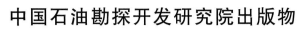
中国石油勘探开发研究院出版物

碳酸盐台地与全球沉积演化

吴因业　张天舒　吴洛菲　郭彬程　赵　霞　周永胜
金春爽　岳　婷　侯宇安　陈瑞银　杜业波　等编译

石油工业出版社

内 容 提 要

本书在分析全球碳酸盐岩沉积特征的基础上,着重阐述了赤道碳酸盐和大陆架环境的碳酸盐沉积相模式,厚层浅水碳酸盐相分布的成因,白垩纪生物与碳酸盐台地,孤立碳酸盐台地的沉积层序演化,碳酸盐台地的淹没与全球缺氧事件,浅水碳酸盐台地的环境变化速率和同步性,地层不整合的沉积演化特征,以及台地沉积物的稳定碳同位素组分与全球旋回重建等科学问题。

本书可作为沉积学与层序地层学、石油地质学研究者、地质与地球物理勘探工作者、油藏工程师以及相关高等院校师生的参考书。

图书在版编目(CIP)数据

碳酸盐台地与全球沉积演化/吴因业等编译.
北京:石油工业出版社,2013.9
ISBN 978-7-5021-9759-9

Ⅰ. 碳…

Ⅱ. 吴…

Ⅲ. 碳酸盐岩油气藏-研究

Ⅳ. TE344

中国版本图书馆 CIP 数据核字(2013)第 215089 号

All Rights Reserved. Authorised translation from the English language edition published by John Wiley & Sons Limited. Responsibility for the accuracy of the translation rests solely with Petroleum Industry Press and is not the responsibility of John Wiley & Sons Limited. No part of this book may be reproduced in any form without the written permission of John Wiley & Sons Limited.

Copyright © 2012 International Association of Sedimentologists

本书经 John Wiley & Sons Limited 授权翻译出版,简体中文版权归石油工业出版社有限公司所有,侵权必究。

著作权合同登记号 图字:01-2013-6696

出版发行:石油工业出版社
(北京安定门外安华里2区1号 100011)
网 址:www.petropub.com.cn
编辑部:(010)64523543 发行部:(010)64523620
经 销:全国新华书店
印 刷:北京中石油彩色印刷有限责任公司

2013年9月第1版 2013年9月第1次印刷
787×1092毫米 开本:1/16 印张:25.75
字数:660千字
定价:140.00元
(如出现印装质量问题,我社发行部负责调换)

版权所有,翻印必究

译者的话

碳酸盐岩沉积在油气勘探领域是十分重要的烃源岩和储层，同时也是重要的非常规储层类型之一。本书除了可应用于常规石油地质勘探领域之外，对非常规资源勘探也可以提供参考，在国民经济的能源建设中将发挥重要作用。本书的学术价值在于：(1)多方法开展碳酸盐岩的沉积环境分析；(2)稳定碳同位素组分与全球旋回重建；(3)碳酸盐岩的不整合特征与沉积层序演化研究。

本书对"十二五"及未来国家科技攻关项目研究、石油石化集团等系统的碳酸盐岩海相重大专项研究、岩性地层油气藏重大专项研究和非常规油气资源勘探研究有重要参考价值。

本书提出了许多全新的学术论点，例如赤道碳酸盐体系特征的认识，建立全球预测模型以掌握环境变化的过去与未来。碳酸盐岩—碎屑岩混积环境的晚始新世珊瑚建造体系中，生物丘随着三角洲复合体的前积，最底部富集骨骼的岩层由具粒泥灰岩到泥粒灰岩基质的苔藓虫浮石构成，它们含有大量浮游有孔虫，缺失与光作用相关的有机生物体。岩石结构、骨骼组分、建造解剖以及相结构的综合分析，揭示了这些珊瑚建造发育在前三角洲环境，在那里三角洲朵体的迁移或者降雨周期变化偶尔会形成适合藻珊瑚生长的水体透明度。其沉积建造的水深位置受依赖光的群落和岩相分布的控制。

在碳酸盐岩层序地层学研究领域，本书也提出了新认识。海相裂谷盆地的半地堑中发育起来的孤立碳酸盐台地，识别出了六个具有不同几何形态、地层结构、层序界面、岩相组成和叠置形式的层序。它们具有复杂的、多重作用的成因，是大地构造作用(区域沉降和断块旋转)叠加到全球海平面升降信号上的结果。另外，伴随沉积中心迁移而改变的碳酸盐生产主体和相对堆积潜力也影响了由大地构造和全球海平面变化混合作用产生的可容空间的充填和内部结构。本研究还证明，在裂谷盆地碳酸盐台地层序发育和台地淹没的过程中，全球海平面的升降和碳酸盐工厂的变化可能与大地构造的活动同等重要。该结果意味着，在其他类似的碳酸盐台地当中，由于人们习惯倾向于用构造解释，因此，全球海平面升降和碳酸盐工厂的作用或许并没有得到正确的认识。

本书的翻译工作主要由中国石油勘探开发研究院、国土资源部油气中心、北京大学、中国矿业大学(北京)等单位教授级高级工程师、高级工程师、博士(生)等总计11名科技工作者完成。中国石油勘探开发研究院研究生部部分博士生和硕士研究生刘群、邓晓娟、陈娅娜、杨沛广、边海关、兰锋、许翠霞和马良涛参加了部分翻译。

主译者名单和翻译任务分工见下表：

《碳酸盐台地与全球沉积演化》译者和翻译分工情况

译者姓名	简　　介	主　要　分　工
吴因业	教授级高级工程师、沉积储层与石油地质专家,工学博士,国际沉积学家协会(IAS)会员,岩相古地理专业委员会委员,北京地质学会理事。科研成果分别获省部级科技进步奖一等奖、二等奖等多项,出版有《油气层序地层学》等多部专著和译著《层序地层学原理》,发表学术论文80余篇	翻译第1章、第13章,编写附录1、附录2,校对初稿2篇。最终审稿与统稿
郭彬程	博士、高级工程师,从事沉积储层与石油地质勘探研究,发表学术论文近20篇	翻译第8章,校对初稿1篇
金春爽	博士、教授级高级工程师,从事沉积储层与油气资源战略研究,发表学术论文20余篇	翻译第7章,校对初稿1篇
陈瑞银	博士、高级工程师,从事有机地球化学与石油地质研究,发表学术论文20多篇	翻译第10章,校对初稿1篇
张天舒	博士(生)、工程师,从事沉积储层与石油地质研究,发表学术论文多篇	翻译第2章、第12章,校对初稿2篇
杜业波	博士、高级工程师,从事沉积储层与海外石油地质勘探研究,发表学术论文10多篇	翻译第9章,校对初稿1篇
侯宇安	理学博士,从事沉积储层与石油地质研究,发表学术论文多篇	翻译第4章,校对初稿1篇
吴洛菲	硕士研究生,从事储层建模与石油工程地质研究,发表学术论文多篇	翻译第3章,校对初稿2篇
岳婷	硕士研究生,从事沉积地质研究,发表学术论文多篇	翻译第6章,校对初稿1篇
赵霞	博士、高级工程师,从事石油地质研究,发表学术论文多篇	翻译第11章,校对初稿1篇
周永胜	德国留学博士、高级工程师,从事海外石油地质勘探研究,发表学术论文10多篇	翻译第5章,校对初稿1篇

翻译工作分三步完成,第一步完成第一稿后自己校译,第二步是相互校译,并请沉积学和层序地层学专家审阅,第三步是统稿,重点对各章节内容和翻译术语的一致性做了大量工作。本译著的完成,得到了中国工程院胡见义院士、中国石油天然气集团公司、中国石油勘探开发研究院有关领导和专家的大力支持,在此一并致谢!

由于本书内容丰富,涉及面广,特别是生物化石方面名词繁杂,加上时间紧迫,不当之处,敬请读者批评指正。

目　　录

第1章　赤道碳酸盐：地球系统方法 …………………………………………………（1）
第2章　大陆架环境的碳酸盐沉积相模式 ……………………………………………（34）
第3章　厚层浅水碳酸盐岩相分布的成因 ……………………………………………（61）
第4章　白垩纪生物厚壳蛤类与碳酸盐台地 …………………………………………（86）
第5章　孤立碳酸盐台地的沉积层序演化 ……………………………………………（122）
第6章　碳酸盐台地的淹没与全球缺氧事件 …………………………………………（158）
第7章　浅水碳酸盐台地的环境变化速率和等时性 …………………………………（190）
第8章　地层不整合的沉积演化特征 …………………………………………………（219）
第9章　侏罗纪斜坡沉积不连续界面的表征 …………………………………………（256）
第10章　海相碳酸盐岩晚古生代冰期的地层特征 ……………………………………（298）
第11章　台地沉积物的稳定碳同位素组分与全球碳循环重建 ………………………（329）
第12章　碳酸盐岩—碎屑岩混积环境的晚始新世珊瑚建造 …………………………（347）
第13章　非海相盆地的大型沉积结构：可容空间与沉积物供给之间相互作用的响应
………………………………………………………………………………………（377）
附录1　中国海相碳酸盐岩的图版：北京及周边长城系蓟县系和寒武系露头 …………（399）
附录2　中国陆相碳酸盐岩的图版：四川盆地侏罗系露头 ……………………………（402）

第1章 赤道碳酸盐:地球系统方法

MOYRA E. J. WILSON 著

吴因业 译,吴洛菲 校

摘 要 本文的假定条件是地球系统方法"从过程到产品"可以用于更好地建立识别和评价未知的赤道碳酸盐体系的预测模型。暖温连同常见碎屑、清水、营养物质的注入,以及赤道热带的盆地背景,都对碳酸盐沉积和成岩作用有重大影响。产生于赤道地区综合沉积作用的碳酸盐体系特征包括:常见光能自养生物和异养生物的出现,文石或钙质为主的矿物,缺少包粒或团粒,常常与碎屑岩伴生,不与蒸发岩伴生,台地类型分散,包括欠透光型碳酸盐台地。此外成岩作用特征包括:常见泥晶化(micritization)和生物侵蚀,缺少海相胶结物,广泛的渗流溶蚀作用和伴生的潜水胶结作用。大气地下水流区域也存在大量的钙质对文石的交代作用,常见埋藏压实作用和渗流作用,以及经由海水或大陆来源的地下水流引起局部的块状白云石化作用。虽然赤道碳酸盐位于暖水的光能生物相组合带(Photozoan Association),许多上面描述的特征却与暖水干旱对应的沉积模式不一致。相反,部分赤道碳酸盐沉积特征显示与冷水形成的相似,因此区分这两种不同气候带的碳酸盐十分困难。建议识别显生宙赤道碳酸盐发育的主要依据有:(1)多种多样的方解石或文石光能自养生物;(2)常常有异能生物相组合因素(Heterozoan Association);(3)独立的(例如同位素的)证据表明暖温(大于22℃)。指向潮湿的赤道背景的其他证据是:(1)合适的古纬度条件;(2)缺少蒸发岩沉积组合,缺少包粒(coated grain)或团粒(aggregate);(3)海水盐度减少或养分上涌的地球化学证据。本文目标是更清楚认识和了解赤道碳酸盐体系,建立全球预测模型以掌握环境变化的过去与未来。

关键词 碳酸盐礁 新生代 碎屑岩 成岩作用 潮湿的赤道气候 养分 东南亚 构造

1.1 概况

温水环境中(热带)和在冷水环境中(温带)中形成的碳酸盐是有很大差别的,这一点大家都很了解(Lees 和 Buller,1972;Nelson,1988;Jones 和 Desrochers,1992;James,1997;James 和 Lukasik,2010)。但是,对于在温水环境中碳酸盐沉积体系的多样性,尤其对湿润的赤道体系中的碳酸盐大家就了解得很少了(图1-1;Lees 和 Buller,1972;Lees,1975;Wilson,2002,2008a)。赤道周围的气候特征为常见碎屑、清水、营养物质的注入,而这些经常是我们容易忽略的,并且可能刚开始我们会认为与碳酸盐群落和系统的多样性是不相协调的(图1-1和图1-2;Fulthorpe 和 Schlanger,1989;Tomascik 等,1997;Wilson,2002,2008a,2011;Park 等,2010)。

提到热带的碳酸盐以及由于"蓝色的海水"所形成的极其多样化的群落,我们脑海中马上会出现贫瘠的营养物质以及热带的礁的成群出现(绿藻动物和光能自养的群落;Lees 和 Buller,1972;James,1997)。在温暖的、受波浪和洋流搅动的浅海中,我们也可以见到鲕粒的沉积和跳跃(Lees 和 Buller,1972;Jones 和 Desrochers,1992)。相比较而言,在冷水的系统中,有洋流刻蚀的小的群落和软体的壳类动物更为突出(Lees 和 Buller,1972;Lees,1975),而在营

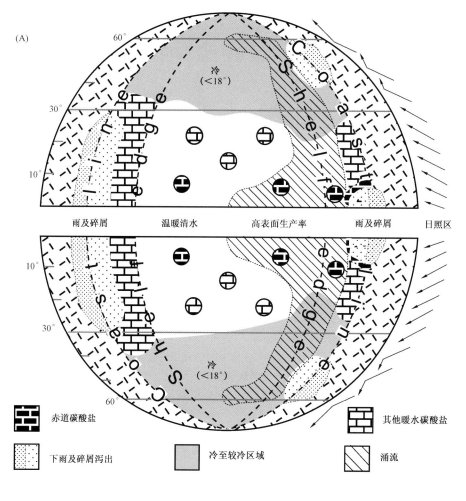

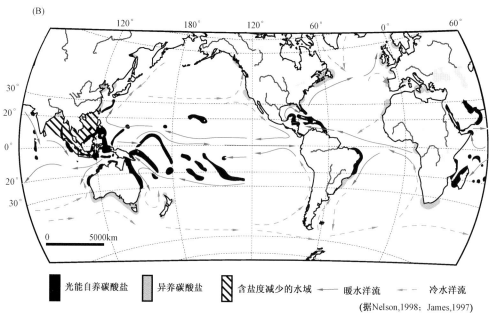

图 1-1 现代碳酸盐全球分布的影响因素

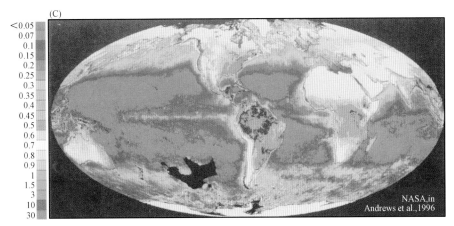

图1-1 现代碳酸盐全球分布的影响因素(续)

(A)浅海碳酸盐沉积的一些重要因素的理想化模型(修改后,Ziegler等,1984;Nelson,1988)。陆架和岛屿上发现的暖水碳酸盐都显示为砖式的排列,赤道碳酸盐(表现为反置的积木式)已经被添加到原始图表,并且它们通常受到地面径流、上升流和温暖温度的影响。异能生物相组合(Heterozoan)碳酸盐(图中没有显示)通常发育在凉爽到寒冷的温度范围内,或在光能生物相组合(Photozoan)系统下(温水)。(B)现代碳酸盐聚集的全球分布(修改后,Nelson,1988;James,1997)。在赤道热带,值得注意的是东南亚重要的光能生物相组合碳酸盐(温水)的发育:一个海洋盐度降低,具有明显的地面径流和上升流的地区。(C)美国宇航局提供的一个全球生物圈的卫星图像(Andrews等,1996),这主要是对以叶绿素为主的光谱的分析(色素浓度的单位是 mg/m^3)。叶绿素含量提供一种对植物量的估计,并代表了植物的生产率,这在海洋中反映了上升流和养分径流

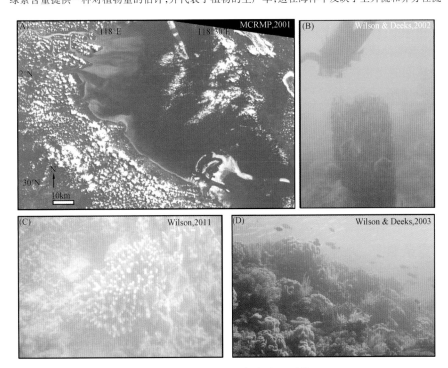

图1-2 东南亚现代碳酸盐系统

(A)现代碳酸盐形成于 Mangkalihat 半岛北部和 Borneo 北东 Berau 三角洲的东部混合碳酸盐碎屑岩陆架上(利用 Landsat-7 卫星所拍摄的遥感影像采用 RGB 321 真彩色合成,轨道参数为116,59。来源于美国宇航局基金赞助的南佛罗里达大学和机构研究的千年珊瑚礁测绘项目,影像获取时间为2001年2月27日)。(B)作者在苏拉威西 Tukang Besi 群岛小于10m水深的地方观察到一个巨大的桶海绵,横向视野是2m。(C)现代珊瑚礁沉积于 Borneo 东海岸浑浊沉积影响区域,水深1~2m的地方(Wilson,2011),横向视野是50cm。(D)位于 Tukang Besi 群岛的现代珊瑚礁边缘主要由软珊瑚构成,横向视野是5m

养物质供应充足的地方,大量的海百合和苔藓虫的草地也会出现(异能生物相组合;Lees 和 Buller,1972;James,1997;Hayton 等,1995;James 和 Lukasik,2010)。浅的近热带的条件下往往会有全球生物礁分布的热点地区(在印度西太平洋存在的"三角礁"),这些地方往往会有一般到极为丰富的营养物质的注入,这很显然与我们上面所说的理论是相互矛盾的。事实上,在很多近赤道的礁体中,很明显的有两种非常不同的光能生物相组合和异能生物相组合系统的出现(Tomascik 等,1997;Wilson,2002,2008a;Wilson 和 Vecsei,2005)。异养型生物比光合自养型生物在体重方面要重(比如说在很多东南亚的石珊瑚的礁体)。另外由于物质的上涌和陆源物质的注入,低的见光度和比水表面更冷的温度是很普遍的,并且它不适用于我们经常所说的"蓝水系统"(Potts,1983;Tomascik 等,1997;Wilson 和 Vecsei,2005;Wilson,2008a,2011)。这些差异就不得不引起大家思考一个问题,对于近赤道的热带的碳酸盐体系以及它们的沉积物来说,有什么是不一样的呢?

在这篇文章中我们用评价环境以及盆地演化过程的地球系统方法来更好地了解赤道碳酸盐体系的沉积和地质产物。这种可以等同为"源到汇"的方法一出现就在碎屑岩沉积中得到了广泛的应用。在碎屑岩体系中,侵蚀(源头)和沉积(沉降)的地形是通过沉积物的沉积路径系统联系起来的,并且受到环境的控制(Allen,2008)。然而在碳酸盐岩系统中,源和汇一般是一体的。有一个共同的认识,那就是"碳酸盐是自己生成的,而不是靠运移的",并且碳酸盐来自于自身的营养供给系统(James 和 Kendall,1992)。碳酸盐受到一系列的复杂环境的影响,生物和化学的影响一般高于物理的。另外,因为沉积后的变化,碳酸盐的沉积物通常更加致密,而这些变化往往也可以反映表面的环境条件和深部盆地的演化进展过程。简而言之,在"过程到产物"的碳酸盐研究中很有必要把碳酸盐的海水环境与相连的地形演化、海洋学、大气、生物圈和地球圈的变化联系起来。

这项研究反映了碳酸盐沉积和相应的变化在多大程度上反映着赤道周边当地的环境情况。在这里我们所说的很多例子都是来自作者在东南亚的新生代地层的研究,这里的生物和地质特征是最为丰富的。这样一个方法同样也具有全球的适用性,在全球很多地方的碳酸盐体系中,都有很强的相似性(Testa 和 Bosence,1998;Gischler 和 Lomando,1999)。在有可能的地方,也会对其他的赤道周边地区作比较,包括其他的气候带以及更早的时期。对于本篇文章内容的理解还可参见另外一本同时出版的著作,其中详细地阐明了东南亚的碳酸盐体系(Wilson,2011)。

1.2 环境条件

按照地球分带,赤道位于 10°~15°之间的热带间聚合带(Inter Tropical Convergence Zone,Lockwood,1974)。尽管在气候学术语上并不一致,许多位于赤道附近 5°范围内的低洼地区都具有高热(大于18℃,通常大于20℃且温度的变化小于2℃),相对湿度高(通常70%~90%)和全年陆上大降雨量(大于1500mm/a;图1-3;Lockwood,1974;Sale,2002)的特点。降雨量的极小值存在于海上,而在赤道大陆地区和东南亚,大降雨量通常伴随着雷雨(图1-3;Lockwood,1974)。多季节性赤道气候的地区是由于低气压带的移动,信风的多变性或者季风的转变引起的,这些地区大多位于赤道附近5°~15°之间(Lockwood,1974;Sale,2002)。这些季节性地区在温度和降雨量上都随年波动,通常在炎热的夏季月份(大于20℃)具有一个明显的潮湿季节(500~>1500mm/a)(图1-3;Lockwood,1974;Sale,2002)。赤道带内,这种"定期潮湿"和"季节性"的气候以及伴随的风模式强烈影响着沿海地区地表温度、径流、碎屑、养分和淡水的流入(图1-4)。东南亚半封闭的海面上,海洋环境中的这种气候条件并不像开放海域一样被全球性海洋特征所"稀释"。事实上,在这种半封闭的环境中,这些因素可能被"放大",比如水流(图1-5;Tomascik 等,1997;Park 等,2010)。赤道地区的温度促进了温水碳酸盐的发育。然而,其他的气候相关的因素影响着赤道地区海洋的盐度、浊度、养分和化学饱和度,这种化学饱和度同时控制着赤道碳酸盐的特征(Tomascik 等,1997;Wilson,2002,2011)。

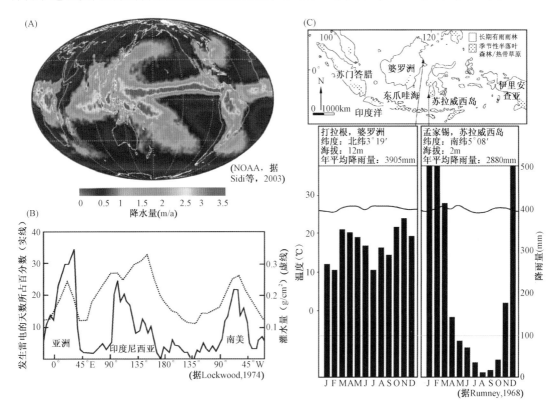

图1-3 东南亚和赤道热带的气候环境

(A)全球各地的年平均降水量,根据特殊传感器微波成像获取的资料(SSM/I)。1987 年开始收集卫星资料。NOAA 资料,来自 Sidi 等(2003)。(B)在北纬10°和南纬10°之间的赤道地区。年雷雨天所占百分比,以5°经度间隔(实线)。500mbar 气压上水蒸气的量,以 g/cm² 为单位,10°经度间隔(据Lockwood,1974)。(C)地图显示东南亚长期有雨雨林和更多季节性森林的分布(据 Rumney,1968;Morley,1999),以及气象站的位置。赤道附近长期有雨雨林(Tarakan)和季节性森林(Makassar)地区的温度和降雨量数据全部来自 Rumney(1968)

东南亚一直是一个具有大量大陆架海洋环境的地区,由大大小小的岛屿组成,海底高地在整个新生界时期被弯曲的海道分隔(Tomascik 等,1997)。这种潜在碳酸盐发育地区的范围和多样性在其他赤道地区是不能相比的,这使东南亚成为研究沉积过程和沉积多样性的理想天然实验室(Fulthorpe 和 Schlanger,1989;Wilson,2002,2008)。相比之下,中美洲、非洲和印度(在新生代时期向北漂移)是典型的大陆架和很小范围的近海浅滩或岛屿(James 和 Glinsburg,1979;Testa 和 Bosence,1998;Gischler 和 Lomando,1999)。尽管太平洋或许与赤道碳酸盐系统的传统观点紧密相关,但其主要还是发育与火山岛和海底山有关的深海碳酸盐(Potts,1983;Camoin 和 Davies,1998;Montaggioni,2005)。东南亚的地区多样性由非常复杂、活跃的构造作用控制,大量的微型大陆断块、盆地和火山岛弧并列发育在欧亚和澳大利亚板块之间的碰撞带(图1-5;Hall,1996,2002;Wilson 和 Hall,2010)。

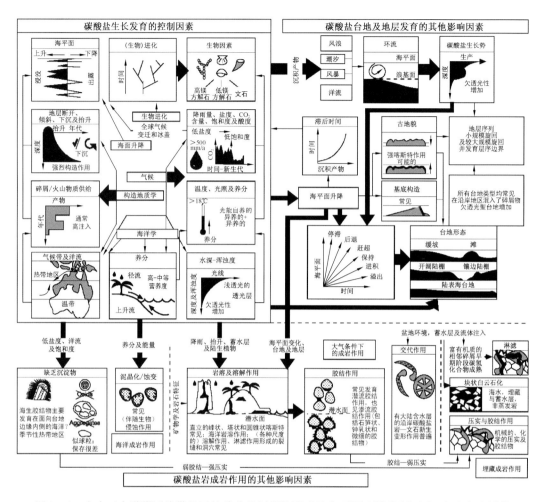

图1-4 东南亚赤道附近热带地区的碳酸盐沉积物形成的主要影响因素(左上)、台地和地层的发育情况(右上)及它们的成岩作用(下)

此图在Jones和Desrochers(1992)的基础上修编而来,包含了更多的影响因素,及其对东南亚地区碳酸盐发育的作用。针对赤道附近热带地区的特殊特征用黑体字加以强调标识

这种复杂构造地区,伴随着大量降雨和丰富的热带植物,通常导致火山碎屑岩、硅质碎屑岩、淡水和营养物质汇聚到这个区域的沿海水域(图1-1、图1-2、图1-4和图1-5;Tomascik等,1997;Wilson和Lokier,2002)。本地的季风气候引起了强烈的季节性地表径流以及风和径流模式的改变(Umbgrove,1947;Park等,2010)。这个区域位于气旋带以外,基本没有强烈的气旋风和波浪(Umbgrove,1947;Tomascik等,1997)。构造断层、沉降和上隆结合冰川—全球海平面变化控制了局部的相对海平面变化,这会影响碳酸盐沉积、珊瑚礁生长、陆上暴露以及洪泛(Fulthorpe和Schlanger,1989;Wilson,2002,2008a;Park等,2010;Wilson和Hall,2010)。

该区内,火山、地震活动和相关的海啸引起陆上和海上景观大的环境变化。从短期来看,这会对生态群落造成破坏,但是从长远来看,可能会带来生态机会(Wilson和Lokier,2002;Satyata,2005;Pandolfi等,2006;Stoddart,2007)。东南亚是赤道附近仅剩的海洋通道,允许太平洋和印度洋之间通过大范围的印度尼西亚穿越洋流进行海水的交换(图1-5;Gordon,2005)。这个地区的气候和水流系统受到全球海洋和大气现象的影响,并且/或与其发生相互作用,包括厄尔尼诺向南震荡,印度洋偶极现象,季风和热带辐合带的波动(Tudhope等,2001;Kuhnt等,

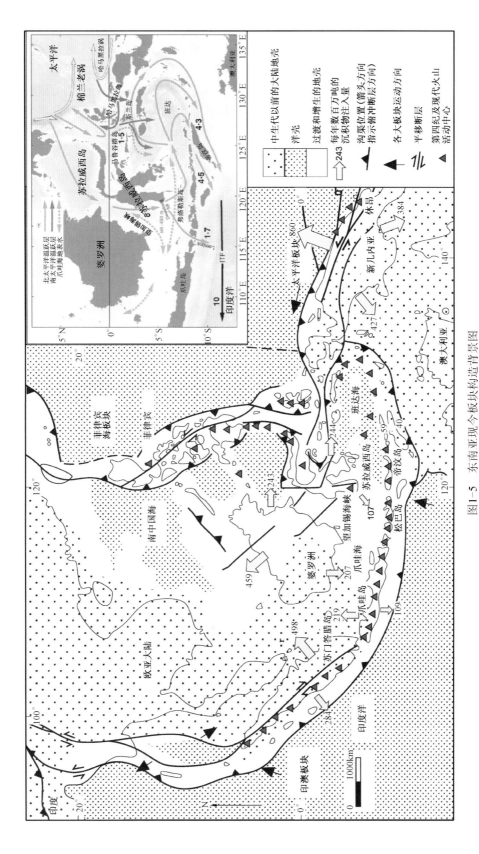

图1-5 东南亚现今板块构造背景图

展示了火山位置及沉积物注入主要来源于六个大的东南亚岛屿（苏门答腊岛、爪哇岛、婆罗洲岛、苏拉威西岛、帝汶岛及新几内亚岛；摘自Wilson和Lokier (2002) 及Wilson (2011) 底图的基础上进行了修编）。插图展示了太平洋与印度洋之间东南亚的一支主要的洋流通路径：印度尼西亚贯通流路径（与上升流相关）及对整个运输量的估计（流量单位：斯维尔德鲁普 $S_v = 10^6 \, m^3/s$；Gordon, 2005）

— 7 —

2004;Wang 等,2005;Abram 等,2009)。这些因素在不同时间尺度上改变海洋表面的温度,使局部比周围环境温暖或变冷有一定的影响(Gagan 等,1998;Penaflor 等,2009)。营养涌入和隆起区也受影响,反过来,引起与浮游生物的水质清澈的变化。长期的海洋(温度、酸度及认知改变)和大气(CO_2)变化在新生代发生,处于由温室气候转变为冰室气候的过程中,在整个新生代中长期的海洋的变化(温度、酸度和组分的变化)和大气(CO_2)的变化也会对海洋生物群和区域系统产生重大的影响(Wilson,2008)。目前认为,影响海洋系统的长期和短期的变化都是全球气候变化的主要驱动力(Gordon 等,2003;Visser 等,2004)。

主要的构造作用、相对海平面频繁变化、低海洋矿化度、碎屑和养分的涌入、所有海洋和温度条件下的变化都强烈地影响着区域和局部的碳酸盐的发育。根据对温暖的、较干旱的亚热带地区的研究,这些条件不同于被认为是珊瑚礁和热带碳酸盐生长的理想条件(Wilson,2002)。虽然没有哪一个因素是东南亚所独有的,但是这些因素的独特组合造成了赤道附近的碳酸盐的特殊性。本书研究评估了碳酸盐沉积以及之后的变化对赤道热带局部环境条件的反映程度(图1-4)。

1.3 赤道热带过程对碳酸盐沉积体系的影响

在新生代,复杂的构造背景以及由此产生的浅水区域的变化对碳酸盐发育的分布和特点有着重大影响(Fulthorpe 和 Schlanger,1989;Wilson 和 Rosen,1998;Wilson,2002;Wilson 和 Hall,2010)。区域构造过程通过板块运动、火山活动、伸展盆地的形成和隆起控制浅海区向热带运动、出现和消失(Wilson,2008;Wilson 和 Hall,2010)。东南亚超过三分之二新生代的浅水碳酸盐岩地层($n=250$)开始具有与陆接触的特征或在碎屑岩陆架之上,而不是孤立的台地(Wilson 和 Hall,2010)。印度洋—太平洋海域的大多数珊瑚礁(53%和20%多来自印度洋)主要集中在东南亚、澳洲及印度洋浅大陆架上,并经受附近陆地的影响(Potts,1983;Tomascik 等,1997)。一般来说,东南亚西部的海洋系统受到径流的强烈影响,而东部群岛的海洋系统主要具有上升流的特征(Tomascik 等,1997)。由此产生的淡水、硅质碎屑岩、火山碎屑岩和受营养影响的碳酸盐系统是赤道热带的主要特征(Potts,1983;Tomascik 等,1997;Wilson 和 Hall,2010;Wilson,2011)。

高降雨和陆上有机物生产导致世界范围内大量的季节性淡水、碎屑和营养物质涌入到沿海地区,以及碎屑陆架地区(图3)。例如,每年婆罗洲某一三角洲的河流输入量和沉积物排入量分别为 $500\sim5000 m^3/s$ 和 $800\times10^4\sim8000\times10^4 m^3/a$(Staub 和 Esterle,1993;Allen 和 Chambers,1998;Woodruffe,2000)。四大东南亚群岛的沉积物输出量是不同的,从 $300\times10^6\sim1650\times10^6 t/a$ 变化(图1-5;Milliman 等,1999)。据保守估计,婆罗洲周围盆地含有长达9km的沉积物来源于该岛(Hamilton,1979),并且新近纪沉积物单位面积的供应量与喜马拉雅山脉相似(Hall 和 Nichols,2002)。这种高径流确实阻碍了一些大陆架上碳酸盐的发育,特别是在东南亚西部和新几内亚(Tomascik 等,1997)。超过80%的与陆相连的碳酸盐系统形成在小规模的岛屿周围(火山的和非火山的),与大规模的岛屿相比,这很可能反映了更有限或周期性的涌入(Wilson 和 Hall,2010)。尽管有这些,有许多现代和新生代碳酸盐的例子与碎屑流涌入同时存在,并且能够适应碎屑涌入(图1-6;Tomascik 等,1997;Wilson 和 Lokier,2002;Wilson,2002,2005)。碳酸盐生产者的类型和其生产的系统将最终取决于碎屑和营养物输入的速率和频率,相关的颗粒粒度以及其他的局部环境条件(Wilson 和 Lokier,2002;Lokier 等,2009)。

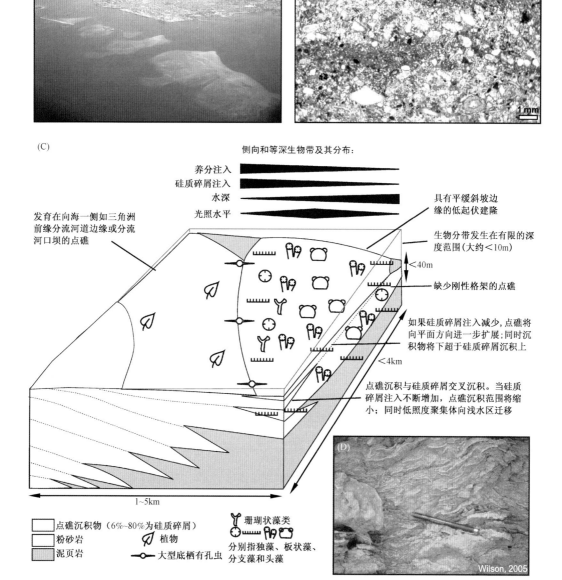

图1-6 东南亚地区碎屑物影响沿岸碳酸盐的例子

(A)苏拉威西碳酸盐—碎屑混合型Spermonde陆棚(offshore Ujung Pandang)的碳酸盐发育和注入河流的概况。观测的水平区域大约小于10km(Wilson和Ascaria,1996)。(B)Manhkalihat peninsula地区的渐新世混合碳酸盐—碎屑沉积的显微影像(Wilson和Evans,2002)。(C)中新世点礁的沉积模式,只强调了主要因素。该区发育在Borneo的Mahakam三角洲向海一侧的边缘(Wilson,2005)。(D)板状珊瑚席状岩(展平后,宽高比>30:1的面状珊瑚,Insalaco,1998)。该席状岩位于某三角洲前缘点礁,该点礁包含60%不可溶的细粒硅质碎屑(Airputih AA剖面,据Wilson,2005)。用作标尺的笔长15cm

— 9 —

在径流不足以阻止碳酸盐发育的地方,一系列混合碳酸盐岩—碎屑岩陆架或者局部碳酸盐岩,包括补丁礁、浅滩或塔礁就发育了(图1-6;例如,东婆罗洲的三角洲前缘;Roberts和Sydow,1996;Wilson,2005;或新几内亚前陆褶皱冲断带的临海位置;Pigram等,1990)。更广泛的碳酸盐岩地层包括裙礁或堡礁,位于主要受碎屑岩控制的陆架的外侧(例如,现代的伯劳系,婆罗洲东北或Borabi的堡礁,巴布亚新几内亚;Leamon和Parsons,1986)。具有隆起的碳酸盐岩地形的近海陆地地区常见广泛的碳酸盐岩陆架(例如,现代的Paternoster台地和南婆罗洲的Berai的石灰岩;Burollet等,1986;Saller等,1993)。

大陆架碳酸盐岩往往产生发育于海底高地位置,包括断层限定的古地形(Wilson和Hall,2010)或地层特征体,如三角洲前缘河口坝(图1-6;Wilson,2005)。洋流、内源沉积中心的迁移(例如,三角洲朵体的废弃)、海侵期及火山或构造运动的相对沉寂时期由于限制了碎屑的注入,都能促进碳酸盐的发育(Roberts和Sydow,1996;Wilson和Lokier,2002;Wilson,2005)。

尽管板状的珊瑚沉积可以包含相当细小的黏土级碎屑,但较大的底栖有孔虫和珊瑚藻相对珊瑚来说对各种粒度沉积物注入都显示出更大的容耐性(Wilson和Lokier,2002;Lokier等,2009)。我们通过推断以下变化来反应不同生物体:(1)迁移率;(2)自洁能力;(3)形态组织;(4)相关的供养机制:① 沉积环境;② 基底;③ 浑浊度;④ 侵蚀;⑤ 能量;⑥ 水深;⑦ 光照度;⑧ 营养度(Wilson和Lokier,2002;Lokier等,2009)。东南亚现代沿岸珊瑚显示了变化的骨架生长密度具有荧光显带,这与雨季洪水、淡水及有机质涌入相关,一些也与厄尔尼诺现象有关(Scoffin等,1989;Tudhope等,1995;Aycliffe等,2004)。生长在火山地区的珊瑚应显示热液同位素特征,富铁纹层,凝灰岩物质以及与喷发事件有关的死亡面(Heikoop等,1996;Pichler等,2000)。就局部而言,生物多样性受限于沉积物和影响整个系统的养分因素。但是,一系列关于浑浊水体碳酸盐的研究表明:该环境下的生物多样性与干净水体的生物多样性大约有三分之二的相似性(Larcombe等,2001)。在经常降水或强降水的区域,动植物群落(种类)主要受控于沉积补给(养分),并向异养型补给转变,而在水体上部几米内主要是光能自养生物(Titlyanov和Latypov,1991)。群落结构上的这些变化,在东南亚海域以珊瑚主导的新近系系统和以大型有孔虫和藻类主导的渐新世(温室效应时期)近赤道海岸碳酸盐系统都有发生。在深度上垂向收缩的生物分带现象造成的一个结果是发育了许多受径流影响的低生长幅度的系统。直立的或倾覆的礁体生物格架一般为碎屑基质包围,同时礁体框架的缺失往往产生缓和的斜坡边缘(Wilson和Lokier,2002)。具有陡峭边缘的高幅度建隆,仅趋向于发生在碳酸盐生产率高于碎屑物累积率,生产速率与相对海平面上升速率相当,发育刚性格架的地方(Wilson和Lokier,2002)。火山碎屑对碳酸盐发展的影响类似于硅质碎屑流,然而棱角状的火山碎片可能对藻类造成伤害(James和Kendall,1992)。在火山休眠期或有遮挡阻止大量火山碎屑注入的区域,边缘礁普遍发育在火山机构周围(Fulthorpe和Schlanger,1989)。不管怎样,来自于火山灰沉积和火成碎屑流的火山碎屑的间隔注入,可能影响与陆源碎屑物质隔离的孤立碳酸盐台地(Wilson和Lokier,2002)。

注入东南亚海域的营养物质流不仅影响着沿岸碳酸盐的发育,同时还影响:(1)区域碳酸盐分带;(2)位于潮水和上升流影响地区的台地结构;(3)新生代时期碳酸盐的发展。总的说来,深度变浅,营养物质增多,大量珊瑚生长,但同时水体透明度下降(Wilson,2008a)。在东南亚海域,珊瑚大量生长的深度大约20m,算得上世界上最浅的(Schlager,1992)。这与营养物质来源为径流和上升流的少营养—中营养区域的光照度很低有关,也与浮游生物生长旺盛有关(Willson,2008a)。现代珊瑚地球化学特征证明:(1)上升流(的存在);(2)浮游生物华;(3)向更多异养供给的转变。这些地球化学变化与某些类型上升流相联系,这些上升流与如下事件有关:(1)IOD(Indian Ocean Dipole,指印度洋偶极)变化;(2)与厄尔尼诺事件相关的闪电;(3)与印度—太平洋暖池(Indo—Pacific Warm Pool)范围和ITCZ(热带间聚合带,译者注)运动相关的强季风(Abram 等,2009)。

区域上,在现代和新生代时期东南亚海域,营养物质的增加和水体透明度的降低,促进了一系列大规模的孤立的和与陆地相连的欠透光台地的产生(图 1-7;比如,Paternoster,Kalulalukung,Berai,Melinau,Spermonde 和 Tonasa Platforms;Wilson 和 Vecsei,2005)。在大部分时间里,欠透光的台地受控于台地顶部的"碳酸盐工厂","碳酸盐工厂"形成于透光区的中深部(也就是 Porma 的欠透光区)。在东南亚海域,台地的欠透光区受控于生物碎屑相,该相富集的是低照度水平下(生长的)扁平状的细长穿孔底栖有孔虫,珊瑚状海藻和仙掌藻属(图1-7;Wilson 和 Vecsei,2005)。这样的有孔虫属可能被一些工作者归类为冷水生物聚集体。但是,虽然多种大型底栖有孔虫和仙掌藻属占优势,它们仍是一个温水条件下的(大于22℃)光能生物相组合(photozoan)。虽然是欠透光的,却是赤道热带区域常见的生物集合体。生成的台地只拥有极小范围的浅水沉积环境,礁体格架发育受限或缺失(可能发生在边缘或作为当地建隆出现),一般具有陡峭的边缘,主要聚集无骨架的欠透光性生物群。赤道附近的东南亚海域所有普遍的特征,比如富营养上升流,地表径流和强潮,都增加了水体的浑浊程度,促进了浮游生物的生长,阻碍了10~20m水深的依赖于高光照水平的造礁生物生长(Wilson 和 Vecsei,2005)。寡光型台地也发育在东南亚海侵期到高位期,例如古近纪造架珊瑚就不发挥重要作用(Wilson 和 Vecsei,2005;Wilson,2008a)。

也有多种理由说明,为什么东南亚从较大的海底有孔虫为主变化为富珊瑚沉积相会出现在渐新世—中新世的边界,似乎滞后于其他暖水带,例如加勒比海(Caribbean)或地中海(Mediterranean)(Wilson,2008a)。这些理由包括:(1)白垩纪—古近纪—新近纪灭绝后珊瑚的缓慢恢复;(2)与板块构造有关的生物地理隔绝,孤立于其他富珊瑚地区(Wilson 和 Rosen,1998;Renema 等,2008)。可是,东南亚潮湿赤道背景有关的其他因素可能是有帮助的,包括:(3)赤道地区大量淡水注入降低了文石饱和度,这是由于大气 CO_2 升高的温室效应(Wilson,2008a;Kleypas 等,1999;Hoegh-Guldberg 等,2007);(4)富养分的印度尼西亚贯穿流沿渐新世—中新世边界带来构造削截(Kuhnt 等,2004);(5)从更多季节性(具有较高的周期性有机质和沉积物注入)变化为更多潮湿持续的情况出现在东南亚渐新世—中新世边界(Wilson,2008a)。

小尺度(几千米)和大尺度(10km 到几百千米)浅水为主的孤立台地广泛形成于东南亚新生代(Fulthorpe 和 Schlanger,1989;Wilson,2002,2008a)。与大多数碳酸盐体系一样,这些发育在早期高地的碳酸盐岩与构造和断层边界有关(Wilson,2002;Wilson 和 Hall,2010)。主要的碎屑注入通常会越过这些高地。碳酸盐生产者一旦建立在光带,如果可以"赶上"海平面上升,就会形成厚的堆积,凸显了台地顶部和周边盆地的深度变化。标志性的不对称是许多台地的共同特征,这是由于:(1)同构造活动旋转断块的差异沉降;(2)迎风面/背风面或洋流的影响;(3)不同沉积环境生产率的变化(Grötsch 和 Mercadier,1999;Wilson 等,2000;Wilson,2002)。构造、海洋地理、相对海平面变化和碳酸盐生产者的类型强烈影响孤立台地的发育和台地形态(Fulthorpe 和 Schlanger,1989;Wilson,2002)。与控制初始发育一样,构造影响单一碳酸盐台地表现在:(1)断层边缘的倒塌和改造;(2)断层的分段;(3)地层的倾斜,沉降,上隆和可容空间的差异产生;(4)内部层序特征的调整和沉积相分布(Wilson 和 Hall,2010;Wilson 和 Bosence,1996;Wilson,1999,2000;Bachtel 等,2004;Wannier,2009;Wilson 等,2000)。季风或洋

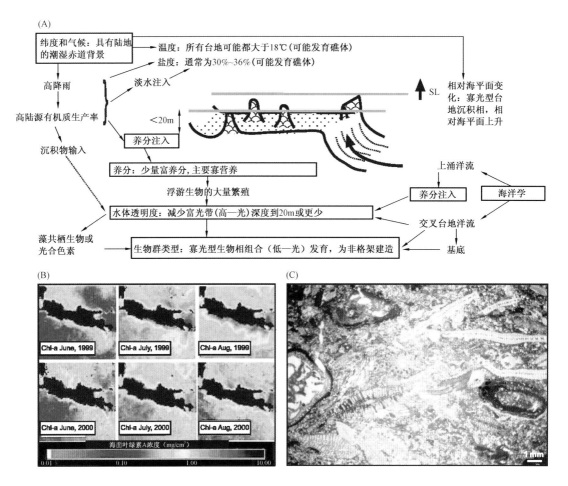

图 1-7 东南亚养分和上涌影响的体系实例

(A)图示总结了赤道地区寡光型沉积相为主的台地发育影响因素(据 Wilson 和 Vecsei,2005)。关键因素是养分、水体透明度、上涌和陆源注入。(B)海面叶绿素 A,伴生季风有关的上涌(1999—2000 年),图像来自 Seawifsimages, Java(据 Basith 等,修改,2011: http://www.google.com.au/imgres? imgurl = http://www.gisdevelopment.net/application/nrm/coastal/wetland/images/ma05_238a.jpg)。(C)薄片显微照片显示多种属平坦的底栖有孔虫和珊瑚藻泥粒灰岩,图片来自 Wonosari Formation,South Java(据 Wilson 和 Lokier,2002)

流,或优势风向导致台地伸长,沉积物的进积和沿台地边缘的相变(Tyrrel 等,1986;Carter 和 Hutabarat,1994;Grötsch 和 Mercadier,1999;Park 等,2010)。

东南亚低海水盐度(大约32‰或更小,相比正常值大于35‰),导致包粒发育的缺少,自然界几乎所有浅水碳酸盐沉积都是生物碎屑(图1-4;Wilson,2002);与缺乏蒸发岩类一样,这是赤道碳酸盐的关键特征(Lees 和 Buller,1972;Wilson,2002)。虽然有球粒形成,但它们很少保存(见赤道热带对碳酸盐成岩作用的影响一文,在后面)。发育叠层石的广泛潮坪没有形成,可能是由于缺少高盐度的条件(Wilson,2002)。可是,海草层,红树林沉积及其伴生物可能会常见。根据局部的环境条件,大量低能到高能浅水沉积会出现在单一台地(Epting,1980;Grötsch 和 Mercadier,1999;Wilson 等,2000)。在洋流影响的或中潮—巨潮汐地区,较高能量沉积相主要发育在孤立台地,包括其内地,尽管有不同大小(Wilson,2008b)。更多泥质的低能沉积相趋向于海侵期发育,或在受保护的环境(包括海草层)内侧高位期,远离礁体镶边的边缘(Epting,1980;Grötsch 和 Mercadier,1999;Fournier 等,2004;

Vahrenkamp 等,2004)。全球海平面和构造会对影响层序发育、沉积相分布和台地几何形态的可容空间变化产生冲击(Epting,1980;Rudolph 和 Lehmann,1989;Vahrenkamp 等,2004;Paterson 等,2006)。高频米级的台地顶部旋回,以及大比例尺碳酸盐岩地层加积、进积和退积,现在已经知道与四级/五级和三级海平面波动有关(Fournier 等,2004;Tcherepanov 等,2008a,b)。新近纪常见的富珊瑚沉积(海平面大波动时期)可以同步或赶上相对海平面上升,相对应有孔虫为主的古近纪则较差(Wilson,2008a)。结果就是,较高幅度的建造和较厚台地序列,或更广泛台地发育建造或塔柱成为新近纪的重要特征,而不是更早的古近纪(Gibson-Robinson 和 Soedirdja,1986;Wilson,2008a)。

1.4 赤道热带作用对碳酸盐成岩作用的影响

三种成岩作用产物[① 海洋的;② 大气的(淡水);③ 埋藏的成岩作用]在赤道热带具有明显的特征。在海洋领域,通常缺少等值线状、葡萄状或放射轴状胶结物(Gischler 和 Lomando,1999;Wilson,2002),良好的资料基础足以和干旱热带区分。东南亚记录有这些胶结物的地方,通常含有较少的组分(Lokier,2000),或在高能边缘中常见(Park 等,1992;Grötsch 和 Mercadier,1999;Wilson 和 Evans,2002;Wilson,2008b)。海相胶结的台地边缘出现在向风面,面向广阔的海洋盆地(图1-8),即在强烈季风区或潮汐驱动的洋流带,或季节性地带,而不是持续潮湿的赤道气候。缺少海洋胶结物或包粒(Milliman,1974;Lees,1975),可能反映了高赤道径流的区域稀释效应,并导致降低海水盐度(Tomascik 等,1997;Jordan,1998;Wilson,2002),同时减少了 $CaCO_3$ 饱和度(Kleypas 等,1999;Hoegh-Guldberg 等,2007)。鲕粒和碳酸盐胶结的团粒已经从法国 Polynesia 和 Cook 岛的现代沉积物中认识到。可是,它们处于赤道带的外面(17°~20°S),属于海洋碱度上升区(Lee 等,2006;Rankey 和 Reeder,2009;Gischler,2011)。

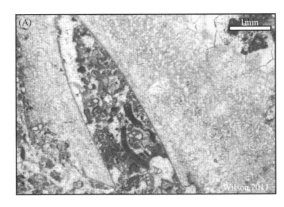
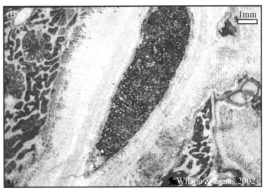

图1-8 东南亚碳酸盐显微照片显示海水成岩作用

(A)平面偏光(PPL)显微照片显示泥晶方解石膜发育在同生变形珊瑚(中心)周围,珊瑚内部(中心)可以看到向地性示踪剂和球粒,显示遮蔽孔隙。(B)平面偏光显微照片反映珊瑚生物黏结灰岩,包括厚的针状到叶片状无二价亮晶铁方解石胶结物,堵塞了原地重结晶分枝珊瑚的孔隙空间(样品、MGA18)。样品来自极高能量面向海洋的礁体镶边的台地边缘沉积,地点在 Mangkalihat Peninsula,Borneo(据 Wilson 和 Evans,2002)

泥晶化主要通过物理和生物作用形成,而不是化学作用,这是东南亚碳酸盐沉积的常见特征(Lees,1975;Wilson,2002)。相似地,颗粒泥晶化和生物侵蚀,会在现代沉积和新生代沉积物中常见(图1-8;Wilson,2002)。微生物有机体和生物膜建造形成的钻孔观察到的泥晶化作用,在低到中等能量的浅水中是常见的,不管有没有海草层(Perry,1999;Reid 和 MacIntyre,2000)。增加的养分会刺激微生物、藻和浮游有孔虫的生长(Hallock 和 Schlager,1986;Perrin 等,1995)。伴生陆源注入(或上涌)的生物礁在东南亚表现出硬壳、生物侵蚀及泥晶化的证据(Tomascik 等,1997;Wilson 和 Lokier,2002;Madden 和 Wilson,印刷中)。球粒的形成,不管是通过整个颗粒泥晶化,还是作为粪球粒(例如鱼或螃蟹),在许多现代潮间带到浅水潮下带都很常见(Wilson,2002)。可是,东南亚地质记录中球粒的保存却很有限,可能是由于缺少早期的海洋胶结作用,常常是埋藏效应(Wilson,2002)。球粒存在的地方,大多保存在生物碎屑或遮蔽孔隙内(图1-8;Wilson,2002)。我们已知赤道地区海洋胶结物通常极少,那么资料丰富的冷水和其他暖水碳酸盐背景的海底溶解的规律是什么(Walter 和 Burton,1990;Sanders,2003;Perry 和 Taylor,2006)?东南亚常常影响新近纪碳酸盐的早期文石溶解和白垩化(部分溶解)归因于喀斯特作用,而海底溶解仍然未知(Wilson,2002)。可是,早期溶解伴生海水成岩作用的地方(Park 等,1995),可能海底溶解起作用(Wilson,2002)。在具有陆源注入的沿岸环境,氧化铁的输入使碳酸盐岩颗粒通过浅部沉积物孔隙水化学的扰动引起的早期成岩溶解减到最小(Perry 和 Taylor,2006)。在东南亚混合的碳酸盐岩—碎屑岩沉积中,通常化石保存良好的界面会反映缺乏早期海洋溶解,或在陆源注入和火山碎屑注入地区,反映快速的碎屑覆盖。相反地,如果海水成岩作用在赤道地区相对有限,大气成岩作用就很有影响(Sun 和 Esteban,1994;Wilson,2002)。高雨量和高陆源有机质形成的微生物生产率,结合构造隆升和陆地的接合,都会产生大量的大气效应,因为这些会影响地下水的流量、活力和酸度(图1-4;Cf. Jennings,1985;Choquette 和 James,1988)。例如,在巴布亚新几内亚的 New Britain 五个地区,计算出碳酸盐溶解剥蚀速率是 270~760mm/ka,这时雨量范围是 5700~12000mm pa(Maire,1981;Ford 和 Williams,2007)。在南苏拉威西 Maros 塔形喀斯特地区,3360mm pa 雨量产生了 $80m^3/km^2$ 的溶解和 $200m^3/km^2$ 的机械剥蚀(图1-9);对比一下匈牙利温带喀斯特地区,溶解和机械的量是 $20m^3/km^2$(650mm pa;Balázs,1973)。可是,在温带和潮湿热带纬度的碳酸盐溶解相对速率方面还有大量争议。水的流量和组成,而不是气候本身,被认为是碳酸盐溶蚀的主控因素(Ford 和 Williams,2007 及其参考文献)。但是,很清楚东南亚潮湿赤道地区是存在大量碳酸盐溶蚀的,具有世界上最大的溶洞群和通道,以及一些最长的溶洞系统(图1-10;Waltham,1997)。喀斯特和溶洞的发育不仅仅依赖外部因素,也有内部因素,例如岩石学(包括组分、结构和渗透性),构造和地层学(Choquette 和 James,1988)。直立的几何形态是赤道热带的典型特征(图1-9;Esteban 和 Klappa,1983;Jennings,1985;Purdy 和 Waltham,1999)。塔形和柱形喀斯特发育形成在不渗透碳酸盐岩地区,溶解集中沿通道或表面径流放射(图1-9)。锥形或舱形喀斯特发育在渗透性地层区(图1-9;Williams,1974;Choquette 和 James,1988;Purdy 和 Waltham,1999)。洞穴通道沿裂隙、层面发育,处于水位成线状或树枝状时(图1-10;Choquette 和 James,1988;Waltham,1997)。溶解和洞穴坍塌受控于流动速率、时间、水的侵蚀性、基准面变化和覆盖层,以及岩石性质(Choquette 和 James,1988;Waltham,1997;Loucks,1999)。渗流带颗粒规模的溶解尤其影响不稳定的文石颗粒,特别是渗透性更好的碳酸盐沉积物(图1-10)。除了上面提到的陆源或大陆喀斯特和洞穴系统外,海洋喀斯特和洞穴的发育常常沿陆上暴露的碳酸盐岩沿岸地区发生(Mylroie 和 Carew,1990;Moore,2001)。在网状的碳酸盐岩"岛翼边缘模式"中,通常不稳定的洞穴网络形成在海洋和淡水混合区(Mylroie 和 Carew,1990)。这些洞穴系统常常发育在东南亚碳酸盐岩岛屿的周围,经常再淹没形成蓝洞(图1-10),并在地质记录上可能显示被忽略的洞穴类型。

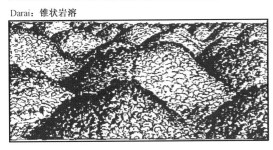

图1-9 东南亚湿润赤道热带地区的表层岩溶发育实例

图片显示了这些岩溶类型草图。(A)发育在新几内亚的三种岩溶类型图(据Williams,1974)。(B)婆罗洲姆鲁山国家公园的剑状岩溶发育图(照片信用保证:Wilson和Khan),尖峰高至10m。(C)苏拉威西岛南部纳沙石灰岩层的塔状岩溶层高达300m(据Wilson,1995,2000)。(D)爪哇岛锥状岩溶发育图(据Lokier,2000),锥状岩溶层高度大约为100m

假定赤道地区碳酸盐沉积常常发生溶解,当溶解流体开始除气、蒸发和停滞不流,就会发生广泛的同生胶结作用(图1-10;James和Choquette,1984)。洞穴堆积物的出现证明渗流沉淀是常见的,就像东南亚出现表面钙化和钟乳石一样。在颗粒规模,渗流胶结物(例如悬挂和吊坠)和潜水胶结物(例如晶簇状和块状)是常见的(图1-10)。东南亚许多碳酸盐岩储层单元含有层状交替的淋滤和胶结物层系,伴随有重复的暴露有关的淋滤(或可能是混合溶解带)、土壤形成作用及潜水胶结作用(Grötsch和Mercadier,1999;Heubeck等,2004;Warrlich等,2010)。这种成岩作用层系仅仅从新近纪孤立台地有区域性报道,与海平面有关,含有易于淋滤的大量文石组分,并受全球海平面波动影响(Wilson,2002)。在东南亚更多受季节性影响地区,大气钙质层的发育(例如凝结纤维状结构,洞室结构;Esteban和Klappa,1983;James和Choquette,1984),会超过喀斯特成为主体,尽管它们可以共同存在(Choquette和James,1988;Wilson,2008b)。

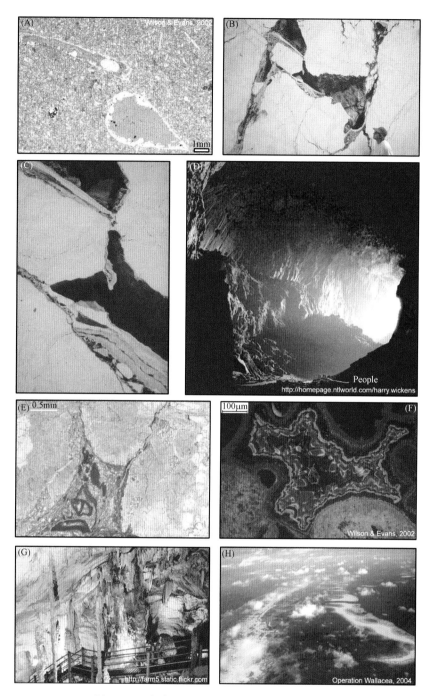

图 1-10　东南亚大气淡水成岩作用影响的实例

(A)婆罗洲 Mangkalihat 半岛的文石双壳类的小尺度渗流溶蚀作用(蓝色表示孔隙,样品:MTR29;据 Wilson 和 Evans,2002)。(B)和(C)喀斯特溶洞和它们的充填:(C)展示了图像(B)中间部分的闭合,(B)是在爪哇西部 Rajamandala 地层的采石场发现的(据 Wilson 和 Lokier)。水平方向 7m 厚的地层。(D)姆鲁山国家公园展示了巨大的溶洞系统,婆罗洲的鹿洞,水平方向大约 100m(图片源自 http://homepage.ntlworld.com/harry.wickens/borneo/borneo-022.jpg,2011 年 3 月 10 号)。(E)马哈坎河三角洲、婆罗洲地下的碎屑岩影响珊瑚礁孔隙间的交代和钙化作用(据 Hook 和 Wilson,2003)。(F)阴极发光显微镜下的块状孔隙充填胶结。晶簇状的胶结展示明亮、灰暗和非阴极发光区,代表了化学沉淀作用的变化(样品:MTM10,据 Wilson 和 Evans,2002)。(G)柱状的钟乳石和石笋,风洞,姆鲁山国家公园,婆罗洲。水平方向 12m。(H)穿过西北方向的空中照片,展示了面向环礁边缘沿东北方向延伸的蓝颜色洞穴,水平厚度大约 10km(图片来源于 Wallacea)

形成于干燥期的南中国海台地钙质地层主要与更新世的低水位相关(Gong等,2005)。有人认为,西太平洋暖池减少的范围与热带间聚合带(ITCZ)向南迁移和冰期时东亚季风强度降低之间有联系(Gong等,2005)。婆罗洲洞穴堆积物的氧同位素数据被用来推知更低的降雨量,也推知与18~20ka前消冰期时热带辐合带向南转移相关的西太平洋对流弱化情况(Partin等,2007)。

越来越多的研究表明,来自岛屿的陆源大气地下淡水流正强烈地影响着海岸、大陆架甚至东南亚的一些孤立碳酸盐台地的成岩作用。在婆罗洲马哈坎三角洲前缘发育的珊瑚点礁显示,文石普遍被方解石和伴生的方解石胶结物早期新生变形交代(图1-10;Madden和Wilson,出版过程中)。点礁含有5%~80%的混合硅质碎屑成分,它与附近持续的陆源碎屑注入同时形成。但没有证据显示陆表暴露或者大气淋滤(Wilson和Lokier,2001;Wilson2005;Madden和Wilson,出版过程中)。新生变形和方解石化是形成于浅埋藏期的早期成岩变化。可能含有高地物源的大气地下水流(James和Choquette定义为蓄水层,1984)冲刷,增加或减少了海水微量元素成分。这被推断为流体变化。这种解释的证据是:(1)早期压实新生变形抑制了原始文石组分的一些结构;(2)气象土壤区指标的缺少(没有负$\delta^{13}C$值);(3)$\delta^{18}O$值与含有高地物源的大气水$\delta^{18}O$值一致,或者与东南亚海水的$\delta^{18}O$值部分相吻合;(4)大约为55℃的温度,这与浅埋深和区域地温相一致。人们推断,与婆罗洲长期潮湿气候相关的大气地下水通过毗邻三角洲而集中于蓄水层,并引起暗礁的广泛早期改变和胶结作用(Madden和Wilson,出版过程中)。一系列的海岸碳酸盐相似成岩作用特征表明,大气地下水流在赤道热带区的影响比之前所认可的更重要。最近的研究显示,在婆罗洲北部和巴布亚新几内亚的褶皱和俯冲带发育区,被追赶的孤立碳酸盐台地也被陆源地下水强烈影响。在这些原始孤立体系区,与陆源地下水流相关的成岩作用显示早期成岩特征(例如,陆表暴露相关的淋滤),并可能引起,(1)广泛的或局部的白云岩化;(2)埋深淋滤(Warrlich等,2010)。

在东南亚,固结成岩较差的海相碳酸盐沉积,除非是通过其他过程胶结作用,一般更倾向于机械和化学压实。依据埋藏深度,压实作用的典型结果是引起原始孔隙度和渗透率大量降低,因为许多碳酸盐岩形成于下陷盆地中,并且其被快速聚集的大套厚层碳酸盐岩、硅质碎屑岩和古近纪(有不同寻常的大气影响)时形成的火山碎屑岩所覆盖了。这些更倾向于是埋藏成岩作用的结果(Wilson和Hall,2010)。虽然埋藏成岩作用经常降低了储层物性,但通过埋深溶解和裂缝,储层的孔隙度和渗透率可能提高。尽管构造成因不是唯一的原因,但东南亚碳酸盐岩裂缝很普遍(Kemp,1992;Grotsch和Mercadier,1999;Wilson和Hall,2010)。如果不是晚期被胶结或被断层封堵充填,裂缝可以是后期改变(例如,白云岩化)或溶解性液流所导致(图1-11;Kemp,1992;Wilson等,2010)。在开放区,典型的裂缝能提高2%~3%的孔隙度,并可能大量地增加成百上千毫达西渗透率(Longman,1985;Wilson和Hall,2010),允许不同储层单元之间的连通(Warrlich等,2010)。一些最近的研究已经强调埋藏淋滤作用的重要性,尤其是对于东南亚区碳酸盐岩台翼、边缘和碳酸盐岩裂缝(Saller和Vijaya,2002;Zampetti等,2003;Sattler等,2004;Pireno等,2009)。这个在区域埋深成岩方面早期被忽略的成因仍有争议。一个有可能的原因是,源于邻近丰富有机质盆地沉积物的烃类在烃类演化的开始阶段产生的有机酸(Moore,2001;Esteban和Taberner,2003)。鉴于高温流或者一些盆地在古近纪—新近纪时的沉积速率,上述这个原因可能在东南亚地区会很重要(Hall,2002b;Hall和Nichols,2002;Wilson和Hall,2010)。然而,其他大量埋藏流体的可能区域来源包括:(1)冷的未饱和海水对台地的冲刷;(2)烃类的生物降解(Heubeck等,2004),或者(3)原地的孔隙流与源自深层更高温度的烃热流的混合作用(Zampetti等,2003)。

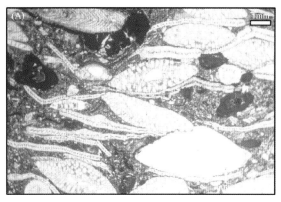 

图1-11 东南亚碳酸盐埋藏成岩作用特征实例

(A)加里曼丹(印尼在婆罗洲岛南部的属地)始新世海底有孔虫和珊瑚藻泥粒灰岩或砾状灰岩的机械压实作用和化学压实作用特征(颗粒破裂和缝合)(据Wilson,1996)。(B)地下大于1000m深普遍发育的白云岩化样品扫描电子显微镜图片。白云岩菱形晶体部分充满了早期破坏的孔隙,同时又受到晚期破坏的影响

(图片来源于Thaariq和Wilson)

在整个东南亚地区,一个地层组规模内的白云石成分和分布变化很大,但是在广泛的碳酸盐沉积物中均有发现(Carnell和Wilson,2004)。东南亚区白云岩的地球化学标志有几个主要的缺失点,能帮助阐明它们的起源(Carnell和Wilson,2003,2004)。然而,部分白云岩化主要伴随有:(1)泥质碳酸盐岩作为替代相,(2)特定的表面或水平层,例如,那些与地表暴露相关的原始孔隙度层,或者(3)构造裂缝和/或压实特征,例如缝合岩面(Carnell和Wilson,2004)。在与地球化学分析相关的局部白云岩化单元研究中,白云化液流是:(1)压实期,与页岩脱水相关(Ali,1995);(2)海水和/或相关的混合区(Mayall和Cox,1988;Park等,1995);(3)源于甲烷的液流(Ali,1995)。最近的研究表明,白云岩化液流使广泛的白云岩化现象发生在浅部至中等埋深处。

白云岩化与以下几方面相关:(1)海水沿着裂缝白云岩化(Wilson等,2007);(2)由于构造的挤压大陆的含水层驱动(图1-11;Warrlich等,2010)。在东南亚这些白云岩化机制不同于其他的地区,有力证据是缺乏蒸发和回注(Sun和Esteban,1994;Carnell和Wilson,2004),这与潮湿的气候相一致。

1.5 讨论

1.5.1 一个地球系统方法:赤道的碳酸盐与其他地区的区别

高温,高降雨量,海洋环境,某种程度的盆地背景共同影响了赤道热带的碳酸盐系统沉积和随后的成岩作用(图1-4和图1-12)。换句话说,赤道热带的沉积成岩作用过程导致特殊的碳酸盐产物,不同于其他的干旱、半干旱和温带的碳酸盐(表1-1;修改自Lees和Buller,1972;Lees,1975;Nelson,1988;James,1997;wilson,2002,2011;Kindler和Wilson,2010)。例如,赤道热带的碳酸盐景观和生物受到碎屑岩和营养物的多少影响,这些碎屑岩主要与陆地的演化,海洋学和板块构造,岩石圈的变化(例如,CO_2水平)有关。其他对碳酸盐生物区的影响受到大气圈的变化,影响了成岩作用的敏感性。溶蚀成岩作用变化和/或者海洋带、大气带和埋藏带的胶结作用会进一步受到高降雨与有机质产率的影响。这些实例会在下面详细讨论,但是很明显"从过程到产品"的方法对于评价赤道地区碳酸盐的发育与变化是适用的。赤道碳

酸盐系统明显与其他气候区有区别,这是由于包含的过程问题,但是"从过程到产品"的方法同样可以用于赤道热带以外地区。新生代碳酸盐岩在东南亚的发育主要集中于这里,因为其记录更加广泛,更加多样,比其他赤道地区更加完整,包括了赤道地区看到的更多变化(James 和 Ginsburg,1979;Fulthorpe 和 Schlanger,1989;Testa 和 Bosence,1998;Gischler 和 Lomando,1999;Wilson,2002,2008a;Montaggioni,2005)。

表1-1 赤道、干旱—半干旱热带、温带区域碳酸盐体系过程和产物对比表(修改自 Leest 和 Buller,1972;Lees,1975;Nelson,1988;James,1997;Wilson,2002,2011;Pascal 和 Wilson,2010)

	赤道—潮湿	(亚)热带—干旱	温带
环境条件			
纬度	北纬15°~20°,南纬15°~20°	北纬/南纬15°/20°~30°	北纬/南纬30°,但未进入副极地纬度
构造	不稳定(到稳定)	稳定—不稳定	稳定—不稳定
海水表面温度	温暖,通常大于22℃	温暖,通常为18~22℃	寒冷,大约为5~10℃
降雨量	大—非常大,通常大于1500mm/a,为常年或季节性潮湿气候	小,通常小于250mm/a	年降雨量为中等到大
盐度	正常到淡水	正常到超咸	正常到淡水
碎屑物输入	对陆架而言全年较高,台地较低	较低(洪水期终止)	低到高
营养	贫营养—富营养;碳酸盐产生区域为中等营养—低贫营养	贫营养(滨岸上升流的地方营养物质较高)	中等—高营养
碳酸盐岩产生			
骨架颗粒	原地或异地——与生物演化、海洋地貌和营养物质无关,藻纹层或叠层石较少,为半透明光照带	受盐度、营养物质及生物地貌影响的原地或异地颗粒	异地颗粒
非骨架颗粒	包壳颗粒和颗粒群较少,鲕粒发育但不宜保存	包壳颗粒和颗粒群常见,鲕粒发育且宜保存	包壳颗粒和颗粒群较少,鲕粒发育但不宜保存
泥晶	由于物理和生物作用,泥晶常见	由于物理、生物和化学作用,泥晶常见	不常见(在外陆架由于物理作用较为常见)
矿物	与微生物有关的文石或方解石	与微生物和沉淀有关的文石或方解石	主要为方解石
沉积速率	中等—高,受生物群落影响,在0.2~1m/ka	中等—高,受生物群落和沉积作用影响,在0.2~1m/ka	中等,约为0.2~1m/ka
台地的发育及岩性组合			
与其他岩石类型有关的组合	与碳酸盐岩—碎屑岩混合沉积有关的岩石组合,硅质碎屑和火山碎屑常见。与海草和红树林有关的沉积也较为常见	主要为与蒸发作用有关的岩石组合	与碎屑岩有关的岩石组合常见,但由于沉积速率中等,所以与赤道的碳酸盐岩相比更易停止生长

续表

	赤道—潮湿	（亚）热带—干旱	温带
台地结构	由原生组合构成的格架——翼部较厚的台地建隆、边缘礁和障壁礁较为常见。靠近陆地，碎屑物质丰富。由原生组合构成的非格架——非边缘礁、浅滩及碳酸盐岩和碎屑岩混合沉积的陆架较为常见	较薄的非翼部陆架、翼部陆架和狭窄的边缘礁较为常见。孤立的边缘或非边缘礁发育。广泛发育潮汐流	非边陆架和斜坡以及浅滩常见，孤立非边缘台地和海岸常见
台地淹没	可能被淹没。与碎屑物输入、富营养化、构造沉降或海平面变化有关	可能被淹没。受构造沉降、海平面变化及盐度增加等环境因素的影响	浅水或饥饿的沉积相——位于光照带以下
后期改造及成岩作用			
微晶化及生物作用	非常常见	常见	生物侵蚀作用常见，微晶化较少？
海相胶结物	除高能量带外，较为普遍	普遍	少见
颗粒溶蚀程度	海底溶蚀作用？文石的淡水淋滤溶蚀作用常见	溶蚀作用较赤道区域弱	海底溶蚀作用？
喀斯特及孔洞发育	两者发育广泛，包括陆上和海域的喀斯特。陆地的喀斯特呈现直立特征	不常见，只是小规模出现	广泛发育。陆地的喀斯特呈现漏斗和落水洞的特征
表生胶结	在渗流带和潜水带较为普遍	发育，但较赤道区域不宜保存	可能常见
与含水层有关的成岩作用	在靠近陆地的滨岸上发育的碳酸盐岩较为普遍——在文石向方解石转化早期稳定期更强	虽然受陆地方向重力的驱动，含水层发育，但滨岸附近碳酸盐岩的表生这样并不普遍	靠近陆地的碳酸盐岩较为普遍，常出现方解石胶结，但文石较少
埋藏成岩作用	由构造和不同压实程度引起的裂缝常见。后期沿着台地边缘常发生渗滤作用。压实效应和埋藏胶结常见，如果较少则出现早期胶结	与赤道一样，发生埋藏成岩作用，但由于早期胶结作用而不易保存	与其他区域一样发生埋藏成岩作用。由于胶结物较少，常发生压实作用
白云石化作用	常发生与海水和埋藏流体有关的强烈的白云石化作用	常发生与蒸发作用和混合作用有关的白云石化作用	白云石化作用较其他区域少见

赤道热带地区高温和均匀的入射光条件促进了依赖于光的暖水碳酸盐生产（光能生物相组合，James，1997）。超高温会阻碍碳酸盐生产（例如珊瑚漂白；Hoegh-Guldberg，1999；Hughes等，2003）。可是，这些条件在东南亚不常常出现，因为有日晒云层效应和洋流的冷却（Wilson，2008a）。东南亚海水真正不同于干旱热带地区的是具有常年的淡水注入、碎屑和养分的注入。降低的赤道海水盐度导致缺少包粒、球粒或耐超盐度的生物群，也缺乏早期海水胶结作用，极少保存粪球粒，文石饱和度低时方解石有机体生长潜力小（Lees 和 Buller，1972；Lees，

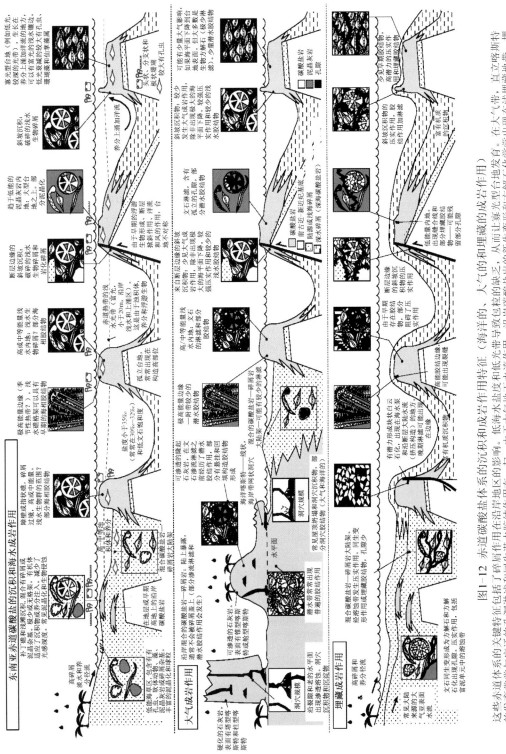

图1-12 赤道碳酸盐体系的沉积和成岩作用特征（海洋的、大气的和埋藏的成岩作用）

这些赤道体系的关键特征包括了碎屑作用在沿岸地区的影响。低海水盐度型光合带号致包括光型合地发育。在大气带，大气型埋藏的发育常与广泛的孔洞淋滤和潜流胶结作用在一起。由于大气（地下）水流作用，沿岸碳酸盐岩常出现石化常见变形和方解石的新生变形环境，沉积物发育。从图上看到的，沉积作用和海水成岩作用的共同特征（顶），埋藏胶结作用发育特征，胶结作用（中）和埋藏成岩作用（底）可以依次看见。例外的是白云石化和方解石，显示了隆起碳酸盐岩人工隆起碳酸盐岩成岩作用的史特征

1975;Wilson,2002,2008a)。这样的连锁反应结果是台地发育趋向于形成广泛加积的生物相关的潮坪。同时,堆积速率会低于干旱热带地区,无镶边的台地不会发育超盐度,而是与狭盐性的生物群类型有关(Wilson,2002)。如果快速埋藏,许多赤道碳酸盐经历明显压实作用,伴随少量早期海洋胶结作用。这些台地发育和成岩作用特征常常与干旱热带不一致,但是与温带有些相似。当然,光能生物相组合的礁生长与海平面有关,在赤道热带形成有镶边的台地,它们不同于温带的标志。

赤道热带与干旱热带相比,沿岸碳酸盐岩的发育和陆上暴露的灰岩有较大的差异,这是由于降雨的变化,陆上有机质的生产率和径流量。赤道热带来自陆地常年的碎屑注入的地方,不足以妨碍碳酸盐生产,会形成混积的碳酸盐岩—碎屑岩大陆架,或近滨的低幅度隆起,无格架补丁礁,这些仅限于极浅水和海侵期生长(图1-12;Wilson和Lokier,2002;Wilson,2005)。与干旱热带相反,岸线附近的碳酸盐岩常常具有"纯"的镶边或无镶边体系,伴生蒸发岩可以被周期性碎屑注入所抑制,或随暴雨事件出现指状交错(Wilson和Lokier,2002)。在温带,沿岸碳酸盐岩发育趋于和碎屑岩混积,就像赤道热带一样。极浅水的温带碳酸盐生产趋于非常有限,主要是由于缺少光依赖的生物群,碎屑的干扰。由于生产率低,温带会比赤道热带具有更多的碎屑干扰。

高侵蚀的大气水的大量生产是赤道热带的特征,这会产生淋滤作用、洞穴和直立的喀斯特地层(Jennings,1985;Ford和Williams,2007)。伴随溶蚀作用,在潜水带会有大量降水,而在渗流带是局部降水(图1-12)。相反,干旱地区雨量少,侵蚀性水也少,只产生少量的溶蚀作用,伴生沉积和大孔洞或喀斯特地层也贫乏。大气水的作用和大规模含水层的发育常常导致沿岸赤道碳酸盐沉积遍布的文石向方解石转化的早期稳定化;干旱地区这一特征不常见(Madden和Wilson,出版中)。长距离的地下水流,部分是构造驱动或海水泵作用,可能是赤道热带广泛白云石化的主要原因,而蒸发或注入机制是干旱地区的关键因素(Sun和Esteban,1994;Carnell和Wilson,2004;Wilson等,2007;Warrlich等,2010)。温带地区的高降雨又能产生大量的洞穴和(隐性的)喀斯特地层(Jennings,1985;Ford和Williams,2007),尽管含水层影响的沉积中大量胶结作用会导致原生的文石生物碎屑比例变化和原生结构的变化(Nelson等,2003)。

养分注入,不管是陆源泻出的还是上涌的,都是东南亚赤道的特征。海的范围从寡营养的到富营养的地区,都有大量碳酸盐产出(每立方米海水大于0.1~1mg叶绿素;Mutti和Hallock,2003)。全球而言,与清水干旱地区相比,养分的区域海拔导致依赖光水平的光能自养生物会变浅(例如珊瑚)(Wilson,2008a)。在更多富营养地区,碳酸盐生产者包括了混合生物、异养生物和光能自养生物,可以转换到异养生物组合。这些组合要有适合的养分,不同于许多干旱热带地区常见的寡营养光能生物群。与海拔养分水平相一致,赤道热带浮游生物大量繁殖(plankton bloom),常见寡营养的(低光)台地发育,水体较深(大于10~20m),内部方解石为主(图1-12)。根据动力学和堆积速率,这些主导型的非骨架欠透光台地显示与温带特征有些相似,尽管形成于暖水生物带(Wilson和Vecsei,2005)。陆源沉积、湖泊和海洋驱动的有机质快速堆积在临近台地区的碎屑单元,或碳酸盐发育的盆地中,这些在赤道热带是常见的。随后烃源岩向烃类的早期成熟可能是晚埋藏淋滤作用大量出现在赤道台地的原因,特别是在其边缘和翼部沉积物(图1-12)。不清楚这种埋藏成岩作用会多大程度影响干旱热带或温带碳酸盐。可是,干旱系统中识别出埋藏淋滤作用的地方,会存在其他原因,例如与蒸发岩有关的侵蚀性流体(Beavington Penney等,2008)。

1.5.2 对地质记录的适用性

尽管每个碳酸盐台地会有特殊性,但区域上会有碳酸盐发育与气候、海洋地理学及盆地背景相关的趋势(Lees 和 Buller,1972;Lees,1975;James,1997)。就赤道热带而言,区域"过程"与上面讨论的碳酸盐"产品"关系明显的是新生界系统,主要是近代的新近纪碳酸盐(图 1-12)。多大程度上赤道系统的特征可以用于较早的地质时期?这时碳酸盐生产的有机体、全球气候、海洋地理及板块构造形态会有很大不同?

在研究全球碳酸盐颗粒组合时,James(1997)定义了术语"光能生物相组合"(Photozoan Association)作为温暖热带的特殊标志。这种相组合被认为是适用于许多显生宙地区,这里碳酸盐生产者推测有光能共栖生物[例如较大的底栖有孔虫,石珊瑚,固着蛤类,F——有孔虫(fusilinid foraminifera)和层孔虫;James,1997]。光能生物相组合"Photozoan Association"是 Lees 和 Buller(1972)的特定生物术语"光合自养生物——Chlorozoan Association"的替代词,还包括了相的内容(例如珊瑚藻灰岩 Coralgal,光合藻灰岩 Chloroalgal 和鲕粒球粒灰岩 Oopeloidal)。James(1997)把光能生物相组合定义为:"一种底栖的碳酸盐颗粒相组合,包括:(1)轻微依赖型有机体骨架,(2)非骨架的颗粒(鲕粒,球粒等等),加上或减去(3)异能生物相组合的骨架"。可是,正如上面所讨论的,赤道热带的新生代沉积不能完全满足这个标准。相反,赤道碳酸盐以底栖的碳酸盐颗粒组合为特征,包括:(1)轻微依赖的有机体骨架,加(2)异能生物相组合骨架。浅水有镶边的大陆架、礁、文石和镁方解石矿物常常出现在新近纪赤道热带,是光能生物相组合的特定标志(James,1997)。可是,较少的海相胶结物,广泛的生物侵蚀和特定的时间段(古近纪)或位于欠透光型台地,方解石矿物为主体,开阔大陆架和斜坡的形成等等,也都可以解释为异能生物,或冷水碳酸盐。赤道碳酸盐的这种"双重人格"大多可以装作看不见(Wilson,2002),尽管认识到某些热带系统难以与温带组合识别区分出来(Lees,1975;Carannante 等,1988;Pascal 和 Wilson,2010)。

Lees(1975)推测,冷水碳酸盐延伸进入低纬度暖水时,替代热带碳酸盐是由于正常盐度的海水被淡水稀释的结果(Nelson,1988)。Carannante 等(1988)首次显示了养分如何在暖水背景下产生相似的冷水碳酸盐。养分的作用和减少光透射显示会促进欠透光型的常常以方解石为主的台地发育(Wilson 和 Vecsei,2005)。换句话说,有可能产生错误解释,把古代赤道碳酸盐当作温带组合。例如,在地中海,曾经被解释为冷水碳酸盐的沉积现在重新解释的结果是潮湿的热带沉积,主要依据来自详细的古生态学研究结合独立的温度资料(Pomar 等,2004)。尽管赤道碳酸盐显示与热带—温度过渡层的相似性(大约 16~22℃;James 的异能生物带,1997;Betzler 等,1997;或 James 和 Lukasik 的异能生物过渡带,2010),但它们在含有大量光能自养生物和一定的暖水单元方面是不同的(Wilson 和 Vecsei,2005)。

识别新生代以外的区域赤道碳酸盐发育的建议是:(1)多种方解石和文石的光能自养生物,加(2)常常见到异能生物相组合单元,加(3)暖温(大于 22℃)的独立证据(例如同位素)。此外指向潮湿赤道背景的依据有:(1)合适的古纬度条件;(2)缺少蒸发岩沉积组合,包粒或球粒;(3)海水盐度减少和养分上升的地球化学证据。但是,与大多数碳酸盐一样,目前还没有识别气候控制评估的可靠标准(James,1997),而且只有通过不同属性的堆积重量评估才能推测潮湿赤道碳酸盐。

方解石为主的暖水赤道碳酸盐发育地区(欠透光型台地),或新生代(古近纪)发育时期,可能会有相似的过去沉积环境条件的线索(Nelson,1988;Wilson,2008a)。在新近纪,许多方解石为主的组合与低海拔养分地区、光透度减少以及或多或少的水流和光能自养生物的欠透光有关。此外,一定海拔不同时期的大气 CO_2 变化连同潮湿地区海水盐度减少,会因文石饱和度降低而促进方解石组分的发育(Kleypas 等,1999;Hoegh-Guldberg 等,2007;Wilson,2008a)。暖水方解石组合的广泛分布可能与海拔、CO_2 的温室效应时期(如古近纪)和增加了海洋酸化有关(Wilson,2008a)。或者,增加了海洋通风的时期,养分上涌或溢出(可能与气候变化及造山有关)会促进方解石光能自养生物,这在过去可以有欠透光型生命模式。显生宙不同的文石和方解石碳酸盐岩生产者的演化和灭绝都会在评价中考虑到,例如固着蛤类和许多珊瑚在白垩纪和古近—新近纪交界处灭绝。

1.6 结论

(1)"从过程到产品"的方法对于主要气候带相关的碳酸盐体系研究是适用的。可是,其他的区域因素或全球因素如生物进化,海洋地理学,全球气候变化和构造形态必须加以考虑。

(2)在赤道热带,暖水作用结合碎屑,清水,养分注入,区域构造作用,是碳酸盐体系发育的主要控制因素。这些因素影响了碳酸盐的产出,最终的沉积物,所建造的碳酸盐台地形态,以及随后的成岩作用,如东南亚实例所示。

(3)潮湿赤道地区产生的碳酸盐产物常常与研究充分的干旱热带暖水碳酸盐模型不一致。

(4)赤道碳酸盐主要是生物碎屑,包括光依赖型和养分依赖型。包粒和团粒几乎不出现,似球粒极少保存,这是由于盐度减少地区早期海相胶结物缺乏。同时代的碎屑注入常常出现在海岸碳酸盐中,在极浅水仅仅发育低起伏特征。与蒸发岩的组合看不见。发育的碳酸盐岩包括了混合的碳酸盐岩—碎屑岩大陆架,孤立的台地,环形大陆架和环形礁。可是,较深水欠透光型沉积为主的非环形大陆架和台地常常处于养分注入区,或在方解石成分为主的沉积时期。高的降雨量和侵蚀性的水域导致大量的溶解作用、喀斯特作用,伴随碳酸盐沉淀发生在陆上暴露区。大气地下水使得沿岸地区和大陆架地区文石向方解石稳定化发展。埋藏压实作用和淋滤作用常常出现,块状白云石化多数与来自陆地的地下水或海水有关。

(5)新生代赤道碳酸盐是暖水光能生物相组合的成员,包括多样的珊瑚,较大的底栖有孔虫(benthic foraminifera)和仙掌藻属(*Halimeda*)。可是,难以从特征上区分冷水碳酸盐(例如,缺少海相胶结物的养分依赖型生物,可能是方解石为主的,在台地结构和成岩作用方面有些相似)。

(6)识别区域显生宙赤道碳酸盐发育的建议是:① 多种方解石和文石的光能自养生物;加 ② 常常见到异能生物相组合单元;③ 加暖温(大于22°C)的独立证据(例如同位素)。此外指向潮湿赤道背景的依据有:① 合适的古纬度条件;② 缺少蒸发岩沉积组合,包粒或球粒;③ 海水盐度减少和养分上升的地化证据。

(7)正如 Nelson(1988)所写引自 Wilkinson(1982)的关于碳酸盐岩沉积学的基础,"很可能我们过于关注现代碳酸盐体系的地区——蓝天、清水和温暖的条件"。但是也要考虑了解温暖却常常阴暗的发育在雷雨天气的赤道碳酸盐的世界。希望本文的贡献可以让人们更好地掌握和发展全球预测模型,以更好地了解过去和未来可能的环境变化。

参 考 文 献

Abram, N. J. , Gagan, M. K. , McCulloch, M. T. , Chappell, J. and Hantoro, W. S. (2003) Coral reef death during the 1997Indian Ocean Dipole linked to Indonesian Wildfires. Science, 301, 952 – 955.

Abram, N. J. , McGregor, H. V. , Gagan, M. K. , Hantoro, W. S. and Suwargadi, B. W. (2009) Oscillations in the southern extent of the Indo – Pacific Warm Pool during the mid-Holocene. Quatern. Sci. Rev. , 28, 2794 – 2803.

Ali, M. Y. (1995) Carbonate cement stratigraphy and timing of diagenesis in a Miocene mixed carbonate-clastic sequence, offshore Sabah, Malaysia: constraints from cathodoluminescence, geochemistry, and isotope studies. Sed. Geol. , 99, 191 – 214.

Allen, P. A. (2008) From landscapes into geological history. Nature, 451, 274 – 276.

Allen, G. P. and Chambers, J. L. C. (1998) Sedimentation in the Modern and Miocene Mahakam Delta. Indonesian Petroleum Association, Jakarta, Indonesia, 236 pp.

Andrews, J. E. , Brimblecombe, P. , Jickells, T. D. and Liss, P. S. (1996) An Introduction to Environmental Chemistry. Blackwell Science, Oxford, 209 pp.

Aycliffe, L. K. , Bird, M. I. , Gagan, M. K. , Isdale, P. J. , Scott-Gagan, H. , Parker, B. , Griffin, D. , Nongkas, M. and McCulloch, M. T. (2004) Geochemistry of corals from Papua New Guinea as a proxy for ENSO ocean-atmosphere interactions in the Pacific Warm Pool. Cont. Shelf Res. , 24, 2343 – 2356.

Bachtel, S. L. , Kissling, R. D. , Martono, D. , Rahardjanto, S. P. , Dunn, P. A. and MacDonald, B. A. (2004) Seismic stratigraphic evolution of the Miocene – Pliocene Segitiga Platform, East Natuna Sea, Indonesia: the origin, growth, and demise of an isolated carbonate platform. In: Seismic Imaging of Carbonate Reservoirs and Systems (Eds G. P. Eberli, J. L. Masafero and J. F. Sarg) , AAPG Mem. , 81, pp. 291 – 308.

Bala′zs, D. (1973) Comparitive morphogenetical study of karst regions in tropical and temperate areas with examples from Celebes and Hungary. Trans. Cave Res. Group Great Brit. , 15, 1 – 7.

Basith, A. , Hendiarti, N. and Syohraeni. (2011) Mapping Chlorophyll-a of Southern Java Waters Using Seawifs and MODIS Aqua. GIS Monitoring. Available at: http: // www. goo gle. com. au/imgres? imgurl = http: // www. gisdevelopment. net/application/nrm/coastal/wetland/images/ma05 _ 238a. jpg (accessed on 3 January, 2011).

Beavington-Penney, S. J. , Nadin, P. , Wright, V. P. , Clarke, E. , McQuilken, J. and Bailey, H. W. (2008) Reservoir qualityvariation on an Eocene carbonate ramp, El Garia Formation, offshore Tunisia: Structural control of burial corrosion and dolomitisation. Sed. Geol. , 209, 42 – 57.

Betzler, C. , Brachert, T. C. and Nebelsick, J. (1997) The warm tempcratc carbonate province. A review of the facies, zonations and delimitations. Courier Forshungs-Institut Senckenberg 201, 83 – 99.

Burollet, P. F. , Boichard, R. , Lambert, B. and Villain, J. M. (1986) The Pater Noster Carbonate Platform. IndonesianPetroleum Association, Proceedings of the 15th Annual Convention, Jakarta, 155 – 169.

Camoin, G. F. and Davies, P. J. (Eds) (1998) Reefs and carbonate platforms in the Pacific and Indian Oceans. Int. Assoc. Sedimentol. Spec. Publ. , 25, 328.

Carannante, G. , Esteban, M. , Milliman, J. D. and Simone, L. (1988) Carbonate lithofacies as palaeolatitude indicators: problems and limitations. Sed. Geol. , 60, 333 – 346.

Carnell, A. J. H. and Wilson, M. E. J. (2003) A review of dolomites in Southeast Asia. Proceedings of the 29th Indonesian Petroleum Association, Jakarta, 413 – 424.

Carnell, A. J. H. and Wilson, M. E. J. (2004) Dolomites in Southeast Asia – varied origins and implications for hydrocarbon exploration. In: The Geometry and Petrogenesis of Dolomite Hydrocarbon Reservoirs (Eds C. J. R. Braithwaite, G. Rizzi and G. Darke) , Geol. Soc. London Spec. Publ. , 235, 255 – 300.

Carter, D. and Hutabarat, M. (1994) The geometry and seismic character of mid-late Miocene carbonate sequences, SS area, offshore northwest Java. Proceedings of the 23rd Annual Convention, Indonesian Petroleum Association, Jakarta, 323 – 338.

Choquette, P. W. and James, N. P. (1988) Introduction. In: Paleokarst (Eds N. P. James and P. W. Choquette), pp. 1 – 21. Springer-Verlag, New York.

Epting, M. (1980) Sedimentology of Miocene carbonate buildups, Central Luconia, offshore Sarawak. Bull. Geol. Soc. Malaysia, 12, 17 – 30.

Esteban, M. and Klappa, C. F. (1983) Subaerial exposure environment. In: Carbonate Depositional Environments (Eds P. A. Scholle, D. G. Bebout and C. H. Moore), AAPG Mem. , 33, 1 – 54.

Esteban, M. and Taberner, C. (2003) Secondary porosity development during late burial in carbonate reservoirs as a result of mixing and/or cooling of brines. J. Geochem. Explor. , 78 – 79, 355 – 359.

Ford, D. C. and Williams, P. C. (2007) Karst Geomorphology and Hydrology. Wiley, Chichester, UK, 562 pp.

Fournier, F. , Montaggioni, L. and Borgomano, J. (2004) Paleoenvironments and high-frequency cyclicity from Cenozoic South-East Asian shallow-water carbonates: a case study from the Oligo-Miocene buildups of Malampaya (Offshore Palawan, Philippines). Mar. Petrol. Geol. , 21, 1 – 21.

Fulthorpe, C. S. and Schlanger, S. O. (1989) Paleo-oceanographic and tectonic settings of early Miocene reefs and associated carbonates of offshore southeast Asia. AAPG Bull. , 73, 729 – 756.

Gagan, M. K. , Aycliffe, L. K. , Hopley, D. , Cali, J. A. , Mortimer, G. E. , Chappell, J. , McCulloch, M. T. and Head, M. J. (1998) Temperature and surface-ocean water balance of the mid-Holocene tropical Western Pacific. Science, 279, 1014 – 1018.

Gibson-Robinson, C. and Soedirdja, H. (1986) Transgressive development of Miocene reefs, Salawati Basin, Irian Jaya. Proceedings of the 15th Annual Convention, Indonesian Petroleum Association, Jakarta, 377 – 403.

Gischler, E. (2011) Sedimentary facies of Bora Bora, Darwin's type barrier reef (Society Islands, South Pacific): the unexpected occurrence of non-skeletal grains. J. Sed. Res. , 81, 1 – 17.

Gischler, E. and Lomando, A. J. (1999) Recent sedimentary facies of isolated carbonate platforms, Belize-Yucatan system, Central America. J. Sed. Res. , 69, 747 – 763.

Gong, S. Y. , Mii, H. S. , Wei, K. Y. , Horng, C. S. , You, C. F. , Huang, F. W. , Chi, W. R. , Yui, T. F. , Torng, P. K. , Huang, S. T. , Wang, S. W. , Wu, J. C. and Yang, K. M. (2005) Dry climate near the Western Pacific Warm Pool: Pleistocene caliches of the Nansha Islands, South China Sea. Palaeogeogr. Palaeoclimatol. Palaeoecol. , 226, 205 – 213.

Gordon, A. L. (2005) Oceanography of the Indonesian Seas and their throughflow. Oceanography, 18, 14 – 27.

Gordon, A. L. , Dwi Susanto, R. and Vrane, K. (2003) Cool Indonesian throughflow as a consequence of restricted surface layer flow. Nature, 425, 824 – 828.

Grötsch, J. and Mercadier, C. (1999) Integrated 3-D reservoir modeling based on 3-D seismic: the Tertiary Malampaya and Camago buildups, offshore Palawan, Philippines. AAPG Bull. , 83, 1703 – 1728.

Hall, R. (1996) Reconstructing Cenozoic SE Asia. In: Tectonic Evolution of Southeast Asia (Eds R. Hall and D. J. Blundell), Geol. Soc. London Spec. Publ. , 106, 153 – 184.

Hall, R. (2002a) Cenozoic geological and plate tectonic evolution of SE Asia and the SW Pacific: computer based reconstructions, models and animations. J. Asian Earth Sci. , 20, 353 – 431.

Hall, R. (2002b) SE Asian heatflow: call for new data. Indonesian Petrol. Assoc. Newslett. , 4, 20 – 21.

Hall, R. and Nichols, G. (2002) Cenozoic sedimentation and tectonics in Borneo: climatic influences on orogenesis. In: Sediment Flux to Basins: Causes, Controls and Consequences (Eds S. J. Jones and L. E. Frostick), Geol. Soc. London Spec. Publ. , 191, 5 – 22.

Hallock, P. and Schlager, W. (1986) Nutrient excess and the demise of coral reefs and carbonate platforms. Palaios, 1, 389 – 398.

Hamilton, W. (1979) Tectonics of the Indonesian region. US Geol. Surv. Prof. Pap., paper number 1078, 345 pp.

Hayton, S., Nelson, C. S. and Hood, S. D. (1995) A skeletal assemblage classification system for non-tropical carbonate deposits based on New Zealand Cenozoic limestones. Sed. Geol., 100, 123 – 141.

Heikoop, J. M., Tsujita, C. J., Risk, M. J. and Tomascik, T. (1996) Corals as proxy recorders of volcanic activity: evidence from Banda Api, Indonesia. Palaios, 11, 286 – 292.

Hendry, J. P., Taberner, C., Marshall, J. D., Pierre, C. and Carey, P. F. (1999) Coral reef diagenesis records pore-fluid evolution and paleohydrology of a siliciclastic basin margin succession (Eocene South Pyrenean foreland basin, northeastern Spain). Geol. Soc. Am. Bull., 111, 395 – 411.

Heubeck, C., Story, K., Peng, P., Sullivan, C. and Duff, S. (2004) An integrated reservoir study of the Liuhua 11-1 field using a high-resolution three-dimensional seismic data set. In: Seismic Imaging of Carbonate Reservoirs and Systems (Eds G. P. Eberli, J. L. Masaferro and J. F. Sarg), AAPG Mem., 81, 149 – 168.

Hoegh-Guldberg, O. (1999) Climate change, coral bleaching and the future of the world's coral reefs. Mar. Freshwat. Res., 50, 839 – 866.

Hoegh-Guldberg, O., Mumby, P. J., Hooten, A. J., Steneck, R. S., Greenfield, P., Gomez, E., Harvell, C. D., Sale, P. F., Edwards, A. J., Caldeira, K., Knowlton, N., Eakin, C. M., Iglesias-Prieto, R., Muthiga, N., Bradbury, R. H., Dubi, A. and Hatziolos, M. E. (2007) Coral reefs under rapid climate change and ocean acidification. Science 318, 1737 – 1742.

Hook, J. and Wilson, M. E. J. (2003) Stratigraphic relationships of a Miocene mixed carbonate-siliciclastic interval in the Badak field, East Kalimantan, Indonesia. Proceedings of the 29th Indonesian Petroleum Association, Jakarta, 398 – 412.

Hughes, T. P., Baird, A. H., Bellwood, D. R., Card, M., Connolly, S. R., Folke, C., Grosberg, R., Hoegh-Guldberg, O., Jackson, J. B. C., Kleypas, J., Lough, J. M., Marshall, P., Nyström, M., Palumbi, S. R., Pandolfi, J. M., Rosen, B. and Roughgarden, J. (2003) Climate change, human impacts, and the resilience of coral reefs. Science, 301, 929 – 933.

Insalaco, E. (1998) The descriptive nomenclature and classification of growth fabrics in fossil scleractinian reefs. Sed. Geol., 118, 159 – 186.

James, N. P. (1997) The cool-water depositional realm. In: Cool-Water Carbonates (Eds N. P. James and J. A. D. Clarke), SEPM Spec. Publ., 56, 1 – 20.

James, N. P. and Choquette, P. W. (1984) Diagenesis 9-limestones-the meteoric diagenetic environment. Geosci. Can., 11, 161 – 194.

James, N. P. and Choquette, P. W. (1988) Paleokarst. Springer-Verlag, New York and Berlin, 416 pp.

James, N. P. and Ginsburg, R. N. (1979) The seaward margin of Belize Barrier and Atoll Reefs. Int. Assoc. Sedimentol. Spec. Publ., 3, 191.

James, N. P. and Kendall, A. C. (1992) Introduction to carbonate and evaporate facies models. In: Facies Models, Response to Sea Level Change (Eds R. G. Walker and N. P. James), pp. 265 – 275. Geological Association of Canada, Ontario.

James, N. P. and Lukasik, J. (2010) Cool-and cold-water neritic carbonates. In: Facies Models 4 (Eds N. P. James and R. W. Dalrymple), pp. 371 – 399. Geological Association of Canada GEOtext 6, St John's, Newfoundland.

Jennings, J. N. (1985) Karst Geomorphology. Blackwell, Oxford, 293 pp.

Jia, G., Peng, P., Zhao, Q. and Jian, Z. (2003) Changes in terrestrial ecosystems since 30 Ma in East Asia: stable isotope evidence from black carbon in the South China Sea. Geology, 31, 1093 – 1096.

Jones, B. and Desrochers, A. (1992) Shallow platform carbonates. In: Facies Models, Response to Sea Level Change (Eds R. G. Walker and N. P. James), pp. 277 – 302. Geological Association of Canada, Ontario.

Jordan, C. J. (1998) The Sedimentation of Kepuluan Seribu: A Modern Patch Reef Complex in the West Java Sea, Indonesia. Indonesian Petroleum Association Field Guide, Jakarta, 81 pp.

Kemp, G. (1992) The Manusela Formation – an example of a Jurassic carbonate unit of the Australian Plate from Seram, Eastern Indonesia. In: Carbonate Rocks and Reservoirs of Indonesia: A Core Workshop (Eds C. T. Siemers, M. W. Longman, R. K. Park and J. G. Kaldi), pp. 11 – 1 – 11 – 41. Indonesian Petroleum Association, Core Workshop Notes No. 1, Jakarta.

Kindler, P. and Wilson, M. E. J. (2010) Carbonate grain associations: their use and environmental significance, a brief review. In: Carbonate Systems During the Oligocene-Miocene Climatic Transition (Eds M. Mutti, W. E. Piller and C. Betzler), Int. Assoc. Sedimentol. Spec. Publ. , 42, 35 – 47.

Kleypas, J. A. , Buddemeier, R. W. , Archer, D. , Gattuso, J. -P. , Langdon, C. and Opdyke, B. N. (1999) Geochemical consequences of increased atmospheric carbon dioxide on reefs. Science, 284, 118 – 120.

Kuhnt, W. , Holbourn, A. , Hall, R. , Zuvela, M. and Käse, R. (2004) Neogene history of the Indonesian Throughflow. In: Continent-Ocean Interactions Within the East Asian Marginal Seas (Eds P. Clift, P. Wang, W. Kuhnt and D. E. Hayes), AGU Geophys. Monogr. , 149, 299 – 320.

Larcombe, P. , Costen, A. and Woolfe, K. J. (2001) The hydrodynamic and sedimentary setting of nearshore coral reefs, central Great Barrier Reef shelf, Australia: Paluma Shoals, a case study. Sedimentology, 48, 811 – 835.

Leamon, G. R. and Parsons, G. L. (1986) Tertiary carbonate plays in the Papuan Basin. SEAPEX Proc. , VII, 213 – 227.

Lee, K. , Tong, L. T. , Millero, F. J. , Sabine, C. L. , Dickson, A. G. , Goyet, C. , Park, G. H. , Wanninkhof, R. , Feely, R. A. and Key, R. M. (2006) Global relationships of total alkalinity with salinity and temperature in surface waters of the world's oceans. Geophys. Res. Lett. , 33, L19605, doi: 10. 1029/2006GL027207.

Lees, A. (1975) Possible influence of salinity and temperature on modern shelf carbonate sedimentation. Mar. Geol. , 19, 159 – 198.

Lees, A. and Buller, A. T. (1972) Modern temperate-water and warm-water shelf carbonate sediments contrasted. Mar. Geol. , 13, 67 – 73.

Lockwood, J. G. (1974) World Climatology an Environmental Approach. Edward Arnold, London, 330 pp.

Lokier, S. W. (2000) The Miocene Wonosari Formation, Java, Indonesia: volcaniclastic influences on carbonate platform sedimentation. PhD Thesis, University of London, London, 648 pp.

Lokier, S. W. , Wilson, M. E. J. and Burton, L. M. (2009) Marine biota response to clastic sediment influx: a quantitative approach. Palaeogeogr. Palaeoclimatol. Palaeoecol. , 281, 25 – 42.

Longman, M. W. (1985) Fracture porosity in reef talus of a Miocene pinnacle-reef reservoir, Nido B Field, the Philippines. In: Carbonate Petroleum Reservoirs (Eds P. O. Roehl and P. W. Choquette), pp. 547 – 561. Springer-Verlag, Berlin.

Loucks, R. G. (1999) Paleocave carbonate reservoirs: origins, burial-depth modifications, spatial complexity, and reservoir implications. AAPG, 83, 1795 – 1834.

Madden, R. H. C. and Wilson, M. E. J. (in press) Diagenesis of delta-front patch reefs: a model for alteration of coastal siliciclastic-influenced carbonates from humid equatorial regions. J. Sed. Res. .

Maire, R. (1981) Karst and hydrology synthesis. Spelunca, Suppl. , 3, 23 – 30.

Mayall, M. J. and Cox, M. (1988) Deposition and diagenesis of Miocene limestones, Senkang Basin, Sulawesi, Indonesia. Sed. Geol. , 59, 77 – 92.

Milliman, J. D. (1974) Marine Carbonates. Springer, New York, 375 pp.

Milliman, J. D. , Farnsworth, K. I. and Albertin, C. S. (1999) Flux and fate of fluvial sediments leaving large islands in the East Indies. J. Sea Res. , 41, 97 – 107.

Montaggioni, L. F. (2005) History of Indo-Pacific coral reef systems since the last glaciation: development patterns and controlling factors. Earth Sci. Rev. , 71, 1 – 75.

Moore, C. H. (2001) Carbonate reservoirs: porosity evolution and diagenesis in a sequence stratigraphic framework. Elsevier, Dev. Sedimentol., 55, 444.

Morley, R. J. (1999) Origin and Evolution of Tropical Rainforests. John Wiley and Sons, Chichester, UK, 362 pp.

Mutti, M. and Hallock, P. (2003) Carbonate systems along nutrient and temperature gradients: some sedimentological and geochemical constraints. Int. J. Earth Sci., 92, 465–475.

Mylroie, J. E. and Carew, J. L. (1990) The flank margin model for dissolution cave development in carbonate platforms. Earth Surf. Proc. Land., 15, 413–424.

Nelson, C. S. (1988) Introductory perspective on non-tropical shelf carbonates. In: Non-tropical Shelf Carbonate: Modern and Ancient (Ed. C. S. Nelson),. Sed. Geol., 60, 3–12.

Nelson, C. S., Winefield, P. R., Hood, S. D., Caron, V., Pallentin A. and Kamp, P. J. J. (2003) Pliocene Te Aute limestones, New Zealand: expanding concepts for cool-water shelf carbonates. NZ J. Geol. Geophys., 46, 407–424.

Netherwood, R. and Wight, A. (1992) Structurally-controlled, linear reefs in a Pliocene Delta-front setting, Tarakan Basin, Northeast Kalimantan. In: Carbonate Rocks and Reservoirs of Indonesia (Eds C. T. Siemers, M. W. Longman, R. K. Park and J. G. Kaldi), pp. 3–1–3–37. Indonesian Petroleum Association Core Workshop Notes 1, Ch. 3.

Pagani, M., Zachos, J. C., Freeman, K. H., Tipple, B. and Bohaty, S. (2005) Marked decline in atmospheric carbon-dioxide concentrations during the Paleogene. Science, 309, 600–603.

Pandolfi, J. M., Tudhope, A. W., Burr, G., Chappell, J., Edinger, E., Frey, M., Steneck, R., Sharma, C., Yeates, A., Jennions, M., Lescinsky, H. and Newton, A. (2006) Mass mortality following disturbance in Holocene coral reefs from Papua New Guinea. Geology, 34, 949–952.

Park, R. K., Siemers, C. T. and Brown, A. A. (1992) Holocene carbonate sedimentation, Pulau Seribu, Java Sea, – the third dimension. In: Carbonate Rocks and Reservoirs of Indonesia, a Core Workshop (Eds C. T. Siemers, M. W. Longman, R. K. Park and J. G. Kaldi), pp. 2–1–2–15. Indonesian Petroleum Association, Jakarta.

Park, R. K., Matter, A. and Tonkin, P. C. (1995) Porosity evolution in the Batu Raja Carbonates of the Sunda Basin-Windows of opportunity. Proceedings of the 24th Annual Convention, Indonesian Petroleum Association, Jakarta, 63–184.

Park, R. K., Crevello, P. D. and Hantoro, W. (2010) Equatorial carbonate depositional systems of Indonesia. In: Cenozoic Carbonate Systems of Australasia (Eds W. A. Morgan, A. D. George, P. M. Harris, J. A. Kupecz and J. F. Sarg), SEPM (Soc. Sed. Geol.), Spec. Publ., 95, 41–77.

Partin, J. W., Cobb, K. M., Adkins, J. F., Clark, B. and Fernandez, D. P. (2007) Millennial-scale trends in west Pacific warm pool hydrology since the Last Glacial Maximum. Nature, 449, 452–455.

Paterson, R. J., Whitaker, F. F., Jones, G. D., Smart, P. L., Waltham, D. and Felce, G. (2006) Accommodation and sedimentary architecture of isolated icehouse carbonate platforms: insights from forward modeling with Carb3D. J. Sed. Res., 76, 1162–1182.

Peñaflor, E. L., Skirving, W. J., Strong, A. E., Heron, S. F. and David, L. T. (2009) Sea-surface temperature and thermal stress in the Coral Triangle over the past two decades. Coral Reefs, 28, 841–850.

Perrin, C., Bosence, D. and Rosen, B. (1995) Quantitative approaches to palaeozonation and palaeobathymetry of corals and coralline algae in Cenozoic reefs. In: Marine Palaeoenvironmental Analysis from Fossils (Eds D. W. J. Bosence and P. A. Allison), Geol. Soc. Spec. Publ., 83, 181–229.

Perry, C. T. (1999) Biofilm-related calcification, sediment trapping and constructive micrite envelopes: a criterion for the recognition of ancient grass-bed environments. Sedimentology, 46, 33–45.

Perry, C. T. and Taylor, K. G. (2006) Inhibition of dissolution within shallow water carbonate sediments: impacts of terrigenous sediment input on syn-depositional carbonate diagenesis. Sedimentology, 53, 495–513.

Pichler, T., Heikoop, J. M., Risk, M. J., Veizer, J. and Campbell, I. L. (2000) Hydrothermal effects on isotope and trace element records in modern reef corals: A study of Porites lobata from Tutum Bay, Ambitle Island, Papua New Guinea. Palaios, 15, 225 – 234.

Pigram, C. J., Davies, P. J., Feary, D. A., Symonds, P. A. and Chaproniere, G. C. H. (1990) Controls on tertiary carbonate platform evolution in the Papuan Basin: new play concepts. In: Petroleum Exploration in Papua New Guinea (Eds G. J. Carman and Z. Carman), pp. 185 – 195. Proceedings of the 1st PNG Petroleum Convention, Port Moresby.

Pireno, G. E., Cook, C., Yuliong, D. and Lestari, L. (2009) Berai carbonate debris flow as reservoir in the Ruby Field, Sebuku Block, Makassar Straits: a new exploration play in Indonesia. roceedings of the 33rd Indonesian Petroleum Association Convention, Jakarta, Indonesia, 15 pp.

Pomar, L. (2001) Types of carbonate platforms: a genetic approach. Basin Res., 13, 313 – 334.

Pomar, L., Brandano, M. and Westphal, H. (2004) Environmental factors influencing skeletal grain sediment associations: a critical review of Miocene examples from the western Mediterranean. Sedimentology, 51, 627 – 651.

Potts, D. C. (1983) Evolutionary disequilibrium among Indo-Pacific corals. Bull. Mar. Sci., 33, 619 – 632.

Purdy, E. G. and Waltham, D. (1999) Reservoir implications of modern karst topography. AAPG Bull., 83, 1774 – 1794.

Rankey, E. C. and Reeder, S. L. (2009) Holocene ooids of Aitutaki Atoll, Cook Islands, South Pacific. Geology, 37, 971 – 974.

Reid, R. P. and MacIntyre, I. G. (2000) Microboring versus recrystallization: further insight into the micritization process. J. Sed. Res., 70, 24 – 28.

Renema, W., Hoeksema, B. W. and van Hinte, J. E. (2001) Larger benthic foraminifera and their distribution patterns on the Spermonde shelf, South Sulawesi. Zoologishe Verhandelingen Leiden, 334, 115 – 149.

Renema, W., Bellwood, D. R., Braga, J. C., Bromfield, K., Hall, R., Johnson, K. G., Lunt, P., Meyer, C. P., McMonagle, L., Morley, R. J., O, Dea, A., Todd, J. A., Wesselingh, F. P., Wilson, M. E. J. and Pandolfi, J. M. (2008) Hopping hotspots: global shifts in marine biodiversity. Science, 321, 654 – 657.

Risk, M. J., Sherwood, O. A., Heikoop, J. M. and Llewellyn, G. (2003) Smoke signals from corals: isotopic signature of the 1997 Indonesian 'haze' event. Mar. Geol., 202, 72 – 78.

Roberts, H. H. and Sydow, J. (1996) The offshore Mahakam Delta: stratigraphic response of late Pleistocene to modern sea level cycle. Proceedings of the 25th Annual Convention, Indonesian Petroleum Association, Jakarta, 147 – 161.

Roberts, H. H., Amaron, P. and Phipps, C. V. (1988) Morphology and sedimentology of Halimeda bioherms from Eastern Java Sea (Indonesia). Coral Reefs, 6, 61 – 172.

Rudolph, K. W. and Lehmann, P. J. (1989) Platform evolution and sequence stratigraphy of the Natuna platform, South China Sea. In: Controls on Carbonate Platform and Basin Development (Eds P. D. Crevello, J. L. Wilson, F. Sarg and J. F. Read), SEPM Spec. Publ., 44, 353 – 361.

Rumney, G. R. (1968) Climatology and the World's Climates. Macmillan, New York, 656 pp.

Sale, C. (2002) Our Wonderful World. Longman/Pearson Education, Melbourne, 424 pp.

Saller, A. H. and Vijaya, S. (2002) Depositional and diagenetic history of the Kerendan carbonate platform, Oligocene, central Kalimantan, Indonesia. J. Petrol. Geol., 25, 123 – 150.

Saller, A., Armin., R., Ichram, L. O. and Glenn-Sullivan, C. (1993) Sequence stratigraphy of aggrading and backstepping carbonate shelves, Central Kalimantan, Indonesia. In: Carbonate Sequence Stratigraphy, Recent Developments and Applications (Eds R. G. Loucks and J. F. Sarg), AAPG Mem., 57, 267 – 290.

Sanders, D. (2003) Syndepositional dissolution of calcium carbonate in neritic carbonate environments: geological recognition, processes, potential significance. J. Afr. Earth Sci., 36, 99 – 134.

Sanders, D. and Baron-Szabo, R. C. (2005) Scleractinian assemblages under sediment input: their characteristics and relation to the nutrient input concept. Palaeogeogr. Palaeoclimatol. Palaeoecol. ,216,139 – 181.

Sattler, U. , Zampetti, V. , Schlager, W. and Immenhauser, A. (2004) Late leaching under deep burial conditions: a case study from the Miocene Zhujiang Carbonate Reservoir, South China Sea. Mar. Petrol. Geol. ,21,977 – 992.

Satyana, A. W. (2005) Oligo-Miocene carbonates of Java, Indonesia: tectonic-volcanic setting and petroleum implications. Proceedings of the 30th Annual Convention, Indonesian Petroleum Association, Jakarta, 217 – 249.

Schlager, W. (1992) Sedimentology and sequence stratigraphy of reefs and carbonate platforms: a short course. AAPG, Continuing Education Course Note Series, 34, 71.

Schlager, W. (2003) Benthic carbonate factories of the Phanerozoic. Int. J. Earth Sci. ,92,445 – 464.

Scoffin, T. P. , Tudhope, A. W. and Brown, B. E. (1989) Fluorescent and skeletal density banding in Porites lutea from Papua New Guinea and Indonesia. Coral Reefs,7,169 – 178.

Sidi, F. H. , Nummendal, D. , Imbert, P. , Darman, H. and Posamentier, H. W. (2003) Tropical deltas of SE Asia – sedimentology, stratigraphy and petroleum geology. SEPM Spec. Publ. ,76,276.

Staub, J. R. and Esterle, J. S. (1993) Provenance and sediment dispersal in the Rajang River delta/coastal plain system, Sarawak, East Malaysia. Sed. Geol. ,89,91 – 106.

Stoddart, D. R. (Ed.) (2007) Tsunamis and coral reefs. Atoll Res. Bull. ,544,164.

Sun, S. Q. and Esteban, M. (1994) Paleoclimatic controls on sedimentation, diagenesis and reservoir quality: lessons from Miocene carbonates. AAPG Bull. ,78,519 – 543.

Tcherepanov, E. N. , Droxler, A. W. , Lapointe, P. , Dickens, G. R. , Bentley, S. J. , Beaufont, L. , Peterson, L. C. , Daniell, J. and Opdyke, B. N. (2008a) Neogene evolution of the mixed carbonate-siliciclastic system in the Gulf of Papua, Papua New Guinea. J. Geophys. Res. ,113, F01S21.

Tcherepanov, E. N. , Droxler, A. W. , Lapointe, P. and Mohn, K. (2008b) Carbonate seismic stratigraphy of the Gulf of Papua mixed depositional system: Neogene stratigraphic signature and eustatic control. Basin Res. ,20,185 – 209.

Testa, V. and Bosence, D. W. J. (1998) Carbonate-siliciclastic sedimentation on a high-energy, ocean-facing, tropical ramp, NE Brazil. In: Carbonate Ramps (Eds V. P. Wright and T. P. Burchette), Geol. Soc. London Spec. Publ. , 149,55 – 71.

Titlyanov, E. A. and Latypov, Y. Y. (1991) Light-dependence in scleractinian distribution in the sublittoral zone of South China Sea Islands. Coral Reefs,10,133 – 138.

Tomascik, T. , Mah, A. J. , Nontji, A. and Moosa, M. K. (1997) The Ecology of the Indonesian Seas. Oxford University Press, Singapore, 1388 pp.

Tudhope, A. W. , Shimmield, G. B. , Chilcott, C. P. , Jebb, M. , Fallick, A. E. and Dalgleish, A. N. (1995) Recent changes in climate in the far western equatorial Pacific and their relationship to the Southern Oscillation; oxygen isotope records from massive corals, Papua New Guinea. Earth Planet. Sci. Lett. ,136,575 – 590.

Tudhope, A. W. , Chilcott, C. P. , McCulloch, M. T. , Cook, E. R. , Chappell, J. , Ellam, R. M. , Lea, D. W. , Lough, J. M. and Shimmield, G. B. (2001) Variability in the El Niño-southern oscillation through a glacial-interglacial cycle. Science,291,1511 – 1517.

Tyrrel, W. W. , Davis, R. G. and McDowell, H. G. (1986) Miocene carbonate shelf margin, Bali-Flores Sea, Indonesia. Proceedings of the 15th Annual Convention, Indonesian Petroleum Association, Jakarta, 124 – 140.

Umbgrove, J. H. F. (1947) Coral reefs of the East Indies. Bull. Geogr. Soc. Am. ,58,729 – 778.

Vahrenkamp, V. C. , David, F. , Duijndam, P. , Newall, M. and Crevello, P. (2004) Growth architecture, faulting, and karstification of a Middle Miocene carbonate platform, Luconia province, offshore Sarawak, Malaysia. In: Seismic Imaging of Carbonate Reservoirs and Systems (Eds G. P. Eberli, J. L. Masafero and J. F. Sarg), AAPG Mem. ,81, 329 – 350.

Visser, K. , Thunell, R. and Goñi, M. A. (2004) Glacial-interglacial organic carbon record from the Makassar Strait,

Indonesia: implications for regional changes in continental vegetation. Quatern. Sci. Rev. ,23,17 – 27.

Walter, L. M. and Burton, E. A. (1990) Dissolution of Recent platform carbonate sediments in marine pore fluids. Am. J. Sci. ,290,601 – 643.

Waltham, T. (1997) Mulu – the ultimate cavernous karst. Geol. Today,13,216 – 222.

Wang, P. ,Clements, S. ,Beaufort, L. ,Braconnot, P. ,Ganssen, G. ,Jian, Z. ,Kershaw, P. and Sarnthein, M. (2005) Evolution and variability of the Asian monsoon system: state of the art and outstanding issues. Quatern. Sci. Rev. , 24,595 – 629.

Wannier, M. (2009) Carbonate platforms in wedge-top basins: an example from the Gunung Mulu National Park, Northern Sarawak (Malaysia). Mar. Petrol. Geol. ,26,177 – 207.

Warrlich, G. , Taberner, C. , Asyee, W. , Stephenson, B. , Esteban, M. , Boya-Ferrero, M. , Dombrowski, A. and Van Kon-ijnenburg, J. -H. (2010) The impact of postdepositional processes on reservoir properties: two case studies of Tertiary carbonate buildup gas fields in Southeast Asia (Malampaya and E11). In: Cenozoic Carbonate Systems of Australasia (Eds W. A. Morgan, A. D. George, P. M. Harris, J. A. Kupecz and J. F. Sarg), SEPM (Soc. Sed. Geol.),Spec. Publ. ,95,99 – 127.

Wilkinson, B. H. (1982) Cyclic cratonic carbonates and Phanerozoic calcite seas. J. Geol. Educ. ,30,189 – 203.

Williams, P. W. (1974) Morphometric analysis of polygonal karst in New Guinea. Geol. Soc. Am. Bull. , 83, 761 – 796.

Wilson, M. E. J. (1995) The Tonasa Limestone Formation, Sulawesi, Indonesia: Development of a Tertiary Carbonate Platform. PhD Thesis, University of London, London 520 pp.

Wilson, M. E. J. (1996) Tertiary Carbonates Study, Mangkalihat Peninsula, Northeast Kalimantan. Confidential Report for MAERSK Oil Maratua Ltd, Indonesia, 63 pp.

Wilson, M. E. J. (1999) Prerift and synrift sedimentation during early fault segmentation of a Tertiary carbonate platform, Indonesia. Mar. Petrol. Geol. ,16,825 – 848.

Wilson, M. E. J. (2000) Tectonic and volcanic influences on the development and diachronous termination of a tropical carbonate platform. J. Sed. Res. ,70,310 – 324.

Wilson, M. E. J. (2002) Cenozoic carbonates in Southeast Asia: implications for equatorial carbonate development. Sed. Geol. ,147,295 – 328.

Wilson, M. E. J. (2005) Equatorial delta-front patch reef development during the Neogene, Borneo. J. Sed. Res. ,75, 116 – 134.

Wilson, M. E. J. (2008a) Global and regional influences on equatorial shallow marine carbonates during the Cenozoic. Palaeogeogr. Palaeoclimatol. Palaeoecol. ,265,262 – 274.

Wilson, M. E. J. (2008b) Reservoir quality of Cenozoic carbonate buildups and coral reef terraces. Proceedings of the 31st Indonesian Petroleum Association, Jakarta,8 pp.

Wilson, M. E. J. (2011) SE Asian carbonates: tools for evaluating environmental and climatic change in the equatorial tropics over the last 50 million years. In: The SE Asian Gateway: History and Tectonics of Australia-Asia Collision (Eds R. Hall, M. A. Cottam and M. E. J. Wilson), Geol. Soc. London Spec. Publ. ,355,347 – 369.

Wilson, M. E. J. and Ascaria, N. A. (1996) IPA Carbonates Field Course, South Sulawesi. Indonesian Petroleum Association Field Guide, Jakarta,77 pp.

Wilson, M. E. J. and Bosence, D. W. J. (1996) The tertiary evolution of South Sulawesi: a record in redeposited carbonates of the Tonasa Limestone Formation. In: Tectonic Evolution of SE Asia (Eds R. Hall and D. J. Blundell), Geol. Soc. London Spec. Publ. ,106,365 – 389.

Wilson, M. E. J. and Evans, M. J. (2002) Sedimentology and diagenesis of tertiary carbonates on the Mangkalihat Peninsula, Borneo: implications for subsurface reservoir quality. Mar. Petrol. Geol. ,19,873 – 900.

Wilson, M. E. J. and Hall, R. (2010) Tectonic influence on SE Asian carbonate systems and their reservoir quality. In:Cenozoic Carbonate Systems of Australasia (Eds W. A. Morgan, A. D. George, P. M. Harris, J. A. Kupecz and J. F. Sarg), SEPM (Soc. Sed. Geol.), Spec. Publ. ,95,13 – 40.

Wilson, M. E. J. and Lokier, S. J. (2002) Siliciclastic and volcaniclastic influences on equatorial carbonates; insights from the Neogene of Indonesia. Sedimentology,49,583 – 601.

Wilson, M. E. J. and Rosen, B. R. R. (1998) Implications of the paucity of corals in the Paleogene of SE Asia: plate tectonics or Centre of Origin. In: Biogeography and Geological Evolution of SE Asia (Eds R. Hall and J. D. Holloway), pp. 303 – 337. Backhuys Publishers, Amsterdam, The Netherlands.

Wilson, M. E. J. and Vecsei, A. (2005) The apparent paradox of abundant foramol facies in low latitudes: their environmental significance and effect on platform development. Earth Sci. Rev. ,69,133 – 168.

Wilson, M. E. J. , Bosence, D. W. J. and Limbong, A. (2000) Tertiary syntectonic carbonate platform development in Indonesia. Sedimentology,47,395 – 419.

Wilson, M. E. J. , Evans, M. J. , Oxtoby, N. , Satria Nas, D. , Donnelly, T. and Thirlwall, M. (2007) Reservoir quality, textural evolution and origin of fault-associated dolomites. AAPG Bull. ,91,1247 – 1272.

Woodruffe, C. D. (2000) Deltaic and estuarine environments and their Late Quaternary dynamics on the Sunda and Sahul shelves. J. Asian Earth Sci. ,18,393 – 413.

Zachos, J. , Pagani, M. , Sloan, L. , Thomas, E. and Billups, K. (2001) Trends, rhythms, and aberrations in global climate 65 Ma to present. Science,292,686 – 693.

Zampetti, V. , Schlager, W. , Van Konijnenburg, J. H. and Everts, A. J. (2003) Depositional history and origin of porosity in a Miocene carbonate platform of Central Luconia, offshore Sarawak. Bull. Geol. Soc. Malaysia, 47, 139 – 152.

Ziegler, A. M. , Hulver, M. L. , Lotts, A. L. and Schmachtenberg, W. F. (1984) Uniformitarianism and palaeoclimates: inferences from the distribution of carbonate rocks. In: Fossils and Climate (Ed. P. J. Brenchley), pp. 3 – 25. Wiley, Chichester.

第 2 章 大陆架环境的碳酸盐沉积相模式

JONE J. G. REIJMER,THORSTEN BAUCH,PRISK SCHAFER 著

张天舒 译,吴洛菲 校

摘 要 本次研究对比了巴拿马境内太平洋一侧巴拿马湾和 Chiriquí 湾的两种碳酸盐沉积环境。这两个海湾的地理纬度邻近,在北纬 7°到 9°之间。巴拿马湾和 Chiriquí 湾的海洋环境截然不同。在 Chiriquí 湾,非上升流全年稳定出现。而在巴拿马湾,旱季(12 月到 4 月)会出现强季节性上升流。上升流的变化仅对碳酸盐的数量造成有限的影响,但是却对生产碳酸盐的生物种群产生重大影响。此外,两个海湾碳酸盐的产生和分布还受到岛屿和陆源物质注入的影响。所发现的陆源物质主要由粒度较小的颗粒构成($<63 \sim 250 \mu m$),可以轻易被水流和波浪搬运。在岛屿周围发现的碳酸盐沉积物(碳酸盐砂屑和混合的碳酸盐—硅质碎屑砂屑)主要由粒度较大的颗粒构成($>500 \mu m$)。巴拿马湾和 Chiriquí 湾发育暖温带碳酸盐生物群,从热带(珊瑚)到热带与冷水混合带(珊瑚红藻)以及冷水藤壶(balanids)环境。Chiriquí 湾以贫养和半自养环境为特征,在岛屿周围的浅水地区发育光养生物(珊瑚)和/或者红藻石相,在向陆架边缘的较深水地区主要发育软体动物相。巴拿马湾季节性上升流导致暂时性富养环境,在岛屿周围发育藤壶(balanids)、棘皮动物和软体动物为主的异养生物相。这里的"冷水"碳酸盐动物群和富养环境发育在易产生季节性上升流的热带地区。在巴拿马太平洋陆架发现的明显的沉积相变指示了上升流和非上升流海洋环境变化的重要性,这种海洋环境的变化决定了热带地区碳酸盐的产量和相应的沉积相模式。

关键词 碳酸盐沉积作用 东太平洋 富养 沉积相模式 非上升流 贫养 巴拿马 上升流

2.1 概况

现代海洋有三个重要的生态梯度(Flügel,2004):(1)深度梯度,与光线强度、温度、氧化作用相关;(2)滨岸到开阔海梯度,营养水平从内陆水道向开阔海逐渐降低;(3)纬度梯度,比如,低纬度地区碳酸盐生产率较高。除了这些大规模的变化以外,一系列环境因素也对海洋相组合的发育起到重要作用:(1)温度;(2)盐度;(3)营养。其他重要的因素还有光照、水深、水动力能量、氧气、浑浊度、基质,以及沉积速率和碎屑沉积物的注入(Lees 和 Buller,1972;Lees,1975;Wilson,1975;Flügel,1982;Carannante 等,1988;Hallock 等,1988;Nelson,1988;James,1997;Schlager,2005)。碳酸盐沉积古环境一般与生物组合相关,从而关联到具体的相组合,这些相组合被认为是某种特定海洋环境的标志。

在美洲太平洋陆架上,由于陆架狭窄,陆源物质注入量大,大陆边缘构造活跃,以及地域性的上升流强劲等因素,碳酸盐沉积稀少(Halfar 等,2000)。近年来暖温带碳酸盐—硅质碎屑混合沉积体系的详细研究(Halfar 等,2000,2004a,b,2006)指出营养和温度变化对碳酸盐组合的影响,并展示了碳酸盐工厂的生物组合和相特征,从而论证了碳酸盐生产率很大程度上受控于营养源。珊瑚为主的浅水碳酸盐工厂在每立方米叶绿素 a 含量为 $0.25 \sim 0.48$ mg 之间的水域最为发育(水深:$5 \sim 20$ m;水体温度:25℃;盐度:30.06psu)。这些数值高于红藻相的水域,每立方米叶绿素 a 含量为 0.71mg(最大 5.62mg)(水深:$10 \sim 25$ m;水体温度:23℃;盐度:

35.25psu)。软体动物—苔藓虫为主的水域,每立方米叶绿素 a 含量为 2.2mg(最大 8.83mg)(水深:20~50m;水体温度:20℃;盐度:35.01psu)。

关于巴拿马太平洋沿岸的研究多数集中在对珊瑚的研究,这些研究包括珊瑚生长速率分析(Glynn 等,1972;Glynn 和 MacIntyre,1977;Wellington 和 Glynn,1983),微孔珊瑚对海水温度变化和紫外线(UV)辐射的反应实验,实验发现上升流地区(巴拿马湾)的珊瑚经紫外线辐射后褪色更早,更严重(D'Croz 等,2001)。Linsley 等(1994)将稳定同位素放射中 Chiriquí 湾珊瑚骨架的变化与东太平洋热带辐合带季节性、年际性的变化联系起来。这种热带辐合带季节性、年际性的变化控制了赤道地区的降水形式和云层覆盖。然而,除了一张整体的巴拿马湾沉积物粒度大小分布地面图(MacIlvaine 和 Ross,1973)以外,缺乏对 Chiriquí 湾和巴拿马湾沉积物的详细研究。本次研究的目的是测试观察海洋学和气候过程的变化对碳酸盐工厂生物组合和相特征的影响。本次研究聚焦于两种截然不同的碳酸盐—硅质碎屑混合沉积环境,位于巴拿马境内太平洋一侧,北纬7°到9°之间,巴拿马湾和 Chiriquí 湾附近。这两个碳酸盐—硅质碎屑混合沉积环境展示了不同的海洋环境,巴拿马湾有强劲的季节性上升流,而 Chiriquí 湾有全年稳定的非上升流。因此,本次研究是一次独特的契机来展示海洋过程变化对热带陆架碳酸盐分布的重要作用。研究表明巴拿马湾的营养条件和强季节性上升流共同导致海湾岛屿周围异养相的发育;这与 Chiriquí 湾的浅水地区发育光养生物(珊瑚)和/或者红藻石相形成鲜明对比。

2.1.1 陆架地形

巴拿马太平洋沿岸(巴拿马海岸线)划分为两个部分,东部的巴拿马湾(GoP)和西部的 Chiriquí 湾(GoC),两者被 Azuero 半岛隔开(图 2-1)。GoP 北部封闭,东部和西部形成逐步加深的陆架斜坡。大部分地区水深小于 75m,平均水深为 65m。海底地形平稳,水体浅,除了一个起始于 Archipelago de Las Perlas 西北部,向南延伸至外陆架的较深的海底峡谷,以及一个 SanMiguel 湾附近较小的西南走向的海底峡谷(MacIlvaine 和 Ross,1973)。Archipelago de Las Perlas (Islas Perlas)位于 GoP 的中央,由 100 个小岛组成,这些小岛主要分布在较大的 Isladel Rey 的西部和北部。

GoC 北部和东部封闭,但是西部仅被半封闭。GoC 的陆架比 GoP 海底狭窄和深很多。平均水深为 110m,在 Hannibal Bank 附近最大水深达到 459m。一条明显的海峡垂直横切这个陆架。这条海峡起始于 Islas Contreras 西北部,向南延伸至 Hannibal Bank。这个地区最大的岛屿是 Coiba,其他大量的岛屿散布在陆架上。在 Azuero 半岛和 Coiba 岛之间水体较浅(30~50m)。

GoC 的海底大地构造(Kolarsky 等,1995)使得在岛屿上和海底形成了一套独特的硅质碎屑岩(大多为晚上新世或更新世)和火山岩露头。相反,GoP 是一个均匀的、和缓倾斜的陆架斜坡,较少的大地构造和火山岩集中分布在 Islas Perlas。

2.1.2 水文地理

近年来的研究表明,巴拿马海岸线的水体循环、温度、盐度和上升流具有明显的季节性(D'Croz 等,1991;Rodriguez-Rubio 等,2003;D'Croz 和 O'Dea,2007)。在雨季,从5月到11月,巴拿马海岸线的水体循环是反气旋型,发育向南的沿岸流,然而在旱季(12月到4月),水体循环倒转为气旋型,发育向北的沿岸流,海洋上升流中心位于海岸线中部。在雨季,巴拿马海岸线主要发育东南信风,而在旱季,北大西洋的东北信风,凭借巴拿马喷气式飞机,经由巴拿马地峡进入巴拿马海岸线(Rodriguez-Rubio 等,2003)。在 GoP 和 GoC 早期的研究中,大多数分析

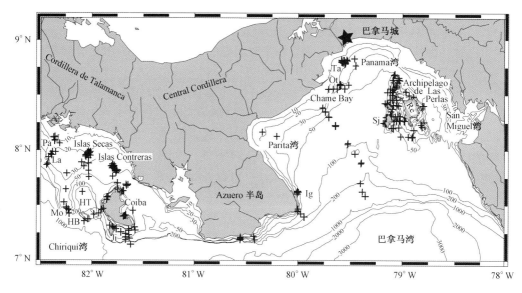

图 2-1 研究区地图(显示采样点位置)

测深数据(以"m"为单位)引自水文图表(英国水文局,UKHO)(Collins,1973;Clark,1988)。HB—Hannibal Bank;HT—Hannibal Trench;Ig—Isla Iguana;Ji—Isla Jicaron;La—Islas Ladrones;Mo—Isla Montuosa;Ot—Isla Otoque;Pa—Isla Padrida;Re—Isla del Rey;Sj—Isla San José;Ta—Isla Taboga

了两个海岸线的水文地理或者观测了 GoP 上升流的相关参数(Schaefer 等,1958;Forsbergh,1963,1969;Smayda,1963,1966;Legeckis,1985;D'Croz 等,1991)。季节性上升流导致了表层水体温度的变化。GoP 温度变化范围从 27.7℃(雨季)到 18.0℃(旱季),而 GoC 在雨季和旱季都稳定在 28℃左右(表 2-1)。GoP 的营养水平在旱季升高,而 GoC 的营养水平则保持相对稳定(表 2-1)(D'Croz 等,1991;D'Croz 和 O'Dea,2007)。在旱季,主导的北风导致了 GoP 较冷水富养水域上升流的形成(Schaefer 等,1958;D'Croz 等,1991;D'Croz 和 O'Dea,2007)。这些富养水域来源于 Humboldt 海流和赤道潜流(EUC)的延展部分(Strub 等,1998)。与 GoP 相反,GoC 受季节变化和上升流影响小得多(表 2-1),这是因为这个海湾被 Cordillera de Talamanca 南部和 Central Cordillera 高耸的火山所保护,这些山脉阻挡了东北信风进入海湾。

表 2-1 巴拿马湾(GoP)和 Chiriquí 湾(GoC)上层海水层(20m)水文特征及相关参数总结
(引自 D'Croz 和 O'Dea,(2007))

参 数	GoP	GoC	GoP	GoC	GoP	GoC	GoP	GoC
	MRS		LRS		EDS		MDS	
温度(℃)	27.7	27.5	27.7	28.8	23.9	28.3	18.0	28.0
盐度(psu)	29.2	30.6	30.2	30.5	33.3	32.5	33.6	32.0
NO_3^- (μm)	0.27	0.34	0.44	0.37	4.59	0.45	14.45	0.75
PO_4^{3-} (μm)	0.14	0.16	0.27	0.24	0.46	0.18	1.20	0.16
叶绿素 a(mg/m³)	0.27	0.34	0.24	0.17	0.83	0.16	1.44	0.29
富养层深度(m)	40.4	37.1	36.8	62.4	16.4	57.5	13.8	54.9

注:斜体数值显示两个海湾相对比在雨季和旱季最大值。MRS—雨季中期(1999 年 8 月 16—22 日;雨季或夏季间隔);LRS—雨季晚期(2004 年 12 月 14—20 日);EDS—旱季早期(2001 年 1 月 15—21 日;旱季或冬季间隔);MDS—旱季中期(2000 年 2 月 11 日—3 月 1 日)

2.2 研究方法

2.2.1 采样

本次研究基于 230 个随机样品的成分和地球化学分析,这些样品采集于 2004 年 5 月(雨季开始)和 2005 年 2 月(旱季)与 RV Urraca(Smithsonian 热带研究院)一起的两次航行。大部分样品由 Van Veen 采泥器采集(最大穿透深度为 15 ~ 20cm)。除了在岛屿周围详细采样以外,沿着海湾地区 3 个横切面也采集了样品,水深从 10 ~ 300m(图 2 - 1)。另外 28 个网捞样品,是在筛选点用容积为 $1.2 \times 100 \times 60cm^3$,网孔为 2mm 的网采集的,为的是将砾石级别的碎片(粒度大于 2mm)和随机采集样品(粒度最大 2mm)做比较。

2.2.2 粒度分析

随机采集的样品,经水洗清除树枝、树叶和鱼类残骸。在清洗过程中,将样品破碎部分还原。样品在 45℃经 48 小时干燥,以便获得样品总重量,包括碎片的重量($<63\mu m$)。接下来,样品经湿筛除去小于 $63\mu m$ 的碎片。在 45℃经 48 小时重新干燥,样品经干筛后,称出 63 ~ $125\mu m$,125 ~ $250\mu m$,250 ~ $500\mu m$,500 ~ $1000\mu m$,1000 ~ $2000\mu m$ 以及大于 $2000\mu m$ 每个粒级段的重量。计算每个粒度段在样品总重所占的百分比。网捞的样品经水洗,干筛后,称出 2 ~ 6.3mm,6.3 ~ 5mm 以及大于 5mm 每个粒级段的重量。

2.2.3 组分分析

从五个粒级段的干筛样品碎片中,清点出碳酸盐岩和陆源碎屑。分析了 230 个样品的颗粒组成,100 个颗粒归到小于 $2000\mu m$ 的各个粒级中。由于颗粒数目有限,平均每个样品只有 60 个颗粒归入大于 $2000\mu m$ 的粒级。这样,每个样品总共清点出 460 个颗粒。有三个样品被清点了 2 次(每个碎片 100 个颗粒对 400 个颗粒),目的是估算每个碎片清点 100 个颗粒的结论的可靠性。高度的正相关($r = 0.999$)证明了每个碎片清点 100 个颗粒可以提供可靠数量的数据集。

在这些族群中,数量显著的是陆源物质、双壳类、腹足类、浮游生物、底栖有孔虫、介形类、藤壶、管虫、棘皮类、珊瑚红藻、苔藓虫、珊瑚和其他生物,包括海绵、鱼类残骸、有机物质(树木和树叶)以及无法辨认的碎片。

清点出每个粒级段中的各个族群的个数。然后经下面的程序计算出总体的分布规律(给出的随机数字,仅用来解释计算程序):

$X\%$　A 粒级段的重量,其中 $N\%$ 的双壳类;
$Y\%$　B 粒级段的重量,其中 $M\%$ 的双壳类;
$Z\%$　C 粒级段的重量,其中 $L\%$ 的双壳类。

这个程序得到了总计的百分比:

$$(X \times N) + (Y \times M) + (Z \times L) = Total\%(wt)(总计质量分数)$$

表 2 - 2 显示的是实际统计结果。

表 2-2 所有清点组的随机抓取样品数点结果总结(包括平均值、最大值和最小值)

生物种类	GoP			GoC		
	平均	最大	最小	平均	最大	最小
陆源物质	*36.6*	92.9	0.0	34.0	*94.4*	0.3
双壳	22.5	*76.8*	0.0	20.5	43.9	2.1
腹足	7.5	34.7	0.0	*11.1*	*64.1*	0.0
藤壶	*11.2*	*51.9*	0.0	2.9	24.4	0.0
浮游有孔虫	3.8	*46.2*	0.0	*6.6*	39.5	0.0
底栖有孔虫	*3.1*	*29.4*	0.0	2.6	24.9	0.0
苔藓虫	1.6	20.1	0.0	*3.9*	*27.9*	0.0
棘皮动物	3.3	*17.3*	0.0	*3.8*	13.7	0.0
龙介虫	1.6	9.4	0.0	*1.7*	*12.2*	0.0
珊瑚红藻	1.6	32.7	0.0	*3.9*	*48.7*	0.0
珊瑚	0.1	5.0	0.0	*4.4*	*37.9*	0.0
鱼类残骸	0.0	0.9	0.0	*0.1*	*2.0*	0.0
介形类	0.4	*8.5*	0.0	*0.8*	4.8	0.0
其他生物	1.7	26.7	0.0	*4.2*	*85.6*	0.0

注:比较两个海湾地区,斜体数值表示最大值。

在第二种方法中,只考虑骨骼颗粒,经计算总计100%(表 2-3 显示实际数据)。在第三种方法中,为校正未被清点粒级段(小于 125μm)的沉积物数量,较小沉积物(未被清点)和大于 125μm 的沉积物(已清点)所占比率的百分比被重新计算(表 2-4 显示实际数据)。

表 2-3 随机抓取样品生物数点结果总结(平均值、最大值和最小值,排除了陆源碎片)

生物种类	GoP			GoC		
	平均	最大	最小	平均	最大	最小
陆源物质	—	—	—	—	—	—
双壳	*37.8*	*88.1*	0.0	31.9	64.4	5.2
腹足	11.9	*69.1*	0.0	*15.9*	64.3	0.0
藤壶	*15.6*	*75.7*	0.0	4.1	34.9	0.0
浮游有孔虫	7.8	59.9	0.0	*14.3*	*70.0*	0.0
底栖有孔虫	*6.0*	*58.2*	0.0	4.3	27.3	0.0
苔藓虫	2.2	25.3	0.0	*5.3*	*36.4*	0.0
棘皮动物	5.5	*50.1*	0.0	5.4	14.4	0.0
龙介虫	2.3	11.3	0.0	*2.6*	*15.5*	0.0
珊瑚红藻	2.0	35.7	0.0	*4.5*	*55.3*	0.0
珊瑚	0.1	6.6	0.0	*4.4*	*44.3*	0.0
鱼类残骸	0.0	1.0	0.0	*0.2*	*3.3*	0.0
介形类	0.8	*17.0*	0.0	*1.3*	9.0	0.0
其他生物	3.0	72.3	0.0	*5.4*	*88.8*	0.0

注:比较两个海湾地区,斜体数值表示最大值。

表2-4 本次研究清点随机样品分析（重新计算的数点结果总结，
关于沉积组分中没有被清点的小于125μm的颗粒与大于125μm的颗粒的比值）

生物种类	GoP			GoC		
	平均	最大	最小	平均	最大	最小
陆源物质	*22.1*	88.8	0.0	21.8	76.5	0.3
双壳	*16.3*	58.1	0.0	15.9	35.5	5.2
腹足	6.1	34.4	0.0	8.6	63.7	0.0
藤壶	9.8	49.6	0.0	2.4	23.5	0.0
浮游有孔虫	1.7	22.5	0.0	*4.1*	24.4	0.0
底栖有孔虫	*1.2*	13.3	0.0	1.7	8.7	0.0
苔藓虫	1.4	19.3	0.0	3.3	25.4	0.0
棘皮动物	2.2	11.1	0.0	2.9	13.2	0.0
龙介虫	*1.4*	9.3	0.0	*1.4*	10.4	0.0
珊瑚红藻	1.4	32.4	0.0	3.5	48.3	0.0
珊瑚	0.1	4.1	0.0	3.4	36.5	0.0
鱼类残骸	0.0	0.9	0.0	*0.1*	*1.4*	0.0
介形类	0.2	4.9	0.0	0.5	3.6	0.0
其他生物	1.5	25.3	0.0	3.2	46.5	0.0

注：比较两个海湾地区，斜体数值表示最大值。

已清点的粒级段 = $A\%$(wt)
未被清点的粒级段 = $B\%$(wt)
以上步骤产生了重新计算的质量分数(R)：

$$(A/100) \times Total\%(wt) = R\%(wt)$$

基于每个样品组分分布，做聚类分析来将样品划分至子集和簇。用树枝形结构关系图来表达各子集各种级别之间的联系，被称为簇树形结构。横坐标表示相似性。当子集在树枝形结构关系图中的趋势沿同一路径，则它们的相似性从右向左增加。做聚类分析使用的是 statistiXL 2006 Vers. 1.6 (Roberts 和 Withers, 2006)，应用沃德法作为链，以及欧几里得距离法来辨别不同的相类型。完整的或者单个的链计算比沃德法更易获得极值解。对网捞样品也做了相同的组分分析。

2.2.4 碳酸盐分析(LECO)

每个采样点的大部分底泥样品经冷冻干燥和玛瑙研钵磨圆后，在海洋地质 IFM-GEO-MAR-Leibniz 研究院 (Kiel, Germany) 使用 LECO C-200 分析仪 (LECO Corporation, St. Joseph, Michigan, USA) 进行 LECO 分析。经对样品热解所产生气态产物的热传导测量，确定一个底泥样品的碳的总含量。基于44个并行测量，分析误差在 ± 0.3% 碳酸钙 ($CaCO_3$) (1θ)。同时，也对酸化样品进行分析，以确定有机碳含量。碳总量的重量百分数 (TC)，有机碳酸盐 (OC)，以及 $CaCO_3$ 用等式相互关联起来：

$(TC - OC) \times 8.33 = CaCO_3$ (碳酸盐总量, TCa)，假设所有的无机碳都绑定在碳酸钙中。

2.2.5 绘图程序

应用 Golden 软件的 Surfer 7(Golden,CO,USA)(1999)绘图,显示出 GoP 和 GoC 的粒度、组分、矿物和相分布。构图应用克里金网格法。

2.2.6 海洋水色卫星数据(NASA,GIOVANNI)

美国国家航空航天局(NASA)在巴拿马湾获得的海洋水色卫星数据,用来获取从 2004 年 5 月到 2005 年 4 月海洋水体温度和叶绿素的数值。来自 SeaWiFS 海洋水色,MODS Aqua 海洋水色和海水表层温度(SST)数据(http：// reason. gsfc. nasa. gov/Giovanni)经基于网络的接口 GES – DISC 互动在线可视化和基本结构分析(GIOVANNI)展现出来。

2.2.7 温度—密度—传导数据(CTD 数据)

在 2004 年 5 月期间应用 Smithsonian 热带研究院(巴拿马)Idronaut Ocean Seven 316 探针(Idronaut Srl,Milan,Italy)测量了 118 个 CTD 站。水体深度测量值在 10m ~ 236m 深。装置了温度、导流能力、氧和叶绿素 a 传感器。

使用了 Leibniz-Institut für Meereswissenschaften IFM – GEOMAR(Kiel,Germany)的 SBE 37 – SMMicroCAT(Sea – Bird Electronics,Washington,USA),用来在 2005 年 2 月做 CTD 剖面。这是一个高精度的传导仪和温度(可选择压力)记录仪,内置电池和记忆。MicroCAT 包括一系列标准接口和非挥发性快闪记忆体,这种设计是为了停泊或其他长期的、固定地点的部署。

2.3 结果

2.3.1 海洋学数据

2.3.1.1 温度

SeaWiFS SST 数据显示 GoC 和 GoP 具有重大的季节性区别。2004 年 5 月,两个海湾地区都有相似的高 SST 值(雨季开始)。GoP 的 SST 平均值是 27.2℃,GoC 是 27.5℃(图 2 – 2,表 2 – 5)。研究测量值显示出 GoP 为 27.9℃(平均)与 GoC 为 28.6℃(平均)稍有差别。在旱季(2005 年 2 月),两个海湾地区的 SST 值差别更为明显。来自卫星影像的 SST 平均值,GoP 为 24.8℃(上升流),GoC 为 28.1℃(非上升流)(表 2 – 5)。在观测中收集的海水表层温度数据显示了两个海湾地区更明显的区别,GoP 为 20.1℃(平均),GoC 为 28.5℃(平均)。

表 2 – 5 海洋学参数和矿物学特征

参　　数	GoP	GoC
	平均	平均
海洋表层温度(雨季早期)(℃)	27.9	28.6
海洋表层温度(旱季)(℃)	20.1	28.5
海洋表层温度(雨季早期)(℃)	27.2**	27.5**
海洋表层温度(旱季)(℃)	24.8**	28.1**
海洋表层盐度(雨季早期)(psu)	33.5	33.5
海洋表层盐度(旱季)(psu)	—	—
叶绿素 a(雨季)(mg/m³)	0.2 ~ 0.7**	0.1 ~ 0.7**
叶绿素 a(旱季)(mg/m³)	0.7 ~ >10**	0.1 ~ 0.4**
温跃层深度(雨季)(m)	50 ~ 60*	60 ~ 70*
温跃层深度(旱季)(m)	0 ~ 20	30 ~ 50

续表

参　数	GoP 平均	GoC 平均
碳酸盐(%)	24.4	*73.2*
文石(%)	19.5	*35.7*
方解石(HMC 和 LMC)(%)	14.4	*16.6*
高镁方解石(HMC)(%)	3.12	*10.1*
低镁方解石(LMC)(%)	*11.2*	6.4

注:对比两个海湾地区,斜体数值为最大值

* 据 D'Croz 和 O'Dea,2007;

** 据 GIOVANNA (http://reason.gsfc.nasa.gov)。

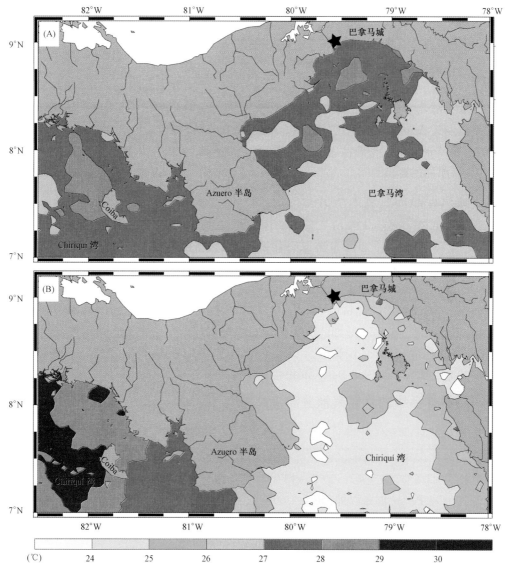

图 2-2　(A)雨季(2004 年 5 月)海洋表层温度(SST)。(B)在 Chiriquí 湾(GoC)和巴拿马湾(GoP)旱季的 SST,展示了两个海湾地区的明显差别

注意到,与 GoC 相比,GoP 旱季的上升流导致较低的 SST 值。数据引自海洋水色卫星数据(NASA,GIOVANNI)

本次研究遵循了 D'Croz 和 O'Dea(2007)等对东热带太平洋温跃层深度的定义,即 20℃等温线的位置。温跃层的位置被认为是透光层主要的营养来源,当上升流冷却海水表层时变浅 D'Croz 和 O'Dea(2007)。在雨季,两个海湾在相似的水体深度有界线清楚的温跃层,GoP 为 50~60m 深度之间,GoC 为 60~70m 深度之间(表 2-5)。在旱季,GoP 的温跃层大幅度变浅到上层水层(到顶部 20m),导致上面提到的表层水体的冷却。在 GoC,变浅幅度有限,仅观测到变浅近 30m(D'Croz 和 O'Dea2007),这样并没有造成表层水体的冷却(表 2-1 和表 2-5)。

2.3.1.2 盐度

两个海湾地区的表层盐度(SSS)在 2004 年 5 月为 33.5 psu,除了在 GoP 中央出现大约 33.0psu 的小透镜体(表 2-5)。由于技术问题,2005 年 2 月没有获得盐度数据。

2.3.1.3 叶绿素 a

从 2004 年 5 月到 11 月,两个海湾地区表层水体的叶绿素 a 值相似,GoP 每立方米叶绿素 a 为 0.1~0.7mg,GoC 为 0.2~0.7mg(图 2-3,表 2-5)。本次研究在两个海湾地区获得的 CTD 最高测量值位于 20~40m 水深,近海岸、岛屿周围以及 Hannibal Bank 的地区(GoC)。

两个研究区表层水体叶绿素 a 值从 2004 年 12 月到 2005 年 4 月有所不同(图 2-3;http://reason.gsfc.nasa.gov)。GoP 每立方米叶绿素 a 值从 0.7mg 到大于 10mg。海湾东南区朝向开阔陆架以及近 Azuero 半岛出现较低值。GoC 每立方米叶绿素 a 值为 0.1~0.4mg,较高值仅出现在 Isla Parida 北部大型海湾近海岸的位置(每立方米叶绿素 a 值>2.5mg),以及沿着 Azuero 半岛西海岸的位置。

2.3.2 随机样品的粒度大小分布

黏土到极细砂(<63~125μm),颜色为深绿灰色,在 GoP 沿大陆海岸,到 Islas Perlas 的西部和东部,以及沿 Azuero 半岛东海岸分布。中砂(250~500μm)出现在 Parita 海湾周围,Chame 海湾北部以及外陆架,从 100~400m 的水体深度范围。较粗粒沉积物仅分布在 Islas Perlas 浅水区。

粗粒沉积物(250~1000μm)主要分布在 GoC Islas Secas 和 Islas Contreras 周围的海岸地区;而这与 GoP 相反,在 GoP 最细的沉积物分布在近岸地区。中部大陆架主要覆盖橄榄绿色到中等灰色泥岩到极细砂(小于 63~125μm)。最细的沉积物出现在 GoC 的中央。在 Isla Montuosa 的东北部和 Hannibal Bank,沉积物又变得较粗(500~1000μm)。

2.3.3 随机样品中的骨骼残骸分布

2.3.3.1 双壳类

双壳类碎片在两个海湾地区几乎所有的沉积类型和水体深度都有分布。GoP 双壳类平均重量百分比为 22.5%(表 2-2—表 2-4)。最高值出现在 Isla Taboga 的西北部(76.8%),以及最北部的 Islas Perlas 群岛以外(50%~60%)(图 2-4)。

GoC 双壳类平均重量百分比是 20.5%,比 GoP 稍低。较高值(30%~43.9%)出现在沿北部海岸的岛屿周围。在海湾中央和朝向 Hannibal Bank 的地区,双壳类丰度降低。

2.3.3.2 腹足动物

腹足动物在 GoP 平均占 7.5%(表 2-2—表 2-4);腹足动物在岛屿周围和浅水区占主体。当水深在 100m 以下,腹足动物数量减少,除了 GoP 一个 300m 深的小区域,测得腹足动物

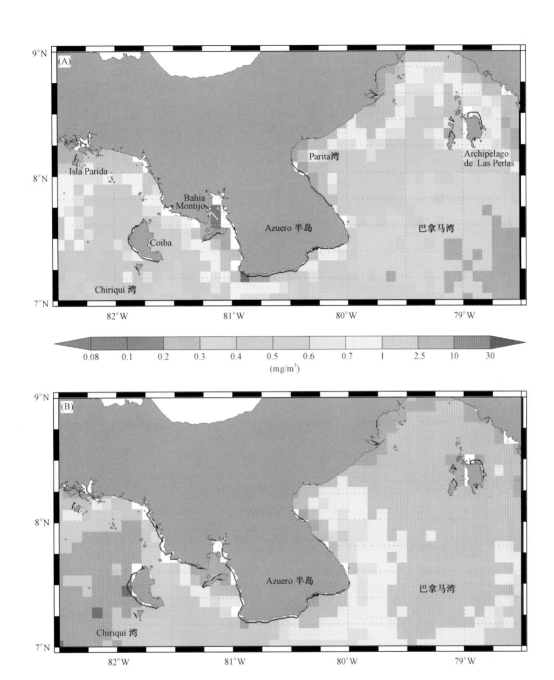

图2-3 (A)叶绿素值,雨季(2004年5—11月)。(B)在Chiriquí湾(GoC)和巴拿马湾(GoP)叶绿素值,旱季(2004年12月—2005年4月),显示明显差异

注意到由于旱季上升流,相比GoC,在GoP叶绿素值更高。数据引自海洋水色卫星数据(NASA,GIOVANNI)

的量多达16.1%(图2-5)。这个地区底泥中主要由壳体碎末,浮游有孔虫和石英颗粒组成。石英颗粒和软体动物碎片揭示了刚性拉张。在Isla San José(Archipelago de Las Perlas)的南部,腹足动物的数量达到最高值(34.7%)。

然而,在GoC,相比GoP,腹足动物数量的平均值较高(11.1%)。同样,水深100m以下没有发现可观数量的腹足动物。腹足动物集中在近岸岛屿周围和Isla Coiba周围的一个狭窄带。在Islas Contreras附近,值达到最大(64.1%)。

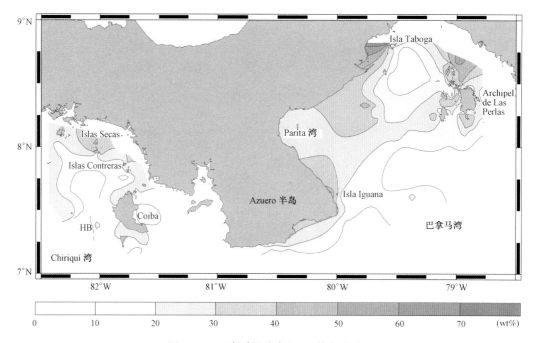

图 2-4 双壳质量分布(wt% 数点碎片)
GoP 和 GoC 没有明显差异;HB—Hannibal Bank

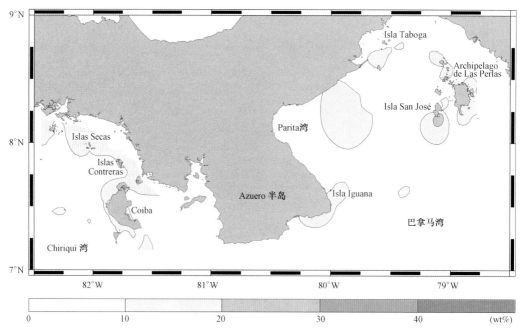

图 2-5 腹足动物质量分布(wt% 数点碎片)
腹足动物在 GoP 和 GoC 屿周围占主体

2.3.3.3 藤壶

在 GoP,藤壶分布普遍,平均值为 11.2%(表 2-2—表 2-4)。较高值(40%~50%)集中出现在 Islas Perlas 周围,最高值在 Isla SanJosé 的南部和 Isla del Rey(表 2-6)。稍低值(30%~

50%)出现在 Isla Taboga、Isla Otoque、Parita 湾和 Isla Iguana 附近。最高值(51.9%)出现在 Isla SanJosé 的南部。

相反,在 GoC,藤壶非常稀少(平均2.9%)(图2-6)。总体上,藤壶仅在 Isla Coiba 西海岸、Islas Contreras 和 Islas Secas 这几个有限的区域分布。最高值(24.4%)出现在 Islas Secas 朝向开阔陆架的地区。

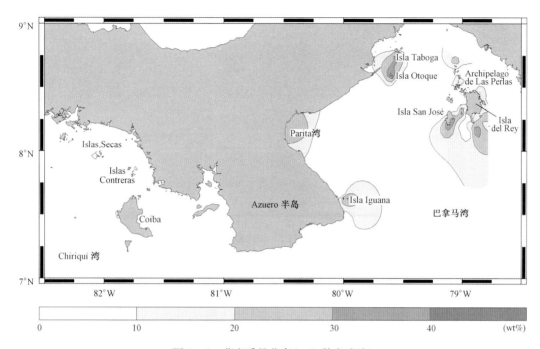

图2-6 藤壶质量分布(wt%数点碎片)
藤壶在 GoP 占主体,但实际上在 GoC 缺失

2.3.3.4 浮游有孔虫

浮游有孔虫在两个海湾地区都普遍分布(表2-2—表2-4)。较高的含量出现在陆源底泥主导的地区;在 GoP 的 Isla Otoque 和 Islas Perlas 之间,以及海湾中央较深水域(300m)普遍分布(平均占3.8%)。最高值(46.2%)在海湾地区的内部中心;然而,这个值是有偏差的,因为数出的粒级段(125~2000μm)仅占整个底泥的2.1%。

浮游有孔虫在 GoC 普遍分布,主要出现在朝向开阔陆架中央。平均值为6.6%。最高值出现在 Islas Secas 附近,在64wt%的底泥中占39.5%。

2.3.3.5 底栖有孔虫

底栖有孔虫在 GoP 陆源底泥地区(海湾中心,Isla Otoque 和 Islas Perlas)普遍分布。在全海湾范围内,平均百分数为3.1%(表2-2—表2-4)。与浮游有孔虫相似,最高值出现在125~500μm 的粒级段中。最高值(29.4%)分布在 GoP 的 Isla Iguana 附近。

底栖有孔虫在 GoC 的平均百分数较低,为2.6%。但有例外的情况出现在 Jicaron 西部的小区域,百分数达到24.9%。

2.3.3.6 苔藓虫

苔藓虫在遍及 GoP 的底泥中平均占1.6%(表2-2—表2-4)。仅在 Islas Perlas 南部的地区有略微升高值(20.1%)。

相比 GoP,在 GoC 苔藓虫平均百分数较高(3.9%)(图 2-7)。苔藓虫在整个岛屿分布普遍。最高值(平均 27.9%)出现在 Isla Coiba 南部,在这里,苔藓虫碎片在 1000~2000μm 的粒级段中占主导,百分数为 44%。

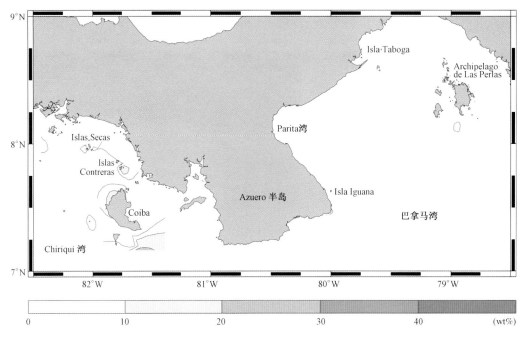

图 2-7 苔藓虫质量分布(wt% 数点碎片)

苔藓虫出现在 GoC,在 GoP 减少

2.3.3.7 棘皮动物

棘皮动物在 GoP 以海胆的棘刺,以及无明显特征的碎片形式出现。所占平均百分数为 3.3%(表 2-2—表 2-4)。棘皮动物碎片分布在 GoC 近岸岛屿周围,Isla Coiba 的东北和东南,Isla Montuosa 周围以及 Hannibal Bank 上。在 GoC 底泥中所占平均百分数为 3.8%。百分数值从 0~13.7%。棘皮动物碎片的分布模式与腹足动物的分布模式非常相似。

2.3.3.8 龙介虫

龙介虫在两个海湾少量出现(没有展示)。在 GoP,龙介虫出现在 Isla Otoque 和 Isla Taboga 之间,Islas Perlas 周围,尤其是南部岛屿西南海岸上,在 Isla Iguana 以及 Azuero 半岛的东南边缘。平均值为 1.6%;没有出现大于 10% 的值(表 2-2—表 2-4)。

在 GoC,最高值出现在 Isla Coiba 南部(12.2%)。所有其他值都在 7% 以下,平均值为 1.7%。

2.3.3.9 珊瑚红藻

珊瑚红藻出现在两个海湾有限的几个地区,占据与珊瑚相同的栖息地。珊瑚红藻出现在 GoP132 个采样点的 21 个,含量在 10% 以下。最高值(32.7%)出现在 Isla del Rey 的东南部(图 2-8)。

在两个海湾都出现了结壳的珊瑚红藻,结带状的和开口分支的红藻石在 GoC 占主导。在 GoC,最高量值的珊瑚碎片(48.7%)出现在 Isla Montuosa 北部,在网捞样品中,以结带状的和开口分支的红藻石的形式出现(表 2-2—表 2-4)。在 Islas Ladrones 周围,红藻石是主要的

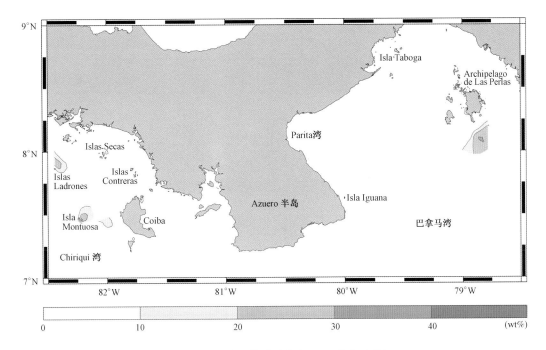

图2-8 珊瑚红藻质量分布(wt%数点碎片)

珊瑚红藻在GoC所有岛屿周围出现；在GoP局限分布在Archipelago de las Perlas，并且和藤壶共同出现

碳酸盐岩生产者，占30.1%。在Islas Secas和Islas Contreras周围也发现了红藻石的高值。在所有的采样点中，红藻石在GoC平均值为3.9%，在GoP为1.6%。叶状体碎片出现在GoC的97个随机样品中的44个，但仅有5个样品的值大于20%。

2.3.3.10 珊瑚

与红藻相似，珊瑚占据了独特的栖息地，虽然在GoC岛屿周围，它们是主要的碳酸盐岩生产者。珊瑚碎片在GoP非常稀少(平均0.1%)，并且仅在Isla Taboga和Isla Iguana周围大于1%。在Islas Perlas周围，含量值在5%以下。

在GoC，珊瑚碎片的平均值较高(3.7%)。造礁石珊瑚碎片出现在GoC所有岛屿周围，最高值(37.9%)出现在Islas Contreras周围，而富含八放珊瑚的沉积物(35.6%)出现在Hannibal Bank 图(2-9)。

2.3.4 网捞样品中骨骼残骸的分布

网捞样品大多在岛屿周围砾级的、碳酸盐岩富集的底泥中收集。网捞样品在两个海湾地区显示出很大不同。在GoP，砾屑主要由双壳(27.8%)和藤壶(21.3%)，其次为腹足动物(13%)和珊瑚红藻(11.5%)构成。岩屑占19%。在GoC，砾屑主要由珊瑚红藻构成(57.7%)，其次为双壳(24.5%)。其他生物，比如腹足和珊瑚以及岩屑，明显表述不清。

两个海湾同样采样点的网捞样品与随机抓取样品最为显著的区别是珊瑚红藻的量。在GoP，网捞样品比随机抓取样品中珊瑚红藻的含量大十倍以上(11.5%比1.8%)，而在GoC，差别是57.7%比10.5%。

实际的网捞样品和随机抓取样品中珊瑚红藻含量最大的差异是，如果是开口分枝 maerl-

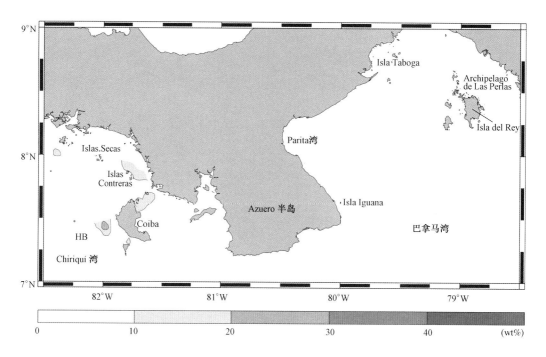

图 2-9 珊瑚质量分布(wt% 数点碎片)

珊瑚主要出现在 GoC 的岛屿周围;在 GoP 珊瑚少量出现在 Archipelago de las Perlas 岛屿周围,包括 Isla del Rey 的南部。HB—Hannibal Bank

type 种类的珊瑚红藻,仅出现在随机抓取样品的细粒沉积物粒级段中(<2mm)。这种情况在 GoP 非常罕见,在 GoP,以结壳生长形式为主。在 GoC,开口分枝菌体的量大得多,导致网捞和随机样品的成分变化较少。

2.3.5 碳酸盐含量

碳酸盐岩富集的沉积物出现在两个海湾的岛屿周围。在 GoP Islas Perlas 周围的受限区域,这种底泥的含量最大值从 70%~80% TCa(图 2-10)。在 Isla Taboga 和 Isla Iguana 周围含量从 60%~70% TCa。

在 GoC,最高的 TCa 值(>90% TCa)出现在 Islas Secas、Islas Contreras 周围,以及 Isla Coiba 的北部周围。Jicaron,Montuosa and Ladrones 群岛附近和 Hannibal Bank 的西部地区含量值从 50%~70% TCa。总体上讲,GoC 碳酸盐平均含量值(73.2% TCa)比 GoP(24.4% TCa)要高,这指示了 GoC 是碳酸盐相对为主体的环境。

2.3.6 相分布

在族群分析基础上,区分出五个相类型(图 2-11),其中有两个相局限于一个或另一个海湾,而三个相在两个海湾都有分布(图 2-12)。相Ⅰ、Ⅱ和Ⅲ是碳酸盐主导的,而相Ⅳ和Ⅴ含有大量的陆源物质(图 2-11)。相分布图(图 2-12)显示陆源沉积物覆盖了 GoP 海底广大的地区(相Ⅳ和Ⅴ),而在 GoC,碳酸盐占优势(相Ⅰ、Ⅱ和Ⅲ)。

相Ⅰ包括了所有光能自养到混合营养组合,而相Ⅱ和Ⅲ表明异养组合占主导。相Ⅰ有两个亚相(图 2-11)。相Ⅰa(珊瑚藻)主要由珊瑚(37%)、红藻(18%)和双壳类(11%)构成。

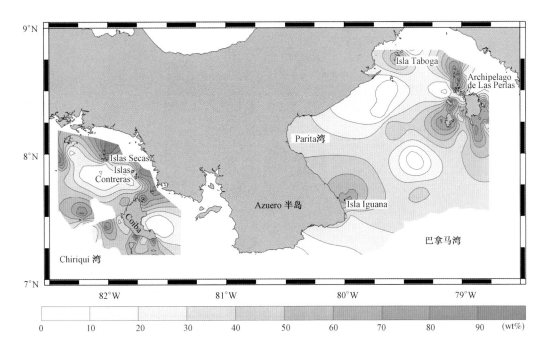

图 2-10 碳酸盐在 GoP 和 GoC 的分布

总碳酸盐含量(wt% TCa)在两个海湾的岛屿周围浅水区较高,表明了总碳酸盐含量与浅水碳酸盐生产和相应的相分布明显关联

相 Ib 红藻—软体组合(rhodmol)由红藻(34%)、双壳类(21%)和珊瑚(6%)构成。网捞样品分析结果强调了这个亚相的细分。相 II 藤壶—软体组合(Balamol)以藤壶为主(36%),其次为双壳类(21%)和腹足类(11%),以及更少量的棘皮动物和龙介虫。相 III(软体动物)以双壳类为主(35%),其次为腹足动物(15%)。相 IV 有孔虫—软体组合(Foramol)的成分和相 III 相似,为双壳类(22%),腹足动物(9%),但是浮游动物(9%)和底栖有孔虫(5%)含量较高,以及较高含量的陆源物质(多达 40%)。相 V 富集双壳类(13%),但以陆源物质占主体(平均 68%)。

相 I(以珊瑚和红藻为主)仅出现在 GoC(图 2-2)。相 Ia(珊瑚藻)主要分布在 Islas Secas 和 Islas Contreras 隐蔽的海湾,以及 Isla Coiba 的北部。相 Ia 出现在 Isla Parida 和 Isla Montuosa 周围(在图 2-11 中并没有显示,因为分布局限,见图 2-1 准确位置)。

相 Ib(rhodmol)出现在相 Ia(珊瑚藻)的附近,但是在更深的水域(图 2-12)。Rhodmol 相在 Islas Ladrones 周围最丰富(见图 2-1 准确位置)。在 Islas Secas,Islas Contreras 周围和 Isla Coiba 的东北部,相 Ib 和珊瑚藻同时出现。

在 GoP,相 II(Balamol)仅出现在 Isla Iguana 周围,Isla Otoque 向北以及 Isla del Rey(Islas Perlas)周围地区。更小的分布地区在 Isla Taboga 和 Islas Perlas 附近(图 2-12)。

相 III(软体动物)在两个海湾广大地区都有分布,显示了软体动物作为碳酸盐主要生产者的重要性。相 IV(Foramol)在两个海湾的中心地区都有广泛分布。相 V(双壳类—陆源相)也在两个海湾地区都有分布,并且主要与细粒沉积物,以及有细粒陆源河流注入的较深水环境相关(比如,Parita 海湾)。

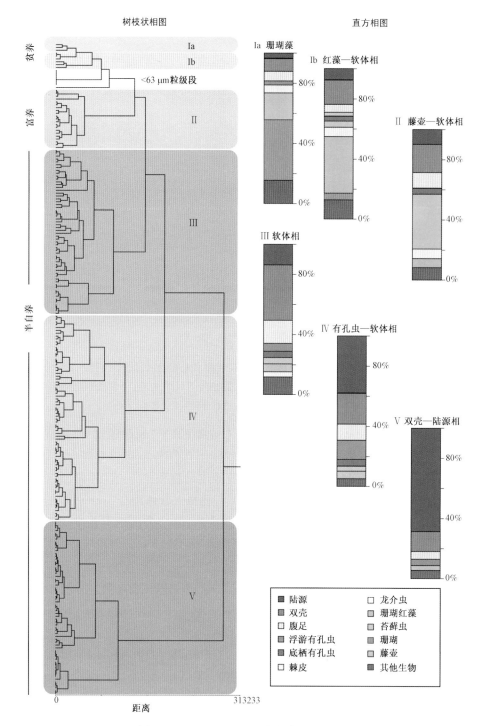

图2-11 GoP 和 GoC 所有相类型聚类图

直方图表示单个相类型中主要生物群落所占的平均百分比。相 Ia 和 Ib（珊瑚藻和红藻—软体相）、相Ⅱ（藤壶—软体相）和相Ⅲ（软体相）代表了碳酸盐主导环境。相Ⅳ（有孔虫—软体相）和相Ⅴ（双壳—陆源相）是以陆源物质为主体。相类型与不同的营养条件相关。相Ⅰ包括两个光养组合，分别指示了贫养或者贫养—半自养环境。相Ⅱ指示了异养组合以及明显的富养环境，而软体动物主导的相（Ⅲ、Ⅳ和Ⅴ）与半自养环境更相关。
颜色环绕的丛集与图2-12中的相的颜色一致

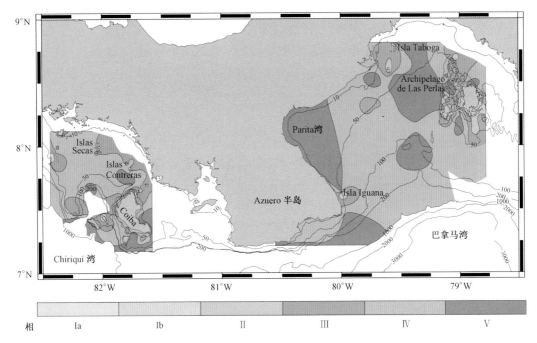

图 2-12 GoC 和 GoP 沉积相分布图

相 Ia(珊瑚藻)和相 Ib(红藻—软体相)仅出现在 GoC,而相Ⅱ(藤壶—软体相)仅分布在 GoP,相Ⅲ(软体相)、相Ⅳ(有孔虫—软体相)和相Ⅴ(双壳—陆源相)在两个海湾都有出现

2.4 讨论

2.4.1 海洋学参数

海洋学参数方面,GoP 和 GoC 在温度、盐度、营养及其他参数有明显差异(图 2-2 和 2-3,表 2-1 和表 2-5)。这些单个参数将在下面内容中表述。

2.4.1.1 温度

GIOVANNI 数据集(图 2-2),D'Croz 和 O'Dea(2007;表 2-1)和 Glynn 和 MacIntyre(1977)表明 GoP 在雨季 SST 值(26 到 28℃)明显比旱季(18 到 24℃)高。在旱季,穿越巴拿马 Isthmus 的北信风取代了表层大量水体,产生了上升流,从而形成了低 SST 值(Fleming,1939;Schaefer 等,1958;Forsbergh,1963,1969;D'Croz 等,1991)。这里展示的数据与这些发现相一致。

在 GoC,观测到的 SST 值在雨季和旱季并没有明显差异(表 2-2,图 2-2)。这个观测结果可以解释为 GoC 的地理位置位于 Cordillera de Talamanca 背风一侧(图 2-1),因此,这个海湾被保护起来,没有受到季节性北风的影响(Schaefer 等,1958;D'Croz 和 O'Dea,2007)。

与温跃层深度变化相关的强季节性温度变化对温度敏感的碳酸盐生产者比如珊瑚产生强烈压力(Halfar 等,2006);这就揭示了为什么珊瑚礁在 GoP 稀少,而在 GoC 相对丰富,因为这里全年温度更稳定。在 GoP,珊瑚礁在 Isla Taboga 和 Isla Iguana 向陆地方向一侧,以及在较少受到寒冷、上升流影响的 Islas Perlas 隐蔽海湾数量减少(Maté,2003;LaVigne 等,2008)。

SST 和相关的参数(盐度和营养)表明年际变化。这些变化由热带辐合带(ITCZ)的转移引起,这些转移改变了上升流的强度、上升流水体的来源以及区域性东太平洋气候(D'Croz

等,1991;Linsley 等,1994)。厄尔尼诺事件对 GoP 和 GoC 的珊瑚礁产生了附加的压力,导致珊瑚褪色((Wellington 和 Dunbar,1995)。D'Croz 等(2001)从 GoP 和 GoC 获取大量的 Porites lobata(团块微孔珊瑚),经实验提高水体温度以及周围环境的 UV 辐射,实验条件与在 1997 或 1998 的 ENSO 事件相似,实验结果表明从 GoP 上升流环境获取的珊瑚比从 GoC 非上升流获取的珊瑚褪色更早、更强烈。

2.4.1.2 盐度

与加利福尼亚湾相比,巴拿马太平洋海岸的两个海湾地区显示出盐度的高季节性变化小于等于30psu(雨季)以及34psu(旱季)(表2-5)。加利福尼亚湾 SSS 值始终保持在35psu(Halfar 等,2006)。D'Croz 等(1991)记录从9月到11月的雨季或夏季盐度最低值小于25psu。GoP 和 GoC 的盐度结构表明34 psu 等密度线在150~190m 范围变化(雨季或夏季),水深在20~40m 范围变化(旱季或冬季)(D'Croz 和 O'Dea,2007)。因此,明显的盐度变化可能仅限于影响狭盐性的种群,阻碍碳酸盐生产,因而控制了碳酸盐生产群落物种的多样性。

2.4.1.3 营养

除了温度、盐度以外,养分被认为是影响碳酸盐环境生物组成的重要因素(Littler 和 Littler,1985;Hallock 和 Schlager,1986;Birkeland,1987;Carannante 等,1988;Hallock,1988,2001;Mutti 和 Hallock,2003;Vecsei,2003;Halfar 等,2004b,2006;Wilson 和 Vecsei,2005)。

养分有效性在两个海湾有明显差异(表2-1,图2-3)。在 GoC,表层养分浓度全年保持低值,而在 GoP,在旱季,随季节性的变化,数值增加(D'Croz 和 O'Dea,2007;LaVigne 等,2008)。养分有效性的主要消耗者浮游植物的数量,由叶绿素 a 测量。D'Croz 等(1991)指出在 GoP 的雨季,叶绿素 a 浓度较低($0.5mg/m^3$),浮游植物的密度较低($20~30cells/m$),而在旱季,叶绿素 a 浓度较高($3mg/m^3$),浮游植物的密度较高($100~300cells/m$)。在 GoP,高叶绿素 a 值和低温度值指示了旱季的富养条件。Smayda(1963,1966)指出在 GoP 浮游植物的出现受控于季节性气候变化。Estrada 和 Blasco(1979)描述了腰鞭毛虫在旱季由于上升流导致养分富集而生长繁盛的情况。

浮游植物的生长,一方面与养分有效性有关,这解释了在两个海湾叶绿素 a 分布的区域性和深度差异,GoP 浮游植物的生产率更高。浮游植物,另一方面,引起水体透明度的降低,$CaCO_3$ 结晶对磷酸盐的抑制作用,以及群落变化和生物侵蚀速率增加(Hallock 和 Schlager,1986)。因此,高叶绿素 a 值,导致了虫黄藻珊瑚和钙性藻类光能自养碳酸盐岩生产的降低。另外,养分促进了浮游植物的生长,浮游动物以此为食。浮游植物和浮游动物是快速生长的种群的食物来源,比如藤壶,藤壶可以占据珊瑚的栖息地,从而在富养环境与珊瑚产生竞争(Birkeland,1987)。沿着一个养分梯度,珊瑚、珊瑚藻,大型藻类到异养底生生物群落出现在从最贫养($0.01mgChl-a/m^3$)到最富养($1.0mgChl-a/m^3$)环境体系(Crossland,1987;Hallock,2001)。珊瑚礁打开或关闭点位于半自养体系内,海岸的上升流区域,比如 GoP,为这个半自养体系提供一个典型的环境。

在 GoP,养分的高度季节性变化有利于低等种群,异养底生生物群落,以快速的生产率循环来适应不稳定或不可预知的环境(Valentine,1973;Birkeland,1987;Hallock,1987)。相反,在 GoC,更低和更稳定的营养水平,和稳定的温度,产生了一个更稳定的环境,这有利于珊瑚的生长和 K 对策的群落。这种在营养体系和环境稳定性之间的负相关也被 Jackson 和 Herrera-Cubilla(2001)在比较两个海湾苔藓虫卵胞富集度和大小时候所得到的发现所证实。在 GoC 发现较大的卵胞证明了在这个海湾 K 对策的盛行。

2.4.1.4　温跃层

D'Croz 和 O'Dea(2007)的研究表明,两个海湾雨季的温度结构相似(表 2-1);这种一致性在这里用数据表示出来。在 1999—2004 年之间,一个界线清楚的温跃层(20°等温线)出现在 0~75m 之间(D'Croz 和 O'Dea,2007)。然而,在 2004 年 5 月的研究中,CTD 数据表明,在 GoP,水体发生了轻微混合,导致温跃层稍微变浅,变得较不明显(表 2-5)。然而在旱季,两个海湾的温度剖面(2005 年 2 月)有明显差异。在 GoC,旱季温跃层位于 20~50m 水深,而在 GoP 20°等温线出现在最大水深 20m 或甚至到达海洋表面。D'Croz 和 O'Dea(2007)解释了在旱季温跃层这种向上运移是一个"对于表层水体附近来说,毫无疑问的营养来源"。

2.4.1.5　透光层

产生微粒有机碳酸盐、溶解有机碳酸盐和通过河流卸载注入的陆源物质的主要生产率,限制了碳酸盐生产。在 GoC,透光层的深度(平均 53m)相对稳定,比 GoP 更深,在 GoP,表现为从 15m(旱季)到 40m(雨季)之间的高度季节性变化(D'Croz 和 O'Dea,2007;表 2-1)。

透光层的深度差异充分解释了光能自养(珊瑚红藻)和混合营养(珊瑚)生物和群落在 GoC 区域性主体分布的差异。这种分泌碳酸盐的生物分布模式更进一步受到陆架结构的限制(Kolarsky 等,1995),在 GoC,海底陆架的尖坡这样更小尺度的地形和广泛分布的岛屿为依赖光能的生物群落富集提供了生长空间。在 GoC 岛屿周围以及海底尖坡,由于水下地形造成光能局限,使得珊瑚红藻胜于珊瑚。相比珊瑚,珊瑚红藻似乎对于缺乏光能的、高营养环境具有更强的忍耐性。因此,珊瑚红藻虽然在热带清澈水体深至透光层较低的基部丰富多产(Foster,2001),但是据报道也出现在极圈更远的地方,在那里它们在极地的冬天遭遇彻底的黑暗(Freiwald 和 Henrich,1994)。

2.4.1.6　陆源影响和基质

在两个海湾陆源注入都很强烈(表 2-2 和表 2-4)。在雨季,当河流卸载悬浮物质进入 GoP 东部和西北部以及 GoC 东部和北部,陆源注入增加。陆源注入的强季节性变化在热带地区很典型,导致了碳酸盐岩和碎屑岩混合沉积。原地混合在 GoP 滨岸和中央以及 GoC 内陆架和深槽水道非常典型。原地混合是由细砂到泥质环境中(Mount,1984),原地生长的和近旁生长—死亡钙质有机生物组合体生产相对连续的碳酸盐沉积物而形成。在 GoC,由于更强的海底构造活动,细粒沉积物混合较少。细粒沉积物通过深槽垂直下切陆架形成水道,在开阔堤岸和岛屿周围留下更纯质的碳酸盐岩沉积。

Mount(1984)指出,沉积物混合发生在两个海湾的岛屿周围,在那里,粗粒碳酸盐砂屑和砾屑与缺乏骨骼颗粒的细粒碎屑岩砂和粉砂交互沉积。此外,偶尔的风暴也使得骨骼碳酸盐和更粗粒的碎屑混合沉积。

基质类型在很大程度上控制了生产碳酸盐的生物群落的分布。软底栖居的生物,比如内底栖双壳类、腹足、独立生长的苔藓虫以及食腐屑的底栖有孔虫,在细粒的粉砂和砂中占主体。硬底栖居的生物,比如腹足、正常海胆、外底栖双壳类和过滤进食的底栖有孔虫,分布在两个海湾岛屿周围碳酸盐富集环境。与固着无柄(fixo-sessile)群落共生,比如藤壶、龙介虫和苔藓虫。

2.4.2　相分布

两个海湾古海洋学上的变化(图 2-2 和图 2-3),导致了清晰的相分布。相 Ia(珊瑚藻)

和相 Ib(Rhodmol)在非上升流 GoC 浅水区域繁盛,而相Ⅱ(Balamol)仅出现在上升流 GoP 的许多岛屿(图 2-12)。相Ⅲ、Ⅳ和Ⅴ在两个海湾都有分布,并且具有非常相似的生物群,与不同百分比的陆源组分相结合。上升流和非上升流条件的对比,产生了特有的碳酸盐相,以适应稳定的(GoC)和变动的(GoP)海洋和气候条件。

相分布图(图 2-12)也展示了,在 GoP,广阔的海底主要覆盖陆源沉积物(相Ⅳ和Ⅴ)。这与 GoC 相反,这里出现更多的碳酸盐(相Ⅰ到Ⅲ)。GoP 和 GoC 在碳酸盐丰度的差异是生物和物理因素联合作用的结果。两个海湾不同的温度和营养体系造成生物群落不同的碳酸盐生产率。在 GoC,全年较高的温度和低营养条件促进珊瑚、珊瑚红藻和软体动物的生长,并且增加了这些有机体的生长速率。这种影响被计算得到的礁生长记录下来,礁从 1.3~4.2m/ka(GoP)和 4.2m/ka(GoC)之间变化,GoP 最大礁厚度 4.2m,GoC 最大礁厚度 8.3m(Glynn 和 MacIntyre,1977)。珊瑚红藻的分支(maerl)也显示了生长速率的差异,在每年最大 0.3mm(GoC)和每年最大 0.15mm(GoC)之间变化(Fortunato 和 Schäfer,2009)。

2.4.3 碳酸盐环境分类

以往过多强调了根据温度体系来定义碳酸盐类型(Lees,1975;Nelson,1988;James,1997)。基于 Lüning(1990)和 Betzler 等(1997)的研究,选取 10°和 20°冬季等温线来区分寒温带和暖温带,暖温带和热带(Halfar 等,2006)。对巴拿马太平洋海岸的温度体系做对比,GoP 落入暖温带,GoC 落入热带。这个分类反映了寒冷的,非热带的海水温度,可以影响 GoP 这样的热带海洋区域。

碳酸盐生产发生在不同的营养环境。光养组合在热带贫养到轻度半自养环境生长。异养碳酸盐在暖温带到极地环境,以及从半自养到完全富养环境中繁盛(Halfar 等,2006)。然而,许多作者展示了重要的碳酸盐生产可以发生在暖温带环境,形成了前面提到的端元之间的转换(Henrich 等,1995,1997;Fornos 和 Ahr,1997;Halfar 等,2000,2001,2006)。古暖温带环境也是一样(Manker 和 Carter,1987;Carannante 等,1988,1995;Gischler 等,1994;Brachert 等,1998)。珊瑚为热带碳酸盐,而在温暖的环境,高百分含量的苔藓虫、软体动物、棘皮动物和有孔虫等异养组合可以代表在透光层以下的环境(James,1997)。另外一种解释是它们代表了典型的浅水"冷水"碳酸盐(Lees,1975;Nelson,1988;James,1997;Halfar 等,2000,2001,2006)。然而,James(1997)明确指出:"冷水碳酸盐一般为异养生物,而异养组合并不意味着冷水沉积环境"。

对在 GoP 和 GoC 发现的相类型的沉积组分按照所出现的环境进行分类。在 GoC,相 Ia(珊瑚藻)和相 Ib(Rhodmol)明显反映出贫养或者贫养—半自养环境(图 2-11 和图 2-12)。在养分增加的环境中,藻珊瑚(Zooxanthellate corals)被珊瑚红藻取代。

GoP 显示半自养到富养环境,由局限在海湾的藤壶—软体相(相Ⅱ)所反映出来的上升流形成。相Ⅱ是典型的异养生物组分,一般解释为狭义的"冷水"陆架碳酸盐(James,1997)。这种由藤壶、棘皮动物和软体动物组合的异养群,与那些由 Halfar 等(2006,图 2-10)在暖温带加利福尼亚湾发现的生物群相似。这些作者也发现藤壶和棘皮动物在富养环境占主体。

2.5　与其他现代和古代碳酸盐对比

当分析 GoP 和 GoC 数据集的时候,出现了以下问题:(1)什么生物可以精确标志热带和冷水碳酸盐工厂;(2)哪一种特征的生物在 GoP 和 GoC 缺失,这种生物标志了碳酸盐工厂的类型?换句话说,所发现的相与现代和古代冷水和热带碳酸盐的数据是否一致?

Nelson(1988)广泛描述了新西兰冷水碳酸盐中的骨骼颗粒组分。观察到的生物有苔藓虫、双壳、软体动物、棘皮动物和底栖有孔虫。此外,藤壶、钙性藻类、腕足动物、单体珊瑚、环节动物、浮游有孔虫、超微化石和海绵骨针也有出现。在 GoP 异养主体的相组合与 Nelson(1988)描述的动物组合极好相匹配,从而被划分为经典的冷水组合。然而,陆架的黏土含量,与其他冷水沉积环境的相分布并不一致。相反,在 GoC,相组合包括数量可观的红藻石(网捞样品:57.7%;随机样品:48.7%),这对现代贫养环境来说是不常见的。而且,在 GoC 沉积物包括结壳的和独立生长的苔藓虫(随机样品:27.9%)(图 2-7,表 2-2—表 2-4)。在加利福尼亚湾,碳酸盐生物的分布(Halfar 等,2004a,b)不仅受到温度和盐度的影响,还受到养分富集的影响。在加利福尼亚湾,营养分布和以下生物的出现存在一种从北向南的趋势:(1)富养的,软体—苔藓虫;(2)以红藻为主;(3)贫养—半自养的珊瑚礁主导的相。这种趋势在其他作者的研究中也被发现(见文献综述,Halfar 等,2004b)。这些相组合在 GoC 贫养—半自养环境出现,包含来自热带和暖温带的动物组分的混合物,例如,珊瑚红藻和珊瑚(红藻主导的相,Halfar 等,2004a)。

南西班牙上中新统碳酸盐沉积环境进行了与 GoC 或 GoP 地区相似的相划分(Gläser 和 Betzler,2002)。这些作者展示了温水红藻碳酸盐发育在无碎屑注入的孤立的海湾,而苔藓虫—软体组合沉积物发育在开阔海湾。Gläser 和 Betzler(2002)提出这种划分方案与由河流注入引起营养供应变化相关,表现为红藻碳酸盐发育在低营养供应区域;或者与水力体系的差异相关,表现为红藻相发育在潮下带,而发育苔藓虫—软体组合相的海藻林分布在浪控区域。观测结果表明,第三种机制,即季节性上升流,产生了多期养分注入,导致碳酸盐生物类型的可对比划分。

Philip 和 Gari(2005)描述了以苔藓虫、红藻和棘皮动物的相组合作为主要骨骼颗粒的南法国下白垩统(Coniacian 到早 Santonian)异养碳酸盐岩。在这些碳酸盐中出现再沉积的厚壳蛤碎片和底栖有孔虫,以及少量碎屑物质。Philip 和 Gari(2005)提出陆源供应和上升流可能增加了陆架上的养分含量,这有利于悬食和半自养生物的发育。此外,这些作者指出,开阔海、相对较深的冷水环境,沉积异养碳酸盐。然而,在 GoP 和 GoC(本次研究)以及加利福尼亚湾发现的相组合(Halfar 等,2001,2004a)表明,异养和光养动物组分代表了暖温带沉积环境。因此,用类比方法,从光养主导的碳酸盐向异养主导的碳酸盐的转换出现在南法国晚白垩世异养碳酸盐,这可能与上升流与非上升流环境的相对变化有关,而不是与 Philip 和 Gari(2005)提出的主要海侵事件相关的海洋环流形式的大尺度变化有关。一系列相似的变化出现在中南部比利牛斯山上白垩统(Santonian)碳酸盐(Pomar 等,2005)。在这里,一系列变化出现在碳酸盐生物的主导类型,厚壳蛤建造和相应的相,这些相与底栖有孔虫富集的颗粒岩或泥粒灰岩相对。这些作者提出所观察到的相变化主要归因于养分来源和温度的变化,但值得注意的是,在这些变化中,很难估计全球因素的影响,比如,碳循环和有机碳埋藏,大气 CO_2 变化,海水中 Ca^{2+} 绝对含量和 Mg/Ca 比值。然而,本次研究展示了季节性上升流可以足够引起碳酸盐生物类型的变化。

由珊瑚和珊瑚红藻主导的温水碳酸盐与以苔藓虫、珊瑚红藻、软体动物和有孔虫主导的"冷水"碳酸盐的混合,被发现在澳大利亚西海岸的 Abrolhos 陆架上(Collins 等,1997)。这些不同相的共生,最开始由早到中更新世向极方向流动的温水环流造成,环流形成被珊瑚生长所需的因素,比如温度、幼虫传送,以及养分含量。在这个转换之前,在中新世和始新世,冷水沉积物占主导地位。James 等(1999)再次确认了 Collins 等(1997)的发现,并且阐述了向极方向流动的、温暖的、贫养的 Leeuwin Current 环流,促进了沉降流和向赤道方向刮的强风,产生了区域性的、季节性的上升流,导致了前面提到的相差异,这与在 GoP 和 GoC 看到的相划分是可比较的。

在澳大利亚 Murray 盆地,与 GoP 和 GoC 发现的相似的季节性变化掌控着渐新世—中新世温水碳酸盐的发育(Lukasik 等,2000)。在这里,干旱气候时期为低营养水平,在多雨气候时期营养水平上升,这种季节性变化导致了包含有孔虫、苔藓虫动物群的光能共生有机体的繁盛。在多雨气候时期,表栖种类和海草减少,导致了内栖种类的海胆和双壳及腹足动物的发育。

在 GoP 和 GoC,对碳酸盐沉积物详细的分析清楚展示了沉积体系对季节性海洋变化的响应。本次研究结果确认了 Hallock 等(1988)的发现,他指出了相对适度的营养来源水平可以抑制珊瑚礁的发育。Mutti 和 Hallock(2003)指出营养和温度变化的影响掌控了碳酸盐相组合的发育。Halfar 等(2000,2004a,b)展示了这种影响对加利福尼亚湾现代沉积的作用,而 Pomar 等(2004)讨论了对一系列地中海盆地的中新世碳酸盐台地的影响作用。

在 GoP,泥质沉积物占主导地位,并缺失苔藓虫,以及在 GoC,红藻石和大量的苔藓虫的出现,是判断 GoP 碳酸盐沉积是一个出现在热带环境的异养相的决定性信息。

本次研究的结果支持了 Nelson(1988)和 James(1997)的理论分析,他们明确指出异养相组合沉积物的出现并不能确凿证明其沉积在经典"冷水"碳酸盐环境。需要找到支持这样一种解释的其他的特征,比如冰川沉积物、地球化学证据以及其他特征的生物。Betzler 等(1997)展示了与热带、寒温带以及冷水碳酸盐相连接的暖温带碳酸盐工厂。作者说明了利用大型底栖有孔虫、钙化绿藻和虫黄藻珊瑚来区分暖温带和冷水碳酸盐。这里展示的研究结果也阐明了对生物含量严密的分析可以区分受冷水影响的热带碳酸盐和"真正"的冷水碳酸盐。

2.6 结论

在巴拿马湾(GoP)季节性上升流对碳酸盐生物的分布产生重要影响。此外,两个海湾碳酸盐生产和分布受岛屿分布和陆源注入影响。陆源物质主要由较小颗粒(小于 63μm 到 250μm)构成,因此,容易被水流和波浪搬运。碳酸盐主导的沉积物(碳酸盐砂屑和混合碳酸盐—碎屑岩砂)在原地堆积,主要由较大颗粒构成(大于 500μm)。

在 GoP,上升流仅对碳酸盐的产量产生有限影响,而对碳酸盐生物群落产生重大影响。在两个海湾的岛屿周围,碳酸盐组合有明显差异。在 GoC,光养组合繁盛,而在 GoP,异养组合更繁盛。两个海湾显示既没有纯的热带碳酸盐,也没有严格意义的冷水碳酸盐。实际上,在 GoP 和 GoC,包含了前面提到的两种沉积环境碳酸盐生物。在 GoC 岛屿周围的浅水区域,显示珊瑚或者红藻石相,指示了贫养到半自养环境。在 GoP,在岛屿周围主要出现藤壶—软体相,表明为与旱季上升流有关的半自养到季节性富养环境(12 月到 4 月)。

本次研究展示了在热带易产生季节性上升流的区域,可发育"冷水"碳酸盐动物群和富养环境。而且,研究也展示了,对比缺乏上升流的,位于相同纬度和陆架结构的巴拿马地峡的太平洋一侧,季节性上升流对表层沉积物成分可能的影响。

参 考 文 献

Betzler,C. ,Brachert,T. C. and Nebelsick,J. (1997) The warm temperate carbonate province: a review of the facies, zonations,and delimitations. Cour. Forsch. Inst. Senckenberg,201,83 – 99.

Birkeland,C. (1987) Nutrient availability as a major determinant of differences among coastal hard-substratum communities in different regions of the tropics. In: Differences between Atlantic and Pacific Tropical Marine Coastal Ecosystems:Community Structure, Ecological Processes and Productivity (Ed. C. Birkeland), pp. 45 – 90. UNESCO Reports in Marine Science,Paris.

Brachert,T. ,Betzler,C. ,Braga,J. C. and Martin,J. M. (1998) Microtaphofacies of a warm-temperate carbonate ramp (Uppermost Tortonian/Lowermost Messinian,Southern Spain). Palaios,13,459 – 475.

Carannante,G. ,Esteban,M. ,Miliman,J. D. and Simone,L. (1988) Carbonate lithofacies as paleolatitude indicators: problems and limitations. Sed. Geol. ,60,333 – 346.

Carannante,G. ,Cherchi,A. and Simone,L. (1995) Chlorozoan versus foramol lithofacies in Upper Cretaceous rudist limestones. Palaeogeogr. Palaeoclimatol. Palaeoecol. ,119,137 – 154.

Clark,J. P. (1988) Panama,Pacific Coast: Gulf of Panama,Nautical Map. Admiralty Charts and Publications,Taunton,UK.

Collins,K. S. B. (1973) Central America,South Coast: Cabo Mala to Punta Burrica,Nautical Chart. Admiralty Charts and Publications,London.

Collins,L. B. ,France,R. E. ,Rong Zhu,Z. and Wyrwoll,K. -H. (1997) Warm-water platform and coo-water shelf carbonates of the Abrolhos Shelf, Southwest Australia. In: Cool-Water Carbonates (Eds N. P. James and J. A. D. Clarke),SEPM Spec. Publ. ,56,23 – 36. SEPM (Society for Sedimentary Geology),Tulsa,Oklahoma,USA.

Crossland,J. C. (1987) Dissolved nutrients in coral reef waters. In: Perspectives on Coral Reefs (Ed. D. J. Barnes), pp. 56 – 68. The Australian Institute of Marine Science,Townsville.

D'Croz,L. and O'Dea,A. O. (2007) Variability in upwelling along the Pacific shelf of Panama and implications for the distribution of nutrients and chlorophyll. Estuar. Coast. Shelf. Sci. ,73,325 – 340.

D'Croz, L. , Del Rosario, J. B. and Gómez, J. A. (1991) Upwelling and Phytoplankton in the Bay of Panama. Rev. Biol. Trop. ,39,233 – 241.

D'Croz,L. ,Maté,J. L. and Ole,J. E. (2001) Responses to elevated sea water temperature and UV radiation in the coral Porites lobata from upwelling and non-upwelling environments on the Pacific coast of Panama. Bull. of Mar, Sci. ,69,203 – 214.

Estrada, M. and Blasco, D. (1979) Two phases of the phytoplankton community in the Baja California upwelling. Limnol. Oceanogr. ,24,1065 – 1080.

Fleming,R. H. (1939) A contribution to the oceanography of the Central America Region. Proceedings of the 6th Pacific Science Congress,3,167 – 176.

Flü gel,E. (1982) Microfacies analysis of limestones. Springer Verlag,Berlin,375 pp.

Flü gel,E. (2004) Microfacies of carbonate rocks. Springer,Berlin,976 pp.

Fornos,J. J. and Ahr,W. M. (1997) Temperate carbonates on a modern, low – energy, isolated ramp: the Balearic Platform,Spain. J. Sed. Res. ,67,364 – 373.

Forsbergh,E. D. (1963) Some relationships of meteorological, hydrographic, and biological variables in the Gulf of Panama. Bull. Inter-Amer. Trop. Tuna Comm. ,7,1 – 109.

Forsbergh, E. D. (1969) On the climatology, oceanography and fisheries of the Panama Bight. Bull. Inter-Amer. Trop. Tuna Comm. ,14,49 – 259.

Fortunato,H. and Schäfer,P. (2009) Coralline algae as carbonate producers and habitat providers on the EasternPacific coast of Panamá:preliminary assessment. Neues Jb. Geol. Paläontol. Abh. ,253,145 – 161.

Foster, M. S. (2001) Rhodoliths: between rocks and soft places. J. Physiol., 37, 217 – 234.

Freiwald, A. and Henrich, R. (1994) Reefal coralline algal build-ups within the Artic Circle: morphology and sedimentary dynamics under extreme environmental seasonality. Sedimentology, 41, 963 – 984.

Gischler, E., Graefe, K. -U. and Wiedmann, J. (1994) The Upper Cretaceous Lacazina Limestone in the Basco-Cantabrian and Iberian Basins of northern Spain: Cold-water grain associations in warm-water environments. Facies, 30, 209 – 246.

Gläser, I. and Betzler, C. (2002) Facies partioning and sequence stratigraphy of cool-water, mixed carbonatesiliciclastic sediments (Upper Miocene Guadalquivir Domain, southern Spain). Int. J. Earth Sci., 91(6), 1041 – 1053.

Glynn, P. W. and MacIntyre, I. G. (1977) Growth rate and age of coral reefs on the Pacific Coast of Panama. In: 3rd International Coral Reef Symposium (Ed. D. L. Taylor), vol. 2: Geology, pp. 251 – 259. Rosenstiel School of Marine and Atmospheric Science, Miami, Florida, USA.

Glynn, P. W., Stewart, R. H. and McCosker, J. E. (1972) Pacific coral reefs of Panama: structure, distribution and predators. Geol. Rundsch., 61, 483 – 519.

Halfar, J., Godinez-Orta, L. and Ingle, J. C. J. (2000) Microfacies analysis of Recent carbonate environments in the Southern Gulf of California, Mexico – A model for warm-temperate to subtropical carbonate formation. Palaios, 15, 323 – 342.

Halfar, J., Godinez-Orta, L., Goodfriend, G. A., Mucciarone, D. A., Ingle, J. C. J. and Holden, P. (2001) Holocene-latePleistocene non-tropical carbonate sediments and tectonic history of the western rift basin margin of the southern Gulf of California. Sed. Geol., 144, 149 – 178.

Halfar, J., Ingle J. C., Jr and Godinez-Orta, L. (2004a) Modern non-tropical mixed carbonate – siliciclastic sediments and environments of the southwestern Gulf of California, Mexico. Sed. Geol., 165, 93 – 115.

Halfar, J., Godinez-Orta, L., Mutti, M., Valdez-Holguín, J. E. and Borges, J. M. (2004b) Nutrient and temperature controls on modern carbonate production. An example from the Gulf of California, Mexico. Geology, 32, 213 – 216.

Halfar, J., Godinez-Orta, L., Mutti, M., Valdez-Holguin, J. E. and Borges, J. M. (2006) Carbonates calibrated against oceanographic parameters along a latitudinal transect in the Gulf of California, Mexico. Sedimentology, 53, 297 – 320.

Hallock, P. (1987) Fluctuations in the trophic resource continuum: a factor in global diversity cycles? Paleoceanography, 2, 457 – 471.

Hallock, P. (1988) The role of nutrient availability in bioerosion: consequences to carbonate buildups. Palaeogeogr., Palaeoclim., Palaeoecol., 63, 275 – 291.

Hallock, P. (2001) Coral reefs, carbonate sediments, nutrients and global change. In: The History and Sedimentology of Ancient Reef Systems (Ed. D. J. Stanley), pp. 387 – 427. Kluwer Academic Press/Plenum Publishers, New York.

Hallock, P. and Schlager, W. (1986) Nutrient excess and the demise of coral reefs and carbonate platforms. Palaios, 1, 389 – 398.

Hallock, P., Hine, A. C., Vargo, G. A., Elrod, J. A. and Jaap, W. C. (1988) Platforms of the Nicaraguan Rise: Examples of the sensitivity of carbonate sedimentation to excess trophic resources. Geology, 16, 1104 – 1107.

Henrich, R., Freiwald, A., Betzler, C., Bader, B., Schäfer, P., Samtleben, C., Brachert, T. C., Wehrmann, A., Zankl, H. and Kühlmann, D. H. H. (1995) Controls on Modern carbonate sedimentation on warm-temperate to Arctic coasts, shelves and seamounts in the Northern Hemisphere: Implications for fossil counterparts. Facies, 32, 71 – 108.

Henrich, R., Freiwald, A., Brachert, T. C. and Schäfer, P. (1997) Evolution of an Arctic open-shelf carbonate platform, Spitzbergen bank (Barents Sea). In: Cool-Water Carbonates (Eds N. P. James and J. A. D. Clarke), SEPM-Spec. Publ., 56, 163 – 181. SEPM (Society for Sedimentary Geology), Tulsa, Oklahoma.

Jackson, J. B. C. and Herrera-Cubilla, A. (2001) Adaptation and constraints as determinants of zooid and ovicell size among encrusting ascophoran cheilostome Bryozoa from opposite sides of the Isthmus of Panama. Proc. 11th Internat. Bryozool. Assoc. Conf. ,2000,249 – 258.

James, N. P. (1997) The cool-water carbonate depositional realm. In: Cool-Water Carbonates (Eds N. P. James and J. A. D. Clarke), SEPM Spec. Publ. ,56,pp. 1 – 20. SEPM (Society for Sedimentary Geology), Tulsa, Oklahoma.

James, N. P. , Collins, L. B. , Bone, Y. and Hallock, P. (1999) Subtropical carbonates in a temperate realm: modern sediments on the southwest Australian Shelf. J. Sed. Res. ,69,1297 – 1321.

Kolarsky, R. A. , Mann, P. and Montero, W. (1995) Island arc response to shallow subduction of the Cocos Ridge, Costa Rica. In: Geologic and Tectonic Development of the Caribbean Plate Boundary in Southern Central America (Ed. P. Mann), GSA Spec. Pap. ,295,235 – 262. Geological Society of America, Boulder, CO.

LaVigne, M. , Field, M. P. , Anagnostou, E. , Grottoli, A. G. , Wellington, G. M. and Sherrell, R. M. (2008) Skeletal P/Ca tracks upwelling in Gulf of Panama coral: Evidence for a new seawater phosphate proxy. Geoph. Res. Lett. ,35, L05604. doi:10. 1029/2007GL031926.

Lees, A. (1975) Possible influence of salinity and temperature on modern shelf carbonate sedimentation. Mar. Geol. , 19,159 – 198.

Lees, A. and Buller, A. T. (1972) Modern temperate-water and warm-water shelf carbonates contrasted. Mar. Geol. , 13,67 – 73.

Legeckis, R. (1985) Upwelling off the Gulfs of Panama and Papagayo in the Tropical Pacific during March 1985. J. Geoph. Res. ,93,15485 – 15489.

Linsley, B. K. , Dunbar, R. B. , Wellington, G. M. and Mucciarone, D. A. (1994) A coral-based reconstruction of intertropical convergence zone variability over Central America since 1707. J. Geoph. Res. ,99,9977 – 9994.

Littler, D. S. and Littler, M. M. (1985) Factors controlling relative dominance of primary producers on biotic reefs. Fifth Int. Coral Reef Congress, Tahiti,4,35 – 39.

Lukasik, J. J. , James, N. P. , McGowran, B. and Bone, Y. (2000) An epeiric ramp: low-energy, cool-water carbonate facies in a Tertiary inland sea, Murray Basin, South Australia. Sedimentology,47,851 – 881.

Lüning, K. (1990) Seaweeds. Their Environment, Biogeography, and Ecophysiology. John Wiley & Sons, New York, 589 pp.

MacIlvaine, J. C. and Ross, D. A. (1973) Surface sediments of the Gulf of Panama. J. Sed. Petrol. ,43,215 – 223.

Manker, J. P. and Carter, B. D. (1987) Paleoecology and paleogeography of an extensive Rhodolith facies from the Lower Oligocene of South Georgia and North Florida. Palaios,2,181 – 188.

Maté, J. L. (2003) Corals and coral reefs of the Pacific coast of Panamá In: Latin American Coral Reefs (Ed. J. Cortés), pp. 387 – 417. Elsevier Science B. V. , Amsterdam, The Netherlands.

Mount, J. F. (1984) The mixing of siliciclastic and carbonate sediments in shallow shelf environments. Geology,12, 432 – 435.

Mutti, M. and Hallock, P. (2003) Carbonate systems along nutrient and temperature gradients: some sedimentological and geochemical constraints. Int. J. Earth Sci. ,92,465 – 475.

Nelson, C. S. (1988) An introductory perspective on nontropical shelf carbonates. Sed. Geol. ,60,1 – 12.

Philip, J. M. and Gari, J. (2005) Late Cretaceous heterozoan carbonates: Palaeoenvironmental setting, relationships with rudist carbonates (Provence, south-east France). Sed. Geol. ,175,315 – 337.

Pomar, L. , Brandano, M. and Westphal, H. (2004) Environmental factors influencing skeletal grain sediment associations: a critical review of Miocene examples from the western Mediterranean. Sedimentology,51,627 – 651.

Pomar, L. , Gili, E. , Obrador, A. and Ward, W. C. (2005) Facies architecture and high-resolution sequence stratigraphy of an Upper Cretaceous platform margin succession, southern central Pyrenees, Spain. Sed. Geol. , 175, 339 – 365.

Roberts, A. and Withers, P. (2006) statistiXL – Statistical Power for Microsoft Excel Vers. 1. 6.

Rodriguez-Rubio, E., Schneider, W. and del Rio, R. A. (2003) On the seasonal circulation within the Panama Bight derived from satellite observations of wind, altimetry and sea surface temperature. Geoph. Res. Lett., 30, 1410.

Schaefer, M. B., Bishop, Y. M. and Landa, G. V. (1958) Some aspects of upwelling in the Gulf of Panama. Bull. Inter-Amer. Trop. Tuna Comm., 3, 77 – 130.

Schlager, W. (2005) Carbonate Sedimentology and Sequence Stratigraphy. SEPM Concepts in Sedimentology and Paleontology, 8. SEPM (Society for Sedimentary Geology), Tulsa, Oklahoma, 200 pp.

Smayda, T. J. (1963) A quantitative analysis of the phytoplankton of the Gulf of Panama I. Results of the regional phytoplankton surveys during July and November, 1957 and March, 1958. Bull. Inter-Amer. Trop. Tuna Comm., 7, 191 – 253.

Smayda, T. J. (1966) A quantitative analysis of the phytoplankton of the Gulf of Panama. III. General ecological conditions, and the phytoplankton dynamics at 8°45′N, 79°23′W from November 1954 to May 1957. Bull. Inter-Amer. Trop. Tuna Comm., 11, 353 – 612.

Strub, P. T., Mesías, J. M., Montecino, V., Rutllant, J. and Salinas, S. (1998) Coastal ocean circulation off western South America – Coastal segment (6, E). In: The Sea (Eds A. R. Robinson and K. H. Brink), 11, pp. 273 – 313. John Wiley & Sons, Inc., New York.

Valentine, J. W. (1973) Evolutionary Paleoecology of the Marine Biosphere. Prentice-Hall, Princeton, NJ, 511 pp.

Vecsei, A. (2003) Nutrient control of the global occurrence of isolated carbonate banks. Int. J. Earth Sci., 92, 467 – 481.

Wellington, G. M. and Dunbar, R. B. (1995) Stable isotope signature of El Nino-Southern Oscillation events in eastern tropical Pacific reef corals. Coral Reefs, 14(1), 5 – 25.

Wellington, G. M. and Glynn, P. W. (1983) Environmental influences on skeletal banding in eastern Pacific (Panama) corals. Coral Reefs, 1, 215 – 222.

Wilson, J. L. (1975) Carbonate Facies in Geological History. Springer Verlag, Berlin, 471 pp.

Wilson, M. E. J. and Vecsei, A. (2005) The apparent paradox of abundant foramol facies in low latitudes: their environmental significance and effects on platform development. Earth-Sci. Rev., 69, 133 – 169.

第3章 厚层浅水碳酸盐岩相分布的成因

PETER M. BURGESS，DAVID A. POLLITT 著

吴洛菲 译，杜业波 校

摘　要　近几年碳酸盐岩地层学最基本的研究结果是人们观察到浅水碳酸盐岩地层岩石相厚度往往具有指数分布的规律。因为它的观察结果不仅可以在各种类型的露头及地下样品中重复验证，同时因为它引出了一个问题：什么样的沉积过程和气候及海洋学背景可以形成特定的岩石相厚度分布。反过来，碳酸盐岩地层如何记录地质时期的气候和海洋学变化这个意义重大的问题也和这个观察结果是密不可分的。本研究应用一种碳酸盐台地地层(Dougal)的简单一维地层正演数值模型来研究相对海平面波动对岩石相分布的控制作用。Dougal 记录了台地顶部的碳酸盐累积量受到透光层、微光层和无光层的沉积产率随水深变化的生产剖面的影响。该剖面中滞后深度控制生产起始点。5 个单个模型的运行结果都凸显了非稳态行为，即抬升地区的地层变化的统计学特性问题，说明确定性建模也可以产生岩石相厚度的指数分布规律。多模型中，生产和可容空间增长率范围跨度很大的总共27200个运算结果证明在这种一维模型中碳酸盐岩岩石相分布受控于三个主要因素：(1)可容空间的复杂变化，这里归因于高频的冰川海平面升降；(2)沉积生产速率在本次模型运算中由透光层沉积生产速率主导；(3)沉积物中自循环振荡是由滞后深度效应驱动，在冰川海平面升降曲线中的高频爬升期中形成的。在多模型运行中，其中只有大约13%产生了指数分布规律，而记载中的露头样品有28%。这说明一些在模型中起重要作用的过程没有考虑进去，我们还需要更深入的研究确定这些过程。这些过程很可能是那些发生在三维空间台地顶部的多种多样的沉积过程，例如与沉积物运移和与生物多样性有关的沉积物生产速率和类型的变化以及其他发生在不同沉积背景下的过程，如斜坡上的沉积物运移过程。

关键词　岩层厚度　碳酸盐岩　指数分布　地层正演模型

3.1　概况

浅水碳酸盐岩地层可以特地质时期的气候和海洋学的微小改变都能记录下来。薄层要比厚层的岩石相厚度分布更容易反映浅水碳酸盐岩地层的基本性质(Wilkinson 等，1997，1999)，显而易见这引出了一个问题：这些厚度能告诉我们关于气候和海洋学变化的什么信息？

Wilkinson 等(1999)以及 Wilkinson 和 Drummond(2004)曾用指数分布来描述这些岩石相厚度分布规律，并通过调用一个随机(泊松)分布累积过程的沉积模型解释观察到的分布情况。可能是因为 Wilkinson 等(1999)及 Wilkinson 和 Drummond(2004)提出的这个模型，这些重要的观测结果及其提出的解释曾引起了一小部分人的详细关注。这次观测似乎和通常无法证实的常识——碳酸盐岩的层序地层学解释中暗示了可容空间的变化作为确定性的外部作用相矛盾。为了进一步了解这种矛盾，已观察到的岩石相厚度分布背后的机理对决定过去全球变化因素显然很重要。

有一种例外出现在 Burgess(2008)的碳酸盐岩岩石相厚度分布问题研究中，他当时通过使用 Kolmogorov – Smirnov(KS)测试比对了露头和理论分布，分析了 56 个露头和表面样品，发现 29% 很显然是指数分布，有半数尽管仍显示出了基本的长宽—厚单位比率，但仍不符合指数分

布。Burgess(2008)强调了控制岩层厚度分布的过程,却很难理解。对于控制过程,一些可能备选在讨论中,但还没有形成任何一种假说。而且同等重要的是,Kolmogorov – Smirnov 测试,尽管一些工作者试图这样用(如 Cooper 和 Smart,2010),但不能从任何感官上展示或者说其缺乏可以计量的周期性。

在之前工作的基础上,很明显会提出下一个问题:除随机积累的泊松过程外,一个重要的备选解释模型本身,其他的沉积过程模型是否可以解释观察到的岩石相厚度分布? 本文的贡献在于使用了碳酸盐积累的一维正演模型来研究简单相对海平面变化振荡的驱动力对岩石相分布类型的作用。

3.2 模型描述

Dougal 是前体模型[Pollitt 在 TED 描述(2008)]衍生出的一个碳酸盐累积的一维数值层序地层学正演模型。Dougal 记录了某一单个地理学点上的理论碳酸盐台地沉积面的垂直运动:

$$h_t = s_{\Delta t} + e_{(wd, \Delta t)} + o_{(wd, \Delta t)} + a_{(wd, \Delta t)} - d_{(wd, \Delta t)}$$

式中 t——时间,Ma;

　　　h——海拔,m;

　　　s——沉降速率,m/Ma;

　　　e——透光层的沉积物生产率,m/Ma;

　　　wd——水深,m;

　　　o——微光层生产率,m/Ma;

　　　a——无透光层的沉积物生产率,m/Ma;

　　　d——剥蚀速率(水下的或地表的),m/Ma;

　　　Δt——固定值,25 年。

测试显示 25 年是使结果数值稳定和精确的最大时间步(Pollitt,2008)。注意到剥蚀速率在本次研究的所有模型案例中都是 0;并需要进一步评估这些结果的侵蚀影响。

因为数值模拟处理的海平面升降变化从本质上说是一维的,而且冰期期间加积地层几何体从常识中认为是贯穿内部地台环境的(例如 Goldhammer 等,1991;Della Porta 等,2004),所以一维模型可以很好地适用于研究这个问题。这种情况下,台内区域假设几乎不显示出了横向相变(Stemmerik,1996;Lehrmann 和 Goldhammer,1999),尽管控制因素中可能存在复杂的变化(见 Burgess 和 Wright,2003)。现代台地显示横向相变大多确实存在(Harris 和 Vlaswinkel,2008)。然而,对于首先理解侧向均质的台地内部是如何受到冰期影响,一个简单的一维模型就能满足我们的需要。

一维模型最大的优势就是其运算时间段短。简单的 Dougal 模型的运算时间只有大约数秒钟,因此可以进行成百上千的单个确定性运算。在这类模型中,必须使用非常短的时间步而不用长时间,来避免在计算水深与产率关系时数值结果导致的不准确性。模型输出结果的定量统计分析要大量的运算,二维或者三维模型使计算机的负荷很大。数千次模拟的结果数据集就可以详细地绘制出模型参数的空间分布,这样就可以理解岩层厚度分布的控制情况。

3.3 模型公式

根据一些单独的简单操作过程,通过模型每次的运算时间变化和重复观察碳酸盐的累积。

3.3.1 沉降速率

沉降是地层积累和长期保存的一种基本控制因素(例如 Burgess 和 Wright,2003)且沉积速率随时间变化。但是,沉降速率的变化,至少是在远离活跃断层的地区,往往发生在数百万年的时间尺度上的(Bosence 等,2009)。这里的模型运算时间跨度可达三百万年,因此所有情况中沉降速率恒定为每百万年 100m。

3.3.2 海平面变化

碳酸盐的早期一维模型采用了对称的正弦曲线海平面波动(Walkden 和 Walkden,1990;Allen 和 Allen,2005)。近期的工作显示应该用非对称曲线来反映冰川的高频非对称的波动和减弱(Clark 等,1999)。受构造影响,如海盆体积的变化,海平面波动也可以出现在百万年的尺度(Kendall 和 Lerche,1988;Goldhammer 等,1991;de Boer 和 Smith,1994;Barnett 等,2002;Artyushkov 等,2003,2002;Artyushkov 等,2003)。在这个模型中,假设长期的海平面变化(周期 >1Ma)是因为冰川海平面变化和板块驱动力变化的共同作用,故模型用的是对称曲线。短期高频海平面变化就假设是冰川海平面变化,由区域性的冰川体积变化驱动,其典型的变化幅度是 35~75m(Crowley 和 Baum,1991;Paterson 等,2006)。

Dougal 模型的海平面非对称变化使用了正弦曲线波动函数:

$$f(x) = \begin{cases} \sin\left(\dfrac{\pi x}{x}\right) & -\alpha/2 < x < \text{ and} \\ \sin\left[\dfrac{\pi}{2} + \dfrac{\pi}{\beta} \times (x - \alpha/2)\right] = \\ \cos\left[\dfrac{\pi}{\beta} \times (x - \alpha/2)\right] & \alpha/2 < x \leq \alpha/2 + \beta \end{cases}$$

这里 α 是周期的相对比例,用每一个海平面波动的整梯度上边缘表示;β 是周期的相对比例,用负梯度上边缘表示。在这个范围之外,$f(x)$ 定义为周期。因为这个函数是奇函数(例如 $f(-x) = f(+x)$),它的傅里叶系数包含 sines,因此:

$$f(x) = \sum b_n \times \sin\left(\dfrac{2\pi n x}{\alpha + \beta}\right)$$

这里:

$$b_n = \dfrac{2}{\pi} \times \left\{ \dfrac{1}{1 + \dfrac{\beta}{\alpha} - 2n} - \dfrac{1}{1 + \dfrac{\beta}{\alpha} - 2n} + \dfrac{1}{1 + \dfrac{\alpha}{\beta} - 2n} - \dfrac{1}{1 + \dfrac{\alpha}{\beta} - 2n} \right\} \times \cos\left(\dfrac{\pi n}{1 + \dfrac{\beta}{\alpha}}\right)$$

3.4 碳酸盐积累速率

为了模拟多种碳酸盐生产者在不同深度以不同速率生产的沉积物,碳酸盐沉积产物用透光层、微光层和无光层单独的产量之后来计算。对每一种产物,经过速率修正的最大生产速率被指定为模型水深函数的一个参数(图 3-1)。

因此,任意时间步长的沉积产物是这三种不同类型产物的总和,其累计值由可容空间极限的沉积物产率计算出来,所以累积物不能发生在海平面以上。根据图 3-1 中的生产剖面,每

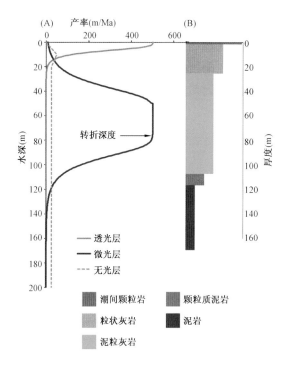

图3-1 (A)模型中使用的碳酸盐水深—产率曲线。注意到转换点在微光层所在的深度,此时产率开始随着水深衰减。(B)模型中用不同碳酸盐工厂单元作用在不同水深产生的不同岩石相的样品。形成这样的模拟实例,水深起始于170m,可容空间允许简单的地层填充。由于模型中不同的工厂作用于不同水深的作用产生不同岩相,前积变浅岩相从泥岩到颗粒岩随着可容空间被填充,水深减少,主要的生产工厂从透光层、微光层到无光层

一个生产剖面都能形成一个不同类型的碳酸盐原料,很可能是同时发生在相同水深范围内(Horbury 和 Adams,1996;Rankey,2004);模型中根据每一个小层中无光层、微光层和透光层的生产原料的相对比例,可以定义岩石相。因此,岩石相厚度是受控于这三个生产剖面的各自的产物相对比例,其次是水深的变化,最后依赖于可容空间的增长与供给速率的比例。例如,如果在给定的迭代次数中无光层产物占了产物的90%~100%,那么岩石相就是泥岩。如果在给定的迭代次数中透光层产物超过了累积厚度的一半且微光层小于80%,无光层小于10%,那么累积的岩石相就是颗粒灰岩。

3.4.1 透光层碳酸盐工厂

透光层生物群是指自养的和异养的依赖于水下显著有光区域,相对占有透光层浅水深度的有机体。透光层的最低限度可以定义为造礁珊瑚虫生长剖面的最大深度(Bosscher 和 Schlager,1993),这样据说在清水中的最大的深度范围是40~50m(Milliman,1974;Hallock 和 Schlager,1986;Pomar,2001b)。基于古老地层解释得到透光层累积速度之前的估计的变化范围比较宽(例如Demicco 和 Hardie,2002;Strasser 和 Samankassou,2003)。这里运行的模型展示了这个值的变化范围。

Bosscher 和 Schlager 曲线是基于主要加勒比海珊瑚虫建造珊瑚(Montastrea annularis)的数据,并假设碳酸盐生产是一条水深的双曲正切函数。基于 Bosscher 和 Schlager 曲线,碳酸盐生产可以用碳酸盐累积速率 e 应用于模型中作为一个离散时间更新规律:

$$e_{(t)} = e_{(m)} \times \tanh[k \times \exp(d \times w_{(t)})]$$

式中 w——水深,m;
t——当前时间步,Ma;
d——腐烂常量;
m——最大生产速率,Ma^{-1};
k——速率常量。

透光层珊瑚的累积的最大速率引自 Bosscher 和 Schlager(1993)是13000m/Ma(13mm每年)。这个速率是明显高于台地内部累积速率估计值的,很可能更低于台地内部背景下长时间海水滞留得到的速率(Demicco 和 Hardie,2002)。例如,Strasser 和 Samankas-sou(2003)引自 Florida, Bahamas 和 Bermuda 的现代系统的累积速率,其变化范围为300~3000m/Ma。来自于

古代系统的经验证据显示出估计得到的沉积速率随着观测时间增加而减少(裂缝和剥蚀估计的误差结果;Sadler,1981)。

Smith 和 Kinsey(1976)及 Bosence 和 Waltham(1990)表示古代台地内部模型累积平均速率是 400~600m/Ma,这与证据相一致。基于这些数据,考察古代碳酸盐台地不同沉积背景的变化下生产的可能速率,250~5000m/Ma 是一个合理的透光层速率变化范围。在给定主要浅水动物群包括来自台地内部的粒状灰岩情况下,假设透光层工厂倾向于生产粒状灰岩相(表3-1)是被认为合理的。

表3-1 模型中使用的岩石相定义

岩石相	最小值			最大值		
	透光层	微光层	无光层	透光层	微光层	无光层
泥岩	0	0	90	10	10	100
颗粒质泥岩	0	0	50	50	50	90
泥粒灰岩	25	0	10	70	70	50
粒状灰岩	50	0	0	100	80	10

注:对于模型中的每一个时间步,根据时间步期间透光层、微光层和无光层矿物沉积的情况,应用这个组合和确定沉积岩石相。对于给定的已生产的岩石相,每个工厂生产的比例一定是介于表中显示的最小值和最大值之间。对于不同岩石相在产量—深度曲线不同部分怎么生产的例子也可参见图 3-1。

3.4.2 微光层碳酸盐工厂

微光层生物群是指自养的和异养的,定居于光线相对不透光层要少些的环境中的有机体。微光层的特点就是光线减弱,有时候温度也会下降(Milliman,1974;Pomar,2001b)。这个层的上下极限取决于海水的光穿透系数,典型的下边界大约在 50~100m 水深范围内(Pomar,2001b)。微光层的沉积速率不像透光层的沉积速率那么明显的取决于沉积的物质(Della Porta 等,2004),但是 Pomar(2001b)和 Schlager(2003)估计它的沉积速率可能是透光层沉积速率的 30%~60%。

给定的时间步长下的微光层生产可以用下式计算:

$$o_t = w_{(t)} < a$$
$$\Rightarrow m \times \tanh[k \times \exp(d_{(u)}) \times (a - w_{(t)})]$$
$$\vee m \times \tanh\{a \times \exp[d_{(l)} \times (w_{(t)} - a)]\}$$

式中 w——水深,m;
t——当前时间步,Ma;
o——碳酸盐累积量,m/Ma;
d——侵蚀常数;
m——生产率的最大值,m/Ma;
k——指数曲线的偏移量。

逻辑运算符 OR 是如果水深超过深度常数转变值 a(图3-1A),那么就会启动。指数曲线的上下部分有不同的对数侵蚀速率,此处分别代表曲线的上下两个部分。微光层曲线倾向生产泥粒灰岩或粒泥状灰岩,并且根据依赖关系,在浅水区部分应该生成泥粒灰岩。

3.4.3　无光层碳酸盐工厂

无光层生物群是指不需要光的异养有机体；它们可以生活在较为宽泛的水深范围内，只是依赖于一些限制因素，例如基质要求，营养物供给，竞争取代，温度，盐度或者水体能量（Pomar，2001b）。无光层生产在给定时间步长的条件下，在模型中代表：

$$p_{(t)} = m\left[w_{(t)} < a \Rightarrow \frac{w_{(t)}}{d} \vee w_{(t)} < j \Rightarrow 1 - \left(\frac{d - w_{(t)}}{d - j}\right) \times 1 - f \vee f\right]$$

式中　w——水深，m；

　　　t——当前时间步，Ma；

　　　m——最大生产率，m/Ma；

　　　d——最大生产深度；

　　　j——平衡生产深度；

　　　f——平衡生产速率，都是作为 m 的一部分。

逻辑运算符 OR 如果水深超过深度转折常数 a，或者平衡生产深度常数 j，那么就起作用。

浅水深度的无光层沉积物的产率是严重受到制约的，并且常用透光层沉积物速率分组（Pomar，2001b）。深水环境的累积更是受到制约，证据就是更新世到全新世数据显示深海沉积的速率一般是 52～66m/Ma（Vollbrecht 和 Kudrass，1990），尽管其他的研究显示无光层碳酸盐沉积物可能更高（Corda 和 Brandano，2003）。当前模型中使用的速率从 50～300m/Ma，在这种制约条件下可以认为是一个合理的范围。无光层产量是主要的，沉积的是泥岩。

3.4.4　滞后深度

滞后深度是指碳酸盐生产开始所需的最小深度，典型情况下是与接近地表暴露面的海侵期间的生产初始化有关的（MacArthur 和 Wilson，1967；Osleger，1991；Tipper，1997）。滞后深度这个概念是基于假设建立一个健康生产的潮下带碳酸盐工厂需要的最小水深，例如，在水体中避免约束导致盐度提高或者碳酸盐饱和度较低。同时也假设复制一个区域海侵后的碳酸盐有机生产可能需要一个最小的时间段。这些多种多样的控制因素可以用滞后深度概括。

Dougal 模型中的滞后深度范围为 0～2m。滞后深度在模型中特别重要，因为在模型中，滞后深度大于 0 时，如果累积发生在海平面处，那么当逐渐上升的海平面使水深达到滞后深度时累积就会停止。在一些特定条件下，例如当沉积物累积速率超过了相对海平面的上升速率时，这就会导致在单一的相对海平面上升期水深和岩石相产生多重的循环（Drummond 和 Wilkinson，1993）。现代碳酸盐体系证明这类对简单外部相对海平面驱动力的复杂响应过程（例如 Vlaswinkel 和 Wanless，2009）是一种自循环形式，并且每次循环的产生对相对海平面扰动都不需任何必要条件（Drummond 和 Wilkinson，1993）。

3.5　岩石相厚度指数分布测试

很多案例都在古碳酸盐岩地层中观察到了岩石相厚度的指数分布，故这种分布模式意义重大（Wilkinson 等，1999；Burgess，2008）。其中一个造就这种分布的可能机理相过渡带之间的等待时间的泊松随机分布过程（Wilkinson 等，1999）。如果累计过程发生在相变化的等待过程中，那么相的厚度应该也是指数分布。根据 Burgess（2008）再次描述和 Press 等（1992）首次描述的情况，将 Dougal 模型中的岩石相分布模式和用相同的平均岩石相单元厚度计算出的理论指数对比，得到：

$$F(t) = 1 - e^{-\mu t}$$

式中　t——岩石相厚度；

　　　μ——平均岩石相单元厚度。

模拟出的理论上的曲线都画在相同的归一化的坐标系中,两个分布之间的最大的偏移值或者差值 D 随后便可以计算出(图3-2)。这个值 D 形成了零假设的 Kolmogorov-Smirnov 测试的基本条件。被调查的分布和一个指数分布不能明显区别出来。这个意义重大的差异 D 的观察值的可能性 P 通过下式计算：

$$P = Q_{ks}(\lambda)$$

式中：

$$Q_{ks}(\lambda) = 2\sum_{j=1}^{\infty}(-1)^{j-1}e^{-2j^2\lambda^2}$$

公式中参数通过下式给定：

$$\lambda = (\sqrt{N} + 0.12 + 0.11/\sqrt{N})D$$

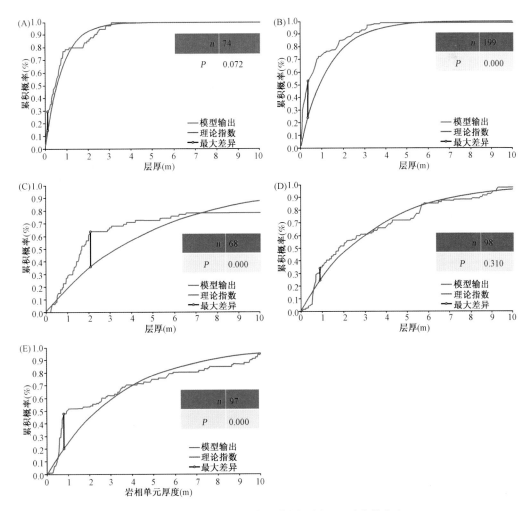

图3-2　岩石相厚度分布和等同的样品尺寸指数分布
(A)模型1;(B)模型2;(C)模型3;(D)模型4;(E)模型5

式中 N 是模拟地区的岩石相单元的总数。

P 是 D 值的可能性,如果分布确实是指数分布,那么观察值至少应该和样品变化的坑内极限一致。因此,P 值接近于 0 意味着观察得到的分布很可能不是指数分布,等同于大于 99% 显著性程度拒绝这种零假设。P 值大于等于 0.1% 意味着没有足够的证据证明合理的拒绝了这种指数分布的解释;这种情况下,一个指数分布可以被认为是展示观察得到的厚度数据的好的模型。对于 P 值在 0.1 到 0.001 之间的解释更难一些,取决于显著性值选取的类型,所以这些案例可以被认为是不确定的。也要注意到这些案例中,样品的尺寸大,Kolmogorov - Smirnov 测试可以为指数曲线相对较小的偏离返回一个显著低的 P 值,但是这种偏离几乎不可能发生在这种随机样品中。

3.6 单模型运算:指数潜力?

正如先前所注意的,在冰期,冰川的波动和减弱引起了大幅度、高频率的海平面升降波动(Clark 等,1999;Rygel 等,2008)。典型的碳酸盐岩地层是这些冰川—海平面波动驱动而形成的,典型的潮下高频序列是受限于发育好的陆上的暴露面(Wright 等和 Vanstone,2001)。为了测试岩石相厚度分布发生在这样的地层中的可能性,一系列的模型,模型案例 1 到 5 都模拟运算过,海平面波动幅度从 10~100m,周期从两万年到一百万年。

3.6.1 模型案例1:0.5Ma 开始案例

运行模型案例 1(MC1;s 参数值的摘要见表 3-2)形成的地层在模型运算中是一般的追赶模式沉积(图 3-3)。这种潮下带追赶模式主导是由于在五千年间 10m 的海平面上升幅度及 500m/Ma 的较低的最大透光层产率,这些导致了相对海平面上升最大速率 6.7m/ka。短期内,海平面上升速率超过了产量。

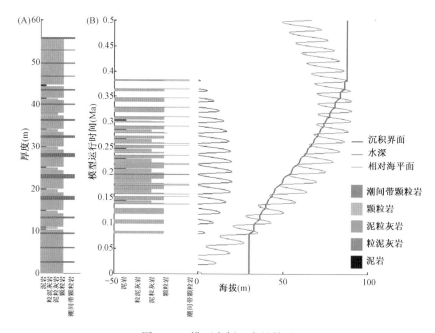

图 3-3 模型案例 1 中的结果

(A)垂直部分和(B)年代地层学图画出了水深、相对海平面、沉积面上升和沉积相。注意到因为(A)中垂直部分的像素分辨率,一些最细的岩石相单元看不见

在每个20ka的相对海平面循环中，水深范围从0~20m，引起了透光层、微光层和无光层系统的沉积，产率的范围从100m/Ma到略超过500m/Ma。水深的变化引起了每一个碳酸盐工厂的产率变化，因而引起了岩石相单元厚度的变化。这些影响因素倒了很多从薄到厚的变化，并且 P 值 0.072 反映出了对于指数曲线缺乏有力的证据（表 3-2 和图 3-2A）。

表 3-2　单个模型案例中使用的参数摘要

模型号	时间(Ma)	海平面变化1时段(Ma)	海平面变化1时段幅度(m)	海平面变化2时段(Ma)	海平面变化2时段幅度(m)	海平面变化3时段(Ma)	海平面变化3时段幅度(m)	不对称比率	最大透光层生产速率(m/Ma)	最大微光层生产速率(m/Ma)	最大无光层生产速率(m/Ma)	滞后深度(m)
1	0.5	0.020	10	N/A	N/A	1.00	50	1:3	500	500	50	0
2	3	0.020	10	N/A	N/A	1.00	50	1:3	500	500	50	0
3	3	0.023	10	0.112	50	1.200	20	1:1	2500	450	125	2
4	3	0.023	10	0.112	50	1.200	20	1:2.5	2500	450	125	2
5	3	0.023	10	0.112	50	1.200	20	1:2.5	2500	450	125	0

给定这个 P 值的前提下，这个分布可以被合理解释为指数分布。相对海平面波动驱动的外部周期影响的完全确定性建模建造的指数分布的岩石相厚度分布的证据是重要的结果。Burgess（2008）提出疑问是否其他除了泊松过程之外的机理也可以产生这样的指数分布。这个模型结果证明其他一个完全不同的机理也可以发生指数分布。但是这个结果也提出了进一步的问题：相对海平面驱动的参数范围内是否存在指数分布？在古地层中相对海平面变化驱动的指数分布是否普遍？

3.6.2　模型案例2:3 个百万年模型运算

在模型案例 2 中（MC2），模拟的持续时间扩展到了 3Ma（表 3-2）。这个时期等同于许多仔细研究的露头样品假设出来的沉积，例如宾夕法尼亚州的 Honaker Trail 地层，犹他州，美国（Goldhammer 等，1991）。因此，这是一个合适的模拟时间长度。除了总的持续时间的增长，模型 MC2 的参数和 MC1 是一样的（见表 3-2）。不止这些，MC2 得到的 P 值为 0，有力地证明了岩石相的指数分布（表 3-3 和图 3-2）。岩石相单元厚度指数曲线的最大偏移值为 0.35m。对比指数曲线可以看出模型案例 2 的地层有很多薄层。但是，模型案例 2 地层的检查显示在图 3-4 的年代地层图中，反映出平均地区厚度是显著受在 1 个百万年周期的海平面波动中海侵到高位阶段时期的水深 20~45m 的微光层形成的两个 50~60m 厚的潮下带泥粒灰岩影响。否则，对比 MC2 和 MC1 的垂直岩石相分布（图 3-3）可以看出它们非常相似。MC2 案例中不同的平均厚度偏移指数函数导致了薄单元的显著不同的频率。

表 3-3　来自于单模型运算案例的 Kolmogorov–Smirnov 测试结果

模型号	n	平均单元厚度（m）	P 值	解释
1	74	0.77	0.072	中间型
2	199	1.26	0.000	非指数型
3	68	4.81	0.000	非指数型
4	98	3.27	0.310	指数型
5	97	3.34	0.000	非指数型

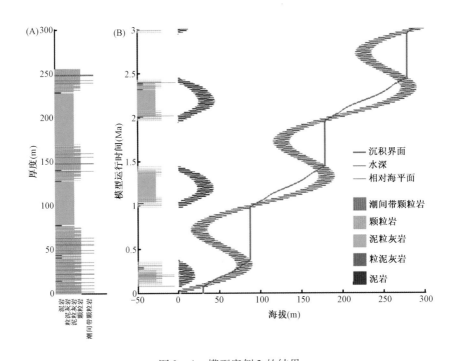

图 3-4 模型案例 2 的结果
(A)垂向剖面和(B)等时地层剖面图,与水深、相对海平面、沉积界面海拔以及沉积相交会

这个结果显示厚度的指数分布可能出现在一个地区的不同部分,但是因为沉积条件随时间的改变,最终的地层从统计学上看不满足统计特性,例如平均厚度和整体厚度分布变化,所以这个地区的一个部分显示出的指数分布(例如,MC1 中的地层)不能持续(例如沉积在 MC2 的附加地层)。在这一案例中,出现了指数的分离,这是因为厚单元沉积在持久稳固的相对较深的潮下带,寡光型生产,时间上处于相对海平面上升赶超生产和堆积时期。一条明显的调查下线可以用来测定较高的生产速率是否对较长时期的模型运算会产生指数厚度分布。

3.6.3 模型案例 3：高生产和对称的全球海平面波动

模型案例 3(MC3)的富光型碳酸盐工厂生产增加到 2500 m/Ma,缓缓减少到寡光型工厂的 450 m/Ma,再增加到无光型工厂的 125 m/Ma(表 3-1)。应用于模拟全球海平面波动幅度的参数符合在冰期气候背景下的米兰柯维奇(Milankovitch)假说(Paterson 等,2006;Rygel 等,2008)。可是要注意,没有证据说明这些参数对更新世之前的地层是现实的,他们完全基于模式驱动下的假定,而更新世之前波动周期性可能会相当的不同(Laskar 等,2004)。重要的是,案例中全球海平面波动是对称的(图 3-5),意味着上升翼和下降翼具有相同的时间。最后,2m 的滞后深度也会对初始水体深度有影响,沉积物生产需要在台地顶部开始。假定合理的滞后深度基于可能的台地顶部沉积时间,可能会影响浅水的效应。对滞后深度的完全敏感性分析是可能的,可以在未来的工作中开展。

由于沉积物生产速率高,堆积作用最可能出现在"保持型"模式中,地层以浅海和潮间带富光型颗粒岩岩相为主(图 3-5)。在 112ka 波动的上升翼上出现 23ka 波动的上升翼时发生生产赶超的情况,例如 MC3 大约 1.48Ma 模型运行时间(EMT)(图 3-6)。在这些时间里,水深在 23ka 高位期可以多次达到 30m,寡光型和无光型生产导致了相对薄层的泥岩(图 3-2C 和图 3-6)。在随后的相对海平面(RSL)下降期,颗粒岩岩相变为主体,这时富光型生产在浅水处得以恢复。

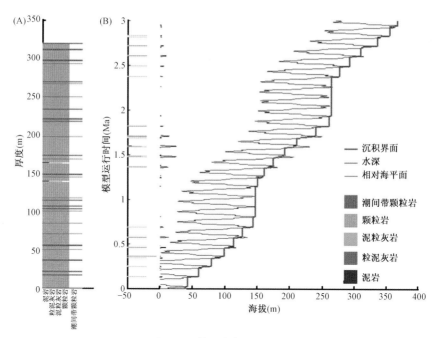

图3-5 模型案例3的结果
(A)垂向剖面和(B)等时地层剖面图,与水深、相对海平面、沉积界面海拔以及沉积相交会

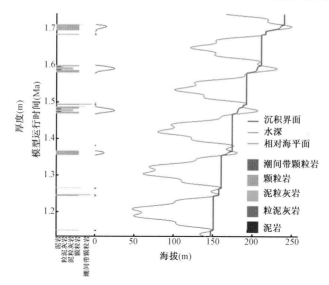

图3-6 模型案例3从1Ma到大约1.7Ma运行时间产生的地层放大图
注意23ka高位期导致水深达30m,具有寡光型和无光型生产,沉积了相对薄层的粒泥灰岩和泥岩。在随后相对海平面下降期,较厚的泥粒灰岩和颗粒岩单元通过浅水富光型生产而沉积

对这一模型,Kolmogorov-Smirnov测试的 P 值计算得出是0.00004(表3-3),强烈证据表明反对指数分布(图3-2C)。与指数曲线最大的离散出现在2.1m厚度处,会有太多的薄层和中间岩相单元分布于1~6m范围,与相对应指数曲线可以对比。

3.6.4 模型案例4:不对称的相对海平面波动

模型案例4(MC4)参数值与MC3相同,只有全球海平面波动的不对称性例外。上升翼与下降翼之比从1:1变到1:2.5(图3-7和图3-8)。这意味着对于112ka时期,上升翼波动是

32ka,而下降翼占80ka,因此海侵比海退更快。冰期海平面波动的典型模型是不对称的,具有较短的上升翼(Goldhammer 等,1991)。基于晚更新世和全新世全球海平面变化史的类比,假定不对称波动是冰期的典型标志,就可以解释不同的冰席堆积和溶化速率(Clark 等,1999)。

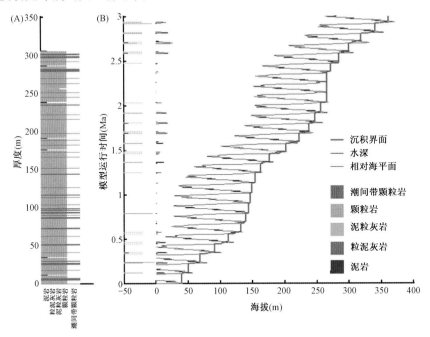

图3-7 模型案例4的结果
(A)垂向剖面和(B)等时地层剖面图,与水深、相对海平面、沉积界面海拔以及沉积相交会

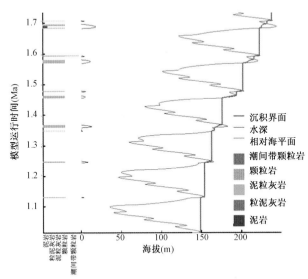

图3-8 模型案例4从1Ma到大约1.7Ma运行时间产生的地层放大图
模拟显示出叠加高级别不对称全球海平面波动曲线驱动的水深具有复杂的变化

MC4 的地层表面上与 MC3 相似,具有富光型泥粒灰岩和颗粒岩,岩相沉积为"追赶型"模式。可是,对比 MC3 时,MC4 的厚岩相单元较少(图3-2D 和图3-6),导致指数厚度分布显示为 Kolmogorov - Smirnov 测试 P 值0.310(表3-3)。两种机制主要对 MC3 和 MC4 的差异负责。第一种是更加复杂的全球海平面曲线,通常具有更加快速的海侵,而且由于叠加了高级别

的不对称波动导致水深更复杂的变化(图3-8)。第二种机制是与非0滞后深度有关的自旋回性及相对高的沉积物生产速率导致多个沉积旋回形成于全球海平面曲线的海侵期。高沉积速率引起全球海平面上升期的快速堆积。在快速堆积赶超相对海平面上升速率的条件下,向上变浅的趋势会引起海平面中断。堆积停留在海平面处,仅仅当相对海平面上升继续,水体加深到2m滞后深度以外的台地顶部才得以恢复。这种向上变浅的开关式堆积过程产生了多个自旋回,中断了MC3中所见到的沉积时期,导致厚颗粒岩岩相单元的形成,从而产生替代的更多中间厚度单元,形成MC4的指数岩相厚度分布。

3.6.5 模型案例5:无滞后深度的不对称相对海平面波动

对比MC3和MC4显示,滞后深度的影响是很大的。而且直观地看,对比另外的同一条件下MC4的2m滞后深度,模型案例5(MC5)0m滞后深度似乎没有对地层模式和岩相厚度分布产生大的差异。可是,对比MC4和MC5的输出(图3-9的2D和2E)显示,MC5的0m滞后深度具有很大差异,相对于MC4产生了较高比例的薄岩相单元,导致MC5的非指数厚度分布($P=0.00$;图3-2E,表3-3),对照一下MC4的指数分布($P=0.310$;图3-2D,表3-3)。

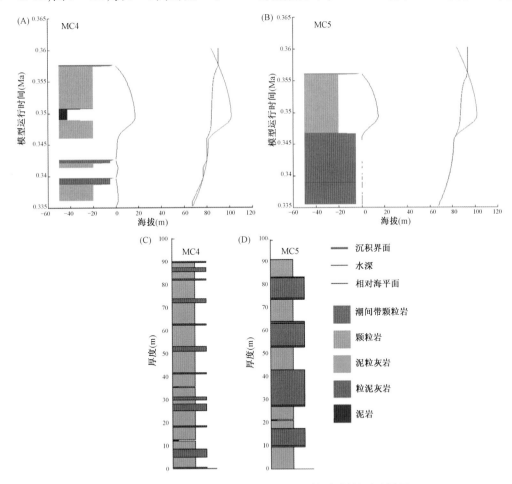

图3-9 从MC4(A)和MC5(B)25ka时间间隔等时地层图

注意由于2m的滞后深度,存在多个水深和岩性旋回,出现在MC4(A)的全球海平面上升翼;MC5(B)由于没有滞后深度则不存在,允许厚颗粒岩单元的持续沉积。滞后深度的差异会对MC4(C)和MC5(D)在同一时间段产生相当不同的垂向序列。注意(D)中的MC5所有厚颗粒岩单元在底部是具有薄层的复合单元,有时也在顶部,被洪泛面或陆上暴露面厚单元分隔

细看 MC5 的结果(图 3-10),97 个岩相单元中超过 50% 厚度小于 1m,与 MC4 非常不同,它只有 35% 小于 1m。可是,MC5 的薄单元优势不代表整个模型剖面都这样。在两个模型剖面的底部 100m(图 3-9C 和 D 所示),实际上 MC4(21 个单元小于 5m)比 MC5(13 个单元小于 5m)具有更多的薄单元。可是,剖面较高时 MC5 以薄颗粒岩单元为主(图 3-10),因此考虑到全部剖面,MC4 具有 25 个小于 1m 厚的颗粒岩单元,而 MC5 有 42 个。模型案例 5 具有更多的薄颗粒岩层,因为在 0m 滞后深度薄层颗粒岩常常作为薄的海侵单元出现在高频层序底部。基于这些结果,滞后深度似乎是岩相厚度分布的重要控制因素。

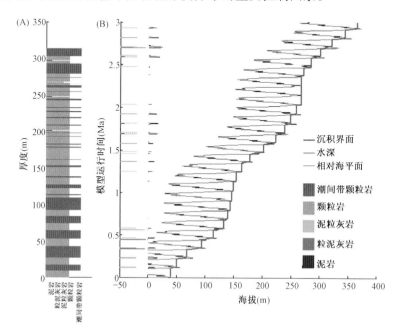

图 3-10 模型案例 5 的结果

(A)垂向剖面和(B)等时地层剖面图,与水深、相对海平面以及沉积相交会

3.7 多模型运行:发现模式

模型案例 1-4 的结果显示,岩相厚度分布的类型依赖于生产速率、全球海平面变化曲线类型、模型时间和沉积条件的变化,例如,MC2 海侵期台地侵没。已知这些控制因素的敏感度时,基于简单的模型运算得出有意义的结论依然困难,这是因为它们代表许多可能的参数值组合当中的一个。因此,从这么少的案例中确定岩相厚度分布的趋势和模式是不容易的。

一维模型的好处在于计算上不昂贵,可以运算上千次,参数范围值很广,可以为每个案例地层计算 Kolmogorov–Smirnov 参数 P 值。这些 P 值可以点在二维或三维参数—空间交会图上,图轴的参数可以是全球海平面波动幅度,或沉积物生产速率。参数—空间交会图标出了岩相厚度分布随多维模型参数空间的变化性质,在划分岩相分布类型时允许更加可靠和有意义的系统评价。MC1 到 MC4 的结果提供了进一步调查岩相单元分布控制因素的起始点。

3.7.1 多次运行组 1:沉积物生产速率控制

图 3-11 显示 MRG1 的参数—空间交会图具有对称的全球海平面波动和 2m 的滞后深度,一定范围的生产速率对应富光型、寡光型和无光型生产曲线(表 3-3)。指数岩相案例限

定于富光型生产速率,范围在1000m/Ma和2000m/Ma之间。对比一下,模型中指数案例出现在寡光型和无光型生产速率范围。

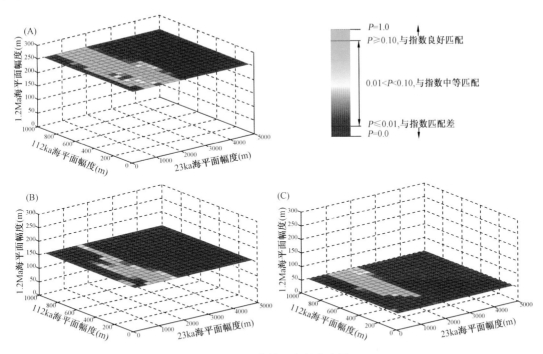

图3-11 参数—空间交会图

颜色代表模型运行组1的KS测试P值,具有对称的全球海平面变化幅度,一定范围的沉积物生产速率(表3-4)和2m的滞后深度。过参数—空间的切片显示,不同的无光生产速率分别是:(A)50m;(B)150m;(C)250m。每个长方形是P值的颜色代码,由岩相厚度分布的KS测试对比产生,模型运算和理论指数分布具有相同的岩相单元数目。每个切片颜色的变化显示出指数和非指数分布怎样通过模型在不同参数值范围产生,详见文中讨论

这一观察表明,在对称的全球海平面波动下,模型中三种生产体系里,富光型生产速率是指数分布出现的首要控制因素。这一结果很可能是由于富光型生产速率相对于其他寡光型和无光型工厂范围更广(表3-4)。案例中单一模拟指数的程度没有用沉积速率测定。富光型生产速率很高时,从2000~2500m/Ma(表3-11),许多厚岩相单元产生了,结果不是指数分布。在低富光型生产速率下,小于500m/Ma,许多薄岩相单元产生了,结果也不是指数分布。

表3-4 多次运行结果模型的总结

模型组数	模型组名称	模型运行次数	生产速率范围(m/Ma)	全球海平面波动幅度(m)	波动幅度的不对称性	滞后深度(m)	模型运行的比例		
							$P \leq 0.01$ 非指数	$0.01 < P < 0.1$ 中间过渡	$P \geq 0.10$ 指数
1	变化的生产、对称海平面	4800	250~5000, 50~1000, 25~300	10,50,20	1:1	2	0.756	0.098	0.146
2	变化的生产、不对称海平面	4800	250~5000, 50~1000, 25~300	10,50,20	1:4	2	0.522	0.377	0.101

续表

模型组数	模型组名称	模型运行次数	生产速率范围（m/Ma）	全球海平面波动幅度（m）	波动幅度的不对称性	滞后深度（m）	模型运行的比例		
							$P \leqslant 0.01$ 非指数	$0.01 < P < 0.1$ 中间过渡	$P \geqslant 0.10$ 指数
3	低生产速率、滞后2m	4400	500	2.5~50,5~100,0~50	1:4	2	0.621	0.153	0.226
4	高生产速率、滞后2m	4400	2500	2.5~50,5~100,0~50	1:4	2	0.498	0.326	0.176
5	低生产速率、滞后0m	4400	500	2.5~50,5~100,0~50	1:4	0	0.856	0.093	0.051
6	高生产速率、滞后0m	4400	2500	2.5~50,5~100,0~50	1:4	0	0.875	0.073	0.052

3.7.2 多次运行组1和2的对比：全球海平面波动对称性的程度

对比单一的MC3，对称的全球海平面波动产生了非指数岩相厚度，而MC4不对称的全球海平面波动产生了指数厚度，表明可能有对岩相厚度分布重要的控制因素是全球海平面波动的对称性程度。这种控制可能是由于不对称曲线的陡升翼导致更高的可容空间产生速率；这可以通过对比MRG1参数空间交会图（图3-11，用对称的全球海平面波动曲线运行）和MRG2参数空间交会图（图3-12，用不对称的全球海平面波动曲线运行，幅度与MRG1相同）。MRG2中指数案例具有相似数量的清晰指数例子（10%对15%），但这会在更广范围的生产速率内出现，并且有更多中间类型的分布（38%对10%），相应地非指数分布较少（52%对76%）。同时，对不对称曲线的响应（MRG2）是非线性的，意味着这种响应不会直接出现，或简单的与输入成正比，而且具有清晰的指数案例出现在参数空间的复杂模式中（例如图3-12B），但相比MRG1会对生产速率范围有更多的限定。

总体上讲，MRG1对称的全球海平面波动曲线生产与MRG2不对称的全球海平面波动曲线生产结果对比显示，全球海平面曲线的对称性对参数空间模型中指数岩相厚度分布在什么地方出现有一定的控制作用。在模型用对称的全球海平面波动运行时，已知高级别波动与低级别曲线的下降翼或上升翼具有相同的可能性。通过模拟，低级别曲线产生的岩相单元有相对一致的厚度，相对于指数分布而言不管是太厚或太薄，意味着仅仅少量参数—空间带可以指出指数分布。不对称全球海平面曲线导致更高级别的波动出现在下降翼，这是由于不对称案例中持续时间比上升翼更长。根据最高级别波动出现在最低级别海平面曲线的位置，就可以得到不同的沉积作用时间量，从而导致岩相厚度的变化和较宽指数分布带或中间类型分布带在参数空间图中的出现。可是，改变全球海平面变化曲线的对称性不会改变指数例子的出现速率，这在MRG1和MRG2中是相似的。这种相似性提示，全球海平面变化曲线的对称性和海平面升降的相对速率几乎控制整个厚度指数分布的出现。

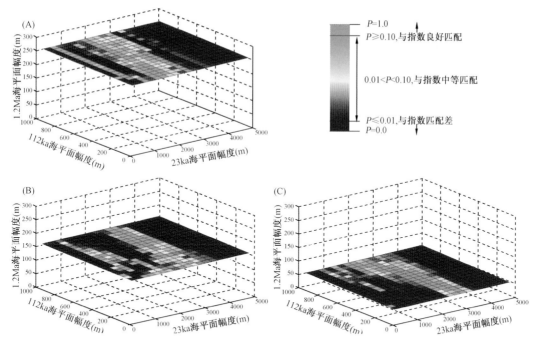

图 3-12 参数—空间交会图

颜色代表模型运行组 2 的 KS 测试 P 值,具有不对称的全球海平面变化幅度,一定范围的沉积物生产速率(见表 3-4)和 2m 的滞后深度。过参数空间的切片显示,不同的无光生产速率分别是:(A)50m;(B)150m;(C)250m

3.7.3 多次运行组 3:全球海平面波动幅度的控制

可容空间产生的速率也受一定时期全球海平面波动幅度的控制,因此会影响指数模拟的出现。多次运行组 3(MRG3)测试了全球海平面波动幅度对岩相厚度分布类型的影响。图3-13 显示 MRG3 的参数—空间交会图;模拟组具有相对低的生产速率 500m/Ma,和不对称的全球海平面波动幅度范围(表 3-3)。整体上讲,MRG3 的 23% 产生了指数岩相厚度分布,非指数占 62%,15.3% 产生了中间类型的曲线(表 3-3)。显示的指数厚度分布在 23ka 时期全球海平面波动幅度低值时最频繁,但在 112ka 时期波动幅度较高(图 3-13)。长期波动幅度相对影响较小,对整体幅度而言具有相似的数目和参数—空间交会图的相似指数分布。

作为对比,MRG3 的参数—空间的岩相厚度指数分布受到 112ka 和 23ka 相对波动幅度的强烈控制。23ka 时期的波动幅度足以影响沉积作用,产生足够多的薄层导致指数分布的发育。在高级别幅度到低级别幅度的低比率条件下的其他地方,高级别旋回不影响足够规律的沉积作用,因此产生足够薄的和中间过渡的厚层,可以认为是指数分布。

本组的多次运行显示,虽然对指数层厚度分布没有先决条件,但如果存在多级别全球海平面波动,它们也会对指数分布的出现有强烈的控制作用。这种结果表明,可容空间和沉积物供给之间的特殊相互作用需要指数分布出现在相对小比例的全球海平面变化曲线参数组合中。

3.7.4 多次运行组 3 和 4 的对比:更加关注生产速率控制

虽然 MRG3 的参数空间交会图(图 3-13)清楚显示了岩相厚度分布受全球海平面波动幅度控制,MRG1 和 MRG2 的结果却表明,考虑全球海平面控制在不同沉积物生产速率下如何变化也十分重要。多次运行组 4(MRG4)在相对高的富光生产速率 2500m/Ma(表 3-4)运行。

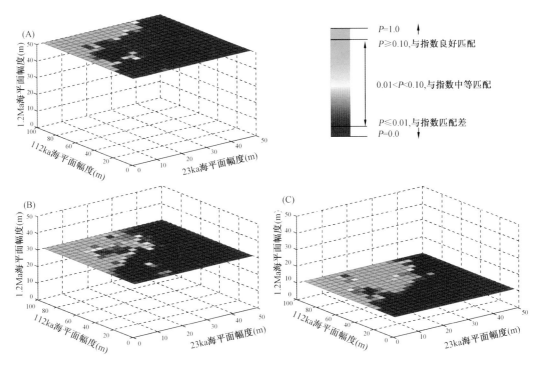

图 3-13 参数—空间交会图

颜色代表模型运行组 3 的 KS 测试 P 值,具有不同的全球海平面变化幅度,低值的生产速率 500m/Ma 和 2m 的滞后深度。全球海平面变化曲线是不对称三单元正弦曲线。过参数空间的切片显示,对于不同的 1.2Ma 时期全球海平面波动幅度分别是:(A)10m;(B)30m;(C)50m。详见文中讨论

与 MRG3 对比,显示 MRG4(图 3-14)产生较少的指数和非指数分布(分别是 18% 和 50%),但是具有更多的中间分布(33%)。可是,指数分布出现在模型运行组高波动幅度 112ka 时期和低波动幅度 23ka 时期(图 3-13 和图 3-14)。高波动幅度 23ka 时期排除了指数分布,这是由于快速海平面上升超过沉积物堆积速率,无光堆积为主,产生了太多的薄层。相反,高波动幅度 112ka 时期产生了指数分布,这是因为波动变化快产生的层厚度,趋向于所需的薄层,但也有一定范围的厚层和中间层产生。就像 MRG3,MRG4 变化较小,波动幅度在 1.2Ma,显示长期波动与高频的全球海平面波动相比,仅仅是次级控制因素。

岩相厚度分布对生产速率敏感的一个原因在于,相对高的生产速率下(例如 MRG4 的较高生产速率和 MRG3 的较低生产速率),得到的可容空间较大程度上会被充填。这一结果出现在大范围的全球海平面参数上,其堆积可以与可容空间产生速率同步,不以厚岩相分布单元为主。换句话说,高生产速率的维持型(keep-up)台地很可能产生指数岩相分布,而不是水淹或追赶型(catch-up)的台地。

3.7.5 多次运行的第五组和第六组:滞后深度的作用

多次运行的第五组和第六组(MRG5 和 MRG6)具有与 MRG3 和 MRG4 同样的参数,只是滞后深度不同,控制碳酸盐开始堆积的水深的参数设为 0m,而不是 2m。就像以前讨论 MC3、MC4 和 MC5 一样,滞后深度的变化对岩相厚度分布有重要控制作用,因为自旋回产生中间的厚度单元,在初始洪泛期没有沉积物抑制了海侵期薄层的发育。MRG4 到 MRG6 运行的多个模型支持这一结论。在 MRG5(图 3-15)和 MRG6(图 3-16)滞后深度为 0,仅仅约 5% 的模型产生了指数分布,对比一下 MRG3 和 MRG4,分别是 23% 和 18%。在滞后深度为 0 时,

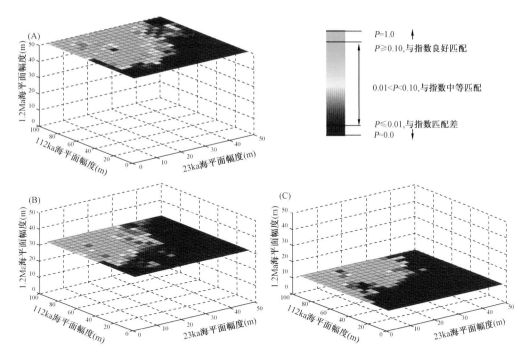

图 3-14 参数—空间交会图

颜色代表模型运行组 4 的 KS 测试 P 值,具有不同的全球海平面变化幅度,高值的生产速率 2500m/Ma 和 2m 的滞后深度。全球海平面变化曲线是不对称三单元正弦曲线。过参数空间的切片显示,对于不同的 1.2Ma 时期全球海平面波动幅度分别是:(A)10m;(B)30m;(C)50m。详见文中讨论

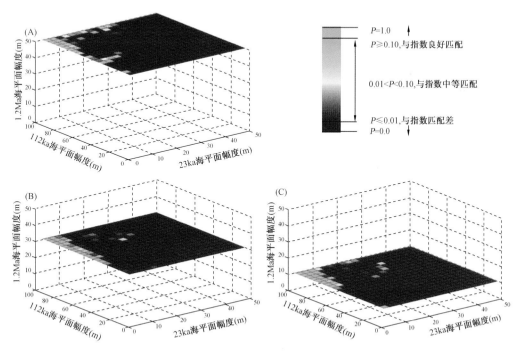

图 3-15 参数—空间交会图

颜色代表模型运行组 5 的 KS 测试 P 值,具有不同的全球海平面变化幅度,低值的生产速率 500m/Ma,没有滞后深度。全球海平面变化曲线是不对称三单元正弦曲线。过参数空间的切片显示,对于不同的 1.2Ma 时期全球海平面波动幅度分别是:(A)10m;(B)30m;(C)50m。详见文中讨论

指数分布只出现在最低幅度的曲线段 23ka 处,这时生产和堆积速率可以和全球海平面变化曲线产生的可容空间速率保持一致。

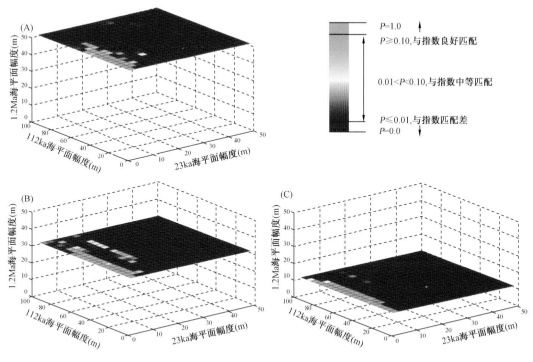

图 3-16 参数—空间交会图

颜色代表模型运行组 6 的 KS 测试 P 值,具有不同的全球海平面变化幅度,高值的生产速率 2500m/Ma,没有滞后深度。全球海平面变化曲线是不对称三单元正弦曲线。过参数空间的切片显示,对于不同的 1.2Ma 时期全球海平面波动幅度分别是:(A)10m;(B)30m;(C)50m。详见文中讨论

3.8 讨论

尽管以前多次说过,本文仍然值得再强调一下:"所有模型都是错误的,但有些模型会是有用的"(Box 和 Draper,1987)。这里提出的模型是一维的,仅仅代表了有限范围的且是已知碳酸盐岩地层形成控制因素的沉积过程;显然这是对真实世界的粗略简化,从这点意义上看模型是错误的。可是,比起许多其他碳酸盐模型来,这一模型不再是错的,尤其是针对定性概念模型而言。在极端案例中,对观察事实的错误解释,没有证据来支持所宣称的沉积过程,那么模型从效果上讲是不可检验的(例如 Spence 和 Tucker,2007)现在的作者会提出争论,说该模型一定比这些例子更有用,在模型中某些合理已知的物理化学过程得到经验数据的支持(Bosscher 和 Schlager,1993),而且最重要的是,它产生了定量结果,可以与碳酸盐岩地层的定量数据进行对比。

就像所有其他模型一样,Dougal 结果的质量依赖于基本的假设。Dougal 用于了解岩相厚度分布的关键假设是水深生产关系,用来代表富光型、寡光型和缺光型的沉积作用,并假定岩相是由于不同生产—深度曲线引起的。如果这些假设是不同的,很明显结果就会不同。所以,这是一个需要进一步调查的领域,什么程度上碳酸盐工厂的性质也会控制厚度分布?为了避免在考虑模型结果时过度解释,这些限定因素应该予以考虑。

已知模型存在的弱点时,岩相厚度分布的指数和非指数结果的成因解释上会有什么建议?

最基本的是这些结果可以用来验证,像所有碳酸盐岩地层模型一样,岩相厚度分布类型可以从沉积物供给和可容空间变化的综合来确定(Schlager,1993)。沉积物生产速率很明显对结果有控制能力,有争议的是模型中测试的图像中它是不是主要控制因素。可容空间变化也起重要作用。注意,对可容空间变量的测试限定于传统的模型即轨道力有关的海平面变化。许多其他的可能性也是存在的,尤其回到地质时期中生代和古生代,轨道力成分的相互作用和作用时间很可能会有差异,需要考虑(Laskar等,2004)。

建模工作中一项特别有趣的发现是,指数岩相厚度分布现在也可以被解释为来自于碳酸盐体系的确定性过程。

过去,指数用于显示随机 Poisson 过程的执行(Wilkinson 等,1999)。Poisson 过程解释可能会用于多个案例,但是这些结果也提高了一些案例确定性过程解释的可能性,认为类似于此处的模拟。可是,结果似乎并不直截了当,因为模型中描述的指数厚度分布的许多案例是在一定的相对海平面曲线上升翼时期因高频自旋回过程的出现而出现。在这些案例中,每一个上升翼的滞后深度和多个自旋回似乎对指数分布的出现都有重要的控制,表明标准的层序地层模型中试图用简单的异旋回来对观察到的地层进行解释是相当乏力。

总的来说,多个运算组中 27200 次运算里面,大约 12% 产生了可以认为是指数分布的岩相厚度分布,19% 产生了中间类型,69% 产生了非指数分布例子。对比来看,56 个露头 Kolmogorov - Smirnov 测试分析案例中,29% 是指数的,21% 是中间的,50% 是非指数分布(Burgess,2008)。基于这一对比,露头案例中指数分布似乎比模型更加普遍。确实,没有更多的模型分析,以及 Burgess(2008)的露头案例中的沉积背景分析,是难以知道怎么解释这种比例关系的,但是一种可能的解释是其他过程参数没有出现在此模型(例如,沉积物搬运和空间上变化的碳酸盐生产类型和生产速率)中,而它对出现在露头中的大部分指数案例起作用。当这一模型集中于再产生台地顶部和台地内部地层时,大概更大比例的指数露头案例会出现,显示出其他过程和沉积背景在 Burgess(2008)考虑的露头案例中占据主导地位。Wilkinson 等(1999)已经建议 Poisson 过程作为解释观察到的岩相厚度分布的候选,但是这些随机 Poisson 过程的正确形式保持了不确定,正如 Burgess(2006)指出的,随机过程和确定性过程的边界可能是模糊的,如本文显示指数分布的形成完全由确定性模型完成。

另一个明显的问题是模型产生的结果中什么类型的分布是非指数厚度分布,多少会类似于露头中观察到的非指数分布(Burgess,2008),而这会告诉我们是什么控制这些案例过程(Sylvester,2007)? 要回答这问题需要做更多的工作,但是这里提出的方法和结果显示出怎样会达到目的。使用 Kolmogorov - Smirnov 测试表明 Dougal 地层模型与观察的厚度分布相似,或许使用自动反转过程就可以得到最优拟合的模型。

最后,由于碳酸盐沉积体系是三维体系,所以考虑一维模型如 Dougal 的限定条件也十分重要。很明显,真正的三维过程,例如碳酸盐沉积物生产中沉积物搬运和空间变化,会在 Dougal 中丢失。可是,它们也在许多层序地层学概念模型中丢失(Spence 和 Tucker,2007;Bosence 等,2009),甚至在许多认为是在三维的数值模型中,在最重要的数学意义上也这样,一系列一维生产模型安排在二维网格里,加上微弱的二维搬运因子(Burgess 和 Wright,2003;Paterson 等,2006;Warrlich 等,2008)。某些模型除此之外就用二维沉积物的生产变量来代表开始移动(Burgess 和 Emery,2004),但是数值模型的下一代需要通过物理过程调查岩相厚度分布的控制因素,例如空间变化的沉积物生产,以及不同的沉积物搬运过程。模型的最终目标是要尽可能了解和定义出岩相厚度分布和三维岩相大小以及体积分布之间的联系。

3.9　结论

(1)观察表明浅海碳酸盐岩地层通常具有指数分布的岩相厚度,这是近年来关于碳酸盐岩最基本的结果之一。其原因在于这种观察可以作普遍适用性的测试,还能提出关于什么沉积作用和气候以及海洋地理背景会导致形成特定岩相厚度分布的问题。这些问题的答案对于正确解释碳酸盐岩地层反映的随地质时间产生的气候变化和海洋地理变化的敏感记录,是十分重要的。

(2)模型案例1(MC1)和模型案例2(MC2)具有同样的参数,只有总运行时间不同(MC1是1Ma,MC2是3Ma),但是MC1产生了指数分布的岩相,MC2为非指数分布。这可能是不稳定特性的实用例子,例子里地层变化剖面的统计特征(Wilkinson等,1999)和进一步分析这种模型结果,可以帮助查明不稳定性怎样使得岩相厚度分布的解释复杂化。

(3)本项研究完成的27205模型运算中三个主要因素可能有利于形成岩相厚度的指数分布。

① 可容空间产生速率的复杂变化。这里由于高频的冰川—全球海平面变化曲线波动,结合沉积物生产和堆积速率,会有利于"追赶"型的沉积作用。在这种条件下,水深变化驱动了碳酸盐堆积类型和速率的变化,从而产生了指数分布所需的许多薄层和相对较少的厚层的混合。例如,薄层常常沉积在快速海侵时期。

② 沉积物生产的速率是关键控制因素。碳酸盐工厂的三个模型中,相对于寡光型和无光型工厂,富光型工厂似乎是主要的控制因素,但这是因为简化了,模型中由于较高的最大速率就有较大范围富光型生产率,因此寡光型工厂也会占主导作用,只要比富光型有较宽范围的生产率。

③ 沉积作用的自旋回波动。这是受深度滞后效应驱动的,形成于冰川—全球海平面变化曲线的高频上升翼的某个时期。这个因素也会有利于薄层和厚层的指数混合,破坏长期的"保持"型颗粒灰岩的沉积作用,进入一系列形成于多期向上变浅旋回的薄层沉积。

(4)岩相厚度的指数分布可以从纯确定性模型中获得。可是,模型运算27200次只有大约13%会产生指数分布,对比露头案例资料则是28%,不包括在本模型中的其他过程起了重要作用。因此需要做更多的工作来确定这些过程会是什么,但是明显的候选因素是陆上暴露时期的剥蚀,台地顶部三维方向上出现的不同沉积过程,以及出现在不同沉积背景如台地斜坡上的其他过程。

参 考 文 献

Allen,P. A. and Allen,J. R. (2005) Basin Analysis,2nd edn. Blackwell Scientific Publications,Oxford,549 pp.

Artyushkov,E. V. ,Chekhovich,P. A. and Tarling,D. H. (2003) The origin of water-depth changes in past epeiric seas. Doklady Earth Sci. ,388,515 – 520.

Barnett,A. J. ,Burgess,P. M. and Wright,V. P. (2002) Icehouse world sea-level behaviour and stratal patterns in late Visean (Mississippian) carbonate platforms:integration of numerical forward modelling and outcrop studies. Basin Res. ,14,417 – 439.

de Boer,P. L. and Smith,D. G. (1994) Orbital forcing and cyclic sequences. In: Orbital Forcing and Cyclic Sequences(EdsP. L. de Boer and D. G. Smith),IAS Spec. Publ. ,19,1 – 14. IAS,Oxford.

Bosence,D. W. J. and Waltham,D. A. (1990) Computer modeling the internal architecture of carbonate platforms. Geology,18,26 – 30.

Bosence, D., Procter, E., Aurell, M., Bel Kahla, A., Boudagher-Fadel, M., Casaglia, F., Cirilli, S., Mehdie, M., Nieto, L., Rey, J., Scherreiks, R., Soussi, M. and Waltham, D. (2009) A dominant tectonic signal in high-frequency, peritidal carbonate cycles? A regional analysis of Liassic platforms from Western Tethys. J. Sed. Res., 79, 389–415.

Bosscher, H. and Schlager, W. (1993) Accumulation rates of carbonate platforms. J. Geol., 101, 345–355.

Box, G. E. P. and Draper, N. R. (1987) Empirical Model-Building and Response Surfaces. Wiley, New York, 424 pp.

Burgess, P. M. (2006) The signal and the noise: forward modeling of allocyclic and autocyclic processes influencing peritidal carbonate stacking patterns. J. Sed. Res., 76, 962–977.

Burgess, P. M. (2008) The nature of shallow-water carbonate lithofacies thickness distributions. Geology, 36, 235–238.

Burgess, P. M. and Emery, D. (2004) Sensitive dependence, divergence and unpredictable behaviour in a stratigraphic forward model of a carbonate system. In: Geological Prior Information (Eds. A. Curtis and R. Wood), Geol. Soc. London Spec. Publ., 239, 77–94.

Burgess, P. M. and Wright, V. P. (2003) Numerical forward modelling of carbonate platform dynamics: an evaluation of complexity and completeness in carbonate strata. J. Sed. Res., 73, 637–652.

Clark, P. U., Alley, R. B. and Pollard, D. (1999) Northern hemisphere ice-sheet influences on global climate change. Science, 286, 1104–1118.

Cooper, K. and Smart, P. L. (2010) Is Milankovitch cyclicity recognisable in carbonate sequences? Numerical experiments using the forward model Carb3D+, in 2010 AAPG annual convention & exhibition; abstracts volume #90104.

Corda, L. and Brandano, M. (2003) Aphotic zone carbonate production on a Miocene ramp, Central Apennines, Italy. Sed. Geol., 161, 55–70.

Crowley, T. J. and Baum, S. K. (1991) Estimating Carboniferous sea-level fluctuations from Gondwanan ice-extent. Geology, 19, 975–977.

Della Porta, G., Kenter, J. A. M. and Bahamonde, J. R. (2004) Depositional facies and stratal geometry of an Upper Carboniferous prograding and aggrading high-relief carbonate platform (Cantabrian Mountains, N Spain). Sedimentology, 51, 267–295.

Demicco, R. V. and Hardie, L. A. (2002) The "carbonate factory" revisited: a re-examination of sediment production functions used to model deposition on carbonate platforms. J. Sed. Res., 72, 849–857.

Drummond, C. N. and Wilkinson, B. H. (1993) Carbonate cycle stacking patterns and hierarchies of orbitally forced eustatic sea-level change. J. Sed. Petrol., 63, 369–377.

Goldhammer, R. K., Oswald, E. J. and Dunn, P. A. (1991) Hierarchy of stratigraphic forcing: example from Middle Pennsylvanian shelf carbonates of the Paradox Basin. In: Sedimentary Modeling: Computer Simulations and Methods for Improved Parameter Definition (Eds E. K. Franseen, W. L. Watney, C. G. S. C. Kendall and W. Ross), Kansas Geol. Surv. Bull., 233, 361–414. Kansas Geological Survey, Lawrence, KS.

Hallock, P. and Schlager, W. (1986) Nutrient excess and the demise of coral reefs and carbonate platforms Palaios, 1, 389–398.

Harris, P. M. and Vlaswinkel, B. M. (2008) Modern isolated carbonate platforms: templates for quantifying facies attributes of hydrocarbon reservoirs. In: Controls on Carbonate Platform and Reef Development (Eds J. Lukasik and J. A. Simo), SEPM Spec. Publ., 89, 323–341.

Heckel, P. H. (1986) Sea-level curve for Pennsylvanian eustatic marine transgressive-regressive depositional cycles along midcontinent outcrop belt, North America. Geology, 14, 330–334.

Horbury, A. D. and Adams, A. E. (1996) Microfacies associations in Asbian carbonates: an example from the Urswick Limestone Formation of the southern Lake District, northern England. In: Recent Advances in Lower Carboniferous Geology (Eds P. Strogen, I. D. Somerville and G. L. Jones), Geol. Soc. Spec. Publ., 107, 221–238. The Geological

Society, London.

Kendall, C. G. S. C. and Lerche, I. (1988) The rise and fall of eustasy. In: Sea-Level Changes – An Integrated Approach (Eds C. K. Wilgus, H. Posamentier, C. A. Ross and C. G. S. C. Kendall), Soc. Econ. Paleontol. Mineral., Spec. Publ., 42, 3–17.

Laskar, J., Robutel, P., Joutel, F., Gastineau, M., Correia, A. C. M. and Levrard, B. (2004) A long-term numerical solution for the insolation quantities of the Earth. Astron. Astrophys., 428, 261–285.

Lehrmann, D. J. and Goldhammer, R. K. (1999) Secular variation in parasequence and facies stacking patterns of platform carbonates: a guide to application of stacking patterns analysis in strata of diverse ages and settings. In: Advances in Carbonate Sequence Stratigraphy: Applications to Reservoirs, Outcrops and Models (Eds A. H. Saller, P. M. Harris and J. A. Simo), SEPM Spec. Publ., 63, 187–225. SEPM, Tulsa, OK.

MacArthur, R. H. and Wilson, E. O. (1967) The Theory of Island Biogeography. Pincetown University Press, Princetown, 203 pp.

Milliman, J. D. (1974) Marine Carbonates. Part 1, Recent Sedimentary Carbonates. Springer-Verlag, New York, 375pp.

Osleger, D. A. (1991) Subtidal carbonate cycles: implications for allocyclic vs. autocyclic controls. Geology, 19, 917–920.

Paterson, R. J., Whitaker, F. F., Jones, G. D., Smart, P. L., Waltham, D. A. and Felce, G. (2006) Accommodation and sedimentary architecture of isolated icehouse carbonate platforms: insights from forward modeling with CARB3D+. J. Sed. Res., 76, 1162–1182.

Pollitt, D. A. (2008) Outcrop and forward modelling analysis of ice-house cyclicity and reservoir lithologies. Unpublished PhD Thesis, Cardiff University, Cardiff.

Pomar, L. (2001a) Ecological control of sedimentary accommodation: evolution from a carbonate ramp to rimmed shelf, Upper Miocene, Balearic Islands. Palaeogeogr. Palaeoclimatol. Palaeoecol., 175, 249–272.

Pomar, L. (2001b) Types of carbonate platforms: a genetic approach. Basin Res., 13, 313–334.

Press, W. H., Teukolsky, S. A., Vetterling, W. T. and Flannery, B. P. (1992) Numerical Recipes in C: The Art of Scientific Computing, 2nd edn. Cambridge University Press, Cambridge, 994 pp.

Rankey, E. C. (2004) On the interpretation of shallow shelf carbonate facies and habitats: how much does water depth matter? J. Sed. Res., 74, 2–6.

Rygel, M. C., Fielding, C. R., Frank, T. D. and Birgenheier, L. P. (2008) The magnitude of Late Paleozoic glacioeustatic fluctuations: a synthesis. J. Sed. Res., 78, 500–511.

Sadler, P. M. (1981) Sediment accumulation rates and the completeness of stratigraphic sections. J. Geol., 89, 569–584.

Schlager, W. (1993) Accommodation and supply – a dual control on stratigraphic sequences. Sed. Geol., 86, 111–136.

Schlager, W. (2003) Benthic carbonate factories of the Phanerozoic. Int. J. Earth Sci., 92, 445–464.

Smith, S. V. and Kinsey, D. W. (1976) Calcium carbonate budget production, coral reef growth and sea-level changes. Science, 194, 937–939.

Spence, G. H. and Tucker, M. E. (2007) A proposed integrated multi-signature model for peritidal cycles in carbonates. J. Sed. Res., 77, 797–808.

Stemmerik, L. (1996) High frequency sequence stratigraphy of a siliciclastic influenced carbonate platform, lower Moscovian, Amdrup Land, North Greenland. In: High Resolution Sequence Stratigraphy: Innovations and Applications (Eds J. A. Howell and J. F. Aitken), Geol. Soc. Spec. Publ., 104, 347–365. The Geological Society, London.

Strasser, A. and Samankassou, E. (2003) Carbonate sedimentation rates today and in the past: Holocene of Florida Bay, Bahamas, and Bermuda vs. Upper Jurassic and Lower Cretaceous of the Jura Mountains (Switzerland and

France). Geol. Croat. ,56,1 – 18.

Sylvester, Z. (2007) Turbidite bed thickness distributions: methods and pitfalls of analysis and modelling. Sedimentology, 54, 847 – 870.

Tipper, J. C. (1997) Modeling carbonate platform sedimentation – lag comes naturally. Geology, 25, 495 – 498.

Vlaswinkel, B. M. and Wanless, H. R. (2009) Rapid recycling of organic-rich carbonates during transgression: a complex coastal system in Southwest Florida. In: Perspectives in Carbonate Geology: A Tribute to the Career of R. N. Ginsburg, (Eds P. K. Swart, G. P. Eberli and J. A. McKenzie), IAS Spec. Publ. ,41,91 – 112.

Vollbrecht, R. and Kudrass, H. R. (1990) Geological results of a pre-site survey for ODP drill sites in the SE Sulu Basin. In: Proceedings of the Ocean Drilling Program, Initial Reports (Eds C. Rangin, E. Silver and M. T. von Breymann), 124, 105 – 111.

Walkden, G. M. and Walkden, G. D. (1990) Cyclic sedimentation in carbonate and mixed carbonate clastic environments: four simulation programs for a desktop computer. In: Carbonate Platforms (Facies, Sequences and Evolution) (Eds M. E. Tucker, J. L. Wilson, P. D. Crevello, J. F. Sarg and J. F. Read), IAS Spec. Publ. ,9, 55 – 78.

Warrlich, G., Bosence, D., Waltham, D., Wood, C., Boylan, A. and Badenas, B. (2008) 3D stratigraphic forward modelling for analysis and prediction of carbonate platform stratigraphies in exploration and production. Mar. Petrol. Geol. ,25, 35 – 58.

Wilkinson, B. H. and Drummond, C. N. (2004) Facies mosaics across the Persian Gulf and around Antigua – stochastic and deterministic products of shallow-water sediment accumulation. J. Sed. Res. ,74, 513 – 526.

Wilkinson, B. H., Drummond, C. N., Rothman, E. D. and Diedrich, N. W. (1997) Stratal order in peritidal carbonate sequences. J. Sed. Res. ,67, 1068 – 1082.

Wilkinson, B. H., Drummond, C. N., Diedrich, N. W. and Rothman, E. D. (1999) Poisson processes of carbonate accumulation on Paleozoic and Holocene platforms. J. Sed. Res. ,69, 338 – 350.

Wright, V. P. and Vanstone, S. D. (2001) Onset of Late Palaeozoic glacio-eustasy and the evolving climates of low latitude areas: a synthesis of current understanding. J. Geol. Soc. ,158, 579 – 582.

第4章 白垩纪生物厚壳蛤类与碳酸盐台地

PETER W. SKELTON, EULALIÀ GILI 著

侯宇安,裴东洋 译,岳 婷 校

摘 要 将阿普弟阶碳酸盐台地生长情况变化及其相关厚壳蛤类的周期性更替现象与气候变化的证据相结合可以产生一个解释性的模型。在阿普弟早期,碳酸盐台地广泛的在整个大西洋、特提斯洋、低古纬度太平洋海山带发育,并且厚壳蛤类具有丰富的多样性,尤其是在台地边缘占优势的文石质的双壳类生物。尽管因果关系仍然有争议,但是在大洋缺氧事件1期达到顶峰的全球碳循环主要扰动的同时,早阿普弟中期发生了广泛的中断。台地增长终止于特提斯北缘,或者美洲地区。同时,*Lithocodium/Bacinella* 和类似的微生物介壳普遍发育在较低的古纬度地区。早阿普弟晚期特提斯台地的恢复仅限于低古纬度地区。含有丰富的双壳类化石的台地边缘相在特提斯中部和南部再次盛行,但是化石类型转变为以伊比利亚半岛周围钙质外壳增厚的厚壳蛤类为主。早阿普弟末期 Lazarus 类的厚壳蛤类亮晶灰岩化石类型消失,尽管在美洲厚壳蛤类一直存在到阿普弟末期,特提斯台地上的厚壳蛤类已经更新为以方解石晶体为主。本研究假设,海水酸化不仅影响中—早阿普弟期台地的分解,还制约了随后的早阿普弟晚期台地恢复的地理范围。起初,由火山衍生的二氧化碳增加了大气中二氧化碳的水平,缓解了由于温室效应造成的低纬度地区热驱逐水溶液 CO_2(这里称为"釜效应")。随后由于有机碳埋藏降低大气中的二氧化碳水平导致的降温使高纬度地区台地水域持续酸化,减少了釜效应效果。双壳类生物对这些因素敏感可能是由于它的矿物学性质和生长环境。在这个过程中方解石和文石比率的升高是由于生物种类周期性的世代交替。

关键词 阿普弟期 碳酸盐台地 气候变化 全球缺氧事件1期 海水酸化 厚壳蛤类

4.1 概况

白垩纪古低纬度地区广泛分布的海相碳酸盐台地其被危机间断的阶段性增长(Dercourt 等,1993,2000;Simo 等,1993;Philip 等,1995;Skelton,2003)模式显然与大洋缺氧事件(OAEs)(Schlanger 和 Jenkyns,1976;Jenkyns,2010)有关(图4-1)。碳酸盐台地由于危机涉及的灭绝的和更替的生物主要包括并不互相排斥的厚壳蛤双壳类生物。无论这些事件的因果关系如何,厚壳蛤类的进化史显然与它们生存的碳酸盐台地有关。阿普弟阶提供了最突出的案例和最彻底的相关记录,尽管双方并不是一致同意。事实上这一阶段是整个白垩纪世界范围的缩影,包含所有时期的标志性事件(图4-2)。除了大规模的台地的发展和消亡,阿普弟阶记录了一个与超地幔喷流柱火山大规模喷发其造成的甲烷水合物和 $\delta^{13}C$ 在碳酸盐和有机碳中的含量大量增加的事实(Jahren 等,2001;Beerling 等,2002;Jenkyns,2003;Renard 等,2005,2009)。这一大灾难与紧接而来的全球缺氧事件以及伴随着的 $\delta^{13}C$ 大量活跃有关。在动荡的历史阶段还出现过低幅度的高频沉积旋回,偶尔幅度较大的沉积旋回以及水进事件,和海洋中与碳酸盐台地上的生物大量钙化以及与湿度温度相关的气候急剧变化(Kemper,1987,1995;De Lurio 和 Frakes,1999;Ruffell 和 Worden,2000;Steuberet 等,2005;Dumitrescu 等,2006;Ando 等,2008)。阿普弟阶提供了一个分析研究以温室效应为主的白垩纪地球系统的相互作用和

反馈情况的案例,与现在二氧化碳含量较少的情况相比,这是巨大碳酸盐台地形成的一个原因。早阿普弟期的结束和中阿普弟期的开始是以第一个系统记录的碳酸盐台地生物的危机作为标志的,这标志还包括文石或方解石骨骼比例的急剧下降,并且有可能与海水中离子类型的化学变化直接相关(由 Steuber 和 Rauch,2005 年探讨的主题)。其他可能的原因包括海侵时期全球缺氧事件相关的碳酸盐台地生态系统的富营养化(Hallock 和 Schlager,1986;Weissert 和 Lini,1991;Föllmi 等,1994;Scott,1995),或者多种类型厚壳蛤类生物与全球环境的扰动的复杂相互作用引起的(Skelton,2000)。

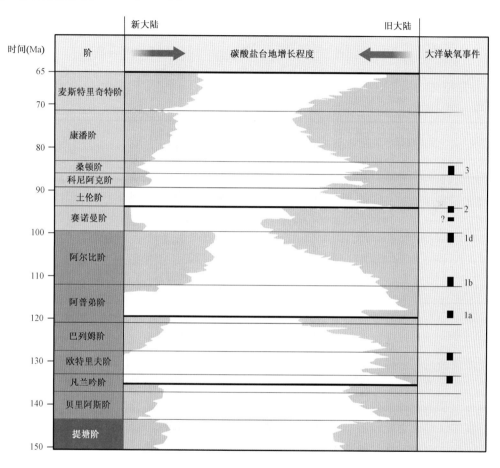

图 4-1 在新世界和旧世界提塘阶和白垩纪中碳酸盐台地(蓝色部分)形成简况,随着时代的推移碳酸盐台地危机(粗体水平线)与大洋缺氧事件的对应关系(Skelton,2003a;Jenkyns,2010)

与此同时,继承 Scholle 和 Arthur(1980)的开创性工作,许多人已经完成该阶段碳同位素曲线的绘制(Weissert 和 Lini,1991;Menegattietal,1998;Kuhntetal,1998;Braloweret 等,1999;Millan 等,2009;Moreno-Bedmar 等,2009),突出和标记全球碳循环扰动的地层标志,尤其是那些与全球缺氧事件有关的特征,尽管有时有不一致的结果。很多人把台地相的显著变化归因于碳循环扰动的影响,特别是特提斯北部边缘地区(Föllmi 等,1994;Weissert 等,1998;Wissler 等,2003;Follmi,2008;Follmi 和 Gainon,2008),常用水侵来形容这些地区的台地消亡。尽管施拉格原来使用的术语提到在:"构造沉降和海平面上升超过碳酸盐堆积,底栖碳酸盐建造停止",但许多后来的文章中往往强调全球性缺氧事件伴随的水体富营养化抑制了碳酸钙建造(Föllmi 等,1994)。稍后将在许多实例中讨论碳酸盐台地上的富厚壳蛤类沉积和富珊瑚沉积

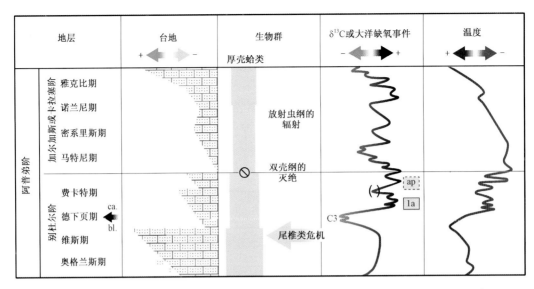

图 4-2　图中显示了在阿普弟期随着地层时代的变化,关键环境指数的变化与碳酸盐台地生长的对应关系。图中从左到右的列分别代表:地层——古特提斯海地层划分应用于此(Reboulet 和 Hoedemaeker,2006),浮游有孔虫区域边界(Moullade 等,1998),台地——总结自古特提斯碳酸盐台地的历史(图 4-4),生物群——厚壳蛤类物种变化的多样性(图 4-10),δ¹³C 或大洋缺氧事件——δ¹³C 变化曲线来自 Menegatti 等(1998)、Renard 等(2005,2009)、Kuhnt 和 Moullade(2009)、Millán 等(2009)和 Herrle 等,ap 为上述的最大负峰值,温度——总结自古气候曲线,摘录自 Weissert 和 Erba(2004)、Steuber 等(2005)、Ando 等(2008)、Dumitrescu 等(2006)、Kuhnt 等(1998)和 De Lurio 和 Frakes(1999)

明显的不连续以及造成多地生物相终结的原因。尽管现在趋于认为碳酸盐台地的消亡、台地上生物种类的灭绝以及全球缺氧事件是相互独立的,但 Weissert 等(1998)发现早阿普弟中期全球碳循环变动之后特提斯中南部地区碳酸盐台地继续生长。作者据此提出这些地区不受陆地风化增强造成的水体富营养化的影响,根据它们的模型,这种富营养化造成了特提斯北部台地的消亡。尽管这样,这个模型还有两个关键的问题不能解释清楚。第一是在孤立的阿普利亚台地以及那些远洋的台地如盖奥特台地,而不仅仅是在靠近陆源的地区的证据表明全球缺氧事件能够影响台地相(Graziano,1999;Jenkyns 和 Wilson,1999);其次,Massee(1998)记录的台地上生物大灭绝是在早阿普弟末期而不是在中期就结束了,这一结论经受了随后的生物地层细化的试验观察(Masse,2003)。

在日益完善的同位素化学地层学和生物地层学,一个单一的事件模型的确是过于简单化(Masse 和 FenerciMasse,2011)。例如 Millan(2009)在西班牙北部巴斯科坎塔布盆地定义的第一期之后从早阿普弟早期到早阿普弟晚期的全球缺氧事件更加受地域限制(图 4-2)。另外在法国东南部 Vocontian 盆地发现了晚阿普弟早期的黑色页岩地层记录(Niveau Noire4 和 Niveau Noire Calcaire2)。这些事件是否为独立产生,或局部复发,或是持续扰动,或是产生全球缺氧事件的条件,仍然是一个有争议的问题。

此外,如上所述,全球缺氧事件对中部和南部特提斯台地并不是完全没有影响的(Graziano,1999,2000,2003;Immenhause 等,2005;Vlahovicetal,2005;Luciani 等,2006;Graziano 和 Ruggiero Taddei,2008;Huck 等,2010),但是与最北端的特提斯台地相比,它们的确是早阿普弟晚期恢复。因此全球缺氧事件可能不是造成阿普弟阶广泛的戏剧性生物事件的唯一原因,

Skelton(2003)根据纬度的组合样式提出气候的影响,包括气候变暖以及之后的变冷,造成早阿普弟期碳酸盐台地的解体。

鉴于早阿普弟期台地和其生物区系的复杂关系,应及时回顾目前已知的有关记录里的古地理变化,从而提供一个详尽、准确、更复杂的解释模型框架。因此本次研究的目标包括以下几点:

(1)通过碳同位素记录的研究来回顾现有的阿普弟阶碳酸盐台地生长及台地相的变化等关于古地理变化的地层上的证据(主要突出特提斯范围,因为这一区域目前数据记录最为广泛)。

(2)伊比利亚地区在古纬度上处于特提斯范围内,记录这些地区厚壳蛤类生物种属的世代更替和形态类型的变化,古地理分布以及外壳的矿物学特征(即文石与方解石比例)。

(3)综合分析气候及其他因素的变化证据,形成有关地球系统可能的相互作用的可行模型,至少给出阿普弟阶碳酸盐台地复杂变化史的主要特点。

4.2 碳酸盐台地地层记录

4.2.1 数据准备

本次研究选取了15个区域内公开发表的有化学或生物资料的碳酸盐台地和继承性斜坡的剖面。欧洲大西洋边缘北部特提斯的九个地区,包括伊比利亚海角,其余六个在特提斯中南部的地区(图4-3)。

这15个地区的碳酸盐台地生长的地层综合记录在图4-4中显示。为了避免中阿普弟期与早阿普弟中期以及英国的二分法和法国的三分法的术语上的混淆,本文中使用应用广泛的法国三分法的别杜尔亚阶(早阿普弟期)、加尔加斯亚阶(中阿普弟期)和卡拉赛阶(晚阿普弟期)。为了对这些地质记录进行比较,这里用的是较广泛的相分类法(图4-4)。本次研究主要强调平台相和斜坡相,同时包括所有其他相类型,为了突出在平台相的地域上的差异。具有厚壳蛤类沉积物和相关的碳酸盐沉积物以及珊瑚类的沉积的 Urgonian 式台地通常情况下在碳酸盐台地顶部外侧发育最好,最突出的例子是绿色的双壳类亮晶生物灰岩为主导形成的台地边缘斜坡,这种情况下蓝色增厚的具有方解石外壳的厚壳蛤类的区别将在下一部分讨论(厚壳蛤类记录)。Follmi(2007)的所谓多生境模式含有丰富的棘皮动物遗迹,苔藓虫和底栖有孔虫和多种硅质碎屑颗粒,而不是鲕屑和缺乏厚壳蛤类、珊瑚的斜坡通道相的沉积。从 Urgonian 型台地继承来的可能是薄的单一的厚壳蛤类生物层,以及潮缘区藻类细纹层和古土壤,这种情况在特提斯中部尤其显著(Luperto Sinni 和 Masse,1986)。小圆片虫为主的泥灰岩和包含结核状 *Lithocodium/Bacinella*(Schlagintweit 等,2010)以及类似微生物结构的独特相类型是对相对异常环境条件的短暂记录(Pittet 等,2002;Rameil,2010;BoverArnal 等,2011)。

在详细研究图4-4中的数据前,别杜尔阶需要细分为早(下)和晚(上)两段。本次研究立足于 Moullade(1998)和 Masse(2003)对菊石的研究,及对标准特提斯的区带划分以及下别杜尔阶的 Deshayesites oglanlensis,上别杜尔阶的 Deshayesites deshayesi 和 Dufrenoyia furcata 的研究。部分台地的沉积物的确受继承性的菊石区带控制,然而,除非继续讨论与菊石方案有关的问题(Reboulet 和 Hoedemaeker,2006;Ropolo 等,2008;Moreno-Bedmar 等,2010;Reboulet 等,2011),否则由于缺乏或者没有合适的化石记录能够有效直接识别其他区域,所以只能建立其他的判别标准来进行代替。

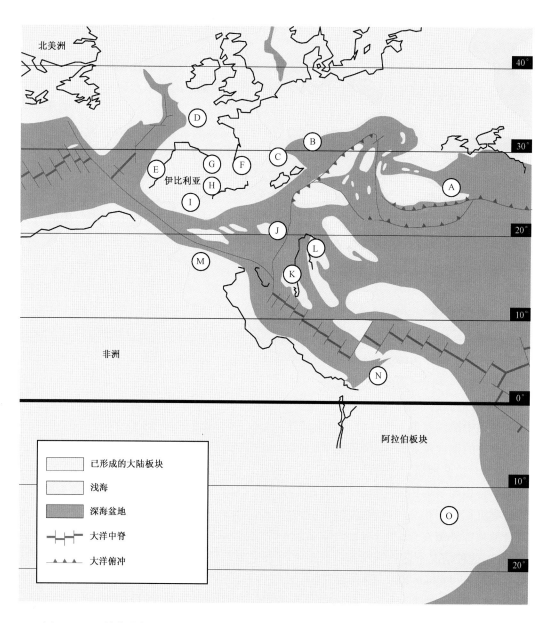

图4-3 阿普弟阶古特提斯东部大西洋区域古地理图显示出15个选定的区域地层记录和台地生长层序的位置(详见图4-4和表4-1)(据Masse等,2000)

例如,全球缺氧事件1期相应的层位的全盆地发育的黑色页岩是建立碳同位素变化曲线的自然目标。特别是那些在威尼斯阿尔卑斯山和瑞士萨特尔的黑色泥岩(Menegatti等,1998)。可是在没有菊石化石记录的地区,生物地层的划分建立在浮游动物和钙质超微化石的基础上,不过一些作者用菊石的区带来校正这种分层(Masse,2003)。然而,图4-4(黑色箭头)显示的是有孔虫和砂科虫属的生物地层边界是稍微在上别杜尔阶之上的,在东南部用菊石区带来确定剖面上别杜尔阶的(Moullade等,1998)。这里必须重申关于这一点的不一致:例如,Weissert和Erba(2004)(图4-2)把生物地层分界等同于菊石化石的区带界限,因此置于下别杜尔阶之内。在许多台地地区识别巴列姆阶和阿普弟阶的界限仍然是复杂的问题。尽管通过地球化学的对比在Cismon地区确定了阿普弟阶的基准,在法国东南部,用菊石化石带

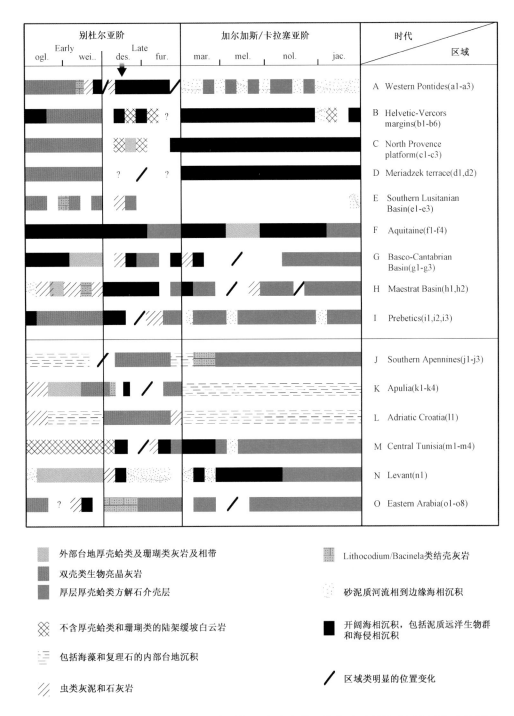

图 4-4 阿普弟阶古特提斯欧洲大西洋区域 15 个选定的区域台地及斜坡地层记录和生长层序。菊石区域的简写及 blowi 或 cabri 边界(黑色箭头)见图 4-2。相带的划分及讨论见下文。图 4-3 显示的是从 A 点到 O 点的位置,参考表 4-1 有详细的说明

来划分了巴列姆阶和阿普弟阶,但在许多地区找到相关的标志仍然很难。由于现阶段的研究主要关注阿普弟阶的变化情况,特别是别杜尔阶中期到晚期,阿普弟阶和巴列姆阶的问题这里不进一步讨论,受这一问题影响的地区将会特别说明。

表 4-1 参考文献及图 4-4 出现的地层概要细节

A	[a1,2] Masse 等（2004b，2009），Öküsmedere 组，Inpiri/Kozlu 段/[a3] Masse 等（2002），Cengellidere 组
B	[b1] Wissler 等（2003），Churfirsten，Alvier 段 + [b2] Föllmi 等（2007），[b3] Föllmi & Gainon（2008），[b4] Föllmi（2008），区域上从 Schrattenkalk 到 Garschella 组地层序列；[b5] Linder 等（2006），Grünten 段 + [b6] Arnaud-Vanneau 等（1982），Vercors
C	[c1] Masse（1976），[c2] Arnaud-Vanneau 等（1982），[c3] Masse（1999）和 Masse 与 Fenerci-Masse（2011），Urgonian 单元从 U1 到 U3，再到 "Gargas Marls"（大多数厚壳蛤在 U2/U3 不整合处终止，但是 U3 含有珊瑚透镜体，以及底部含有丰富的虫类化石）
D	[d1] Pastouret 等（1974），样品采自倾斜的平地，含有 Offneria rhodanica，暗示可能是下 Bedoulian 地层（见 Masse，2003）/[d2] Roberts 等（1980），岩心位置在 402A
E	[e1,2] Rey（1979，2006），[e3] Burla 等（2008），Crismina 组（Ponta Alta 组，包含 Lithocodium/Bacinella 的重要单元—厚壳蛤灰岩之间的薄壳状珊瑚碎片）
F	[f1] Masse & Philip（1981），[f2] Masse & Gallo Maresca（1997），[f3] Bourrouilh（2000），Arudy 段；[f4] Masse & Chartrousse（1997），Arudy Caprina parvula 地层 = 上部 Bedoulian（见 Masse，2003）
G	[g1] García-Mondéjar 等（2009），Aralar 地层序列：[g2] Millán（2009），Madotz 台地；[g3] Millán 等（2009），Sarastarri 组
H	[h1] Bover-Arnal 等（2009，2010），Galve 次盆地的 Barranco de las Calzadas 段（Lithocodium/Bacinella 单元—发育在下别杜尔亚阶上部的结壳的珊瑚和厚壳蛤碎屑）/[h2] Tomás 等（2008），Benicàssim-Orpesa 地区
I	[i1] Castro 等（2001，2008），Llopis 组/[i2] Masse 等（1998a），[i3] Vilas 等（2003），"Caroch" 台地（尤其是 Sierra del Carche 地层序列）
J	[j1] D'Argenio 等（1992），[j2] Carannante 等（2009），[j3] Masse 等（1993），Matese，Serra Sbregavitelli 和 Serra della Macchiatelle 段
K	[k1] Graziano（1999，2000），[k2] Luciani 等（2006），[k3] Graziano & Ruggiero Taddei（2008），Gargano，Montagna degli Angeli 灰岩（Fucoid 泥灰岩）/[k4] Masse（1992a,b），Calcari de Bari，Livello Corato
L	[l1] Masse 等（2004a），Tounj 地区
M	[m1] Lehmann 等（2009），[m2] Heldt 等（2008），Hamada 组外斜坡/[m3] Masse（1984），[m4] Heldt 等（2010），Orbata-Serdj 组（包括 Gafsa-Kasserine-Kairouan 地区，在 Jebel Serdj 旁边）
N	[n1] Bachmann & Hirsch（2006），Ein el Assad 和 Hidra 组

近年来常用地球化学方法进行对比碳同位素曲线的显著波动变化（图 4-2）。这种做法引起了关于如何与生物地层的菊石化石进行对应的讨论，部分原因是因为上面提到的其他相关问题，另一部分原因是作者概念的使用范围，比如全球 1 期缺氧事件的确切意义。自从在某些富含碳地层（Schlanger 和 Jenkyns，1976）上建立的全球缺氧事件的概念出现以及认识到全球缺氧子事件（Arthur 等，1990）的主要实例后，全球缺氧事件这一概念慢慢的发生改变。因此，当 Menegatti 等（1998）讨论 Selli 事件时认为是 "第一次和第二次 C_{13} 活跃漂移（C_4 段和 C_5 段）和相应中心段（C_5）的时间间隔"，由于对这一偏移的不断增长的兴趣，一些研究者，包括 Erba（1999）等把全球缺氧事件 1 期定义为更广泛的概念。

除了这种不一致的定义，有关生物地层相关的碳同位素曲线图形，甚至在部分能提供丰富的菊石化石的地区，也有分歧。例如，分析了法国东南部别杜尔阶的典型剖面的团队（Kuhnt 等，1998；Renardet 等，2009）认为这个负向偏移预示着全球缺氧事件 1 期处在 deshayesi 化石的双峰的区带内。相比之下,Moreno-Bedmar(2009,2010)等在西班牙东部 Maestrat 盆地工作的研究,把碳元素的负漂移,以及整个全球缺氧事件1期,认为是等同于 forbesi 菊石区带的上部(Reboulet 等,2011)。Millan 等(2009)在西班牙西北部巴斯科—卡特布赖恩盆地的工作认识到 weissi 化石区带存在一个急剧的负尖峰,等同于全球缺氧事件1期中的较长间隔。这一争论包括生物地层学上的菊石化石的分类,而全球缺氧事件的定义不在本次讨论的范围之内。相反,如前所述,本次讨论主要分析阿普弟阶特别是别杜尔阶古地理的分布和碳酸盐台地的特征。

因此,鉴于这些矛盾和随之而来的潜在的地层混乱,全球缺氧事件没有在图4-4中表示出来,而是在一些能够具体识别的相关的案例中进行讨论。最后,这里避免为每一个地层界限确定绝对的年代,因为在同位素年代学文献仍然存在互相矛盾的结果(Gradstein 等,2004;Fiet 等,2006)。

4.2.2 结论

图4-4显示别杜尔早期碳酸盐台地的大幅增长,以及中期台地相发育的间断。沿特提斯北部台地边缘的剥蚀面就是 weissi/deshayesi 的边界(参见表4-1)。

别杜尔期晚期特提斯北部和紧邻大西洋东部边缘地区(图4-3位置A到I)碳酸盐台地的恢复被限制在伊比利亚海角地区(图4-3位置E到I),虽然时间性差异可能反映了区域构造或其他的影响。例如,在巴斯科坎塔布东部 Sarastarri 组台地淹没在盆地分叉的间隔水面下,在比利亚的罗亚德洛斯西部 Maestrat 盆地台地发展发生在该年代间隔的结束时期。在这两种情况下,仍然在 Prebetic 地区,阿普弟阶台地仍能继续的发育。在 Lusitanian 盆地南部,别杜尔期晚期 orbitoline 泥灰岩表面经受淋滤侵蚀之后(图4-5),早期台地由于构造的因素影响,产生了沉积间断,这一间断一直持续到晚阿普弟期(图4-4)。

图4-5 (A)Lusitanian 盆地南部,下部别杜尔亚阶 Ponta Alta 段厚壳蛤类灰岩(顶部可见侵蚀面,r字母处为厚壳蛤类铸模)与上部别杜尔亚阶 Praia da Lagoa 段 orbitoline 泥灰岩接触(见图4-4位置E)。(B)相同位置处的 orbitoline 泥灰岩,可见 Rhizocorallium 遗迹化石。(A)图与(B)图的比例尺是欧元硬币,直径23mm(详述见 Rey,2006;Burla 等2008)

相比之下,沿着韦科尔到海尔维边缘的佛朗哥瑞士 Urgonian 台地(图4-3C 位置)在 weissi 或 deshayesi 界面及附近时仍被水体淹没,其证据包括在海尔维断层和冲断带浓缩的磷酸盐床和 Grunten 组的砂质泥灰岩覆盖和侵蚀上 Schrattenkal 组的表面。相应地,北普罗旺斯台地(图4-3C 位置)下别杜尔阶厚壳蛤类灰岩覆盖有不连续的具有外源性质的燧石碎屑灰岩,这些碎屑灰岩用代表了缺氧事件的 C_{13} 负偏移指数和含菊石化石的泥灰岩来标定了时代（图 4-4，位置 C；ArnaudVanneauet 等，1982；Masse 等，1999；Masse 和 Fenerci-Masse，2011）。

Öküsmedere 组下别杜尔阶双壳蛤类和珊瑚灰岩上局部覆盖着珊瑚，下部与 weissi 组的含菊石绿泥石泥灰质灰岩（图 4-4A；Masse 等，2009），然而在一些地方的珊瑚床上有上别杜尔阶的小圆片虫记录，阿普弟晚期如 Cengellidere 组碳酸盐台地重新发育（Masse 等，2002）。在阿基坦阶的记录，反映了与比斯开湾扩展有关的活跃的沉降（Bourrouilh，2000），但这里一个显著特点是如伊比利亚半岛，晚别杜尔期时厚壳蛤类的发育。

遗憾的是，Meriadzek 斜坡的地质记录问题（图 4-3D 位置）不能很好地解决。然而，下阿普弟阶台地灰岩样品包括双壳类生物亮晶灰岩和其变种类型（Masse 和 Chartrousse，1997）构成了别杜尔阶下部典型的组合类型（Masse，2003）。

特提斯台地中部记录了整个阿普弟阶碳酸盐台地相的发展，尽管这些孤立的台地不可避免地在别杜尔期受到全球缺氧事件 1 期的影响，这些在介绍部分已经提过。例如加尔加诺岬的综合部分能够识别出别杜尔阶台地时的层序类型，含有丰富的粗碎屑的双壳类生物灰岩，内部有双壳类的碎片，上覆有高岭土蒙脱石结壳，下伏上别杜尔阶 Mattinata 组丰富的腕足层。此外，古近—新近纪台地滑塌产生的泥石流沉积在帕纳萨斯台地上，但被认为是来自西部边缘地区，包含有已经灭绝的蛤类和珊瑚（Baron-Szabo 和 Steuber，1996），否则只有沙特东部上别杜尔阶有这样的双壳类生物（Skelton 和 Steuber，1999；Skelton，2004）。

沿特提斯南部边缘（图 4-3 位置 M 到 O）可以发现中别杜尔阶台地相发育的中断，尽管可能包含了晚别杜尔期的沉积。突尼斯中北部的杰贝尔（图 4-4 位置 M），Heldt（2003）和 Lehmann（2009）等都发现了 Hamada 组下段水体富营养化和海底由于全球缺氧事件 1 期造成的还原环境的古地理证据，灰岩和泥灰岩变为主要是泥质灰岩相。这一区域代表了半深海的中央突尼斯南部碳酸盐台地和深海突尼斯北部斜坡沉积。在上面提到的古地理变化方面的证据得到 Heldt 等（2008）得出的结论："全球缺氧事件 1 期的碳酸盐建造扰动并没有影响南特提斯台地边缘"。加尔加斯亚阶（Heldt 等，2010）继承了 Serdj 组典型特征，体现了台地相厚壳蛤类的发育，Masse（1984）记录了更南部台地剖面别杜尔阶小圆片虫和蛤类灰岩。后者顶部 Masse 记录了 Polyconites 软体动物的出现，当时认为属于 Gargasian 组底部，而现在认为是属于上别杜尔阶的物种（Skelton 等，2010）。再往南，在 Chotts 地区，Marzouk 和 Ben Youssef（2008）记录了 Orbata 组底部的灰岩和泥岩内的丰富的海胆、厚壳蛤类和有孔虫动物，这一层位学者认为属于别杜尔阶上部。Marzouk 和 Ben Youssef（2008）描述了这些层位上白云质灰岩和泥灰岩通过传递向上蒸发，上覆了含 Deshayesites weissi 和 Dufrenoyia furcata 以及其他类型菊石化石的砂岩和泥岩。如果得到证实，这样的分类限制了别杜尔阶底部（如果不是 Barremian 最顶部）Orbata 厚壳蛤类台地相的研究，尽管在厚壳蛤类和相关类群的鉴定可以提供对这项建议的一个测试。

围绕地中海东部海岸（地中海东部，图 4-4N），Bachmann 和 Hirsch（2006）认为含丰富有孔虫的层位介于中别杜尔阶碳酸盐台地阿萨德组和 Hidra 组碳酸盐岩和碳酸盐岩碎屑混合物之间，与全球缺氧事件密切相关，由于海平面上升造成了这一中断现象。在 Hidra 组的中间部分，研究者也指出，河流砂岩和铁质结壳的插值，标志着早阿普弟期台地的出现。

世界上最大规模的碳氢化合物贮存之一位于东部阿拉伯板块（涵盖东部沙特沙特阿拉伯，阿拉伯联合酋长国和阿曼，地点在图 4-3）阿普弟阶剖面，经过穿层的，特别是层序地层学方面的分析，最近由 Van Buchem 等（2010）大量编辑并研究。尽管这一研究很专业，但因为周围用来判断一些关键层位的生物地层标志相对贫乏，年代归属的问题仍没有解决。

这些问题涉及 Kharaib 组顶部含丰富厚壳蛤类的台地灰岩。在 Van Buchem 等(2010)的综合论述中,Kharaib 组的上界是别杜尔阶和阿普弟阶的分界(在剖面上部存在的底栖有孔虫 Montseciella arabica),虽然这些作家承认:"不能排除的是这个剖面顶部可能含有早阿普弟期的地层"(Van Buchem 等,2010)。然而,在考虑碳同位素和锶同位素的相关性的基础上,Strohmenger 等(2010)通过分析图 4-4 的位置 O,由于台地相持续发育把别杜尔阶和阿普弟阶的分界定在 Kharaib 组。作者也认为 Hawar 是别杜尔阶下部超覆的不整合界面,这一解释也被 Granier 认同(2008)。Strohmenger 等(2010)展示 Shuaiba 组底部 C_{13} 同位素曲线,并将这前面的负漂移曲线放在 Hawar 段底部。在图 4-4H 中,后者剖面的开始可能与中别杜尔阶台地相的中断有关。平台的阶段 Shuaiba 组上覆的建造[Van Buchem 等(2010)记录的,图 4-2]毫无疑问在上别杜尔阶到达顶峰,含生物碎屑灰岩的地层延伸到亚阶(Shuaiba 序列 3,Strohmenger 等,2010)的顶部;厚壳蛤类相确实在 Shuaiba 序列 4 重新出现,尽管与早 Gargasian 期蛤类属于不同的族群(Yose 等,2010)。在 Al Hassanat 组,年轻的阿普弟阶台地继续作为阿拉伯板块东南缘的低位楔(Masse 等,1997)。

阿拉伯 Bedoulian 继承的一个显著特征是 Lithocodium 或 Bacinella 结壳(图 4-6A)在原地充分发育,形成庞大的规模(Hillgartner,2010)。再往南部,在阿曼,Qishn 组(图 4-6B)相当于上 Kharaib 组和下 Shuaiba 组,在此处 Immenhauser 等(2005)和 Rameil 等(2007)总结出两个不同的时间间隔是全球缺氧事件在浅水区的表现。虽然这种发展的例子确实已经从其他地方描述(如 Huck 等,2010;Bover-Arnal 等,2011),但其他的都是不代表地层意义的缺氧间隔。无论是从地层较低的区间内,例如 Lusitania 盆地南部下别杜尔阶 Ponta Alta 段厚壳蛤类和珊瑚灰岩(图 4-4E;图 4-6C 和 D),Pontides 西部 weissi 区地层(图 4-4A),还是在意大利东北部的"BA2-AP1"间隔,以及更高的层位,例如南亚平宁 Sbregavitelli 剖面可能是 Gargasian 期的内陆台地相(图 4-4J;图 4-6E 和 F)。台地相的发育状态因不同地方的条件而不同,只要后者出现(Masses 等,2009),而不是完全的地层承压,在一些特定的时间间隔时,比如全球缺氧的间隔内,虽然这种情况下台地范围被更频繁的、范围更广的侵蚀(Huck 等,2010)。

最近在位于阿拉伯板块的东北部伊朗西南部扎格罗斯山脉 Kazhdumi 盆地(Van Buchem 等,2010;Vincent 等,2010)的研究工作与阿曼和阿拉伯联合酋长国的南部的剖面进行了对比。特别要说明的是,中别杜尔阶是以 Kazhdumi 组底部放射洪水区以及 deshayesi 段菊石化石为标志的,这中断了 Dariyan 组含丰富牡蛎和鲕粒的碳酸盐台地沉积。台地边缘相对 Kharaib 组受到周边泥质入侵,但与 Hawar 的含丰富鲕粒的台地相相比有一些类似之处。但是在晚别杜尔阶 Shuaiba 组含蛤类的石灰岩在盆地边缘周围继续发育。

对南部进一步考虑,埃塞俄比亚东南部在河流相的碎屑岩剖面中有薄层含丰富鲕粒灰岩 Graua 灰岩(Bosellini 等,1999)。由于内部含有 Praeorbitolina 生物,Cormyi 同样把层位认定为上别杜尔阶。

为了进一步的考虑,特提斯地区之外大部分的阿普弟阶碳酸盐台地地层记录的分辨率需要进一步细化,虽然可以强调一些有关意见。横跨加勒比海省有丰富蛤类动物群的地方存在广泛的从巴列姆阶到别杜尔阶的岩石记录,地域范围从得克萨斯州北部(Sligo 组;Scott 和 Hinote,2007)的地下,通过大安的列斯群岛和墨西哥,委内瑞拉南部(Barranquin 组;Masse 和 Rossi,1987)。同时在深海半深海的剖面上可以明显地看到全球缺氧事件的影响,例如墨西哥北部的马德雷(Bralower 等,1999)。

但是新大陆的巴列姆阶到别杜尔阶间隔内沉积的精确地层范围仍然不确定。牙买加和古巴的含蛤类灰岩可能不会超出巴列姆阶的范围(Chart 和 Masse,1998),墨西哥西南地区(Hu-

图 4-6 阿普弟阶 Lithocodium 或 bacinela 及相似的生物结壳的例子。(A)是结核边缘薄层显微照片,包含 Bacinella 类的构造,如"b"点所示,"p"所示是耳壳红藻类,产自阿布扎比的滨岸带,比例尺为 1mm,(B)是中别杜尔阶 Qishn 组地层,在阿曼 Haushi-Huqf 南端 Wadi Baw,箭头所示低隆起处地质学家为比例尺(1.8m 高),(C)和(D)是下别杜尔阶 Crismina 组 Ponta Alta 段地层,产自葡萄牙 Forte da Crismina(见图 4-4,E 观察点),(C)是结壳层上覆在生物碎屑的河道上,如"ch"所示(红色圆圈长 35cm),(D)为垂直的结壳层剖面,"c"所示为珊瑚,被"1b"Lithocodium 或 Bacinella 类包壳,比例尺为 1 欧元硬币(直径 23mm),(E)和(F)是意大利 Apennines 南部 Lago di Matese 西端公路剖面(见图 4-4,观察点 J),(E)为风化的似核形石被腹足类和厚壳蛤类介壳包裹,(F)为核形石的放大照片,箭头所示为分离的壳瓣,右边为厘米刻度比例尺

etamo 区)原始状态的豆粒灰岩 Palorbitolina"显示了早期阿普弟期族群的特征"(Alencaster 和 PantojaAlor,1996;以 Rolf Schroeder 和 Antonietta Cherchi 的研究为基础)。此外,得克萨斯州地下的 Sligo 组上覆开放的海相 Pearsall 组沉积,包括有孔虫 Globuligerina hoterivica 和 Nannoconus bucheri,都属于 Hauterivian 到下阿普弟阶的范围(Scott 和 Hinote,2007)。墨西哥东北部的 LaPena 组菊石同样受碳酸盐台地上别杜尔阶 Cupido 组沉积范围的限制,至少 Dufrenoyia justinae 族群被认为等同于特提斯 furcata 地区(Barragan 和 Maurrasse,2008)。Scott 和 Hinote (2007)认为"LaPena 页岩是全球缺氧事件的地球化学记录"。如果正确的话,这意味着台地底部的灰岩不会超过上别杜尔阶的底部(Menegatti 等,1998;Renard 等,2009;接近有孔虫 blowi 或 cabri 的界面)。然而还有一种可能性需要检验,上述给出别杜尔阶年代的 LaPena 菊石 (Scott 和 Hinote,2007),被质疑并不对应全球缺氧事件,而是 Millan 等识别出的稍微早一点的别杜尔阶 Aparein 段(2009)。然而无论确切的年代是什么,目前出现的图片是与特提斯中南部区域相比,在别杜尔阶结束前加勒比台地的发育逐渐消失。此外,新大陆碳酸盐台地在阿普弟期的恢复生长也是有限的(Skelton,2003;Scott 和 Filkorn,2007)。

现在(如 Chartrousse 和 Masse,1998;Jenkyns 和 Wilson,1999;Masse 和 Shiba,2010)太平洋台地记录的情况大部分是关于海洋深处的海山的,显然限制了可用的地层数据,虽然双壳蛤类灰岩在海底平顶山终止已被证明是等同于前 Selli 段的碳同位素负漂移,他们通过鲕粒灰岩及富含有机质的黏土的碳循环识别出这个规律(Jenkyns 和 Wilson,1999)。在大陆边界处可以获得额外的露头信息。Iba 和 Sano(2007)研究了在日本和特提斯类似的碳酸盐台地相在中东阿尔比期与晚阿普弟期的逐步解体。这些作者指出可能属于巴列姆期早期到阿普弟期的碳酸盐台地相插大阪组的浅海砂岩以及中日本的其他地方,九州四国等地也有类似的发展变化,尽管这些物种的年代等结构性的问题仍需要进一步的研究和修正。日本北部 Hokkaido 岛的 Yezo 群下部晚阿普弟期 olistolith 灰岩包含有 polyconitid 蛤类和辐射蛤类,通过 Orbitolina 有孔虫以及浮游有孔虫来确定年代。类似的,相对较早的多足科和放射科蛤类在巴基斯坦北部的岛弧剖面 Yasin 群灰岩中被发现(Pudsey 等,1985)。与特提斯北部边缘的发展一道考虑(Montenat 等,1982;Marcoux 等,1987),这些实例指出晚阿普弟期返大陆周围台地的广泛发育能够到达日本的范围(Takashima 等,2007)。

4.2.3 台地发展史总结

根据上面研究的地层数据,在古地理范围内早阿普弟期台地发展的顶点是在早别杜尔期:图 4-3 和图 4-4 中所有考虑到的地区都记录了这段时间内碳酸盐台地的发展。

这一阶段在大约中别杜尔期(晚 weissi 到 deshayesi)被各种不同的方式中断了(图4-4)。同期全球碳循环的扰动,以及造成 C3 的负向漂移以及对应的全球缺氧事件(图 4-2)开始被认为是造成这一中断的原因,影响了全世界范围内碳酸盐台地相的发展(Weissert 和 Erba,2004;Burla 等,2008;Follmi,2008),但是通过不同岩性来表达。一些研究者认为这一中断事件有更复杂的线索,需要关注时间上的差异(Masse 和 Fenerci-Masse,2011)。例如在北普罗旺斯台地识别出的"U2"或"U3"的连续性(图 4-4C),以及北部亚高山白云岩伏于前 Selli 的 $\delta^{13}C$ 漂移之下(Masse 等,1999 年)。在 La Bedoule 与此相对应的中断是轻微的不整合(129),不整合下面有第一次 $\delta^{13}C$ 负漂移的记录(Renard 等,2005 或 2009)。这一偏移让人质疑沿着特提斯北部终止了 Urgonian 台地的侵蚀不整合面的起源,以及可能发挥的作用,后面将讨论碳循环造成的台地的困境。与此同时,阿拉伯台地东南部地层中有丰富的 Lithocodium 或 Bacinella 或类似的微生物结壳的蛤类灰岩(Immenhauser 等,2005;Rameil 等,2010;图 4-6A 和 B)。

晚别杜尔期台地相的发展有很大的地域性(图4-7)。特提斯北部边缘较高古纬度地区的碳酸盐台地没有能够恢复(图4-4A到D),而新大陆晚别杜尔期furcata区受到泥质海相沉积的入侵。碳酸盐台地的建造还受到普利亚的Gargano周围水退水进的动荡影响(Graziano,2000,2003)。一些准大陆边缘(伊比利亚半岛,北非和地中海东部地区)的海侵沉积含有大量的碎屑以及鲕粒(图4-5)。尽管伊比利亚海角和阿基坦(图4-4F到I)周围含蛤类台地的确恢复了,在一定程度上突尼斯(图4-4M)处在低纬度地区,虽然受到碎屑涌入的影响,仍有多种的相类型(图4-4,J到L和O)。然而,分类学和矿物学上一些有趣的古地理差异造成晚别杜尔期的蛤类群体的差异显著(图4-7),将在下一节讨论。Gargasian到Clansayesian之间(尤其是较晚的部分)有台地广泛的发育,向东延伸超出特提斯北缘,远到日本。

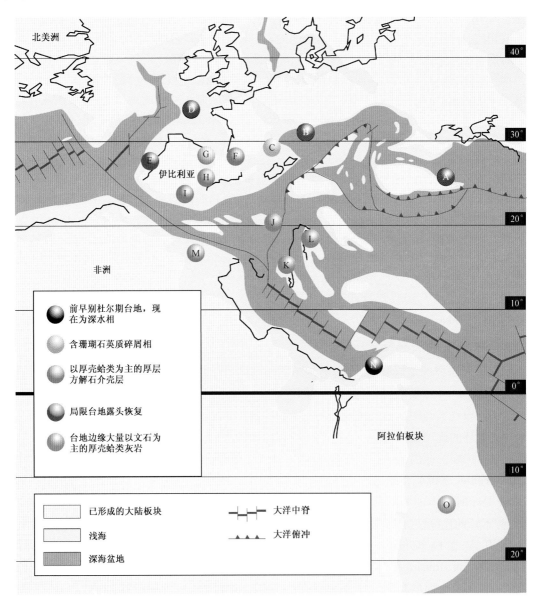

图4-7 如图4-3所示古地理图,显示了晚别杜尔期碳酸盐台地沉积条件
(具体位置见图4-1和表4-1)

4.3 蛤类记录

4.3.1 分类单元和生活习性

阿普弟期蛤类产生三个主要的生态形态型(Skelton,1991,2003;Ross 和 Skelton,1993;Gili 等,1995)。直立管状的插在泥沙淤积的沉积物中的蛤类数量大量增长(图 4-8A 和 B),并不同程度的相互附着缠绕。螺旋盘绕的类型长在硬质的基底上(图 4-8C 和 D),而个体较大,弯曲的种类长在不稳定的泥沙频繁通过的基底上(图 4-8E)。根据不同的沉积环境可以对比不同族群的蛤类生物生存的台地环境(如 Masse,1979;Masseet 等,1998;Hughes,2000,2004;Fenerci-Masse,2006;Masse 和 Fenerci-Masse,2008)。

台地的标准断面或者陡坡(图 4-9)能够作为衡量阿普弟期蛤类物种变化的参考模型。在台地剖面上可以看见多样性组合类型经常性的循环出现,开始是斜坡来源的泥质相,局部含板状到穹隆珊瑚(Rosen 等,2002;Tomas 等,2008),然后过渡到电梯状蛤类、斜卧状蛤类的碎屑的沉积体,后者现今生长在台地边缘的底部(Hughes,2000)。相比之下台地顶部内的蛤类族群,主要是珍惜的种类或细长的微小的种类占主导地位。

不同蛤类分支的结构特征倾向于演化成使其个体成为有利的特殊形态型。因此新种类的出现以及旧种类的灭绝伴随着台地组合性质的变化。考虑到这些构造特征的影响,图 4-10 显示了现今已知的晚巴列姆期到早阿普弟期蛤类物种的数量。

最具戏剧性变化的是 Caprinidae 族群(图 4-10 顶部所示)。许多这些松散盘绕拉长的管状类型采用了横卧生长习性(图 4-8E;Skelton 和 Masse,1998),尽管个体小的通常长成电梯状直立,通常情况下它们生存在台地外缘的环境(Masse,1998;Hughes,2000;Fenerci-Masse,2006;Masse 和 Fenerci Masse,2011)。这些种类的内壳由文石构成能够抵御外力冲击,个别种类发育较厚的外壳(图 4-11A;Skelton,1991)。相比之下,方解石的外壳一般总是较薄(图4-11B;通常小于 1mm;Skelton 和 Smith,2000),已成为最终外壳外附着的多余壳体(见下文讨论的 requieniids 种),因此这些种类外壳成分主要以文石为主。晚巴列姆期到早别杜尔期的蛤类多样性之后,晚别杜尔期种类出现部分下降,特别是美洲台地 caprinuloideinids 种的缺失(图 4-10;Skelton,2003)。别杜尔阶顶部已知地层记录中 Caprinids 消失(如后面将讨论的 Shuaiba 组),在美洲以及旧大陆的阿尔比、阿普弟阶地层中这些种类又重新出现,这证明 Lazarus 类在太平洋等未知区域保存(Skelton,2003)。

Requieniids 种的遗传多样性(图 4-10 底部)同样在早别杜尔期达到顶峰,此时能够很好地进行分类直到晚阿普弟期晚阿尔比期种类重新减少。尽管都是几何级的增长(图 4-8C;Skelton,1985),图卡斯蛤属凭借它们增厚的方解石外壳很好地适应了新环境(Skelton 和 Smith,2000),并且能够很好地固着在基底上生长(图 4-8D)。因此 requieniids 种在台地顶部无论内外都找到合适的基底兴旺生长(Masse,1979;Masse 和 Fenerci-Masse,2006)。

Polyconitids 和 radiolitids 的对比是一个明显的记录(图 4-10 中部到底部)。前者代表的早别杜尔期单一特异性种类,而后者则是由别杜尔期单一种类变化而来(Skelton 和 Smith,2000)。从晚别杜尔期 polyconitids(Skelton 等,2010)和加尔加斯放射性蛤类(Masse 和 Gallo Maresca,1997;Masse 等,1998;Fenerci-Masse 等,2006)的情况来看,两种种类的蛤类伴随着分类学上的多样化出现的丰度都显著提高。此外,两种类型的个体都进化出增厚的钙质外壳,这又促进了它们保持聚集起来直立向外生长的生活习性;甚至别杜尔阶的 Polyconites 有相对较

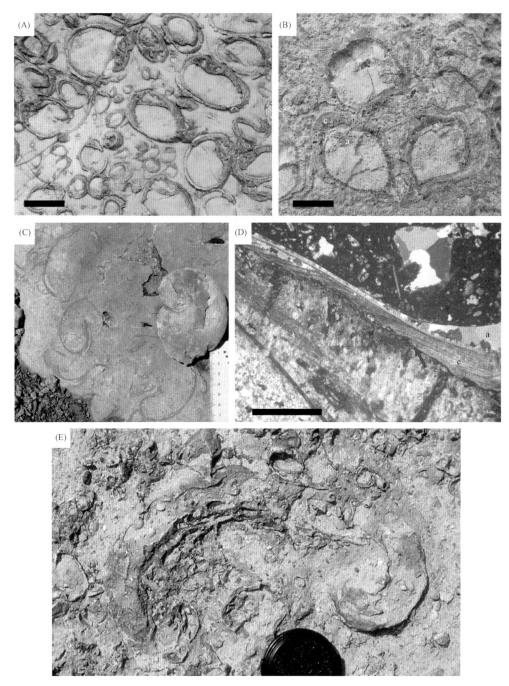

图4-8 阿普弟阶厚壳蛤类古生态类型的例子,(A)和(B)是直立状,(C)和(D)是缠绕状,(E)是平卧状,(A)是 Mathesia 类松散族群横断面,产自意大利 Apennines 南部早阿普弟期 Machierelle 段(见图4-4,观察点 J),比例尺为10mm,(B)是 Archaeoradiolites 类族群横断面,产自西班牙东部 Benassal 组 La Venta 段(见图4-4,观察点 H),"C"是方解石介壳层,"a"是文石质内部层,比例尺是10mm,(C)是 Toucasia 族群,俯看的角度拍摄,右边是脱落的样本,比例尺是厘米刻度尺,产自 Maestrat 盆地西部 Pinares 组(见图4-4,观察点 H),(D)是 Toucasia 类壳瓣的一张偏光显微照片,并有藻类壳质和珊瑚类(下别杜尔阶,Ponta Alta 段,见图4-6D),"a"是亮晶胶结文石质内部介壳,"c"是方解石质壳外壳,箭头所示的是基底铸模,比例尺是10mm,(E)是横卧的 Pachytraga 类,俯看的角度拍摄,产自葡萄牙 Lusitanian 盆地南部下别杜尔阶 Crismina 组 Ponta Alta 段,镜头盖直径是55mm

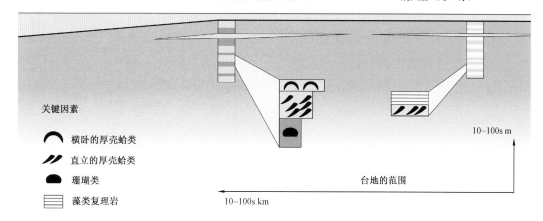

图4-9 白垩纪碳酸盐台地的典型模式,典型的旋回式沉积包括从外部台地边缘到内部台地顶部的厚壳蛤类灰岩(淡蓝色部分)。Clinger 厚壳蛤类可以在这些不同的地方找到(Skelton,2003)

厚的方解石外壳。像 requieniids 一样,两种类型都有在台地顶部无论内外的基底上发育的习性(Masse 等,1997;Bover-Arnal 等,2010)。

除了上面提到的四种主要的类型外,阿普弟阶的 monopleurids(图4-10中部)主要是小型个体,包括细长的管状(图4-8A;Masse 和 Fenerci-Masse,2009)以及更复杂的帽状集合体。如上所述,一些 monopleurid 表明有方解石外壳增厚的趋势,特别是蛤类派生的旁支 Agriopleura 种比较明显。另一类直立生长的族群 Caprinulidae(图4-10底部)以及未知起源的 Himeraelites 形成了更大更厚的外壳。

4.3.2 古地理变化

特提斯范围内早别杜尔期台地灰岩的特点与台地外部 caprinids 种的丰富数量和多样性密切相关(特别是边缘高能量带),它们可能贡献了数量可观的粗碎屑碎片(图4-12;Burla 等,2008;Fenerci-Masse,2006)。在范围更广的台地内部,主要的生物层是蛤类而不是 requieniids 类(Fenerci-Masse,2006),而在阿拉伯地区,requieniids 受地域的局限,有可能是从 Glossomyophorus 衍生而来(Masse,1998;Skelton 和 Masse,2000)。

晚别杜尔期台地边缘一些蛤类族群已经演化出纵向的变化(图4-7)。特提斯中部南部的低纬度地区台地上,在台地边缘相上 caprinids 一般能够保持主导地位,特别是 Offneria 种,经常产生大量生物碎片,例如在 Shuaiba 组和 Gargano 台地边缘比较明显(Hughes,2000;Skelton,2004;Al-Ghamdi 和 Read,2010)。相比之下,伊比利亚略高纬度台地(图4-7)泥灰岩沉积越来越厚,突尼斯(图4-4M)caprinid 的发展被限制,polyconitid 和 requieniid 种大量发育。

例如 Galve 盆地,西部 Maestrat 盆地,东部伊比利亚(图4-4H)台地晚别杜尔期碳酸盐沉积包含了大量种类丰富的蛤类以及珊瑚(Bover-Arnal 等,2010)。Camarillas-Loma del Morron 台地和 Las Mingachas 继承性台地边缘的剖面显示了台地顶部从厚变薄的厘米级别的变化周期(Bover-Arnal 等,2009,2010)。两个不同的种群现在都能识别出来。第一,台地顶部相以 Toucasia carinata 蛤类为主(图4-13A),以及直立型蛤类特别是 P. hadriani(Skelton 等,2010)以及 Monopleurid(可能是 Mathesia;图4-13B)和一些珊瑚。在台地边缘灰岩中,通常紧密排列着 P. hadriani(图4-13C 和 D)以及丰富的分枝珊瑚,Chondrodonta(图4-13D)和 nerineid。

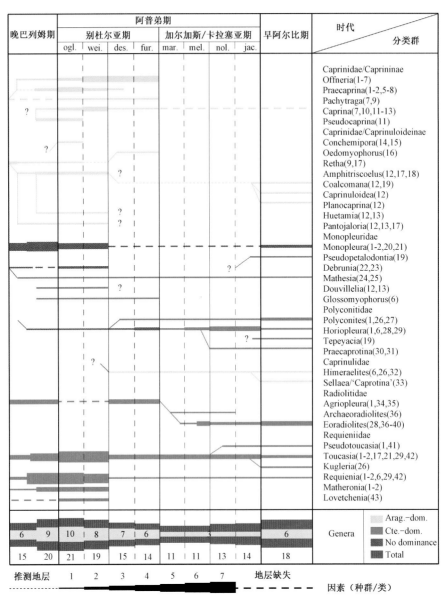

图4-10 从晚巴列姆期到早阿尔比期全球厚壳蛤类种群变化表,由SkeltonheSmith(2000)系统归纳,其引用的信息如下。缩写Arag-dom代表方解石外部介壳层减少,转变为文石质,缩写Cte-dom代表厚层的外部方解石介壳。各个分类群的参考文献如下:1. Masse(1976);2. Masse et al.(1999);3. Chartrousse&Masse(1998a);4. Chartrousse & Masse(1998c);5. Masse et al.(1998c);6. Skelton & Masse(2000);7. Masse(2003);8. Masse & Steuber(2007);9. Skelton & Masse(1998);10. Masse & Chartrousse(1997);11. Chartrousse&Masse(2004);12. Scott&Filkorn(2007);13. Scott&Hinote(2007);14. Chartrousse & Masse(1998b);15. Jenkyns&Strasser(1995);16. Skelton(2004);17. Pantoja-Aloret al.(2004);18. Alencaster & Pantoja-Alor(1996);19. Masse et al.(2007a);20. Masseet et al.(1998d);21. Coogan(1977);22. Masse & Fenerci-Masse(2009);23. Fenerci-Masse&Masse(2010a);24. Fenerci-Masse & Masse(2010b);25. Fenerci-Masseet et al. 2011);26. Masse et al.(1998e);27. Skelton et al.(2010);28. Pudsey et al.(1985);29. Masse et al.(1998a);30. Sano(1995);31. Iba & Sano(2007);32. Camoin(1983);33. Masse et al.(1997);34. Hughes(2000);35. Masse & Fenerci-Masse(2008);36. Fenerci-Masse et al.(2006);37. Masse & Gallo Maresca(1997);38. Steuber(1999);39. Masse et al.(2007b);40. Pons et al.(2010);41. Masse et al.(2010);42. Masse(1994);43. Masse(1992b)

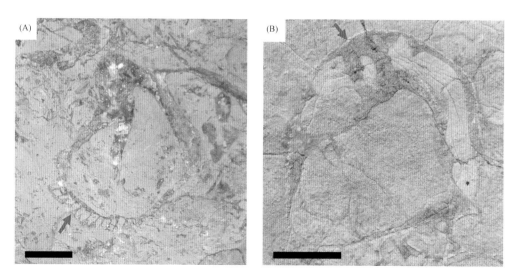

图4-11 Caprinid类的壳瓣如下照片:(A)左边的箭头所示是壳瓣外套膜的沟槽印迹,来自于上别杜尔阶Sarasrarri地层,在西班牙的北西部,Aralar山脉(位置见图4-4,G观察点),(B)右侧是壳瓣,箭头所示的位置是薄层的外部方解石介壳层,来自于上别杜尔阶Pinares地层,在西班牙东部西Maesrtat盆地(位置见图4-4,H观察点),比例尺都是10mm的横条

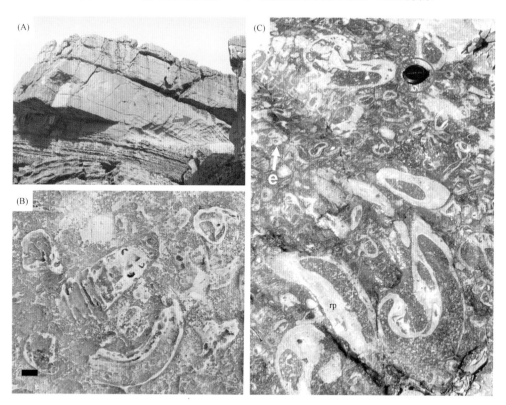

图4-12 台地外别杜尔亚阶富Caprinid类沉积,(A)和(B)照片所示是Crismina组下Ponta Alta段,位于Forte da Crismina(见图4-4,E观察点),(A)箭头所示的是3m厚的斜坡滨岸沉积,(B)是(A)图中箭头所示的滨岸沉积顶部放大照片,从上俯看,卷曲的caprinids类和大量nerineid gasrtopod类壳瓣散乱分布,比例尺是10mm,(C)上部表层是Comburindio组,在墨西哥南西部的Huetamo,显示了内源的caprinid类,包括了小的簇状直立壳瓣,如图"e"字母所示,其下是平卧生长的pantojaloia类,如字母"rp"所示(PantojaAlor等,1994),镜头盖的直径是55mm

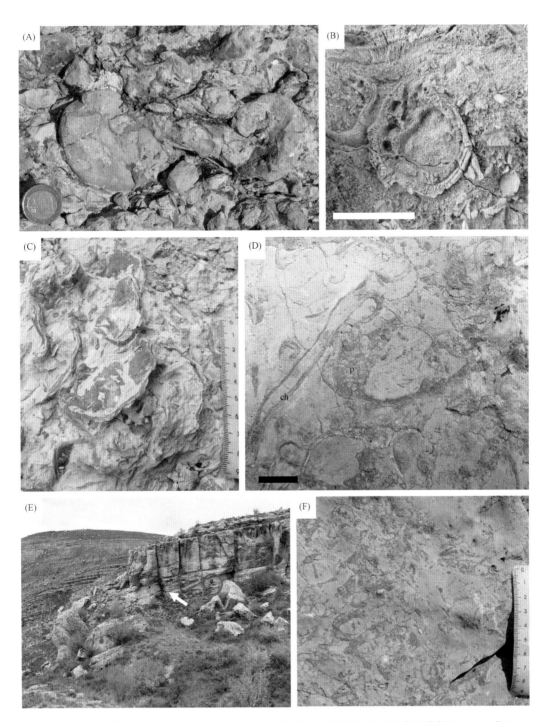

图4-13 上别杜尔亚阶Villarroya de los pinares组厚壳蛤类种群,位于西班牙东部Maestart盆地西部(见图4-4,观察点H,Bover-Arnal,2010),(A)和(B)是俯看台地顶部的种群,(A)是Toucasia种群(欧元硬币作为参考比例尺,直径为25mm),(B)是Monopleurid或是Mathesia类的种群(比例尺为10mm),(C)到(F)是以Polyconitid为主的台地边缘相,(C)是紧密排列的Polyconites类盖在斜坡沉积边缘,位于Las Mingachas,比例尺在其右边,(D)是浮蛋白石,"ch"代表chondrodonta类,"p"代表Polyconites类(比例尺为10mm),(E)是"Las Mingachas"台地边缘,大量的台地边缘灰岩进积在斜坡灰岩上(箭头所示是地质学家靠近的地方,环状是比例尺,高1.8m,(F)caprinid类浮蛋白石在台地中隔离出来,靠近(D)

Caprina parvula 和 Offneria 作为 requieniids 和少数 caprinids 的附属也有出现。Las Mingachas 台地上含有丰富蛤类和珊瑚的台地灰岩横向过渡为含更多泥质的台地边缘斜坡相（图 4-13E），那里 Polyconites 特别丰富，含有一些板状珊瑚，而且有大量的生物碎屑。这些台地斜坡相的沉积有含有菊石、有孔虫和不规则的小海胆的泥质石灰岩。这些区域 Caprinid 灰岩仍然存在，但局限于一些与碳酸盐岩相关的孤立的台地内（图 4-11B 和 4-13F）。

西班牙西北部晚别杜尔期 Sara-starri 组蛤类生物群也有明显的类似现象（图 4-4G；Millan 等，2009 年），类似于 Bilbao 附近的 Galdames 组。在这两种情况下 caprinid 生物碎屑的数量与发育程度与 requieniids 和 polyconitids 等主导台地剖面的情况相比明显受到限制。Masseet 等（1998 年）在西班牙东南部（图 4-4Ⅰ）再次记录到 Urgonian 组 requieniid 相的主导地位，尽管他们也认为还含有丰富的 Pachytraga 和 Offneria。目前这些比较仅仅是定性的，建立在现有研究者的观察以及文献资料的基础上。进一步的工作，显然需要对这些量化标准来测试比较。

别杜尔阶晚期 caprinids 种的崩溃和阿普弟期 polyconitids 种以及放射性蛤类广泛的增殖（图 4-10 和图 4-14A 到 D），伴随着 requieniids 继续存在以及蛤类外壳的方解石/文石的比例改变，这一改变造成应该关注的外壳结构的改变（Steuber，2002）。然而，无论这种变化是对海洋化学直接改变的回应，或者干脆是由其他原因推动了分类学上的变化仍然是一个有争议的问题，下个部分将讨论这个问题。值得注意的大趋势中的一个例外是含丰富文石的 Himera-elites 种蛤类的适度扩展，尤其是在特提斯台地中部。

4.4 台地发展史与气候数据的综合讨论

根据在泥岩的光谱伽马射线数据基础上的风化情况分析早别杜尔期台地发展的顶峰（图 4-4）正值北特提斯台地西部边缘处于温暖和半干旱的环境条件（Ruffell 和 Worden，2000）。这种情况下显然有利于干净的碳酸盐台地以及含有丰富 caprinid 种的台地边缘相的广泛发展（图 4-12）。氧同位素曲线数据（Steuber 等，2005；Ando 等，2008）和生物分子代黏物（Dumitrescu 等，2006）指示了全球缺氧事件之前的海面极端变暖的现象。Mehay 等（2009）和 Tejada 等（2009）认为这种变暖的趋势是由于大量火山活动增加了大气中 CO_2 的含量的温室效应的影响，根据前者的估计，能够解释全球缺氧事件之前的 C_3 明显的负漂移（图 4-2）。

用来解释这些扰动和全球碳循环以及中别杜尔阶台地发育中断的解说已经提出来了，特别是考虑了全球缺氧事件间隔期内的温度变化趋势。Follmi（2008）引用本人 1994 年和 Weissert 等（1998）的原始模型"作为水体中磷酸的来源，化学风化作用的加强使气候条件向更温暖和潮湿的方向转变"，这种转变驱使台地富营养化，与全岩碳酸盐分析中 C_{13} 记录同步。相比之下，Weissert 和 Erba（2004），Ando 等（2008）和 Bover-Arnal 等（2010）把预示全球缺氧事件的碳同位素 C_3 的负漂移（图 4-2）和温度最高值联系起来，认识到同位素正漂移涉及到有机碳的埋存，有可能涉及到大气中二氧化碳的减少造成气温下降。后者的解释同样被 Karl Follmi 和 Stein 等（2011）证实，他们认为 Sell 层位本身记录了一次变冷的阶段。Keller 等（2010）通过分析特提斯中部孢粉报告说最热气候带是紧随碳同位素负漂移之后，在 $200 \times 10^4 a$ 之后开始变冷。Keller 等（2011）的氧同位素曲线仍然显示出在全球缺氧事件间隔期内的积极变化趋势（Menegatti 等，1998 年的多数 C_4 和 C_5 碳同位素段）。另一个不相互排斥的对生物钙化作用的解释是假设在全球缺氧事件期碳酸酸化的海水引起的大气效果（Weissert 和 Erba，2004；Burla 等，2008；Mehay 等，2009）。作为对中别杜尔期台地危机的解释，这种假设比水体表面富营养

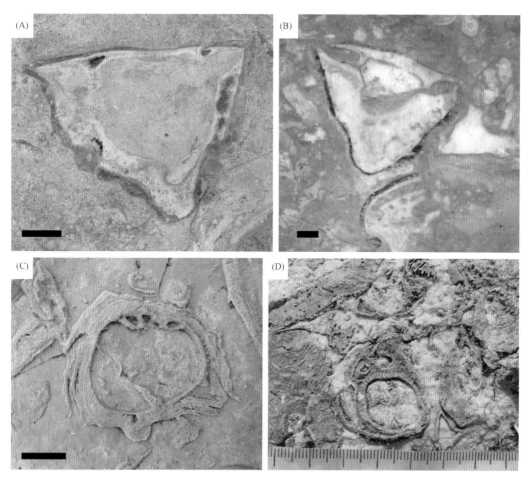

图4-14 典型的晚阿普弟期厚壳蛤类，(A)和(B)是polyconitids类种群，(A)是Polyconites类，产自西班牙东部Benassal组的"La Venta段"(见图4-4,观察点H)，(B)是Horiopleura类，产自Oman的Wadi al Assyi的Hassanat组(见图4-4,观察点O)，(C)是Radiolitid,Eorradiolites类，照片显示了厚层的外部方解石壳瓣，产自意大利Apennines南部Cava Uria地区(比例尺是10mm)，(D)是大量文石质caprinulid类，下部中间是Himeraelites类，并有radiolitid类的碎屑，产自意大利Apennines南部Lago di Matese西端阿普弟阶和阿尔比阶的Machietelle段(见图4-4,观察点J)，毫米比例尺在下部

化的优点在于能够在现代的浅海环境看到富营养化的程度。通过实验室得知CO_2浓度增加导致海水pH值的降低造成了海洋贝类生物生长速率降低和外壳的钙化程度减弱(Doney等，2009；Iesias-Rodri-guez等，2008)，特别是具有特殊意义的双壳蛤类(Berge等；2006；Lannig等，2010)。

在这一点上再次强调Erba(1994)的观察："成岩尾椎类多样性的减少在地磁变化前被记录下来并在中别杜尔阶危机内短期达到最强烈，但明确在代表全球缺氧事件的黑色页岩沉积之前"(图4-2)。Cismon和Erba等的太平洋深海钻探计划网站最新的曲线(图4-1；2010)的确显示尾椎类界限从上别杜尔阶开始直到下阿普弟阶前Selli段C_{13}负漂移出现前达到危机的程度。这种模式对上述提到的台地的地层记录有有趣的比较作用，并且从下别杜尔阶到中别杜尔阶都有发育(图4-2，图4-4)。那么为何尾椎类的危机出现在中别杜尔阶台地危机之前？在此背景下，值得高度关注早阿普弟期特提斯或大西洋台地矛盾蔓延的程度，单独估计

以每年 $0.07 \times 10^{12} kgC$ 的速度进行埋存,相当于现今全球的29%(Skelton,2003),这是全球大气中二氧化碳埋存在海水并造成海水酸化的证据。Skelton(2003)把这个明显的物理化学上的复杂矛盾称为釜效应(图4-15)。

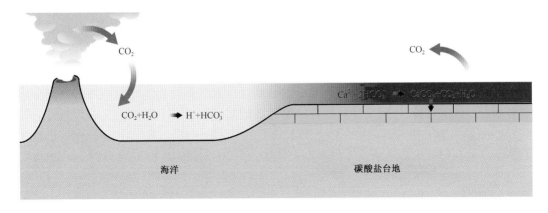

图4-15　如图示的是"釜效应"过程,由于受温暖气候影响 CO_2 逸出,碳酸盐台地上的水蒸发并覆盖其上,尽管 CO_2 在空气中的浓度很高(红色部分),相反,在开阔较冷的海洋环境, CO_2 酸化并溶解其中

"水覆辽阔的浅水台地和陆架在极端温室阶段可能已经经历强烈的热量和蒸发,它的温度和盐度有可能与周边的开放水体相关联……通过钙化来驱替 CO_2(图4-15),这样的条件下可以避免相反的情况,受到大气中过高的 CO_2 含量的影响。"由于缺少阿普弟阶海水表面古温度数据特别是开放水体浅海台地的数据,一个现实的类比模型有助于支持这一假设。Kauffman 和 Thomson(2005)对西部巴拿马加勒比海沿岸的浅水潟湖进行了五年的纵向研究,他们发现"离岸海水表面的温度模式类似于我们监测的三个潟湖,近岸水域往往有更高的高点,普遍较低的低点(Kauffman 和 Thomson,2005)导致离岸和近岸水域有 0.2~0.8℃ 的长期差距"。尽管只有 20~30km,Bahh1a Almirante 是对中白垩纪时期巨大台地和陆架的缩微模型,更大的温度和盐度偏差有可能已经发生,后者的确能够作为台地内部蒸发的实例(Lehmann 等,2000)。同样,法国东南部 Orgon(图4-3C 附近)上巴列姆阶台地内的沉积预示了早阿普弟期温度最大值的氧同位素古温度记录,暗示季节性的 16~34.6℃ 的温度变化(Steuber 和 Rauch,2005)。

因此,在早别杜尔期温室效应时期相对较冷的海水中(图4-15左)可以推测尾椎类在台地受到影响之前首当其冲的承受海水酸化的影响(Heldt 等,2008;Erba 等,2010)。这种情况下可能的说法就是釜效应作用使得中低纬度地区变暖,使中部南部特提斯台地免受酸化的影响。然而台地仍然受到不仅仅是极端温度,还有其他因素的影响,而不是 Lithocodium 或 Bacinella 或类似微生物的繁殖记录(Graziano,2003;Rameil 等,2010)。

同时,如上所述,特提斯北部很多 Urgonian 型台地终止于 weissi 或 deshayesi 界面附近的不规则的侵蚀面。在某些情况下,侵蚀不整合与露头和构造的因素有关(Renard 等,2009)。然而 Follmi 和 Gainon (2008)在上 Schrattenkalk 段与 Grunten 段(图4-4B)的接触面的详细讨论中表示台地表明是石灰岩但却没有明确提及露头的证据。这些案例中值得进一步讨论的一个问题是以后有没有发生酸化使水下台地解体。无论哪种方式,在 Follmi 等(1994)和其他人的富营养化模型中,Urgonian 型台地终止后上覆地层的营养条件有可能与其原始的灭亡无关。

这需要一个独特的解释,尽管有明显的动荡,然而晚别杜尔期(图4-7)高纬度地区 Urgonian 型台地在 C_3 的负尖峰之后仍没有能够恢复,这是由于有机碳的广泛埋存(Danelian 等,2004)增进了大气中 CO_2 含量的下降导致的温度下降造成的(Dumitrescu 等,2006;图 4-2)。纬度限制的台地恢复模式,以及在台地边缘由 Caprinids 种主导的模式更受到空间的限制(图 4-7),是对气候限制的暗示,涉及整个伊比利亚半岛气温变冷温带已经向南移动的的过程(Burla 等,2008,2009;Bover-Arnal 等,2010)。即便削减大气中的 CO_2,减弱了釜效应的效果,那么在较冷的水体中抑制生物碳酸盐的生长也是个问题,尤其是涉及了较易溶解的文石的时候。因为海水的温度与其 PH 值的关系比与 CO_2 浓度关系弱,文石的溶解度对这两个因素是很敏感的(Cao 等,2007;图 4-1)。Flushing 台地边缘以及陆架受周围相对较冷的开阔水体影响能够加强原住的种群,特别是针对高古纬度地区的 Caprinids 类(见前述的蛤类剖面记录)。这些 Caprinids 可能觉得自己在形态学上的亲和力在错误的时间出现在错误的地点,虽然在当时有利于方解石形成的海水条件下(Steuber 和 Rauch,2005;Masse 和 Fenerci-Masse,2008),大型文石外壳只能增加他们的痛苦。随后的海侵,无论是海平面的升降还是构造上的变化,都让早已负重不堪的台地困扰不已。

晚别杜尔期特提斯北部台地边缘 Urgonian 型台地受到抑制的富营养化模型的替代模型与上述提到的解释并不互相矛盾。然而除了不适于对这时期全球台地发展做解释外,另外在这部分的问题是关于蛤类及其相关生物群的营养生态环境过于依赖了不受支持的假设。在 Follmi、Jenkyns 和 Wilson 等人的叙述中除了一些特殊种类没有找到,蛤类与 photosymbiotic 共生的证据来应用他们提出的贫营养区这个概念,确切地说大多数蛤类是悬浮物摄食者(Gili 等,1995;Scott,1995)。此外,甚至一些通常居住在斜坡环境附近的片状珊瑚如 microsolenids 类显示出最适于在浑浊的水环境下捕食的形态特征(Rosen 等,2002)。就像 Masse 和 Fenerci-Masse(2008)通过分析蛤类万花筒般多样的分布特点得到的启发性的解释是蛤类生物多样性的重要证据。因此像上述那样引用不明物种的蛤类物种是帮倒忙的简单化。这并不是说,没有蛤类是容易富营养化,相反,更具体的记录蛤类的族群生活地点能够用来知道古地理环境标志的变化,这已经被 Masse 和 Fenerci-Masse 解决了(2008)。

从特提斯北部的西边到大西洋区域,晚杜尔期转换到了更潮湿的气候条件(Ruffell 和 Worden,2000),从陆地径流输入了更多的泥质碎屑,使干旱带向南撤退(Burla 等,2008,2009)。如前所述 Maestrat 和其他伊比利亚各地丰富的灰质沉积伴随台地边缘相对较低的能量带对新的 polyconitid 种出现以及向外扩展有利。这类的不对称的锥形底瓣和近似水平的上瓣使这些个体能像牡蛎那样紧密的排列(图 4-13C;Skelton 等,2010)。polyconitid 坚硬的外壳和壳之间的空间为后来者提供了生存的空间。这些 polyconitid 在台地边缘和斜坡上的确找到了充足的食物供给如浮游植物等。这些 polyconitids 和 requieniids 与 caprinids 相比既有相对增厚的钙质外壳,又能忍受特提斯北部更冷的水体。

通过上述气候趋势的顶点和 Steuber 等(2005)的氧同位素古温度曲线(图 4-2),加尔加斯加的大规模灭亡和阿普弟期蛤类经历的古环境变化与 Cenomanian 的规模的确能够相比(Skelton,2003)。新大陆蛤类记录仍然仅限于阿普弟阶最上部(Scott 和 Filkorn,2007),caprinids 类从已知的地层记录中消失,而不是旧大陆的阿布尔阶上部。这些变化符合 Kemper 建立在德国西北部浅色富含碳酸盐岩和暗色富含泥质沉积基础上对阿普弟中期到晚期变冷阶段的假设,他把这种变化归因于海进条件下高位域温暖干燥的环境以及低位域潮湿凉爽的气候环境。对极地冷却的支持来自于 Arctic 盆地(Kemper,1987)和 Eromananga 盆地,澳大利亚(De Lurio 和 Frakes,1999)的滴石和鹦鹉螺化石,以及相关的大幅度的层序地层学意义上的海平面升降(如 Bover-Arnal 等,2009,2010)。相反,Heimhofer 等(2008)报告了上阿普弟阶 Algarve 剖面泥灰岩的风化情况和植被的稳定性,并暗示与中低古纬度气候变化没有直接的关系。

然而,图 4-2 和图 4-4 以及前面几节的讨论显示,晚阿普弟期推测的古温度和台地建造都在此达到一个新的高峰,并伴随着蛤类多样化再次出现(图 4-10),预示着旧大陆晚阿尔比期至赛诺曼期之后新大陆阿尔比阶台地发展的黄金时代(图 4-1)。

4.5 结论

(1)早别杜尔期(最早的阿普弟期)整个大西洋、特提斯洋、太平洋低古纬度带经历了碳酸盐台地发展的极盛期,并且伴随着蛤类高产的生物多样性,特别是占据台地边缘环境的横卧的文石为主的 caprinids 类。

(2)中别杜尔期(早阿普弟中期)全球碳循环的波动以及全球缺氧事件 1 期影响了全球碳酸盐岩台地相的发育。含丰富蛤类的台地在特提斯北部边缘以及新大陆的低古纬度地区 Lithocodium 或 Bacinella 的界面微生物大量繁殖的地方终止。

(3)晚别杜尔期(早阿普弟晚期)台地的恢复仅限于低古纬度地区,并且 caprinids 类台地相在特提斯中南部地区再次盛行。同时,伊比利亚海角周围灰质剖面有增厚钙质外壳的蛤类(主要为 polyconitids 和 requieniids)很大程度上取代了主要为文石质的 caprinids。

(4)特提斯区域台地重新增长发生在 Gargasian 或 Clansayesian(晚阿普弟期),caprinids 类从全球已知的地层记录中消失到一些未知地方避难直到新大陆晚阿普弟期以及旧大陆阿布尔阶才出现。新大陆蛤类直到晚阿普弟期仍然数量稀少,晚阿普弟期台地发育加快,超越特提斯北缘向日本东部的部分延伸。

(5)晚别杜尔期纬度对台地恢复的限制以及直到晚阿普弟期蛤类的稀缺,都与同位素曲线和全球缺氧事件 1 期从晚别杜尔期到中阿普弟期的气候变冷的沉积证据相对应。

(6)这些变化的解释对比大气的因素强调海水酸化早阿普弟期中期台地以及随后台地恢复的地域限制的影响。步骤 2 归因于火山喷发造成大气中 CO_2 浓度上升,以及一些低纬度地区由于温室效应造成的热量使水中的 CO_2 气化(这被称为釜效应)。另外,步骤 3 高纬度地区台地周围水体随后的变冷是由于大气持续提供 CO_2 使水体继续酸化减弱了釜效应。

(7)对这些因素与其他族群比敏感的多的富含文石的 caprinids 类,由于它们的矿物学成分以及占据了台地外边缘的生活环境,首先受到了影响。从整个阿普弟阶蛤类化石记录中观察到方解石或文石比率的大幅度增加是由于生态环境的变化。

参考文献

Alencaster, G. and Pantoja-Alor, J. (1996) The rudist Amphitriscoelus (Bivalvia-Hippuritacea) in the Lower Cretaceous of southwestern Mexico. J. Paleontol. ,70,399-407.

Al-Ghamdi, N. and Read, F. J. (2010) Facies-based sequencestratigraphic framework of the Lower Cretaceous rudist platform, Shu' aiba formation, Saudi Arabia. In: Barremian-Aptian Stratigraphy and Hydrocarbon Habitat of the Eastern Arabian Plate (Eds F. S. P. van Buchem, M. I. Al-Husseini, F. Maurer and H. J. Droste), Vol. 2, pp. 367-410. Gulf PetroLink, Bahrain.

Ando, A., Kaiho, K., Kawahata, H. and Kakegawa, T. (2008) Timing and magnitude of early Aptian extreme warming: unraveling primary $\delta^{18}O$ variation in indurated pelagic carbonates at deep sea drilling project site 463, central pacific ocean. Palaeogeogr. Palaeoclimatol. Palaeoecol. ,260,463-476.

Arnaud-Vanneau, A. , Arnaud, H. , Cotillon, P. , Ferry, S. and Masse, J. -P. (1982) Caractères et Évolution des Plates-formes Carbonatées Peéivocontiennes au Creéacé Inférieur (France Sud-Est). Cretaceous Res. ,3,3 – 18.

Arthur, M. A. , Jenkyns, H. C. , Brumsack, H. -J. and Schlanger, S. O. (1990) Stratigraphy, geochemistry, and paleoceanography of organic carbon-rich Cretaceous sequences. In: Cretaceous Resources, Events and Rhythms – Background and Plans for Reserach (Eds R. N. Ginsburg and B. Beaudoin), NATO ASI Series C: Mathematical and Physical Sciences, 304, 75 – 119, Kluwer Academic Publishers.

Bachmann, M. and Hirsch, F. (2006) Lower Cretaceous carbonate platform of the eastern Levant (Galilee and the Golan Heights): stratigraphy and second-order sea-level change. Cretaceous Res. ,27, 487 – 512.

Baron-Szabo, R. and Steuber, T. (1996) Korallen und Rudisten aus dem Apt in tertiären Flysch des Parnass-Gebirges bei Delphi-Arachowa (Mittelgriechenland). Berl. Geowiss. Abh. ,E 18, 3 – 75.

Barragan, R. and Maurrasse, F. J. -M. R. (2008) Lower Aptian (Lower Cretaceous) ammonites from the basal strata of the La Peña Formation of Nuevo León State, northeast Mexico: biochronostratigraphic implications. Rev. Mex. Ciencias Geológicas, 25, 145 – 157.

Beerling, D. J. , Lomas, M. R. and Gröcke, D. R. (2002) On the nature of methane gas hydrate dissociation during the Toarcian and Aptian oceanic anoxic events. Am. J. Sci. ,302, 28 – 49.

Berge, J. A. , Bjerkeng, B. , Pettersen, O. , Schaaning, M. T. and Ønevad, S. (2006) Effects of increased sea water concentrations of CO_2 on growth of the bivalve Mytilus edulis L. Chemosphere, 62, 681 – 687.

Bosellini, A. , Russo, A. and Schroeder, R. (1999) Stratigraphic evidence for an early Aptian sea – level fluctuation: the Graua limestone of south-eastern ethiopia. Cretaceous Res. ,20, 783 – 791.

Bourrouilh, R. (2000) CHAPTER 23, Mud-mounds on divergent extensional and transform margins: devonian and Cretaceous examples from southern France. In: Geological Exploration in Murzuq Basin (Eds M. A. Sola and D. Worsley), pp. 463 – 483. Elsevier Science B. V, Amsterdam, Netherlands.

Bover-Arnal, T. , Salas, R. , Moreno-Bedmar, J. A. and Bitzer, K. (2009) Sequence stratigraphy and architecture of a late Early-Middle Aptian carbonate platform succession from the western Maestrat Basin (Iberian Chain, Spain). Sed. Geol. ,219, 280 – 301.

Bover-Arnal, T. , Moreno-Bedmar, J. A. , Salas, R. , Skelton, P. W. , Bitzer, K. and Gili, E. (2010) Sedimentary evolution of an Aptian syn-rift carbonate system (Maestrat Basin, ESpain): effects of accommodation and environmental change. Geol. Acta, 8, 249 – 280.

Bover-Arnal, T. , Salas, R. , Martín-Closas, C. , Schlagintweit, F. and Moreno-Bedmar, J. A. (2011) Expression of an oceanic anoxic event in a neritic setting: lower Aptian coral rubble deposits from the western Maestrat Basin (IberianChain, Spain). Palaios, 26, 18 – 32.

Bralower, T. J. , CoBabe, E. , Clement, B. , Sliter, W. V. , Osburn, C. L. and Longoria, J. (1999) The record of global change in mid-Cretaceous (Barremian-Albian) sections from the Sierra Madre, northeastern Mexico. J. Foramin. Res. ,29, 418 – 437.

Burla, S. , Heimhofer, U. , Hochuli, P. A. , Weissert, H. and Skelton, P. W. (2008) Changes in sedimentary patterns of coastal and deep-sea successions from the North Atlantic (Portugal) linked to early Cretaceous environmental change. Palaeogeogr. Palaeoclimatol. Palaeoecol. ,257, 38 – 57.

Burla, S. , Oberli, F. , Heimhofer, U. , Wiechert, U. and Weissert, H. (2009) Improved time control on Cretaceous coastal deposits: new results from Sr isotope measurements using laser ablation. Terra Nova, 21, 401 – 409.

Camoin, G. (1983) Plates-formes carbonates et récifs à rudistes du Crétacé de Sicile. Thèse doctorale, Université de Provence. Travaux du Laboratoire de Gélogie historique et de Paléontologie, no. 13, Université de Provence, Marseille, pp. 244 .

Cao, L. , Caldeira, K. and Jain, A. K. (2007) Effects of carbon dioxide and climate change on ocean acidification and carbonate mineral saturation. Geophys. Res. Lett. ,34, L05607. doi:10. 1029/2006GL028605.

Carannante, G. , Cherchi, A. and Simone, L. (1995) Chlorozoan versus foramol lithofacies in Upper Cretaceous rudist limestones. Palaeogeogr. Palaeoclimatol. Palaeoecol. ,119,137 – 154.

Carannante,G. ,Cherchi,A. ,Graziano,R. ,Ruberti,D. and Simone,L. ,(2008) Post-Turonian rudist-bearing limestones of the Peri-Tethyan region: evolution of the sedimentary patterns and lithofacies in the context of global versus regional controls. In: Controls on Carbonate Platform and Reef Development (Eds. J. Lukasik and J. A. Simo), SEPM Spec. Publ. ,89,255 – 270.

Carannante,G. ,Pugliese,A. ,Ruberti,D. ,Simone,L. ,Vigliotti,M. and Vigorito,M. (2009) Evoluzione cretacica di un settore della piattaforma apula da dati di sottosuolo e di affioramento (Appennino campano-molisano). Boll. Soc. Geol. It. ,128,3 – 31.

Castro,J. M. ,Company,M. ,de Gea,G. A. and Aguado,R. (2001) Biostratigraphy of the Aptian-Middle Cenomanian platform to basin domain in the prebetic zone of Alicante,SE Spain: calibration between shallow water benthonic and pelagic scales. Cretaceous Res. ,22,145 – 156.

Castro,J. M. ,de Gea,G. A. ,Ruiz-Ortiz,P. A. and Nieto,L. M. (2008) Development of carbonate platforms on an extensional (rifted) margin: the Valanginian-Albian record of the Prebetic of Alicante (SE Spain). Cretaceous Res. , 29,848 – 860.

Chartrousse, A. and Masse,J. -P. (1998a) Offneria simplex nov. Sp. (rudiste,Caprinidae) du Barrémien du Sud-Est de la France et de Cuba. Implications sur la biostratigraphie et l'évolution du genre offneria. Bull. Soc. Géol. Fr. , 169,841 – 850.

Chartrousse, A. and Masse,J. -P. (1998b) Coalcomaninae (Rudistes,Caprinidae) nouveaux de l'Aptien inférieur des mid pacific mountains. Géobios,mémoire spécial,22,87 – 92.

Chartrousse,A. and Masse,J. -P. (1998c) Offneria arabica nov. Sp. (rudiste,Caprinidae) de l'Aptien inférieur du Jebel Madar (Sultanat d'Oman). Cretaceous Res. ,19,827 – 841.

Chartrousse,A. and Masse,J. -P. (2004) Revision of the early Aptian Caprininae (Rudist Bivalves) of the new world. evolutionary and palaeobiogeographic implications. Cour. Forsch. -Inst. Senckenberg,247,19 – 34.

Coogan,A. H. (1977) Early and middle Cretaceous Hippuritacea (rudists) of the Gulf coast. In: Cretaceous carbonates of Texas and Mexico (Eds. D. G. Bebout and R. G Loucks) Report of Investigation,89,32 – 70,University of Texas at Austin,Bureau of Economic Geology.

Danelian,T. ,Tsikos,H. ,Gardin,S. ,Baudin,F. ,Bellier,J. -P. and Emmanuel,L. (2004) Global and regional palaeoceanographic changes as recorded in the mid-Cretaceous (Aptian – Albian) sequence of the Ionian zone (NW Greece). J. Geol. Soc. London,161,703 – 709.

D'Argenio,B. ,Ferreri,V. and Ruberti,D. (1992) Cicli,ciclotemi e tempestiti nei depositi carbonatici aptiani del Matese (Appennino Campano). Soc. Geol. Ital. Mem. ,41 (for 1988),761 – 773.

De Lurio,J. L. and Frakes,L. A. (1999) Glendonites as a paleoenvironmental tool: implications for early Cretaceous high latitude climates in Australia. Geochim. Cosmochim. Acta,63,1039 – 1048.

Dercourt,J. ,Ricou,L. E. and Vrielynck,B. (1993,eds.) Atlas Tethys,palaeoenvironmental maps. Gauthier – Villars, Paris,France,307 pp,14 maps,1 pl.

Dercourt,J. ,Gaetani,M. ,Vrielynck,B. ,Barrier,E. ,Biju-Duval,B. ,Brunet,M. F. ,Cadet,J. P. ,Crasquin,S. and Sandulescu,M. (2000,eds.) Peri-Tethys Palaeogeographical Atlas,Commission for the Geological Map of the World (CCGM/CGMW),Paris,France;24 maps and explanatory notes: I – XX,1 – 269.

Dinis,J. L. ,Rey,J. ,Cunha,P. P. ,Callapez,P. and Pena dos Reis,R. (2008) Stratigraphy and allogenic controls of the western Portugal Cretaceous: an updated synthesis. Cretaceous Res. ,29,772 – 780.

Doney, S. C. , Fabry, V. J. , Feely, R. A. and Kleypas, J. A. (2009) Ocean acidification: the other problem. Annu. Rev. Mar. Sci. ,1,169 – 192.

Dumitrescu,M. ,Brassell,S. C. ,Schouten,S. ,Hopmans,E. C. and Sinninghe Damsté,J. S. (2006) Instability in tropical Pacific sea-surface temperatures during the early Aptian. Geology,34,833 – 836.

Erba, E. (1994) Calcareous nannofossils and Mesozoic oceanic anoxic events, Mar. Micropaleontol., 52, 85 – 106.

Erba, E.. (and 23 others) (1996) The Aptian stage. Bull. Inst. Roy. Sci. Nat. Belg., Sci. Terre, 66 Supp., 31 – 43.

Erba, E., Channell, J. E. T., Claps, M., Jones, C., Larson, R., Opdyke, B., Premoli Silva, I., Riva, A., Salvini, G. and Torricelli, S. (1999) Integrated stratigraphy of the Cismon Apticore (southern Alps, Italy): a "reference section" for the Barremian-Aptian interval at low latitudes. J. Foramin. Res., 29, 371 – 391.

Erba, E., Bottini, C., Weissert, H. J. and Keller, C. E. (2010) Calcareous nannoplankton response to surface-water acidification around oceanic anoxic event 1a. Science, 329, 428 – 432.

Fenerci-Masse, M. (2006) Les communautés à rudistes du Créacé inférieur de la marge ouest Européenne de la Téthys. Thèse doctorale, Université de Provence (Aix-Marseille I), Centre de Sédimentologie et Paléontologie, Marseille, France, pp. 436.

Fenerci-Masse, M. and Masse, J. -P. (2010a) Debrunia occitanicanov. sp. (Monopleuridae) from the early Aptian of SE France. Turk. J. Earth Sciences, 19, 573 – 581.

Fenerci-Masse, M. and Masse, J. -P. (2010b) Mathesia Mainelli (Hippuritoidea, Monopleuridae) from the late Aptian – Albian of the Mediterranean region: a revision. Turk. J. Earth Sciences, 19, 543 – 556.

Fenerci-Masse, M., Masse, J. -P., Arias, C. and Vilas, L. (2006) Archaeoradiolites, a new genus from the Upper Aptian of the Mediterranean region and the origin of the rudist family Radiolitidae. Palaeontology, 49, 769 – 794.

Fenerci-Masse, M., Masse, J. -P., Kołodziej, B., Ivanov, M. and Idakieva, V. (2011) Mathesia darderi (Astre) (Bivalvia, Hippuritoidea, Monopleuridae): morphological, biogeographical and ecological changes in the Mediterranean domain during the late Barremian-Albian. Cretaceous Res., 32, 407 – 421. doi: 10.1016/j.cretres.2010.10.005

Fiet, N., Quidelleur, X., Parize, O., Bulot, L. G. and Gillot, P. Y. (2006) Lower Cretaceous stage durations combining radiometric data and orbital chronology: towards a more stable relative time scale? Earth Planet. Sci. Lett., 246, 407 – 417.

Föllmi, K. (2008) A synchronous, middle Early Aptian age for the demise of the Helvetic Urgonian platform related to the unfolding oceanic anoxic event 1a ("Selli event"). Comment on the article 《ur la présence de grands foraminifères d'âge aptien supérieur dans l'Urgonien de la Nappe du Wildhorn (Suisse centrale). Note préliminaire》 by R. Schroeder, K. Schenk, A. Cherchi & B. Schwizer, Revue de Paléobiologie, 2007, 665 – 669. Rev. Paléoobiol., Genève, 27, 461 – 468.

Föllmi, K. B. and Gainon, F. (2008) Demise of the northern Tethyan Urgonian carbonate platform and subsequent transition towards pelagic conditions: the sedimentary record of the Col de la Plaine Morte area, central Switzerland. Sed. Geol., 205, 142 – 159.

Föllmi, K. B., Weissert, H., Bisping, M. and Funk, H. (1994) Phosphogenesis, carbon-isotope stratigraphy, and carbonate-platform evolution along the Lower Cretaceous northern Tethyan margin. Geol. Soc. Am. Bull., 106, 729 – 746.

Föllmi, K., Bodin, S., Godet, A., Linder, P. and van de Schootbrugge, B. (2007) Unlocking paleo-environmental information from Early Cretaceous shelf sediments in the Helvetic Alps: stratigraphy is the key. Swiss J. Geosci., 100, 359 – 369.

García-Mondéjar, J., Owen, H. G., Raisossadat, N., Millán, M. I. and Fernández-Mendiola, P. A. (2009) The Early Aptian of Aralar (northern Spain): stratigraphy, sedimentology, ammonite biozonation, and OAE1. Cretaceous Res., 30, 434 – 464.

Gili, E., Masse, J. -P. and Skelton, P. W. (1995) Rudists as gregarious sediment-dwellers, not reef-builders, on Cretaceous carbonate platforms. Palaeogeogr. Palaeoclimatol. Palaeoecol., 118, 245 – 267.

Gradstein, F., Ogg, J. and Smith, A. (2004, eds.) A Geologic Time Scale 2004. Cambridge University Press, Cambridge, UK, pp. 589.

Granier, B. R. C. (2008) Holostratigraphy of the Kahmah regional Series in Oman, Qatar, and the United Arab Emirates. Carnets de Géologie/Notebooks on Geology, Brest, 33 p, Online article 2008/07 (CG2008_A07). http://paleopolis. rediris. es/cg/CG2008_A07/(accessed on 13 October, 2011).

Graziano, R. (1999) The Early Cretaceous drowning unconformities of the Apulia carbonate platform (Gargano Promontory, southern Italy): local fingerprints of global palaeoceanographic events. Terra Nova, 11, 245 – 250.

Graziano, R. (2000) The Aptian – Albian of the Apulia Carbonate Platform (Gargano Promontory, southern Italy): evidence of palaeoceanographic and tectonic controls on the stratigraphic architecture of the platform margin. Cretaceous Res. , 21, 107 – 126.

Graziano, R. (2003) The Early Cretaceous drownings of Tethyan carbonate platforms: controlling mechanisms and palaeoceanography. Insights from the Apulia record. In: Temperate-type (foramol facies) carbonate platforms versus tropical-type (chlorozoan facies) carbonate platforms: tridimensional arrangement of lithofacies, benthic associations and evolution of the related depositional systems (Eds. L. Simone, D. Ruberti and R. Graziano), Proceedings of the COFIN 2000 Workshop Italia – Napoli – Pozzuoli, 25 – 27 February 2003, pp. 55 – 62, De Frede, Napoli.

Graziano, R. and Ruggiero Taddei, E. (2008) Cretaceous brachiopod-rich facies of the carbonate platform-to-basin transitions in southern Italy: stratigraphic and paleoenvironmental significance. Boll. Soc. Geol. It. , 127, 407 – 422.

Hallock, P. and Schlager, W. (1986) Nutrient excess and the demise of coral reefs and carbonate platforms. Palaios, 1, 389 – 398.

Heimhofer, U. , Adatte, T. , Hochuli, P. A. , Burla, S. and Weissert, H. (2008) Coastal sediments from the Algarve: low-latitude climate archive for the Aptian-Albian. Geol. Rundsch. , 97, 785 – 797.

Heldt, M. , Bachmann, M. and Lehmann, J. (2008) Microfacies, biostratigraphy, and geochemistry of the hemipelagic Barremian – Aptian in north-central Tunisia: influence of the OAE 1a on the southern Tethys margin. Palaeogeogr. Palaeoclimatol. Palaeoecol. , 261, 246 – 260.

Heldt, M. , Lehmann, J. , Bachmann, M. , Negra, H. and Kuss, J. (2010) Increased terrigenous influx but no drowning: palaeoenvironmental evolution of the Tunisian carbonate platform margin during the Late Aptian. Sedimentology, 57, 695 – 719.

Herrle, J. O. , Kössler, P. and Bollmann, J. (2010) Palaeoceanographic differences of early Late Aptian black shale events in the Vocontian Basin (SE France). Palaeogeogr. Palaeoclimatol. Palaeoecol. , 297, 367 – 376.

Hillgärtner, H. (2010) Anatomy of a microbially constructed, high-energy, ocean-facing carbonate platform margin (earliest Aptian, northern Oman Mountains). In: Barremian-Aptian stratigraphy and hydrocarbon habitat of the eastern Arabian Plate (Eds F. S. P. van Buchem, M. I. Al-Husseini, F. Maurer and H. J. Droste), Vol. 1, pp. 285 – 300. Gulf Petro-Link, Bahrain.

Huck, S. , Rameil, N. , Korbar, T. , Heimhofer, U. , Wieczorek, T. D. and Immenhauser, A. (2010) Latitudinally different responses of tethyan shoal-water carbonate systems to the early Aptian oceanic anoxic event (OAE 1a). Sedimentology, 57, 1585 – 1614.

Hughes, G. W. (2000) Bioecostratigraphy of the Shu'aiba formation of Saudi Arabia. GeoArabia, 5, 545 – 578.

Hughes, G. W. (2004) Palaeoenvironments of selected Lower Aptian rudists from Saudi Arabia. Cour. Forsch. - Inst. Senckenberg, 247, 233 – 245.

Hughes, G. W. (2005) Micropalaeontological dissection of the Shu'aiba Reservoir, Saudi Arabia. In: Recent Developments in Applied Biostratigraphy (Eds A. Powell and J. B. Riding), Geological Society of London, for the British Micropalaeontological Society, Bath, UK, 69 – 90.

Husinec, A. and Jelaska, V. (2006) Relative sea-level changes recorded on an isolated carbonate platform: tithonian to Cenomanian succession, southern Croatia. J. Sed. Res. , 76, 1120 – 1136.

Iba, Y. and Sano, S. (2007) Mid-Cretaceous step-wise demise of the carbonate platform biota in the Northwest Pacific and establishment of the North Pacific biotic province. Palaeogeogr. Palaeoclimatol. Palaeoecol. , 245, 462 – 482.

Iglesias-Rodriguez, M. D. , Halloran, P. R. , Rickaby, R. E. M. , Hall, I. R. , Colmenero-Hidalgo, E. , Gittins, J. R. ,

Green, D. R. H., Tyrrell, T., Gibbs, S. J., von Dassow, P., Rehm, E., Armbrust, E. V. and Boessenkool, K. P. (2008) Phytoplankton calcification in a high CO_2 world. Science, 320, 336 – 340.

Immenhauser, A., Hillgärtner, H. and van Bentum, E. (2005) Microbial-foraminiferal episodes in the Early Aptian of the southern Tethyan margin: ecological significance and possible relation to oceanic anoxic event 1a. Sedimentology, 52, 77 – 99.

Jahren, A. H., Arens, N. C., Sarmiento, G., Guerrero, J. and Amundson, R. (2001) Terrestrial record of methane hydrate dissociation in the early Cretaceous. Geology, 29, 159 – 162.

Jenkyns, H. C. (2003) Evidence for rapid climate change in the Mesozoic – Palaeogene greenhouse world. Phil. Trans. Roy. Soc. London A, 361, 1885 – 1916.

Jenkyns, H. C. (2010) Geochemistry of oceanic anoxic events. Geochemistry Geophysics Geosystems, 11, Q03004. doi: 10. 1029/2009GC002788.

Jenkyns, H. and Strasser, A. (1995) Lower Cretaceous oolites from the Mid-Pacific Mountains (Resolution Goyot, Site 866). Proc. ODP Sci. Results, 143, 111 – 118.

Jenkyns, H. and Wilson, P. A. (1999) Stratigraphy, paleoceanography, and evolution of Cretaceous Pacific guyots: relics from a greenhouse earth. Am. J. Sci., 299, 341 – 392.

Kauffman, K. W. and Thomson, R. C. (2005) Water temperature variation and the meteorological and hydrographic environment of Bocas del Toro, Panama. Carib. J. Sci., 41, 392 – 413.

Keller, C. E., Hochuli, P. A., Weissert, H., Bernasconi, S. M., Giorgioni, M. and Garcia, T. I. (2011) A volcanically induced climate warming and floral change preceded the onset of OAE1a (early Cretaceous). Palaeogeogr. Palaeoclimatol. Palaeoecol., 305, 43 – 49. doi: 10. 1016/j. palaeo. 2011. 02. 011

Kemper, E. (1987) Das Klima der Kreide-Zeit. Geol. Jb., 96A, 5 – 185.

Kemper, E. (1995) The causes of the carbonate and colour changes in the Aptian of NW Germany. Neues Jb. Geol. Paläontol. Abh., 196, 275 – 289.

Kühnt, W. and Moullade, M. (2009) The Gargasian (Middle Aptian) of La Marcouline section at Cassis-La Bédoule (SE France): stable isotope record and orbital cyclicity. Ann. Mus. Hist. Nat. Nice, 24, 251 – 262.

Kuhnt, W., Moullade, M., Masse, J. -P. and Erlankeuser, H. (1998) Carbon isotope stratigraphy of the lower Aptian historical stratotype at Cassis-La Bédoule (SE France). Géol. Mediterr., 25, 63 – 79.

Lannig, G., Eilers, S., Pörtner, H. O., Sokolova, I. M. and Bock, C. (2010) Impact of ocean acidification on energy metabolism of oyster, Crassostrea gigas – changes in metabolic pathways and thermal response. Mar. Drugs, 8, 2318 – 2339.

Larson, R. L. (1991) Geological consequences of superplumes. Geology, 19, 933 – 966.

Lehmann, C., Osleger, D. A. and Montañez, I. P. (1998) Controls on cyclostratigraphy of lower Cretaceous carbonates and evaporates, Cupido and Coahuila platforms, northeastern Mexico. J. Sed. Res., 68, 1109 – 1130.

Lehmann, C., Osleger, D. A. and Montañez, I. P. (2000) Sequence stratigraphy of lower Cretaceous (Barremian-Albian) carbonate platforms of northeastern Mexico: regional and global correlations. J. Sed. Res., 70, 373 – 391.

Lehmann, J., Heldt, M., Bachmann, M. and Negra, M. El. H. (2009) Aptian (lower Cretaceous) biostratigraphy and cephalopods from north central Tunisia. Cretaceous Res., 30, 895 – 910.

Linder, P., Gigandet, J., Hüsser, J. -L., Gainon, F. and Föllmi, K. B. (2006) The early Aptian Grünten member: description of a new lithostratigraphic unit of the helvetic Garschella formation. Eclogae Geol. Helv., 99, 327 – 341.

Luciani, V., Cobianchi, M. and Lupi, C. (2006) Regional record of a global oceanic anoxic event: OAE1a on the Apulia platform margin, Gargano Promontory, southern Italy. Cretaceous Res., 27, 754 – 772.

Luperto Sinni, E. and Masse, J. -P. (1986) Données nouvelles sur la stratigraphie des calcaires de plate-forme du Crétacé inférieur du Gargano (Italie méridionale). Riv. Ital. Paleontol. Stratigr., 92, 33 – 66.

Luperto Sinni, E. and Masse, J. -P. (1993) Biostratigrafia dell' Aptiano in facies di piattaforma carbonatica delle

Murge baresi (Puglia – Italia meridionale). Riv. Ital. Paleontol. Stratigr. ,98,403 – 424.

Marcoux, J., Girardeau, A., Fourcade, E., Bassoulet, J. -P., Philip, J., Jaffrezo, M., Xuchang, X. and Chengfa, C. (1987) Geology and biostratigraphy of the jurassic and lower cretaceous series to the north of the Lhasa block (Tibet, China). Geodin. Acta,1,313 – 325.

Marzouk, L. and Ben Youssef, M. (2008) Relative sea-level changes of the lower Cretaceous deposits in the chotts area of Southern Tunisia. Turk. J. Earth Sci. ,17,835 – 845.

Masse, J. -P. (1976) Les calcaires urgoniens de Provence Valanginien-Aptien inférieur. Stratigraphie, Paléontologie, les Paléoenvironnements et leur évolution. Thèse, Docteur ès Sciences. Université d'Aix-Marseille II, U. E. R. des Sciences de la mer et de l'environnement, Marseille, France, pp. 511,60 pls.

Masse, J. -P. (1979) Les rudistes (Hippuritacea) du Crétacé inférieur. Approche paléoécologique. Geobios, Mém. Spéc. ,3,277 – 287.

Masse, J. -P. (1984) Données nouvelles sur la stratigraphie de l'Aptien carbonaté de Tunisie centrale, conséquences paléogéographiques. Bull. Soc. Géol. Fr. ,(sér. 7) 26,1077 – 1086.

Masse, J. -P. (1989) Relations entre modifications biologiques et phénomènes géologiques sur les plates-formes carbonatées du domaine périméditerranéen au passage Bédoulien-Gargasien. Geobios, Mém. Spéc. ,11,279 – 294.

Masse, J. -P. (1992a) Les Rudistes de l'Aptien Inférieur d'Italie continentale: aspects systématiques, stratigraphiques et paléobiogeographiques. Geol. Romana,28,243 – 260.

Masse, J. -P. (1992b) Systématique, stratigraphie et paléobiogeographie du genre Lovetchenia (Requieniidae) du Crétacé inférieur méditerranéen. Géobios,26,699 – 708.

Masse, J. -P. (1994) L'évolution des Requieniidae (Rudistes) du Crétacé inférieur: caractères, signification fonctionnelle adaptative et relations avec les modifications des paléoenvironnements. Géobios,27,321 – 333.

Masse, J. -P. (2003) Integrated stratigraphy of the Lower Aptian and applications to carbonate platforms: a state of the art. In: North African Cretaceous carbonate platform systems (Eds E. Gili, H. Negra and P. W. Skelton), NATO Science Series, IV. Earth and Environmental Sciences,28,215 – 227, Kluwer Academic Publishers.

Masse, J. -P. and Chartrousse, A. (1997) LesCaprina (rudistes) de l'Aptien inférieur d'Europe occidentale: systématique, biostratigraphie et paléobiogéographie. Geobios,30,797 – 809.

Masse, J. -P. and Fenerci-Masse, M. (2008) Time contrasting palaeobiogeographies among Hauterivian – lower Aptian rudist bivalves from the Mediterranean Tethys, their climatic control and palaeoecological implications. Palaeogeogr. Palaeoclimatol. Palaeoecol. ,269,54 – 65.

Masse, J. -P. and Fenerci-Masse, M. (2009) Debrunia, a new Barremian genus of petalodontid Monopleuridae (Bivalvia, Hippuritoidea) from the Mediterranean region. Palaeontology,269,54 – 65.

Masse, J. -P. and Fenerci-Masse, M. (2011) Drowning discontinuities and stratigraphic correlation in platform carbonates. The late Barremian-early Aptian record of southeast France. Cretaceous Res. 32,659 – 684. doi: 10. 1016/ j. cretres. 2011. 04. 003

Masse, J. -P. and Gallo Maresca, M. (1997) Late Aptian Radiolitidae (rudist bivalves) from the Mediterranean and Southwest Asiatic regions: taxonomic, biostratigraphic and palaeobiogeographic aspects. Palaeogeogr. Palaeoclimatol. Palaeoecol. ,128,101 – 110.

Masse, J. -P. and Philip, J. (1981) Cretaceous coral-rudist buildups of France. In: European fossil reef models (Ed. D. F. Toomey). SEPM Spec. Publ. ,30,399 – 426.

Masse, J. -P. and Philip, J. (1986) L'évolution des rudistes au regard des principaux évènements géologiques du Crétacé. Bull. Centres Rech. Explor. -Prod. Elf-Aquitaine,10,437 – 445.

Masse, J. -P. and Rossi, T. (1987) Le provincialisme sud-caraibe à l'Aptien inférieur. Sa signification dans le cadre de l'évolution géodynamique du domaine Caraibe et de l'Atlantique central. Cretaceous Res. ,8,349 – 363.

Masse, J. -P. and Shiba, M. (2010) Praecaprotina kashimae nov. sp. (Bivalvia, Hippuritacea) from the Daiichi-Kashima seamount (Japan Trench). Cretaceous Res. ,31,147 – 153.

Masse,J. -P. and Steuber,T. (2007) Strontium isotope stratigraphy of early Cretaceous rudist bivalves. In: Cretaceous Rudists and Carbonate Platforms: Environmental Feedback (Ed. R. W. Scott),SEPM Spec. Publ. ,87,159 – 165.

Masse,J. -P. ,Gallo Maresca,M. and Luperto-Sinni,E. (1993) Aptian rudists from Lago di Matese (Central Italy) and their stratigraphic and palaeoenvironmental framework. In: Third International Conference on Rudists,Proceedings (Eds G. Alencaster and B. E. Buitrón), pp. 42. Universidad Nacional Autónoma de México, Instituto de Geología,Mexico,DF.

Masse,J. -P. ,Borgomano,J. and Al Maskiry,S. (1997) Stratigraphy and tectonosedimentary evolution of a late Aptian-Albian carbonate margin: the northeastern Jebel Akhdar (Sultanate of Oman). Sed. Geol. ,113,269 – 280.

Masse,J. -P. , Arias, C. and Vilas, L. (1998a) Lower Cretaceous rudist faunas of southeast Spain: an overview. Géobios,mémoire spécial,22,193 – 210.

Masse,J. -P. ,Borgomano,J. and Al Maskiry,S. (1998b) A platform-to-basin transition for lower Aptian carbonates (Shuaiba Formation) of the northeastern Jebel Akhdar (Sultanate of Oman). Sed. Geol. ,119,297 – 309.

Masse,J. -P. ,Chartrousse,A. and Borgomano,J. (1998c) The Lower Cretaceous (Upper Barremian-Lower Aptian) caprinid rudists from northern Oman. Géobios,mémoire spécial,22,211 – 223.

Masse,J. -P. , Gourrat, A. , Orbette, D. and Schmuck, D. (1998d) Hauterivian rudist faunas of southern Jura (France). Géobios,mémoire spécial,22,225 – 233.

Masse,J. -P. ,Gallo Maresca,M. and Luperto-Sinni,E. (1998e) Albian rudist faunas from southern Italy: taxonomic, biostratigraphic and palaeobiogeographic aspects. Géobios,31,47 – 59.

Masse,J. -P. ,El Albani,A. and Erlenkeuser,H. (1999) Stratigraphie isotopique (δ^{13}C) de l' Aptien inférieur de Provence (S. E. de la France). Applications aux corrélations plate-forme/bassin. Eclogae Geol. Helv. , 92, 259 – 263.

Masse,J. -P. ,Bouaziz,S. , Amon, E. O. Baraboshkin, E. , Tarkowski, R. , Bergerat, F. , Sandulescu, M. , Platel, J. P. , Canerot,J. ,Guiraud,R. ,Poisson,A. ,Ziegler,M. and Rimmele,G. (2000) Early Aptian. (112114 Ma) In: Peri-Tethys Palaeogeographical Atlas (Eds J. Dercourt, M. Gaetani, B. Vrielynck, E. Barrier, B. Biju-Duval, M. F. Brunet,J. P. Cadet,S. Crasquin and M. Sandulescu),CCGM/CGMW,Paris,France,Map 13.

Masse,J. -P. ,Fenerci-Masse, M. and Özer,S. (2002) Late Aptian rudist faunas from the Zonguldak region,western Black Sea,Turkey (taxonomy, biostratigraphy, palaeoenvironment and palaeobiogeography). Cretaceous Res. ,23, 523 – 536.

Masse,J. -P. ,Fenerci-Masse, M. ,Korbar,T. and Velic,I. (2004a) Lower Aptian Rudist Faunas (Bivalvia, Hippuritoidea)from Croatia. Geol. Croatica,57,117 – 137.

Masse,J. -P. ,Özer,S. and Fenerci,M. (2004b) Upper Barremian-lower Aptian rudist faunas from the western Black sea region (Turkey). Cour. Forsch. -Inst. Senckenberg,247,75 – 88.

Masse,J. -P. ,Beltramo,J. ,Martinez-Reyes,J. and Arnaud-Vanneau, A. (2007a) Revision of Albian polyconitid and monopleurid rudist bivalves from the New World. In: Cretaceous Rudists and Carbonate Platforms: Environmental Feedback (Ed. R. W. Scott),SEPM Spec. Publ. ,87,221 – 230.

Masse,J. -P. ,Fenerci-Masse, M. , Vilas, L. and Arias, C. (2007b) Late Aptian-Albian primitive Radiolitidae (bivalves,hippuritoidea) from Spain and SW France. Cretaceous Res. ,28,697 – 718.

Masse,J. -P. ,Tüysüz,O. ,Fenerci-Masse, M. ,Özer,S. and Sari,B. (2009) Stratigraphic organisation,spatial distribution,palaeoenvironmental reconstruction,and demise of Lower Cretaceous (Barremian-lower Aptian) carbonate platforms of the Western Pontides (Black Sea region,Turkey). Cretaceous Res. ,30,1170 – 1180.

Masse,J. -P. ,Fenerci-Masse,M. ,Işintek,I. and Güngör,T. (2010) Albian Rudist Fauna from the Karaburun Peninsula,izmir Region,Western Turkey. Turk. J. Earth Sciences,19,671 – 683.

Méhay,S. ,Keller,C. E. ,Bernasconi,S. M. ,Weissert,H. ,Erba,E. ,Botín,C. and Hochuli,P. A. (2009) A volcanic

CO_2 pulse triggered the Cretaceous Oceanic Anoxic Event 1a and a biocalcification crisis. Geology, 37, 819 – 822.

Menegatti, A. P., Weissert, H., Brown, R. S., Tyson, R. V. and Farrimond, P. (1998) High-resolution $\delta^{13}C$ stratigraphy through the early Aptian "Livello Selli" of the Alpine Tethys. Paleoceanography, 13, 530 – 545.

Millán, I. (2009) Palaeoceanographic changes record during the Early Aptian of Aralar (N. Spain). Tesis Doctoral, pp. 157. Universidad del País Vasco/Euskal Herriko Unibertsitatea, Leioa.

Millán, M. I., Weissert, H. J., Fernández-Mendiola, P. A. and García-Mondéjar, J. (2009) Impact of Early Aptian carbon cycle perturbations on evolution of a marine shelf system in the Basque-Cantabrian Basin (Aralar, N Spain). Earth Planet. Sci. Lett., 287, 392 – 401.

Montenat, C., Moullade, M. and Philip, J. (1982) Le Crétacé inférieur à Orbitolines et Rudistes d'Afghanistan central. Géol. Méditerr., 9, 109 – 122.

Moreno-Bedmar, J. A., Company, M., Bover-Arnal, T., Salas, R., Delanoy, G., Martínez, R. and Grauges, A. (2009) Biostratigraphic characterization by means of ammonoids of the lower Aptian Oceanic Anoxic Event (OAE 1a) in the eastern Iberian Chain (Maestrat Basin, eastern Spain). Cretaceous Res. 30, 864 – 872.

Moreno-Bedmar, J. A., Company, M., Bover-Arnal, T., Salas, R., Delanoy, G., Maurrasse, F. J.-M. R., Grauges, A. and Martínez, R. (2010) Lower Aptian ammonite biostratigraphy in the Maestrat Basin (Eastern Iberian Chain, Eastern Spain). A Tethyan transgressive record enhanced by synrift subsidence. Geologica Acta, 8, 281 – 299.

Moullade, M., Masse, J.-P., Tronchetti, G., Kuhnt, W., Ropolo, P., Bergen, J. A., Masure, E. and Renard, M. (1998) Le stratotype historique de l'Aptien inférieur (région de Cassis-La Bédoule, SE France): synthèse stratigraphique. Géol. Mediterr., 25, 289 – 298.

Pantoja-Alor, J., Schroeder, R., Cherchi, A., Alencaster, G. and Pons, J. M. (1994) Fossil assemblages, mainly foraminifers and rudists, from the Early Aptian of southwestern México. Palaeobiogeographical consequences for the Caribbean region. Rev. Esp. Paleontología, 9, 211 – 219.

Pantoja-Alor, J., Skelton, P. W. and Masse, J.-P. (2004) Barremian rudists of the San Lucas Formation around San Lucas, Michoacan, SW Mexico. Cour. Forsch. -Inst. Senckenberg, 247, 1 – 17.

Pastouret, L., Masse, J.-P., Philip, J. and Auffret, G.-A. (1974) Sur la présence d'Aptien inférieur à faciès urgonien sur la marge continentale armoricaine. Conséquences paléogéographiques. CR Acad. Sci. Paris, D., 278, 203 – 207.

Philip, J. and Airaud-Crumière, C. (1991) The demise of the rudist bearing carbonate platforms at the Cenomanian/Turonian boundary: a global control. Coral Reefs, 10, 115 – 125.

Philip, J., Masse, J.-P. and Camoin, G. (1995) Tethyan carbonate platforms. In: The Ocean Basins and Margins, Volume 8: The Tethys Ocean (Eds A. E. M. Nairn et al.), pp. 239 – 265. Plenum Press, New York.

Pittet, B., van Buchem, F. S. P., Hillgärtner, H., Razin, P., Grötsch, J. and Droste, H. (2002) Ecological succession, palaeoenvironmental change, and depositional sequences of Barremian-Aptian shallow-water carbonates in northern Oman. Sedimentology, 49, 555 – 581.

Pons, J. M., Vicens, E., Chikhi-Aouimeur, F. and Abdallah, H. (2010) Albian Eoradiolites (Bivalvia: Radiolitidae) from Jabal Naïmia, Gafsa region, Tunisia, with revisional studies on the Albian forms of the genus. J. Paleontol., 84, 321 – 331.

Pudsey, C., Schroeder, R. and Skelton, P. W. (1985) Cretaceous (Aptian/Albian) age for island-arc volcanics, Kohistan, N. Pakistan. In: Geology of Western Himalayas (Eds V. J. Gupta et al.), Contributions to Himalayan Geology, 3, 150. 168. Hindustan Publishing Corp., India.

Rameil, N., Immenhauser, A., Warrlich, G., Hillgärtner, H. and Droste, H. J. (2010) Morphological patterns of Aptian Lithocodium – Bacinella geobodies: relation to environment and scale. Sedimentology, 57, 883 – 911.

Reboulet, S. Rawson, P. F., Moreno-Bedmar, J. A., Aguirre-Urreta, M. B., Barragán, R., Bogomolov, Y., Company, M., González-Arreola, C., Idakieva Stoyanova, V., Lukeneder, A., Matrion, B., Mitta, V., Randrianaly, H., Vašíček, Z., Baraboshkin, E. J., Bert, D., Bersac, S., Bogdanova, T. N., Bulot, L. G., Latil, J.-L., Mikhailova,

I. A., Ropolo, P. and Szives, O. (2011) Report on the 4th International Meeting of the IUGS Lower Cretaceous Ammonite Working Group, the "Kilian Group" (Dijon, France, 30th August 2010). Cretaceous Res., 32, 786 – 793, doi:10.1016/j.cretres.2011.05.007

Reboulet, S. and Hoedemaeker, P. (reporters) and 19 others (2006) Report on the 2nd International Meeting of the IUGS Lower Cretaceous Ammonite Working Group, the "Kilian Group" (Neuchâtel, Switzerland, 8 September 2005). Cretaceous Res., 27, 712 – 715.

Renard, M., Raféli s, M. de., Emmanuel, L., Moullade, M., Masse, J.-P., Kühnt, W., Bergen, J. A. and Tronchetti, G. (2005) Early Aptian δ^{13}C and manganese anomalies from the historical Cassis-La Bédoule stratotype sections (S. E. France): relationship with a methane hydrate dissociation event and stratigraphic implications. Carnets de Géologie/Notebooks on Geology, Brest. Online article 2005/04 (CG2005_A04), 18 p, http://paleopolis.rediris.es/cg/CG2005_A04/index.html. (accessed on 13 October, 2011).

Renard, M., Raféli s, M. de., Emmanuel, L., Moullade, M., Masse, J.-P., Kühnt, W., Bergen, J. A. and Tronchetti, G. (2009) Early Aptian δ^{13}C and manganese anomalies from the historical Cassis-La Bédoule stratotype sections (S. E. France): relationship with a methane hydrate dissociation event and stratigraphic implications. Ann. Mus. Hist. Nat. Nice, 24, 199 – 220. [Published version of Renard et al. (2005)]

Rey, J. (1979) Les formations bioconstruites du Crétacé inférieur d'Estremadura (Portugal). Géobios, Mémoire spécial, 3, 89 – 99.

Rey, J. (2006) Stratigraphie séquentielle et séquences de dépôt dans le Crétacé inférieur du Bassin Lusitanien. Ciências da Terra, Volume Especial, 6, Departamento de Ciências da Terra, Faculdade de Ciências e Tecnologia, pp. 120 Universidade Nova de Lisboa Lisbon, Portugal.

Roberts, D. G., Montadert, L., Thompson, R. W., Auffret, G. A., Lumsden, D. N., Kagami, H., Timofeev, P. P., Muller, C., Bock, W. D., DuPeuble, P. A., Schnitker, D., Hailwood, E. A., Harrison, W. and Thompson, T. L. (1980) Geological Setting and Principal Results of Drilling on the Margins of the Bay of Biscay and Rockall Plateau during Leg 48. Phil. Trans. Roy. Soc. London, A, 294, 65 – 75.

Rojas, R., Iturralde-Vinent, M. and Skelton, P. W. (1996) Stratigraphy, composition and age of Cuban rudist-bearing deposits. Rev. Mex. de Ciencias Geológicas, 12 (for 1995), 272 – 291.

Ropolo, P., Moullade, M., Conte, G. and Tronchetti, G. (2008) About the stratigraphic position of the Lower Aptian Roloboceras hambrovi (Ammonoidea) level. Carnets de Géologie/Notebooks on Geology, Brest, Letter 2008/03 (CG2008_L03), 7 p, http://paleopolis.rediris.es/cg/CG2008_M03/ (accessed on 20 October, 2011).

Rosen, B. R., Aillud, G. S., Bosellini, F. R., Clack, N. J., Insalaco, E., Valldeperas, F. X. and Wilson, M. E. J. (2002) Platy coral assemblages: 200 million years of functional stability in response to the limiting effects of light and turbidity. In: Proceedings of the Ninth International Coral Reef Symposium, Bali, Indonesia. (Eds. M. K. Moosa, S. Soemodihardjo, A. Soegiarto, K. Rominmohtarto, A. Nontji, S. Soekarno and L. Suharsono). Ministry of Environment, Indonesian Institute of Sciences and International Society for Reef Studies, 1, 255 – 264.

Ross, D. J. and Skelton, P. W. (1993) Rudist formations of the Cretaceous: a palaeoecological, sedimentological and stratigraphic review. In: Sedimentology review (Ed. V. P. Wright), Sedimentology review, 1, 73 – 91. Blackwell Scientific Publications, Oxford.

Ruffell, A. and Worden, R. (2000) Palaeoclimate analysis using spectral gamma-ray data from the Aptian (Cretaceous) of southern England and southern France. Palaeogeogr. Palaeoclimatol. Palaeoecol., 155, 265 – 283.

Sano, S. (1995) Lithofacies and biofacies of Early Cretaceous rudist-bearing carbonate sediments in northeastern Japan. Sed. Geol., 99, 179 – 189.

Sartorio, D., Tunis, G. and Venturini, S. (1997) The Iudrio Valley section and the evolution of the northeastern margin of the Friuli Platform (Julian Prealps, NE Italy-W Slovenia). Memorie di Scienze Geologiche (Padova), 49, 163 – 193.

Schlager, W. (1981) The paradox of drowned reefs and carbonate platforms. Geol. Soc. Am. Bull. ,92,197 – 211.

Schlagintweit, F. ,Bover-Arnal, T. and Salas, R. (2010) Erratum to: new insights into Lithocodium aggregatum Elliott 1956 and Bacinella irregularis Radoičić 1959 (Late Jurassic-Lower Cretaceous): two ulvophycean green algae (? Order Ulotrichales) with a heteromorphic life cycle (epilithic/euendolithic). Facies 56,635 – 673.

Schlanger, S. O. and Jenkyns, H. C. (1976) Cretaceous oceanic anoxic events: causes and consequences. Geol. Mijnbouw,55,179 – 184.

Scholle, P. and Arthur, M. A. (1980) Carbon isotope fluctuations in Cretaceous pelagic limestones: potential stratigraphic and petroleum exploration tool. AAPG Bull. 64,67 – 87.

Scott, R. W. (1995) Global environmental controls on Cretaceous reefal ecosystems. Palaeogeogr. Palaeoclimatol. Palaeoecol. ,119,187 – 199.

Scott, R. W. and Filkorn, H. F. (2007) Barremian-Albian rudist zones, U. S. Gulf coast In: Cretaceous Rudists and Carbonate Platforms: Environmental Feedback (Ed. R. W. Scott),SEPM Spec. Publ. ,87,167 – 180.

Scott, R. W. and Hinote, R. E. (2007) Barremian-Early Aptian rudists, Slgo Formation, Texas, U. S. A In: Cretaceous Rudists and Carbonate Platforms: Environmental Feedback (Ed. R. W. Scott),SEPM Spec. Publ. ,87,237 – 246.

Simo, J. A. T. , Scott, R. W. and Masse, J. -P. (1993) Cretaceous carbonate platforms: an overview. In: Cretaceous carbonate platforms (Eds J. A. T. Simo, R. W. Scott and J. -P. Masse), AAPG Mem. ,56,1 – 14.

Skelton, P. W. (1985) Preadaptation and evolutionary innovation in rudist bivalves. Spec. Pap. Palaeont. , 33, 159 – 173.

Skelton, P. W. (1991) Morphogenetic versus environmental cues for adaptive radiations. In: Constructional Morphology and Evolution (Eds N. Schmidt-Kittler and K. Vogel),pp. 375 – 388. Springer Verlag, Berlin.

Skelton, P. W. (2000) Rudists and carbonate platforms – growing together, dying together. In: Crisi biologiche, radiazioni adattative e dinamica delle piattaforme carbonatiche (Eds A. Cherchi and C. Corradini), Accad. Naz. Sci. Lett. Arti di Modena, Collana di Studi,21,231 – 235.

Skelton, P. W. (2003a) ,ed. The Cretaceous World. Cambridge University Press and The Open University, Cambridge, UK, pp. 360.

Skelton, P. W. (2003b) Rudist evolution and extinction – a North African perspective. In: North African Cretaceous carbonate platform systems (Eds E. Gili, H. Negra and P. W. Skelton), NATO Science Series, IV. Earth and Environmental Sciences,28,215 – 227, Kluwer Academic Publishers.

Skelton, P. W. (2004) Oedomyophorus shaybahensis, a new genus and species of caprinid (?) rudist from the Lower Aptian Shu'aiba Formation of eastern Saudi Arabia. Cour. Forsch. -Inst. Senckenberg,247,35 – 47.

Skelton, P. W. and Masse, J. -P. (1998) Revision of the Lower Cretaceous Rudist Genera Pachytraga Paquier and Retha Cox (Bivalvia: Hippuritacea),and the origins of the Caprinidae. Géobios, mémoire spécial,22,331 – 370.

Skelton, P. W. and Masse, J. -P. (2000) Synoptic guide to the Lower Cretaceous rudist bivalves of Arabia. SEPM Spec. Publ. ,69,85 – 95.

Skelton, P. W. and Smith, A. B. (2000) A preliminary phylogeny for rudist bivalves: sifting clades from grades. In: The evolutionary biology of the Bivalvia (Eds. E. M. Harper, J. D. Taylor and J. A. Crame), Geol. Soc. London Spec. Publ. ,177,97 – 127.

Skelton, P. W. and Steuber, T. (1999) Aptian rudists from Arachowa near Delphi) Parnassis Mts.), central Greece. In: Fifth International Congress on Rudists, Abstracts and Field Trip Guides (Eds R. Höfling and T. Steuber), Erlanger Geol. Abh. ,3,67 – 68. Institut für Geologie der Universität Erlangen-Nürnberg, Erlangen.

Skelton, P. W. ,Gili, E. ,Bover-Arnal, T. , Salas, R. and Moreno-Bedmar, J. A. (2010) A new species of Polyconites from the Lower Aptian of Iberia and the early evolution of polyconitid rudists. Turk. J. Earth Sciences, 19, 557 – 572.

Stein, M. , Föllmi, K. B. , Westermann, S. , Godet, A. , Adatte, T. , Matera, V. , Fleitmann, D. and Berner, Z. (2011)

Progressive palaeoenvironmental change during the Late Barremian – Early Aptian as prelude to Oceanic Anoxic Event 1a: evidence from the Gorgo a Cerbara section (Umbria-Marche basin, central Italy). Palaeogeogr. Palaeoclimatol. Palaeoecol. ,302,396 – 406. doi:10.1016/j.palaeo.2011.01.025

Steuber, T. (1999) Cretaceous rudists of Boeotia, central Greece. Spec. Pap. Palaeont. ,61,229.

Steuber, T. (2002) Plate tectonic control on the evolution of Cretaceous platform carbonate production. Geology,30, 259 – 262.

Steuber, T. and Löser, H. (2000) Species richness and abundance patterns of Tethyan Cretaceous rudist bivalves (Mollusca: Hippuritacea) in the central-eastern Mediterranean and Middle East, analysed from a palaeontological database. Palaeogeogr. Palaeoclimatol. Palaeoecol. ,162,75 – 104.

Steuber, T. and Rauch, M. (2005) Evolution of the Mg/Ca ratio of Cretaceous seawater: implications from the composition of biological low-Mg calcite. Mar. Geol. ,217,199 – 213.

Steuber, T., Rauch, M., Masse, J.-P., Graaf, J. and Malkoc, M. (2005) Low latitude seasonality of Cretaceous temperatures in warm and cold episodes. Nature 437,1341 – 1344.

Strohmenger, C. J., Weber, L. J., Ghani, A., Al-Mehsin, K., Al-Jeelani, O., Al-Mansoori, A., Al-Dayyani, T., Vaughan, L., Khan, S. A. and Mitchell, J. C. (2006) High-resolution sequence stratigraphy and reservoir characterization of Upper Thamama (Lower Cretaceous) Reservoirs of a giant Abu Dhabi oil field, United Arab Emirates. In: Giant hydrocarbon reservoirs of the world: From rocks to reservoir characterization and modeling (Eds. P. M. Harris and L. J. Weber), AAPG Memoir 88/ 139 – 171.

Strohmenger, C. J., Steuber, T., Ghani, A., Barwick, D. G., Al-Mazrooei, S. H. A. and Al-Zaabi, N. O. (2010) Sedimentology and chemostratigraphy of the Hawar and Shu'aiba depositional sequences, Abu Dhabi, United Arab Emirates. In: Barremian-Aptian stratigraphy and hydrocarbon habitat of the eastern Arabian Plate (Eds F. S. P. van Buchem, M. I. Al-Husseini, F. Maurer and H. J. Droste), Vol. 2, pp. 341 – 365. Gulf PetroLink, Bahrain.

Takashima, R., Sano, S.-I., Iba, Y. and Nishi, H. (2007) The first Pacific record of the Late Aptian warming event. J. Geol. Soc. London,164,333 – 339.

Tejada, M. L. G., Katsuhiko, S., Kuroda, J., Coccioni, R., Mahoney, J. J., Ohkouchi, N., Sakamoto, T. and Tatsumi, Y. (2009) Ontong Java Plateau eruption as a trigger for the early Aptian oceanic anoxic event. Geology, 37, 855 – 858.

Tomás, S., Löser, H. and Salas, R. (2008) Low-light and nutrient-rich coral assemblages in an Upper Aptian carbonate platform of the southern Maestrat Basin (Iberian Chain, eastern Spain). Cretaceous Res. ,29,509 – 534.

Van Buchem, F. S. P., Al-Husseini, M. I., Maurer, F. and Droste, H. J. (Eds.) (2010a) Barremian-Aptian stratigraphy and hydrocarbon habitat of the eastern Arabian Plate. GeoArabia Special Publication,4,2 Vols, pp.614 Gulf PetroLink, Bahrain.

Van Buchem, F. S. P., Al-Husseini, M. I., Maurer, F., Droste, H. J. and Yose, L. A. (2010b) Sequence-stratigraphic synthesis of the Barremian-Aptian of the eastern Arabian Plate and implications for the petroleum habitat. In: Barremian-Aptian stratigraphy and hydrocarbon habitat of the eastern Arabian Plate (Eds F. S. P. van Buchem, M. I. Al-Husseini, F. Maurer and H. J. Droste), Vol. 1, pp. 9. 48. Gulf PetroLink, Bahrain.

Van Buchem, F. S. P., Baghbani, D., Bulot, L. G., Caron, M., Gaumet, F., Hosseini, A., Keyvani, F., Schroeder, R., Swennen, R., Vedrenne, V. and Vincent, B. (2010c) Barremian – Lower Albian sequence-stratigraphy of southwest Iran (Gadvan, Dariyan and Kazhdumi formations) and its comparison with Oman, Qatar and the United Arab Emirates. In: Barremian-Aptian stratigraphy and hydrocarbon habitat of the eastern Arabian Plate (Eds F. S. P. van Buchem, M. I. Al-Husseini, F. Maurer and H. J. Droste), Vol. 2, pp. 503 – 548. Gulf PetroLink, Bahrain.

Vilas, L., Masse, J.-P. and Arias, C. (1995) Orbitolina episodes in carbonate platform evolution: the early Aptian model from SE Spain. Palaeogeogr. Palaeoclimatol. Palaeoecol. ,119,35 – 45.

Vilas, L., Martín-Chivelet, J. and Arias, C. (2003) Integration of subsidence and sequence stratigraphic analyses in the Cretaceous carbonate platforms of the Prebetic (Jumilla-Yecla Region), Spain. Palaeogeogr. Palaeoclimatol. Palaeoecol. ,200,107 – 129.

Vincent, B. , Van Buchem, F. S. P. , Bulot, L. G. , Immenhauser, A. , Caron, M. , Baghbani, D. and Huc, A. Y. (2010) Carbonisotope stratigraphy, biostratigraphy and organic matter distribution in the Aptian – Lower Albian successions of southwest Iran (Dariyan and Kazhdumi formations). In: Barremian-Aptian stratigraphy and hydrocarbon habitat of the eastern Arabian Plate (Eds F. S. P. van Buchem, M. I. Al-Husseini, F. Maurer and H. J. Droste), Vol. 1, pp. 139 – 197. Gulf PetroLink, Bahrain.

Vlahović, I. , Tiš. ljar, J. , Velić, I. and Matičec, D. (2005) Evolution of the Adriatic Carbonate Platform: palaeogeography, main events and depositional dynamics. Palaeogeogr. Palaeoclimatol. Palaeoecol. ,220,333 – 360.

Weissert, H. and Erba, E. (2004) Volcanism, CO_2 and palaeoclimate: a Late Jurassic – Early Cretaceous carbon and oxygen isotope record. J. Geol. Soc. London, 161, 695 – 702.

Weissert, H. and Lini, A. (1991) Ice age interludes during the time of Cretaceous greenhouse climate? In: Controversies in modern Geology (Eds D. W. Müller, J. A. McKenzie and H. Weissert), pp. 173. 191. Academic Press, London.

Weissert, H. , Lini, A. , Föllmi, K. B. and Kuhn, O. (1998) Correlation of Early Cretaceous isotope stratigraphy and platform drowning events: a possible link? Palaeogeogr. Palaeoclimatol. Palaeoecol. ,137,189 – 203.

Wissler, L. , Weissert, H. , Masse, J. -P. and Bulot, L. (2002) Chemostratigraphic correlation of Barremian and lower Aptian ammonite zones and magnetic reversals. Geol. Rundsch. ,91,272 – 279.

Wissler, L. , Funk, H. and Weissert, H. (2003) Response of Early Cretaceous carbonate platforms to changes in atmospheric carbon dioxide levels. Palaeogeogr. Palaeoclimatol. Palaeoecol. ,200,187 – 205.

Wissler, L. , Weissert, H. , Buonocunto, F. P. , Ferreri, V. and d'Argenio, B. (2004) Calibration of the Early Cretaceous time scale: a combined chemostratigraphic and cyclostratigraphic approach to the Barremian-Aptian interval, Campnaia Apennines and southern Alps (Italy). In: Cyclostratigraphy: approaches and case histories (Eds. B. d'Argenio, A. G. Fischer, I. Premoli Silva, H. Weissert and V. Ferreri), SEPM Spec. Publ. ,81,123 – 133.

Yose, L. A. , Ruf, A. S. , Strohmenger, C. J. , Schuelke, J. S. , Gombos, A. , Al-Hosani, I. , Al-Maskary, S. , Bloch, G. , Al-Mehairi, Y. and Johnson, I. G. (2006) Three-dimensional characterization of a heterogeneous carbonate reservoir, Lower Cretaceous, Abu Dhabi (United Arab Emirates). In: Giant Hydrocarbon Reservoirs of the World: From Rocks to Reservoir Characterization and Modeling (Eds. P. M. Harris and L. J. Weber), AAPG Memoir 88,173 – 212.

Yose, L. A. , Strohmenger, C. J. , Al-Hosani, I. , Bloch, G. and Al-Mehairi, Y. (2010) Sequence-stratigraphic evolution of an Aptian carbonate platform (Shu'aiba Formation), eastern Arabian Plate, onshore Abu Dhabi, United Emirates. In: Barremian-Aptian Stratigraphy and Hydrocarbon Habitat of the Eastern Arabian Plate (Eds F. S. P. van Buchem, M. I. Al-Husseini, F. Maurer and H. J. Droste), Vol. 2, pp. 309 – 340. Gulf PetroLink, Bahrain.

第 5 章　孤立碳酸盐台地的沉积层序演化

ÓSCAR MERINO-TOMÉ,GIOVANNA DELLA PORTA,
JEROEN A. M. KENTER,KLAAS VERWER,PAUL (MITCH) HARRIS,
ERWIN W. ADAMS,TED PLAYTON,DIEGO CORROCHANO 著

周永胜 译，吴因业 校

摘　要　布达哈山的孤立碳酸盐台地是于早侏罗世在大阿特拉斯的海相裂谷盆地的半地堑中发育起来的。其中识别出了六个具有不同几何形态、地层结构、层序界面、岩相组成和叠置形式的层序。它们具有复杂的、多重作用的成因，是大地构造作用（区域沉降和断块旋转）叠加到全球海平面升降信号上的结果。另外，伴随沉积中心迁移而改变的碳酸盐生产主体和相对堆积潜力也影响了由大地构造和全球海平面变化混合作用产生的可容空间的充填和内部结构。层序Ⅰ和Ⅱ呈板状几何形态，主要反映了均匀的构造沉降和全球海平面升降的组合作用，二者共同作用形成了具有潟湖和潮坪岩相的缓坡到陆架碳酸盐体系。大致在洛林期伊始，一期裂谷的启动和断块旋转促成了与下伏层序Ⅱ的角度不整合，进而在洛林期和普林斯巴赫期有浅水孤立台地和楔形沉积层序Ⅲ到Ⅵ的逐步发育。古海洋水文状况和碳酸盐工厂的同时变化，导致台地的几何形态向平顶台地转变，并在台地边缘发育高能包粒和次光带海绵微生物沉积。层序Ⅲ、Ⅳ和Ⅵ都具有由加积到退积的旋回特点推断是原地碳酸盐生产外加上盘可容纳空间迅速增加联合作用的结果。环境对碳酸盐生产的影响最有可能是导致布达哈山台地在托尔期淹没的原因，同时它还影响了层序Ⅳ到Ⅵ的充填和几何形态。本项研究表明，以往普遍接受的半地堑"脉冲驱动"式层序发育模式可能并不完全适用于布达哈山；因为在这里，与布达哈山断块沉降脉冲相伴随，还有层序级的更加连续的旋转活动看来也与全球海平面升降和碳酸盐生产进行了互动来生成层序；连续旋转的标志有：发散地层的规律分布、层序Ⅴ和Ⅵ潟湖沉积中有无数的次级暴露面。此外，本研究还证明，在裂谷盆地碳酸盐台地层序发育和台地淹没的过程中，全球海平面的升降和碳酸盐工厂的变化可能与大地构造的活动同等重要；该结果就意味着，在其他类似的碳酸盐台地当中，由于人们习惯倾向用构造解释，因此，全球海平面升降和碳酸盐工厂的作用或许并没有得到正确的认识。

关键词　碳酸盐台地体系　碳酸盐生产　早侏罗世　全球海平面升降　大阿特拉斯　摩洛哥　同沉积的大地构造活动

5.1　概况

碳酸盐沉积体系的地质记录展示了丰富多样的碳酸盐序列，呈现出各式各样的相组成、结构、沉积几何形态和生长形式，并且在整个显生宙随着时空变化而变化。内在的控制因素，比如碳酸盐工厂的类型及其生长潜力，还有外在的控制因素，比如大地构造背景、全球海平面升降、气候和海洋水文参数，都作用于和影响到碳酸盐沉积体系的成核、发育和消亡。高度多样化的碳酸盐工厂（在时空尺度上的），受全球海平面升降和区域构造活动驱使的可容空间变化，都极大地限制了普适性沉积模式的应用范围，增加了地震图像解释、层序地层学概念应用和油气勘探的复杂性（Wright 和 Burgess，2005；Markello 等，2008）；对于一些特定的构造背景来说尤其如此，比如裂谷盆地，其中的碳酸盐台地是在旋转的断块上、随着可容空间的变化发育

的,而可容空间的变化是由拉张构造活动和全球海平面升降相互作用产生的。构造产生的可容空间变化,如果其速率与碳酸盐生产速率相当的话,就可以超越全球海平面升降的影响(Bosence 等,1998 及其参考文献)。此外,时间量级为数千年到一百万年的平均海平面波动速率与构造活动产生的可容空间变化相当,从而就产生了复杂的相互作用(Dorobek,2008)。因此,除非有很好的年代标定并且能够定出三维地层关系,否则的话,很难把构造活动的影响从全球海平面升降的信号中区分出来(Dorobek,2008)。尽管如此,在裂谷半地堑发育的碳酸盐台地还是拥有一些共同的特点(Leeder 和 Gawthorpe,1987),如:(1)下盘区具有悬崖式的、过路或增生的台地边缘,后由珊瑚藻礁镶边,并有从下盘输入的再沉积相;(2)上盘的倾向坡具有碳酸盐缓坡,其后演化成镶边的陆架边缘(Bosence,2005;Dorobek,2008)。

露头研究、基于地震的地下研究以及构造—地层正演模拟表明,在同裂谷阶段拉张断裂活动主要控制的是:(1)碳酸盐台地的几何形态和样式,(2)不整合和沉积层序的发育(Bosence 等,1998 及其参考文献)。由断块旋转产生的、具有以不整合为界沉积层序的碳酸盐序列,不同于经典的被动边缘模式(Handford 和 Loucks,1993)。断块旋转同时引发上盘下降(相对海平面上升)和下盘上升(相对海平面下降)。因此,下盘层序的界面(剥蚀不整合和陆上暴露)就与上盘的海泛和淹没事件相互对应。这种楔形沉积层序是半地堑断块旋转所特有的、并且可以在地震剖面上查看到的(Prosser,1993);可是,要在几公里大小的露头上连续观测到从下盘的暴露、再顺倾向到上盘的海泛以及其中的层序界面和岩相特点,机会甚少。例外的情况要算红海的中新统了,断块台地模式正是由此提出的(Bosence 等,1998;Cross 等,1998;Bosence,2005)。尽管此后继续有断块台地模式的发表(Bosence,2005;Cross 和 Bosence,2008;Dorobek,2008),然而,仍有一些关于此类背景下碳酸盐台地演化的问题悬而未决,包括:(1)沿着断层的不同段落,碳酸盐台地的几何形态和所形成的层序是如何变化的;(2)海平面的相对迅速上升和淹没是否可以单单归因于裂谷盆地的沉降,或者说,可容空间的迅速增加是否就需要与环境驱使的碳酸盐工厂生产无效相配合;(3)全球海平面的升降和碳酸盐工厂的特点是否在断块碳酸盐序列演化和层序地层发育当中同样也起着根本性的作用。

本研究聚焦于一处出露绝佳的下侏罗统碳酸盐台地露头[即摩洛哥大阿特拉斯山脉的布达哈山(DBD),图 5-1]。该台地发育形成在一个下盘隆起上,现今作为一个孤立碳酸盐台地几乎原封完整的出露。布达哈山(DBD)能够提供空间上连续的信息,不仅是关于以不整合面和相应海泛面为界的沉积层序中的相特征和相架构,也关乎其成因,即断块旋转、全球海平面升降和碳酸盐生长潜力变化的复合作用。对本研究来说特别重要的地质背景和事件有:特提斯洋的拉张构造体制、在早—中侏罗世发生的两次全球重大生物危机。在西特提斯洋,有许多碳酸盐台地曾受到了拉张构造作用的影响,并经历了以凝缩序列沉积为标志的淹没和远洋沉积的覆盖(Winterer 和 Bosellini,1981;Bosellini,1989;Santantonio,1993,1994;Cobianchi 和 Picotti,2001;Ruiz-Ortiz 等,2004;Jadoul 和 Galli,2008)。三叠纪与侏罗纪的分界以生物灭绝事件为标志(Hesselbo 等,2002;Tanner 等,2004),而早侏罗世末期又经受了生物危机和随后物种多样化的影响(Little 和 Benton,1995;Pálfy 和 Smith,2000;Wignall 等,2005,2006),并伴随出现了早托尔期大洋缺氧事件中所呈现的古海洋状况的改变(Jenkyns,1988)。

Agard 和 Du Dresnay(1965)、Du Dresnay(1976,1977)最初将布达哈山(DBD)作为下侏罗统的生物礁露头进行了描述,提供了重要的沉积学和地层学信息。继此之后,有许多人员专门针对布达哈山(DBD)的台地露头进行了研究(Crevello,1991;Kenter 和 Campbell,1991;

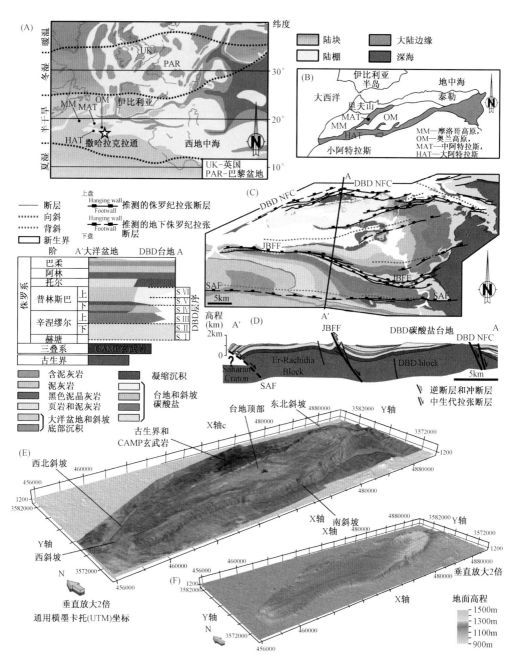

图5-1 (A)西特提斯区在托尔期的岩相古地理图(据 Thierry 等,2000 修改),图中标示出了布达哈山(DBD)的位置(五角星),古气候带是依照 Dera 等(2009a)重绘。(B)摩洛哥主要地质构造大区图(据 Teixel 等,2003 修改)。(C)大阿特拉斯山脉东南部综合地质图[位置见图(B)中的小红框]。该地质图展示了所推测的主要侏罗纪伸展大地构造(据 Du Dresnay,1976 修改),A—A′代表图(D)所示剖面的位置(其中的 DBD NFC = 布达哈山北侧断裂带)。(D)大阿特拉斯东部的综合地质剖面图,图中显示布达哈山(DBD)所处的位置(据 Agard 和 Du Dresnay,1965 重绘);SAF—南阿特拉斯断裂,JBFF—Bou Ferma 山断裂,Saharan Craton—撒哈拉克拉通,Er - Rachidia Block—拉希迪耶断块,DBD Block—布达哈山断块。(E)快鸟卫星影像投在数值高程模型(DEM)上的斜视图,图像显示布达哈山露头完好保存的原始形态以及文中所提及的沉积带,卫星影像的水平分辨率是0.7m,数值高程模型的水平分辨率为15m×15m。(F)反映露头地形起伏的数值高程模型

Campbell 和 Stafleu,1992;Blomeier 和 Reijmer,1999,2002;Scheibner 和 Reijmer,1999）。尽管如此,前人的研究成果还没有涉及对整个碳酸盐体系及其从成核到最后淹没演化的探讨。本文介绍一种新方法,即通过综合沉积相的沉积剖面和空间分布、地层结构,建立以 GIS 为基础的数字露头模型（DOMs）。研究初期只是集中在布达哈山（DBD）个别空间有限的研究窗口（Verwer 等,2009a,b）。Verwer 等（2009a）研究的目的是要揭示布达哈山（DBD）南部在普林斯巴赫末期由台地顶部到台地边缘复杂的三维沉积相关系。Verwer 等（2009b）的研究窗口位于布达哈山（DBD）东南端,约有 2km² 大小,其中描述了布达哈山（DBD）南斜坡的岩相、斜坡架构和演化,并在上辛涅缪尔阶—普林斯巴赫阶的地层中识别出四个地层单元。然而,本项研究却是从整个碳酸盐台地着眼,将台地生长的各个不同阶段中沉积相的内部架构和横向变化予以定量化标定。其中,数字化的露头模型覆盖了布达哈山（DBD）台地的整个范围,包括在穿越整个布达哈山进行详细勘查所收集到的野外数据资料（见附图 S1❶;关于本研究的范围,Verwer 等（2009a,b）描述的研究窗口位置请参见支持信息）。通过数字露头模型的方法,能够把揭示可容空间变化、全球海平面升降和碳酸盐生产之间相互作用所必需的沉积层序、相叠置形式和构造—地层格架予以空间界定。这些成果可能也适用于其他半地堑的断块碳酸盐台地。本文旨在:（1）展示在整个受不同断滑控制的断块上,碳酸盐沉积体系和所观测层序的几何形态变化;（2）评价针对台地淹没所提出的多重作用成因方案;（3）将识别出的沉积层序与全球海平面变化曲线进行对比,从全球海平面升降的信号中把构造因素的作用区分出来。

5.2 地质背景

5.2.1 大阿特拉斯山脉

摩洛哥的大阿特拉斯山脉是中生代裂谷经由阿尔卑斯回返运动形成的（Beauchamp 等,1996;Teixel 等,2003,2007;Arboleya 等,2004）。在阿尔卑斯回返运动过程中,构造缩短总量由西向东增大,最大值达到 24%（Teixel 等,2007）。由于该构造缩短作用通过早期拉张断层的后期复活和薄皮构造的有限发育而得到了调节,因此,中生代裂谷活动所产生的断块显示较小的内部构造变形（Teixel 等,2007;Tesón 和 Teixel,2008）。由此而来,早期裂谷盆地的结构和演化也就得到较好的恢复和再现（Beauchamp 等,1996;Teixel 等,2003;Arboleya 等,2004;Lachkar 等,2009）。在侏罗纪时期,大阿特拉斯盆地属于南特提斯被动边缘上一个夭折的海相裂谷;该被动边缘分别在三叠纪、早侏罗世和中侏罗世经历了大的裂谷活动（Ellouz 等,2003;Teixel 等,2007;Lachkar 等,2009）;其中,三叠纪、早侏罗世的裂谷期以拉张体制为特征,中侏罗世的裂谷期则以张扭体制为主（Ellouz 等,2003）;而后者又与中大西洋开始洋壳增生有关（Ellouz 等,2003）,根据最近的年代鉴定结果,其时代为托尔期或托尔期—阿林期（Steiner 等,1998;Davison,2005）。Lachkar 等（2009）将早侏罗世裂谷期（辛涅缪尔期—普林斯巴赫期）进一步划分成两个阶段:（1）早裂谷期（辛涅缪尔期）,以广泛的、低应变正断裂活动为特征,断距小、上盘下降速率小;（2）裂谷顶峰期,出现在普林斯巴赫期（卡瑞辛期—多米尔期;Jamesoni-Gibbosus 化石亚带）,以高应变正断裂活动为标志,断距大、上盘下降速率大。

早侏罗世（赫塘期?—辛涅缪尔期）,随着海侵大阿特拉斯盆地的中部和东部变成海相环境（图 5-1A）;一个开阔碳酸盐体系在盆地的南缘成核。由于裂谷盆地构造发育的结果,在洛林期和普林斯巴赫期,该体系演化成了窄的碳酸盐台地（Warme,1988;Crevello,1991;Wilms-

---

❶ 译者注:此为作者内部资料目录,未在本文中出现,后同。

en 和 Neuweiler,2008;Lachkar 等,2009)。在早托尔期,大多数台地被淹没(Agard 和 Du Dresnay,1965;Warme,1988;Elmi 等,1999),随后又被托尔期—巴柔期的大洋盆地沉积序列所覆盖(Elmi 等,1999)。

5.2.2 布达哈山

布达哈山(DBD)碳酸盐岩台地矗立在冲积平原当中,长 35～40km,宽 4～15km,山顶平坦,起伏落差在 100～400m 之间(图 5-1E、F 和图 5-2)。该平顶山的西北侧、西侧、南侧和东北侧为斜坡,倾角 20°～35°,出露的陡倾灰岩岩层(10°～32°)恰与台地斜坡对应(图5-1E、附图 S3);平坦的山顶以呈水平的或近水平层状的灰岩序列为特征,又与台地顶部地层相吻合(图 5-1E、附图 S3)。碳酸盐序列成核的基岩有:志留系—奥陶系变质的硅质碎屑地层、由与中大西洋岩浆省(CAMP)玄武岩相关的拉斑玄武质熔岩和基性岩床组成的火山地层单元(Marzoli 等,2004)。这些岩石在中北部形成了一条 WSW—ENE 向的基岩带(图 5-1E 和图5-2A)。在布达哈山(DBD)周围冲积平原出露的则是时代属托尔期、阿林期、巴柔期和局部属普林斯巴赫期的绿色泥灰岩、页岩和深色泥晶灰岩(图 5-2);它们代表的是与布达哈山(DBD)碳酸盐台地属同时代或较年轻的深水大洋盆地沉积。

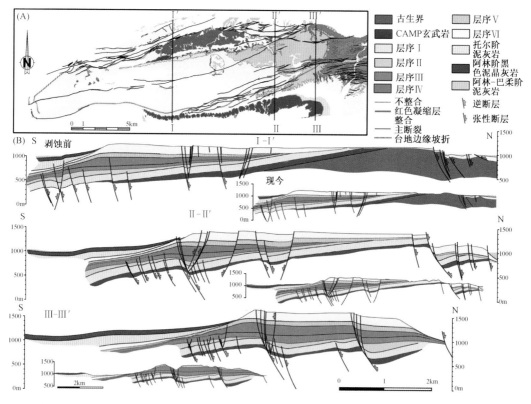

图 5-2 (A)布达哈山(DBD)平面图展示了六个层序及其他岩性地层单元的分布。黑杠线代表三条地质剖面(图 B)所穿过的位置。(B)穿切布达哈山(DBD)碳酸盐台地的三条地质剖面,每条剖面又分别包括复原后的剖面和现今的剖面,剖面的绘制是基于:实测剖面的空间信息、野外分散观察点记录的空间观测数据以及在快鸟(QuickBird)卫星影像上做的遥感地理信息系统(GIS)填图,后者就是通过将卫星影像投放在 ASTER 数字高程模型(DEM)上(图 5-1E)完成的

5.3 研究方法

利用先进的定位技术对所有相关的野外观测(采样点、实测剖面、层面、层序界面和构造特征)都进行了登记。主要利用的是精度达到 1cm 的实时动态全球定位技术(RTK GPS),在没有服务支持时,就采用 1~10m 精度的手持 GPS 接收机(并且用 RTK GPS 系统进行校对)。利用 GIS 软件(Microstation V.8 软件,由总部设在美国宾夕法尼亚州埃克斯顿的奔特力公司开发),其他或附加的信息被直接记录到快鸟正射卫星影像(0.7m 的像素大小)上,以及 1:50000 和 1:20000 的航片上。所有空间相关数据都被整合到数字露头模型(DOM;Bellian 等,2005)当中。该模型包括:由 ASTER 卫星影像(空间分辨率 15m×15m×8m)提取的区域数字高程模型(DEM),辅以用实时动态全球定位(RTK GPS)断面图和激光雷达影像制成的高分辨率 DEMs,和叠在 DEM 上的快鸟正射卫星影像。该数字露头模型为本研究提供了一个强有力的手段,用它可以提取有关台地几何形态和架构(如斜坡的坡度、掀斜活动和地层结构)、岩相组合的体量和分布等方面的定量信息。

出露于布达哈山(DBD)台地顶部、斜坡和邻近冲积平原的赫塘期?—普林斯巴赫期地层包含有各种不同的岩相。根据所含化石、层理样式、沉积构造和结构,鉴定出了 35 种岩相(图 5-3,附图 S2)。所观察到的特征以及相应提出的沉积环境解释已总结到了附表 1 当中,附图 S4 到附图 S8 则提供了有关实录。在台地顶部的岩相类型当中,包括有陆相沉积、潮间到潮上带沉积、浅水潮下带低能沉积、和以潮下带为主的高能到中能颗粒灰岩和砾屑灰岩。生物建造岩相绝大多数是在台地斜坡产出的,局部出现在靠近台地坡折的台地顶部。再沉积和远洋沉积是台地斜坡和邻近大洋盆地区的特征岩相。凝缩相则代表台地淹没期间欠补偿沉积的层段。

对超过 80 条剖面进行了测录(附图 S1 和 S9)、定位(用实时动态全球定位技术——RTK GPS)、取样。岩石样品用来切磨(近 700 个)薄片和抛光板以做进一步的岩石学分析;通过对这些薄片进行室内研究,同时结合露头观测记录,促成了对主要岩相类型的识别和描述。

利用公开发表的资料数据(Agard 和 Du Dresnay,1965;Du Dresnay,1976;Crevello,1991;Campbell 和 Stafleu,1992)和新的菊石样本时代鉴定结果,我们建立起了一个生物地层格架(图 5-3)。对采自布达哈山(DBD)碳酸盐层系下面的玄武岩、辉长岩和微辉长岩(岩床和岩墙)样品,进行了化学分析[即用电感耦合等离子体质谱(ICP-MS)分析高场强元素(HFSE),如铪(Hf)和镥(Lu)]。根据分析获得的结果,能够把布达哈山(DBD)的玄武岩与大阿特拉斯山的中大西洋岩浆省(CAMP)熔岩层序(Marzoli 等,2004)做比对,以此进一步推断层序 I 开始沉积的时代(赫塘期 p.p. 菊石亚带)。新近发表的数据表明,中大西洋岩浆省(CAMP)的岩浆事件始于大约距今 201.75Ma(三叠纪末,几乎与三叠纪和侏罗纪的分界线对应),持续了 58.0 到 61.0 万年(Olsen 等,2002,2003;Marzoli 等,2004;Whitesite 等,2007)。

5.4 沉积层序、几何形态、岩相和地质年代

布达哈山(DBD)碳酸盐台地地层和与其邻接的同时代大洋盆地沉积可以被进一步划分为六个可填图的成因地层单元(图 5-2 和图 5-3);这些地层单元,即本文所称的"层序",等同于 Mitchum(1977)所描述的、并由 Bosence 等(1998)用于断块台地的以不整合为界的沉积层序。各个层序以不同的内部架构(地层结构)和沉积相为特征(表 5-1)。在层序内部,各层序(除层序 V 外)是由加积到退积的地层单元构成。岩相的纵向序列和外台地相带的轨迹表明,这些层序大多是在可容空间增加—减少的旋回过程当中生成的,因此也与 Catuneanu 等

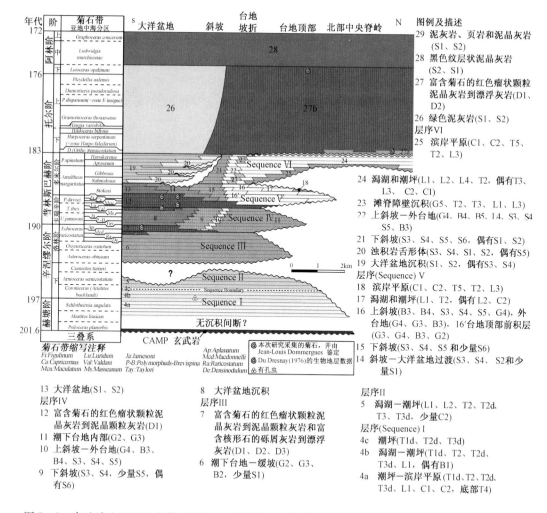

图 5-3 布达哈山(DBD)年代地层图,生物地层方面的数据来自于菊石样品和 Du Dresnay(1976)发表的数据,年代地层和磁极性年代根据 Gradstein 等(2004),绝对年龄依照 Walker 和 Geissman (2009),菊石亚带根据 EL Hariri 等(1996);附表 S1 有对岩相的详细描述,附表 S2 有对层序 V 和 VI 相组合更详细的解释

(2009)引入的"地层层序"概念相匹配。尽管如此,传统的体系域却难于区分出来。以下是对每个层序的一个简要描述,内容包括它们的分界面、岩相构成和厚度变化。

表 5-1 布达哈山(DBD)层序界面特征和层序几何形态的总结

层序	层序的底 (和层序 VI 的顶)	层序的几何形态	台地的几何形态/台地的幅度(相对于邻近洋盆)	台地生长与南部地层结构
层序 VI	底界面:在布达哈山中心区为角度剥蚀不整合、具有陆上暴露特征;在台地斜坡为水下侵蚀面,具有下切沟与峡谷。顶界面:在台地顶部上边是岩溶面,在台地斜坡是淹没面	楔形几何形态,分平顶台地区和斜坡区;在南边缘区地层厚度最大(达 200m),向布达哈山中心区地层厚度逐渐减小;由台地边缘到斜坡地层厚度减小,到大洋盆地而增大	前沿变陡的孤立台地,台地的幅度达 600m	退积型(南台地边缘退积约 250m),内台地地层上超底部不整合和古地貌高地;台地顶部地层由布达哈山中心区向台地边缘发散;在布达哈山中心区,台地顶部地层内局部有小型不整合

续表

层序	层序的底（和层序Ⅵ的顶）	层序的几何形态	台地的几何形态/台地的幅度（相对于邻近洋盆）	台地生长与南部地层结构
层序Ⅴ	具有陆上暴露标志的角度剥蚀不整合（布达哈山中心区），到硬底发育的整合海侵面遍布台地顶部；在台地斜坡和大洋盆地区是红色的凝缩段	楔形几何形态，分平顶台地区和斜坡区；在南缘区地层厚度最大（达100m），由台地边缘向台地内和大洋盆地，层序厚度均减小；布达哈山中心区层序逐渐尖灭	前沿变陡的孤立台地，台地的幅度达470m	加积型（3~6km）；前积层由布达哈山中北部向台地边缘和斜坡带递进；台地顶部地层由布达哈山中心区朝向台地边缘发散；在布达哈山中心区，台地顶部地层内局部有小型不整合
层序Ⅳ	在中北部，推测为角度剥蚀不整合；在台地斜坡和大洋盆地区为红色凝缩层	楔形几何形态，分平顶台地区和斜坡区；在南缘区地层厚度最大（达200m），由台地内和大洋盆地，层序厚度均减小，向布达哈山中心区层序逐渐尖灭	前沿变陡的孤立台地，台地的幅度达400m	退积型（南台地边缘退积约250m），台地顶部地层由布达哈山中心区朝向台地边缘发散（在布达哈山东部露头区）
层序Ⅲ	角度剥蚀不整合	楔状几何形态；在布达哈山南侧斜坡和在布达哈山东部的生长向斜当中，地层厚度最大（到100m）；向布达哈山中心区层序逐渐尖灭	低幅台地，北缘偏陡，台地的幅度达130~150m	退积型（台地南边缘退积约400m）；横向构成连续的楔形地层单元；地层沿上盘上超，并且向下方倾角增大，在东部露头观察到生长地层
层序Ⅱ	以广泛发育的潮上和潮间带白云岩相为层序分界	由于受后期剥蚀削减而呈楔形；推测层序的原始几何形态为板状（沿布达哈山）；初始地层厚度>150m	区域碳酸盐体系（缓坡或碳酸盐陆架?）	加积型，平行层状的台地顶部地层
层序Ⅰ	与CAMP玄武岩整合接触；局部发育古土壤	由于受后期剥蚀削减而呈楔形；推测层序的原始几何形态为板状（沿布达哈山）；初始地层厚度>130m	区域碳酸盐体系（缓坡或碳酸盐陆架?）	加积型，平行层状的台地顶部地层

5.4.1 层序Ⅰ和Ⅱ

层序Ⅰ覆盖在中大西洋岩浆省（CAMP）玄武岩之上，至少有126m厚；层序Ⅱ则厚达150m。层序Ⅰ和层序Ⅱ之间的分界标志为7m厚、横向连续的、由窗格状和叠层状白云岩、鲕粒白云岩构成的地层单元，其中常见原地角砾岩化和干裂（图5-4和图5-5）。层序Ⅰ和层序Ⅱ都是由板状平行成层的、米级互层构成；互层属浅水潮下带灰岩（L1、少量B1和L2）、潮间到潮上带灰岩（T2、T3）和白云岩（G1d、T1d、T2d和T3d，附图S4）的交互。这两个层序的岩相类型和垂向叠置在空间上没有显著变化，说明该碳酸盐体系的相带延伸广阔，可能都超过了布达哈山（DBD）的现今宽度。层序Ⅰ和层序Ⅱ都有一个海侵—海退（T—R）旋回，其中，最大海侵段由潮下带沉积(岩相 B1、L1 和 L2,图 5-5)大量出现为标志。而保存原始结构的白云岩(岩相 T1d、T2d、T3d 和 G1d)推断是由早期白云岩化作用造成的。

图 5-4 野外露头照片说明层序Ⅰ、层序Ⅱ以及下伏大西洋岩浆省(CAMP)玄武岩之间的地层产出关系。(A)布达哈山(DBD)的中北部,层序Ⅰ底部的白云岩层覆盖在 CAMP 玄武岩之上,层序Ⅴ又覆盖在层序Ⅰ之上,SP1 的剖面线也标在上面;(B)全景照片显示:布达哈山(DBD)中北部,层序Ⅴ以不整合上覆于层序Ⅰ的白云岩地层;(C)在盖斯尔—穆盖勒谷(KMG)出露的层序Ⅰ和Ⅱ,KMG1 的剖面线也标在上面(位置参见附图 S2)

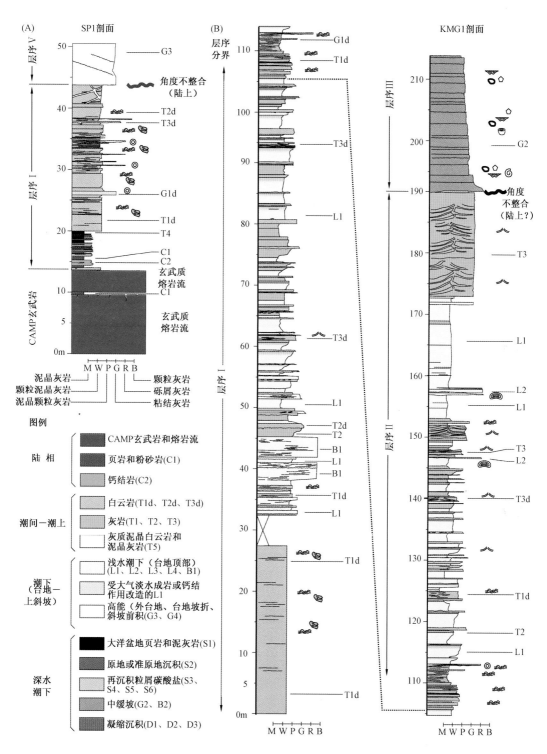

图5-5 层序Ⅰ和Ⅱ的地层剖面图,其中的图例也作为本文其他地层剖面的岩相图例和说明。(A)SP1剖面的地层柱状图,展示的是布达哈山(DBD)中北部区层序Ⅰ和Ⅴ的特点(图5-4A),关于相标志符号,在图5-7当中有相应的图例说明;(B)综合地层剖面图,展示的是在盖斯尔—穆盖勒谷(KMG)出露的层序Ⅰ和Ⅱ(位置参见附图S1和S2)

层序Ⅰ和层序Ⅱ的地层发育、岩相组成、地层厚度酷似时代属赫塘期 p.p. 菊石亚带到洛林期的 Idikel 组（Mehdi 等，2003；Wilmsen 和 Neuweiler，2008，图 5-2C）。层序Ⅰ可以与 Idikel 组（Wilmsen 和 Neuweiler，2008）的下段对比，而层序Ⅱ可以与 Idikel 组的上段对比。与 Idikel 组相似，层序Ⅰ上段潮下带沉积当中的生物群指示的时代也是辛涅缪尔期（图 5-3）。这些生物群中包含有属辛涅缪尔期的代表性组合：粗枝藻（Palaeodasycladus mediterraneous）和底栖有孔虫（Siphovalvulina gibraltarensis、Amijiella amiji 和 Riyadella prearegularis）（Barattolo 和 Romano，2005；BouDagher-Fadel 和 Bosence，2007）。

5.4.2 层序Ⅲ

层序Ⅲ是作为一个加积—退积型、低幅碳酸盐台地而发育形成的（图 5-6）。台地南侧的斜坡缓倾（1°~5°），北侧斜坡则较陡（7°~12°），记录了断块的掀斜活动。在布达哈山（DBD）东北侧和东侧的 ENE—WSW 向生长向斜当中，该层序的地层厚度最大（约 100 m）。在南部，层序Ⅲ有 50m 厚（图 5-7），向布达哈山（DBD）的中心区，逐渐变薄。

由包粒骨屑泥晶颗粒灰岩、泥晶颗粒灰岩或颗粒灰岩及少量颗粒泥晶灰岩（局部含海绿石颗粒）构成了板状到楔状的基本层（G2，附图 S5A），由这些基本层组合进而形成了米到十米厚的、横向连续（数百米）的地层单元（图 5-8A）。这些地层单元的分界通常是厘米到分米厚的泥灰岩，或是红色薄层状（分米厚）到瘤状的颗粒泥晶灰岩，常含有六射海绵和普通海绵。另有米到分米厚、分米宽的透镜状海绵生物丘（B2，图 5-8B、附图 S5B）常出现在该层序的底部附近以及东北侧的斜坡带（图 5-6）。

层序Ⅲ与 Foum Zidet 组具有许多相似的特征。Foum Zidet 组的底界属盆地级的主要等时面（Wilmsen 和 Neuweiler，2008），局部呈角度不整合（Mehdi 等，2003）。Agard 和 Du Dresnay（1965）把该地层界面划定为洛林阶的底，并且在后来通过菊石动物群予以了印证（Wilmsen 和 Neuweiler，2008）。

5.4.3 层序Ⅳ

层序Ⅳ的几何形态为楔状，具有一个明显的平顶台地相带和下倾的斜坡相带。层序Ⅳ的最大地层厚度（约 200m）出现在布达哈山（DBD）的南缘区，越靠近布达哈山（DBD）的中心区，厚度越小，在中心区，层序Ⅳ没有沉积（图 5-2 和图 5-9）；在层序Ⅳ沉积的同时，布达哈山（DBD）的中心区则以层序Ⅰ和Ⅱ遭受陆上暴露和岩溶改造为特征（图 5-9）。在南侧、东侧和东北侧斜坡，层序Ⅳ的底以一段 3~5m 厚的红色板状层系为标志，该层系的主要岩相类型为富菊石的颗粒泥晶灰岩、颗粒泥晶灰岩或泥晶颗粒灰岩（D1，图 5-7 和图 5-8A，附图 S5C），次要岩相类型为海百合泥晶颗粒灰岩或颗粒灰岩（D2，附图 S5D）和核形石漂浮灰岩（D3，附图 S5E）。向斜坡上方，层序Ⅳ的斜坡沉积由与该层系交互，到取而代之。在纵向上，层序Ⅳ的大洋盆地沉积则覆盖在该层系之上，从而记录了该碳酸盐体系一次重要的退积过程。

层序Ⅳ的斜坡带是由向洋盆方向变薄、向上呈退积叠置的楔形地层单元所构成（图 5-9B 和 C）；这也导致了沉积层的倾角逐步增大，从而造成斜坡变得越来越陡（由层序Ⅳ沉积早期到晚期，坡度由 8°增加到 19°，Verwer 等，2009b）。珊瑚—海绵—微生物粘结灰岩单元（B3）在台地坡折到之下大约 70~100m 水深的范围发育；朝生地方向，则与包粒颗粒灰岩（G4）交互（Verwer 等，2009b）。再到斜坡的下方，珊瑚—微生物粘结灰岩被海绵—微生物粘结灰岩

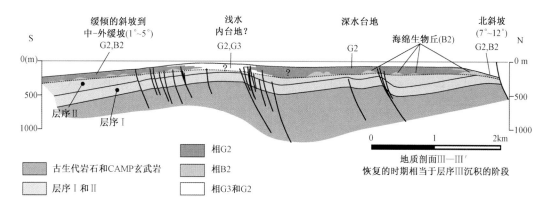

图5-6 针对地质剖面Ⅲ所做的复原,展示了层序Ⅲ沉积期布达哈山东部的几何形态、沉积相分布(附表S1有对岩相的详细描述)

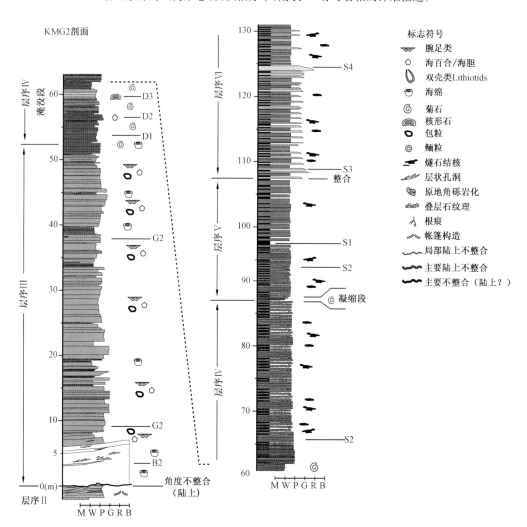

图5-7 盖斯尔—穆盖勒谷(KMG)层序Ⅲ、Ⅳ、Ⅴ及Ⅵ的综合地层剖面
(位置参见附图S1和S2,图5-5有岩性柱的图例及说明)

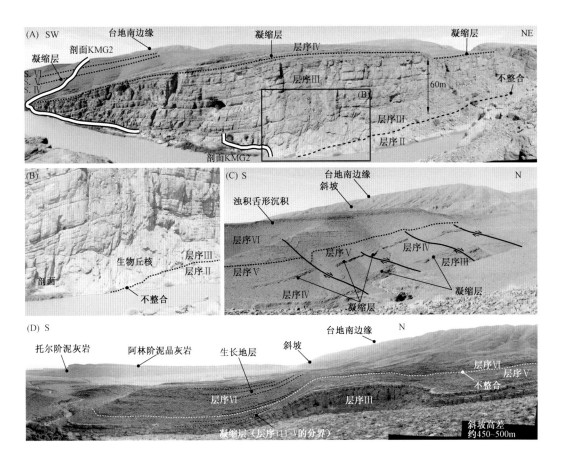

图5-8 布达哈山(DBD)南侧,层序Ⅲ到Ⅵ地层序列的野外露头照片。(A)盖斯尔—穆盖勒谷(KMG)南端的全景照,其中可以看到主要地层界面和层序都是斜倾的,碳酸盐台地的南缘出现在后方,盖斯尔—穆盖勒谷剖面(KMG2)的位置也标示在该全景照片上;(B)野外露头照片显示,在层序Ⅲ底界附近有一个高15m的海绵生物丘;(C)布达哈山(DBD)南侧的露头显示,层序Ⅳ、Ⅴ和Ⅵ的大洋盆地沉积层系覆盖在层系Ⅲ之上,主要层系分界线也被标示在该露头照片上,其中注意有一系列隐伏的张性生长断层改变了地层的展布;(D)盖斯尔—穆盖勒附近布达哈山南斜坡的远景照,注意单斜背斜和上覆层系Ⅵ的生长地层,在此位置,从层序Ⅲ沉积末期到层序Ⅴ的大洋盆地沉积开始一直都处于沉积欠补偿状态

(B4)取代,该岩相出现在距台地坡折140m的水深。微生物粘结灰岩呈楔形与再沉积的颗粒灰岩、泥晶颗粒灰岩及少量漂浮灰岩层(S3、S4和少量S5)交互。下斜坡包括再沉积的颗粒灰岩和泥晶颗粒灰岩(S3);朝洋盆方向,它们与含燧石结核的细粒似球粒颗粒泥晶灰岩(S2)夹泥灰岩(S1)交互(Verwer等,2009b)。在东侧和东北侧,粘结灰岩稀少,而最常见到的岩相是砂粒级包粒颗粒灰岩、颗粒灰岩或泥晶颗粒灰岩(S3、S4和G2)。

布达哈山(DBD)东部台地顶部地层为一个加积序列;其单元层0.1~1.0m厚、板状平行成层、由包粒颗粒灰岩或泥晶颗粒灰岩组成,常见有海百合、腹足类、似球粒和层孔虫(G2和G3);局部,在其内部也显示有槽状、板状交错层理,层系厚0.1~1.0m;其中尚未识别到潮间到潮上带的岩相类型。

层序Ⅳ、Ⅴ和Ⅵ在时代上与被划归为辛涅缪尔末期—普林斯巴赫期的Aganane、Choucht、Aberdouz和Ouchbis组(Mehdi等,2003;Wilmsen和Neuweiler,2008)相当。在层序Ⅳ底部红色

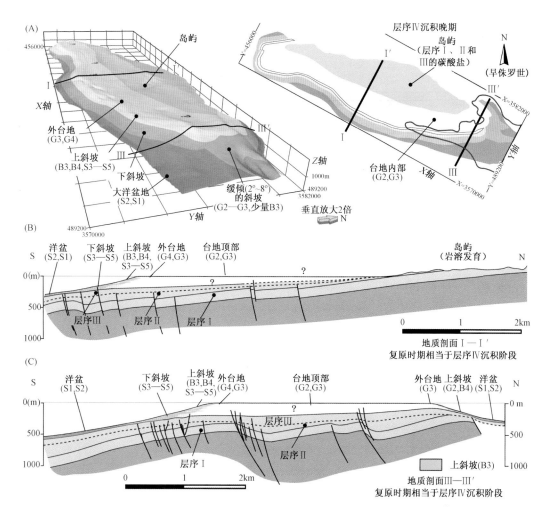

图5-9 (A)层序Ⅳ沉积晚期的相带分布和3D沉积模式斜视图,3D沉积模式是从数字露头模型恢复得到的,其中,用实线圈出的范围代表有直接露头数据来完成相带恢复的地方。(B)和(C)分别为南北向综合地质剖面Ⅰ和Ⅲ在层序Ⅳ沉积期的几何形态、沉积相分布(附表S1有对岩相的详细描述)

凝缩段采集到的菊石生物群属于Raricostatum带(可能是Aplanatum亚带),即对应于辛涅缪尔末期(图5-3)。

5.4.4 层序Ⅴ

层序Ⅴ为一楔状地层单元,在布达哈山(DBD)的南边缘、东边缘和东北边缘,厚度达到最大(70~100m)(图5-2);从中可以区分出一个台地顶部区和周围的斜坡区。在布达哈山(DBD)的中心区,层序Ⅴ的海相沉积覆盖在层序Ⅰ和Ⅱ经过岩溶改造的碳酸盐岩之上(图5-4A和B、图5-10和图5-11)。

层序Ⅴ属于一个进积型前积序列;由于在层序Ⅳ以前的浅海台地发生海泛,上凹的前积层递进并充填该海泛产生的可容空间,形成的前积层系高达50~70m,倾角达20°(图5-4A和图5-10);在平面上,由布达哈山(DBD)的中心区向南、东南、东和东北进积。到了布达哈山(DBD)的边缘带,前积层向外扩展并且推进到层序Ⅳ的斜坡沉积上面,形成高达29°的倾角,达到450m的落差高度。

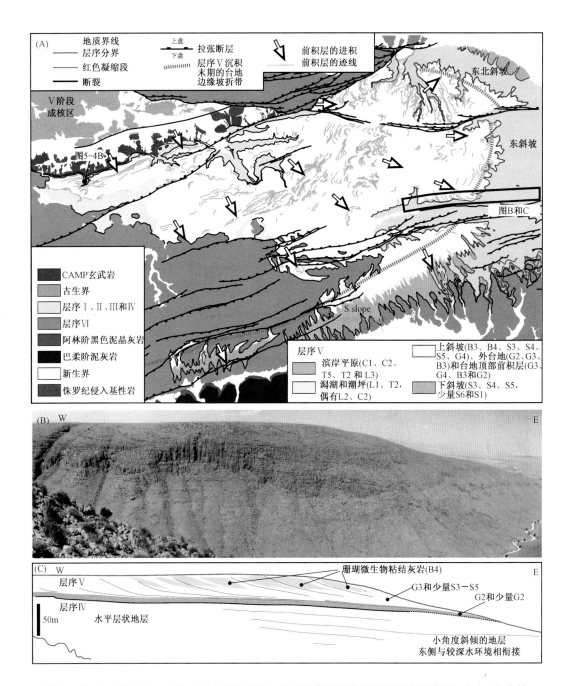

图5-10 (A)在布达哈山(DBD)东部地质图上显示层序Ⅴ的前积层轨迹线及其进积方向(白色箭头,附图S2),图上也标注了(B)和(C)以及图5-4B的取景范围,有关岩相、相组合和相带的详细描述参见附表S1和S2;(B)布达哈山(DBD)东部出露的层序Ⅳ和Ⅴ的野外全景照,层序Ⅴ由约50m厚的进积型前积层系组成,覆盖在层序Ⅳ的水平层状地层上;(C)露头照片(图B)的地质素描解释

前积层的顶部地层以水平为主,由浅水潮下带和潮间到潮上带的岩相交互组成(L1和T2,图5-11和图5-12,附表S2),常见陆上暴露标志,包括钙结岩(C2)以及一些层序内的不整合,反映出在前积层递进的过程当中,可容空间有限。朝边缘方向,这些沉积横向上被向洋盆斜倾达5°的沉积层代替,其岩性为分选中等到好的包粒颗粒灰岩,包括常见的潮上带鲕粒、海

— 136 —

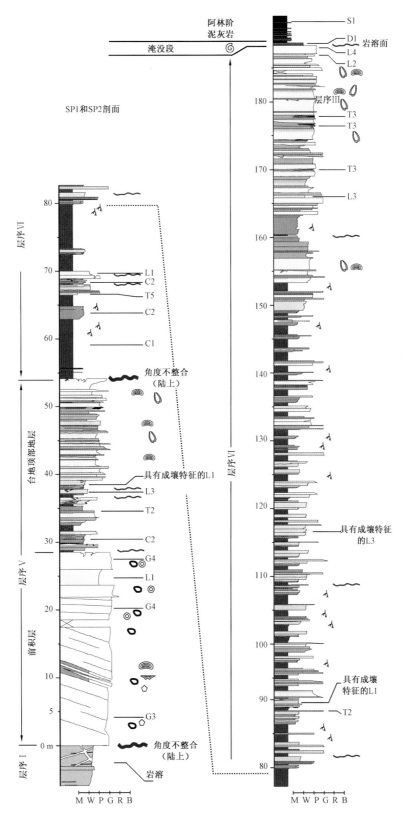

图 5-11 布达哈山(DBD)中心区层序Ⅴ和Ⅵ的综合地层剖面;位置参见附图 S1 和 S2,岩性柱和相标志符号分别参见图 5-5 和图 5-7 的图例及说明

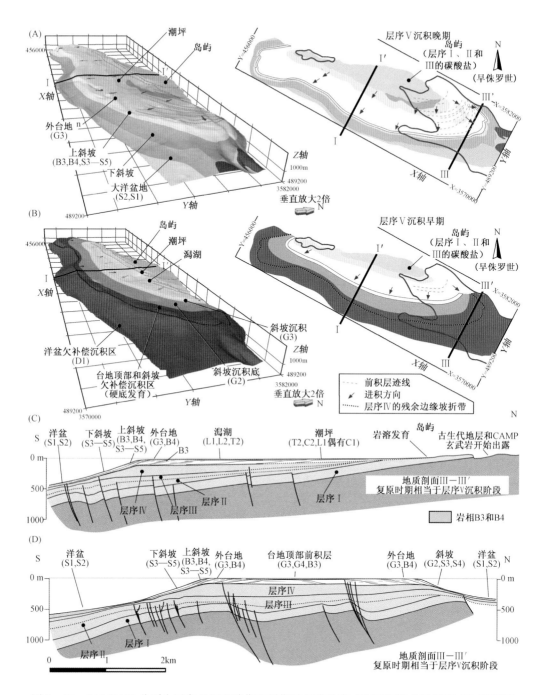

图5-12 (A)和(B)分别为层序Ⅴ沉积晚期和早期的相带分布、3D沉积模式斜视图,3D沉积模式是从数字露头模型恢复得到的;其中,用实线圈出的范围代表有直接露头数据来完成相带恢复的地方,所用色带的图例描述参见图5-3;(C)和(D)南北向综合地质剖面展示层序Ⅴ沉积期的几何形态和沉积相分布(有关岩相、相组合和相带的详细描述参见附表S1和S2)

百合、泥晶化颗粒和核形石(G3)。发育在布达哈山(DBD)内部的前积层(即沉积在层序Ⅳ台地顶部地层上面的前积层),其岩相构成有以下特点和变化:在前积层最陡(10°到20°)的部位,由包粒颗粒灰岩和颗粒灰岩—泥晶颗粒灰岩(局部富含似球粒)(G3;附图S5F)构成,层厚几米、呈缝合型接触。向下到前积层的底部(地层倾角 <5°~7°),则以包粒和似球粒泥晶颗粒灰岩层为主,含有潮下开阔海到洋盆的层孔虫,比如 Involutina liassica(G2)。

在布达哈山(DBD)的南缘、东缘、东北缘,前积层的上部以丰富的生物建造岩相为特征,包括产出在 10~60m 古水深的珊瑚—微生物粘结灰岩(B3,图 5-10B 和 C,附图 S6A 和 B)。此外,在布达哈山(DBD)的南斜坡,位于珊瑚—微生物黏结灰岩的下倾方还出现有海绵—微生物粘结灰岩(B4,附图 S6C 和 D);在该部位,随着坡度角增大(由约 20°增大到 29°),生物建造相和粗粒(砾到巨砾)再沉积岩相(砾级到巨砾级砾屑灰岩和角砾岩,S5,附图 S6E 和 S6)(Verwer 等,2009b))也相应增多。在布达哈山(DBD)的斜坡带,前积层的远端部分由成层性好的粒屑沉积组成(颗粒灰岩、颗粒灰岩或泥晶颗粒灰岩,S3,附图 S6F 和 S4;砾级到巨砾级砾屑灰岩,S5 和 S6)。向洋盆方向粒屑沉积尖灭、相变为大洋盆地沉积。后者则是由互层的绿色泥灰岩(S1)、细粒似球粒和生屑泥晶颗粒灰岩到颗粒泥晶灰岩(S2,附图 S6G 和 H)组成,灰岩当中常含有燧石结核和放射虫。

在大洋盆地区标志层序 V 开始沉积的凝缩段中,采集到的菊石类表明,沉积欠补偿状况始于 Ibex 带,经过 Davoei 带(由 Valdani 亚带到 Figulinum 亚带)一直持续,并且可能持续到 Stokesi 亚带(Margaritatus 带),所对应的时代为卡瑞辛期(图 5-3)。不过,如从大洋盆地到南斜坡各个不同地点采集的菊石样本所证明的,该凝缩段顶面的时代在朝台地方向上趋于更早一些。

5.4.5 层序Ⅵ

层序Ⅵ为楔形地层单元,台地顶部区和斜坡区分辨明显。在位于中心的台地内部区,层序Ⅵ上超了由层序Ⅰ和Ⅱ的岩溶化沉积和古生界岩石或中大西洋岩浆省(CAMP)玄武岩所构成的暴露古地貌,因此层序Ⅵ的底界属角度不整合(附图 S10A)。布达哈山(DBD)现今的地形地貌贴切地代表了在层序Ⅵ沉积结束时的台地几何形态(图 5-13)。其中可细分出九条主要相带,它们具有特征的相组合(附表 S2),依次环绕着布达哈山(DBD)中北部区的暴露古地貌高地、呈同心状分布(图 5-13)。首先是陆相沉积围绕着中北部区分布,并形成宽泛的陆源沉积楔,这些陆相沉积包括含根系有机质的红色页岩和粉砂岩(C1)和常见的钙结岩(C2,附图 S7A 和 B),属滨岸平原环境沉积的产物,记录了从中大西洋岩浆省(CAMP)玄武岩和下伏古生界基底剥蚀而来的硅质碎屑沉积物的输入。台地顶部地层为一加积序列,由几米厚、板状层理和平行层理的组合单元所构成。组合单元的岩相有:潮下带沉积(L1、L2、L3 和 L4,附图 S7A、C 和 D),和潮间—潮上带沉积(T2、T3、T4 和 T5,附图 S7E)。接下来是一个宽 300~800m 的相带—滩脊障壁,其特征性的岩相有:具交错层理的砾屑灰岩和颗粒灰岩(G5;在 Verwer 等 2009a 中的 CGG 岩相;附图 S7F)、具窗格构造和帐篷构造的潮间到潮上带沉积(T3、T2;在 Verwer 等 2009a 中分别为 PPG 和 FPP 岩相);该砾状包粒沙坝体系位于台地内,距离实际台地坡折有 150~400m,并且在南缘更为发育。潮下带沉积从台地坡折向台地内延伸 200~500m,岩相有:包粒、似球粒和生屑颗粒灰岩和砾屑灰岩(G4 和 G6,附图 S8A 和 B),核形石砾屑灰岩—漂浮灰岩(L2,附图 S8C)及双壳类 lithiotid 滩(L4,附图 S8D),珊瑚—微生物粘结灰岩(B3 和 B5,附图 S8E)和米级四射珊瑚丛状体。在台地坡折的下方是一断续分布的上斜坡礁,它一直延伸到古水深 60~100m 处;该上斜坡礁当中发育块状到成层性差的珊瑚微生物粘结灰岩(B3 和 B5)及少量富海绵微生物粘结灰岩(B4);粘结灰岩单元与包粒颗粒灰岩(G4)

交互,再往下,则横向相变为来自礁和台地顶部的再沉积相(S5、S4 和 S3);后者向洋盆内斜倾,倾角达 19°到 32°。

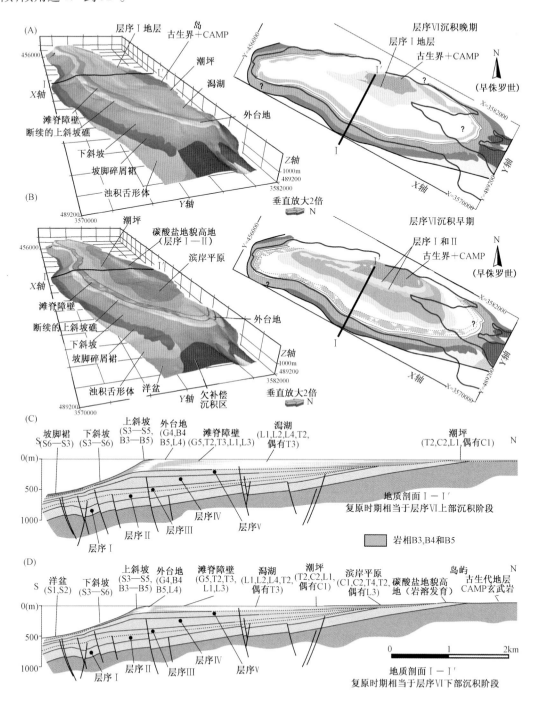

图 5-13 (A)和(B)分别为层序Ⅵ沉积晚期和早期的相带分布图、3D 沉积模式斜视图,3D 沉积模式是由数字露头模型恢复得到的,其中,用实线圈出的范围代表有直接露头数据来完成相带恢复的地方,所用色带的图例描述参见图 5-3;(C)和(D)南北向综合地质剖面分别展示层序Ⅵ沉积晚期和早期的几何形态、沉积相分布(有关岩相、相组合和相带的详细描述参见附表 S1 和 S2)

下斜坡和斜坡底部呈现出一个上凹形的剖面，上部倾斜15°到19°，朝大洋盆地方向，变缓至小于8°。来自台地和台地边缘的沉积物（S3、S4、S5 和 S6）形成规模几千米大的扇体，分布在南斜坡的坡脚，在横向上联成一个断续分布的带（Verwer 等，2009b）。这些沉积的出现证明了南缘上部经历了一系列块体崩坏（mass-wasting）事件。大洋盆地则以下列岩相互层为特征：泥灰岩（S1）、夹含燧石结核的细粒似球粒颗粒泥晶灰岩（S2）、板状薄层骨屑泥晶颗粒灰岩或颗粒灰岩（S3）。

由于在层序Ⅴ和Ⅵ的过渡带缺乏生物地层资料，因此不能准确鉴定这两个层序分界的时代。从层序Ⅴ底部大洋盆地区凝缩段采集到的最年轻的菊石组合（为 Margaritatus 带的 Stokesi 亚带）属早多米尔期（图 5-3），我们以此来间接锁定该层序分界所属时代的范围。在布达哈山（DBD）台地顶部区该层序分界以陆上暴露为特征，它可能与在撒哈拉台地记录到的早多米尔期大范围台地暴露事件相对应（Wilmsen 和 Neuweiler，2008）。在撒哈拉台地，台地海泛发生在上 Margaritatus 带和下 Spinatun 带（多米尔期的 p.p. 菊石亚带），导致了又一期台地加积（Wilmsen 和 Neuweiler，2008）。这期加积可能与层序Ⅵ属同期（图 5-3）。据研究表明，层序Ⅵ最年轻地层的时代是普林斯巴赫末期（Spinatun 带；Du Dresnay，1976；Crevello，1991；Campbell 和 Stafleu，1992）。

5.4.6　淹没期的和淹没期后的地层

在台地顶部，层序Ⅵ之后的沉积是分米到米级厚的地层，在该地层的底，局部可见明显不整合面，且常有岩溶特征（附图 S8F）。这些地层包含有：（1）红色颗粒泥晶灰岩或泥晶颗粒灰岩——富含腕足类、海百合以及属 Tenuicostatum 带（时代为托尔期伊始，Du Dresnay，1976）的菊石；（2）局部产出的海绵微生物黏结灰岩层——在布达哈山（DBD）最北边的露头中，层几米厚，分米级宽度。最常见的岩相是红色、富菊石的瘤状颗粒泥晶灰岩，其中含磷酸盐豆粒（D1）。沿台地斜坡见有红色海百合颗粒灰岩或泥晶颗粒灰岩，富含菊石、腕足类和泥晶化颗粒（D2）。在掩蔽的内部孔洞中，有纤维状和放射轴状纤维状方解石胶结物局部围绕骨屑颗粒形成毫米厚的壳。所有这些类型的岩相还填充了水下岩墙；这些厘米到米宽、近垂直的岩墙交叉穿切早期形成的碳酸盐沉积。它们记录了碳酸盐台地淹没和欠补偿沉积一直到早阿林期（Du Dresnay，1976；Blomeier 和 Reijmer，1999；Elmi 等，1999）。

托尔阶绿色粉砂质页岩和灰岩或泥灰岩的互层（与下托尔阶的 Tagouditte 组相当，Lachkar，2000）、阿林阶—下巴柔阶泥晶灰岩上超于布达哈山（DBD）的斜坡带，表明此时布达哈山（DBD）台地俨然为一个欠补偿沉积的海底隆起。

5.5　沉积层序的演化与构造控制

5.5.1　层序内的几何形态和沉积相演化

层序Ⅰ和Ⅱ具有以下特征：地层结构为平行层状、没有明显的内部斜交接触关系，无同沉积断裂或掀斜活动的迹象（图 5-14）。它们与区域广泛分布的 Idikel 组相对应说明，层序Ⅰ和Ⅱ是当时那个依托撒哈拉克拉通成核的宽阔碳酸盐体系的一部分。它们与大阿特拉斯盆地的其他地方的 Idikel 组厚度相近（Mehdi 等，2003；Wilmsen 和 Neuweiler，2008），表明裂谷盆地的广大区域拥有相对统一的沉降速率。

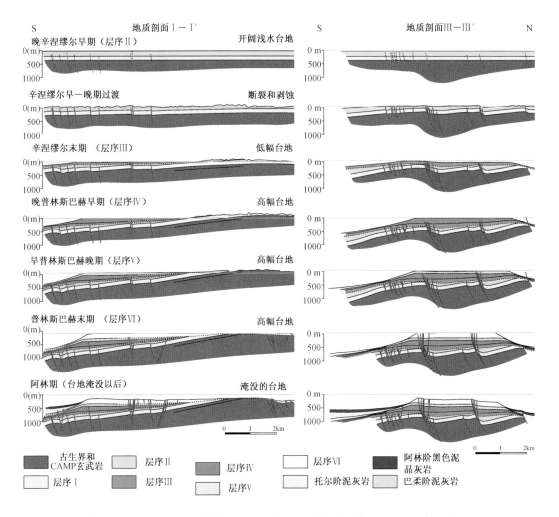

图 5-14 通过对地质剖面 Ⅰ—Ⅰ′和Ⅲ—Ⅲ′进行分层复原,再现了布达哈山(DBD)碳酸盐台地的发育过程

地质记录显示,大约在洛林期伊始,普遍有拉张断裂开始活动。以前的层序Ⅰ和Ⅱ地层被断切成许多小的、WSW—ENE 向断块,并且这些断块沿着北倾的正断层逐渐向南旋转。此时,北部的一条大滑脱断层确立和界定了布达哈山(DBD)大断块的下盘。在布达哈山(DBD)中北部,断块旋转引起下盘上升,并且形成了一个 WSE—ENE 向延伸的岛屿,成为该碳酸盐岩台地演化的一大特色。洛林期(层序Ⅲ沉积时),布达哈山(DBD)成为了一个加积—退积型、低幅碳酸盐台地,仅在局部有较陡的地势(图 5-6 和图 5-14)。在此阶段,下盘隆起的东端似乎仍浸没在海水里,因此缺乏浅水、浪基面以上的碳酸盐生产,而以具有海绵生物丘的中—外缓坡相为主。

层序Ⅳ沉积期(卡瑞辛期),台地的生长以台地加积和边缘退积为特征,因此导致了台地到大洋盆地的落差加大。沿着南部边缘,随着一条断续的珊瑚—海绵礁镶边筑起,斜坡带变得越来越陡。相反,在东部,由台地顶部向大洋盆地的过渡带却是数公里宽、局部受欠补偿沉积影响的缓降型斜坡(图 5-14)。最后,发生大退积、碳酸盐沉积体系几乎完全消亡和相对海平面上升,宣告了层序Ⅳ的终结;整个布达哈山(DBD)斜坡带及其附近,凝缩序列沉积和欠补偿沉积遍布,记录了该事件的发生。由于相对海平面上升,台地顶部和大面积原先出露的岛屿又被淹没了。

层序Ⅴ沉积伊始,在浅水中心区产出的碳酸盐砂屑形成"S"形前积层,它们由布达哈山(DBD)的中北部、呈放射状向南、向东、向东北,迅速前积在下伏层序Ⅳ被淹没和欠补偿沉积的台地顶部和斜坡带上(图5-12和图5-14)。沿着台地边缘还构筑起了珊瑚—海绵—层孔虫礁并伴有大量微生物集群。随着层序Ⅵ开始沉积,下盘区不断遭受剥蚀,导致已沉积的碳酸盐层系被剥走,露出中大西洋岩浆省(CAMP)玄武岩和古生界基底。同时,剥蚀而来的大量黏土质泥和粉砂在陆地滨岸环境堆积,它们绕岛形成大范围分布的陆源沉积楔(图5-13B和图5-14)。

层序Ⅵ沉积期,台地顶部地层的加积导致了高幅(达450~600m)碳酸盐岩台地的发育。这种地势促成斜坡不稳定、引起边缘垮塌(可能由地震活动触发)。在普林斯巴赫期末,相对海平面下降造成了陆上暴露及岩溶改造,这在布达哈山(DBD)台地顶部的大部分区域都有记录。该事件过后,台地开始淹没。托尔期和早阿林期,布达哈山(DBD)成了一个欠补偿沉积的海底隆起(图5-14);隆起又被复杂的网格状水下岩墙和断裂切割。这些断裂构造是在层序Ⅵ沉积末期开始发育,进而控制了布达哈山(DBD)的淹没和淹没后的演化;但是,这些构造对早期阶段的台地发育可能只起了轻微的控制作用。

5.5.2 构造对层序的控制

通过对布达哈山(DBD)断裂与断块活动、碳酸盐岩台地内部架构及其演化的分析,揭示出了一些有助于弄清拉张大地构造活动在台地发育中所起作用的重要特征。布达哈山(DBD)东端的露头、特别是在盖斯尔—穆盖勒谷(KMG),出露极好,清楚地揭示了与早侏罗世拉张活动相关的构造,以及这些构造与布达哈山(DBD)层序和淹没后地层的关系。最普遍的构造是向北倾、走向介于70°~100°N之间的正断层,以及向南倾的次级共轭正断层。在布达哈山(DBD)的东北端,三条断距大(250~400m)的主干断裂或断裂带部分地切割出宽两公里、向南旋转的断块群(图5-2B)。在断块的内部,无论是布达哈山(DBD)断块,还是前述几公里宽的次断块,都具有由正断裂活动形成的高度内部破裂;这些正断裂主要影响到的是层序Ⅰ到Ⅲ的地层。尽管如此,沿着沉积序列自下而上,断裂活动的程度逐渐减弱。因此,上普林斯巴赫阶和托尔阶的地层只显示出有限的断裂作用和相关的变形。在布达哈山(DBD)断块和前述几公里宽的次断块的范围内,层序Ⅰ和Ⅱ的地层被切割成数目众多窄的(数十到数百米宽)、向南旋转的断块。穿过层序Ⅲ的断层显示小的断距,从几米到数十米,并且越往上,断层错动越小,位移被断裂相关的单背斜所调节(图5-8C),和/或者被上覆地层给封固住了。另外还可以明显注意到的是,在地层序列厚的地方,上部地层(层序Ⅳ、Ⅴ和Ⅵ)中的断裂发育非常有限,可能是由于沉积速率超过了断滑的速率。

层序Ⅲ到Ⅵ总体呈楔状几何形态,向南缘、西缘和东缘变厚,并且与旋转断块上发育的碳酸盐台地中所描述的"以不整合面分界的沉积层序"相当(Bosence,2005)。层序Ⅴ和Ⅵ的水平层状台地顶部地层,向布达哈山(DBD)的中北部呈现收敛的地层几何形态,并且越靠地层下部,倾角越大;此外,这些层序还显示有:层序内的角度不整合、钙结层频繁与潮下潟湖沉积交互。所有这些观测结果说明,沿着布达哈山(DBD)WSW—ENE向北边缘下盘上升,同时沿着东、南和西边缘上盘下降,正像在旋转断块上发育的沉积体系中所描述的那样(Gawthorpe等,1997;Bosence等,1998;Cross等,1998;Cross和Bosence,2008;Dorobek,2008)。这表明:自洛林期开始,布达哈山(DBD)碳酸盐台地演变成了一个生长在向南旋转断块上的孤立碳酸盐台地。布达哈山(DBD)断块呈 WSW—ENE 向拉长形,以两条主要滑脱断层为界;滑脱断层沿着在大阿特拉斯山脉所观察到的区域走向(即 WSW—ENE 向;Ellouz 等,2003;Wilmsen 和 Neuweiler,2008)延伸过来。其中,位于布达哈山(DBD)南边的 Bou Ferma 山断裂(JBFF)把布达哈山(DBD)断块与拉希迪耶断块分割开来(图 5-1D 中的 JBFF)。布达哈山断块的下盘由一条向台地东西两侧延伸的复杂断裂带所界定,布达哈山(DBD)由东到西在几何形态和岩相架构上有明显的差异,反映出沿着该断裂带断层的滑动和构造驱使的可容空间都有变化(图 5-14)。

本研究基于倾角测量给出了层序Ⅲ到Ⅳ沉积期的平均同沉积断裂旋转估值;倾角测量的对象包括:台地顶部地层层面(例如,潮下到潮间的过渡层,被作为大体水平的古标志)和层序的几何形态(图 5-15、附图 S11 和 S12、附表 S3 到 S10)。基于观测的构造旋转和 DBD 南缘产生的可容空间,我们还推算出了布达哈山(DBD)断块下盘的构造抬升值。

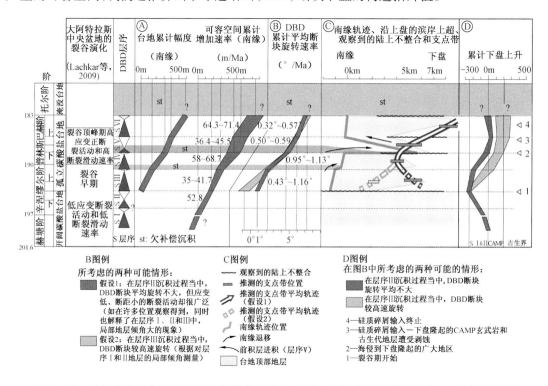

图 5-15 以年代地层柱为标尺对比分析布达哈山(DBD)南缘台地幅度的演化、台地地层的演化、南缘可容空间增加的变化、DBD 断裂旋转的估测值以及中北部区的相对构造抬升;此外,图中也标注了中北部记录到的一些主要地质事件(具体解释参看附图 S11 和 S12)

针对在层序Ⅲ沉积期的构造旋转,提出有两种可能的情形。第一种情形推断认为,广泛发育的正断层就像在布达哈山(DBD)和盖斯尔—穆盖勒谷(KMG)东部露头看到的那样:断滑速率小、断距小(介于几米到几十米)。较小断块的构造旋转(就像对层序Ⅰ、Ⅱ和Ⅲ的地层进行倾角测量得到的结果:多数 >2°)只是让整个布达哈山(DBD)断块发生一个向南平缓的旋转(<2°)(附表 S4 到 S6)。此假设与本文介绍的大部分地质观测结果一致,也与 Lachkar 等(2009)提出的模式相吻合,因此我们认为是最有可能的情形。第二种情形假定,布达哈山(DBD)断块在层序Ⅲ沉积期的旋转速率较高(附表 S7 到 S10),且要求 Bou Ferma 山断裂(JBFF)断滑速率也高。

图 5-15 提出的数据说明：至晚从层序Ⅳ开始沉积起，布达哈山（DBD）断块的平均旋转速率随着时间在减小，支点带（即零位移点）的位置由上盘斜坡带在向下盘隆起迁移。这些数据提供了有关整个布达哈山（DBD）断块沉降及其构造演化方面的信息，据此可以区分出四个不同的演化阶段：

（1）层序Ⅲ沉积期，尽管其岩相的特点是较深水潮下带，但是，可容空间的生成速度很可能比在层序Ⅰ和Ⅱ沉积期要低。这看似与其正处在新一期裂谷开始的构造背景相矛盾，但可以从其所处的位置来解释，即布达哈山（DBD）台地是绕着布达哈山（DBD）断块的下盘分布。

（2）层序Ⅳ沉积期，可容空间的生成速度加快，布达哈山（DBD）断块的旋转速率增高，可能伴随着支点带向南迁移。这种效应也许是由于：Bou Ferma 山断裂（JBFF）上盘下降加速（断裂及其上盘位置见图 5-1C 和 D），同时那些在层序Ⅲ沉积期切割布达哈山（DBD）断块的许多小型正断层逐渐停止活动。这刚好与大阿特拉斯盆地的裂谷高峰期（即裂谷坳陷期）开始相吻合（Lachkar 等，2009）。

（3）层序Ⅴ沉积期，整个布达哈山（DBD）断块的沉降速率显著减小，从而减少了台地顶部区可容空间的生成，有利于前积层进积。

（4）层序Ⅵ沉积期，沉降速率再度加大，致使支点带向下盘方向迁移。这次沉降速率的加大可能与在 Bou Ferma 山断裂（JBFF）、和／或者与在大阿特拉斯盆地东部的南阿特拉斯断裂（SAF）上盘大幅下降有关（断裂及其上盘位置见图 5-1C 和 D）。

支持上述构造演化的沉积学方面的证据和露头观察结果包括：（1）布达哈山（DBD）中北部层序Ⅰ和Ⅱ陆上暴露和岩溶改造，再有岛屿的形成；（2）在布达哈山（DBD）中北部（即向支点带的下盘方向）可容空间迅速减小，层序Ⅴ的前积序列陆上暴露，同时在上盘引发前积层顺坡进积，导致构造强迫型海退；（3）层序Ⅴ和Ⅵ的台地内部地层以广泛发育大气淡水成岩改造和钙结层为特征；（4）在层序Ⅴ到层序Ⅵ的过渡时期，由于布达哈山（DBD）中心区整个碳酸盐序列被剥蚀，硅质碎屑开始输入到布达哈山（DBD）的内部区域；（5）在层序Ⅵ沉积晚期，下盘隆起的大部分地方被淹没。

5.6 讨论：控制碳酸盐台地几何形态和演化的因素

尽管构造体制显然对布达哈山（DBD）台地体系的发育施加了重要的影响。然而，大地构造活动、全球海平面升降和环境—古海洋水文条件变化是三个基本控制因素，只有进行细致的分析才能够揭示三者对于促成几何形态、地层结构、岩相类型和叠置形式所起作用的相对大小。在以下两节我们将介绍：（1）所观测的沉积层序与全球海平面升降曲线、与区域相对海平面变化记录的对比；（2）区域环境—古海洋水文条件变化的记录与布达哈山（DBD）主要颗粒类型、建隆类型和台地样式观察结果的综合和对比。由于对层序Ⅰ和Ⅱ缺少准确的生物地层学界定，因此，讨论主要聚焦在布达哈山（DBD）碳酸盐台地发育的晚期阶段，即从层序Ⅲ到层序Ⅵ。

5.6.1 可容空间：大地构造活动与全球海平面升降

对于在同裂谷期发育的断块台地来说，构造活动对于台地的架构、对于不整合面和沉积层序的发育起主要的控制作用（Bosence 等，1998；Dorobek，2008）。Bosence 等（1998）证明过，在这种类型的碳酸盐台地中，楔形、以不整合为界的层序可以经由断块的连续脉冲式旋转与构造平静期交替而产生。在布达哈山（DBD）碳酸盐台地，构造旋转被证明是层序Ⅲ到Ⅵ楔状几何形态发育的动因，因为这些层序与用 Bosence 等（1998）模型预测的层序类型非常相似。尽管如此,布达哈山(DBD)所呈现出的复杂性比前述模式更大。一个关键的差别就是,在布达哈山(DBD),构造旋转看起来是一个相当连续的、而不是间断的过程。此外,有些观察到的层序边界也不能用快速脉冲式断块旋转来解释。层序V的底对应的是一个海侵面。在推测的支点带以外(向北)好几千米的地方,该底界以硬底发育为特征。假如当时发生的是快速脉冲式构造旋转的话,那么北边这些地方本应该受到构造抬升和陆上暴露的影响。再就是层序VI顶部广泛发育的陆上暴露,同样存在类似的问题,因为该陆上暴露还影响到了支点带的南侧一方。如果是脉冲式旋转的话,这些地方本应该发生沉降。显然,在引起可容空间变化、进而控制布达哈山(DBD)沉积层序发育的过程中,除了构造旋转作为动因外,还应涉及到其他别的机制;其中有以下两个主要的影响因素:

(1)整个布达哈山(DBD)断块沉降速率的波动,起因是 Bou Ferma 山断裂(JBFF)或者南阿特拉斯断裂的上盘下降的变化,例如层序Ⅳ和Ⅵ沉积期都对应一次重大沉降,但在层序Ⅴ沉积期,整个布达哈山(DBD)断块的净沉降却减少了(图5-15)。

(2)时间量级为数千年到一百万年的平均海平面波动速率与构造活动产生的可容空间变化相当(Dorobek,2008)。侏罗纪,海平面变化的总形式是在一个长期上升的大趋势(参见Hallam,1988,2001;Haq 等,1988;Surlyk,1991;Li 和 Grant-Mackie,1993;Hardenbol 等,1998)上,不时穿插以低幅的波动(图5-16),这些三级波动要么是是由海平面下降引起的(Haq 等,1988;Surlyk,1991;Li 和 Grant-Mackie,1993),要么是由海平面静止期形成的(Hallam,1988,2001)。

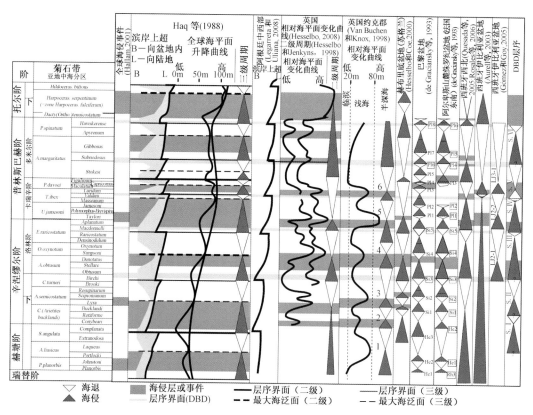

图5-16 在年代地层图上将布达哈山(DBD)层序与不同作者所建立的早侏罗世全球海平面升降、相对海平面变化曲线、上超曲线、海侵—海退周期进行对比

图 5-16 将布达哈山(DBD)层序的时代划分与已发表的资料进行对比,包括:早侏罗世的海平面变动、滨岸上超曲线和海进—海退旋回。对比效果尚存不足,主要是由于时代界定的不确定性,即布达哈山(DBD)层序分界的时代和全球侏罗纪海平面升降的时代。尽管在不同沉积盆地当中识别出的层序分界和海进—海退旋回的数目和时代可能存在差异,但是,许多重大海侵却是广为人知的(图 5-16,表 5-2),并且在布达哈山(BDB)也记录到了,包括:那些标志着层序Ⅲ、Ⅳ和Ⅴ开始的海侵,和标志着布达哈山(DBD)碳酸盐台地淹没的海侵(图 5-16,表 5-2)。此外,广泛被记录到的晚 Pleuroceras spinatum(带)海平面下降同样在布达哈山(DBD)层序Ⅵ的顶部也识别到了。最后,布达哈山(DBD)层序的各个分界(层序Ⅲ以上)与以下区域变化有显著的对应关系:(1)由 Haq 等(1988)建立的三级层序;(2)由 Legarreta 和 Uliana(1996)专为阿根廷中西部建立的滨岸上超曲线;(3)由 Gómez 和 Goy(2005)描述的西班牙伊比利亚盆地晚辛涅缪尔期—普林斯巴赫期海进—海退旋回。

表 5-2　在辛涅缪尔期、普林斯巴赫期和托尔早期广泛识别到的海侵或海水变深事件和海退或海水变浅事件

地质年代	菊石带	海侵或海水变深;海退或海水变浅	盆地和地理分区	布达哈山
托尔早期	Dactylioceras tenuicostatum to Harpoceras Serpentinum (= zone Harpoceras Falciferum) Zones	海侵或海水变深	阿根廷(Legarreta 和 Uliana,1996),英国(Hesselbo 和 Jenkyns,1998,Van Buchen 和 Knox,1998,Hesselbo,2008),巴黎盆地(Graciansky 等,1998),阿尔卑斯山麓侏罗纪盆地(Graciansky 等,1993),北海(Jacquin 等,1998),西班牙(阿斯图里亚斯和巴斯克-坎塔布连盆地,以及伊比利亚盆地,Aurell 等,2003、Gómez 和 Goy,2005;Quesada 等,2005,Rosales 等,2006),意大利(南阿尔卑斯山,Cobianchi 和 Picotti,2001),大阿特拉斯盆地(摩洛哥,Mehdi 等,2003;Wilmsen 和 Neuweiler,2008),突尼斯(Soussi 和 Ben Isma?,2000),沙特阿拉伯,德国,北美北极圈地区,阿尔伯达,马达加斯加,巴基斯坦,(Hallam,1988)	在层序Ⅵ顶部有海侵;台地淹没
多米尔末期	Pleuroceras spinatum Zone (hawskerense subzone)	海退或海水变浅	阿根廷(Legarreta 和 Uliana,1996),英国(Hesselbo 和 Jenkyns,1998;Hesselbo,2008),北海(Jacquin 等,1998),巴黎盆地(Graciansky 等,1998),阿尔卑斯山麓侏罗纪盆地(Graciansky 等,1993),西班牙(阿斯图里亚斯和巴斯克—坎塔布连盆地,以及伊比利亚盆地,Aurell 等,2003;Gómez 和 Goy,2005;Quesada 等 2005,Rosales 等,2006)	在层序Ⅵ顶部有海退
多米尔早期	Mid Amaltheus margaritatus Zone (Subnodosus subzone)	海侵或海水变深	阿根廷(Legarreta 和 Uliana,1996),英国(Hesselbo 和 Jenkyns,1998;Hesselbo 2008),阿尔卑斯山麓侏罗纪盆地(Graciansky 等,1993),北海(Jacquin 等 1998),西班牙最大海侵(阿斯图里亚斯和巴斯克—坎塔布连盆地,Quesada 等,2005;Rosales 等,2006)	在层序Ⅵ底部有海侵

续表

地质年代	菊石带	海侵或海水变深；海退或海水变浅	盆地和地理分区	布达哈山
卡瑞辛晚期	Late Tragophylloerans ibex Zone（Valdani?，Luridum subzones）to early Prodactylioceras davoei	海侵或海水变深	格陵兰（Surlyk，1991），阿根廷（Legarreta 和 Uliana，1996），英国（Hesselbo 和 Jenkyns，1998；Van Buchen 和 Knox，1998；Hesselbo，2008），巴黎盆地（Graciansky 等，1998），阿尔卑斯山麓侏罗纪盆地（Graciansky 等，1993），瑞士南部，德国，波兰，内华达和加拿大北极圈（Hallam，1988）；西班牙最大海侵（阿斯图里亚斯和巴斯克—坎塔布连盆地，以及伊比利亚盆地，Aurell 等，2003；Gómez 和 Goy，2005；Quesada 等，2005；Rosales 等，2006）	在层序Ⅴ底部有海侵
辛涅缪尔末期—普林斯巴赫初期	Late Echinoceras raricostatum Zone（aplanatum subzone）to Uptonia jamesoni Zone	海侵或海水变深	格陵兰（Surlyk，1991），苏格兰（Stephen 和 Davies，1998），阿根廷（Legarreta 和 Uliana，1996），英国（Hesselbo 和 Jenkyns，1998；Van Buchen 和 Knox，1998；Hesselbo，2008）和西班牙（阿斯图里亚斯和巴斯克—坎塔布连盆地，以及伊比利亚盆地，Aurell 等，2003；Quesada 等，2005；Rosales 等，2006），伊比利亚盆地（Gómez 和 Goy，2005）	在层序Ⅳ底部有海侵
辛涅缪尔早期与晚期的过渡	Oxinoticeras obtusum Zone	海侵或海水变深	阿根廷（Legarreta 和 Uliana，1996），英国（Hesselbo 和 Jenkyns，1998；Van Buchen 和 Knox 1998，Hesselbo 2008），西班牙（阿斯图里亚斯和巴斯克—坎塔布连盆地，Aurell 等，2003；Quesada 等 2005，Rosales 等，2006；伊比利亚盆地，Gómez 和 Goy，2005；以及西班牙南部贝蒂克山脉，O'Dogherty 等，2000），大阿特拉斯盆地（摩洛哥，Mehdi 等，2003，Wilmsen 和 Neuweiler，2008）	在层序Ⅲ底部有海侵

如果考虑到所有露头研究都存在的明显限制，以及布达哈山（DBD）层序分界时代确定方面存在的不确定性，会不难发现由布达哈山（DBD）层序所记录的可容空间旋回有复杂的成因。比如，套用 Bosence 等（1998）提出的模型简单地以布达哈山（DBD）断块的构造旋转为动因，不能解释层序Ⅲ到Ⅵ。因此就提出另外一种模式（图 5-17），该模式考虑下列过程的互动：(1) 布达哈山（DBD）断块的连续旋转；(2) 由于断滑速率的变化以及 Bou Ferma 山（JBF）断裂和南阿特拉斯断裂的上盘下降，造成整个布达哈山（DBD）断块沉降速率的变化；(3) 广泛记录到的（全球？）海平面升降，如：洛林期伊始（Obtusum 带）、洛林末期到早卡瑞辛期（Aplanatum 亚带到 Jamesoni 带）、晚卡瑞辛期（Valdani?—Luridum 亚带到下 Davoei 带）和早多米尔期（Subnodosus 亚带）的海侵，和晚多米尔期（Hawskerense 亚带）海退事件；(4) 该体系的碳酸盐沉积物生产量的变化。

5.6.2 碳酸盐工厂的作用

碳酸盐工厂对岩相特征具有主导控制作用，它影响碳酸盐体系的几何形态、进—退积趋势

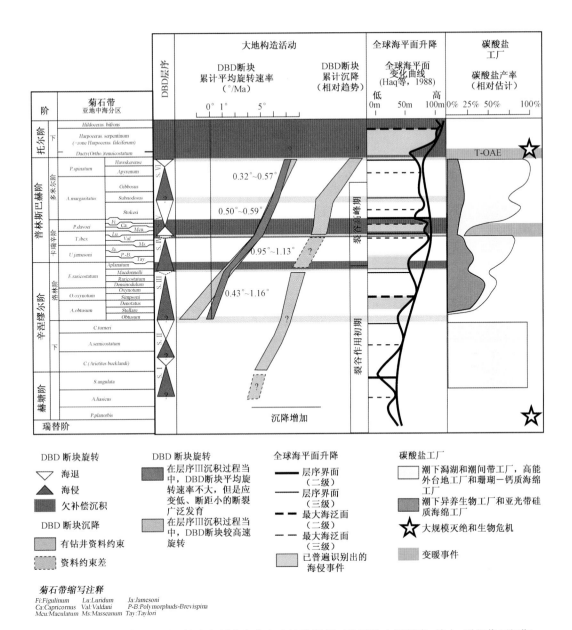

图 5-17 在年代地层图上综合与层序发育和台地淹没相互作用的主要因素,其中,裂谷作用初期和裂谷高峰期参照 Lachkar 等(2009)识别的裂谷期,辛涅缪尔末期和卡瑞辛晚期的欠补偿沉积主要影响大洋盆地和台地斜坡

和消亡(Pomar,2001;Schlager,2003,2005;Kenter 等,2005;Pomar 和 Hallock,2008)。在布达哈山(DBD),通过分析构成各层序的岩相以及它们在不同时段的相对比例,可以识别出不同类型的碳酸盐工厂及其与台地发育有关的演化(图 5-18)。碳酸盐工厂的变化影响了台地斜坡的几何形态、台地边缘的构型以及台地最终的淹没(图 5-18)。

层序Ⅰ和Ⅱ的沉积物大部分由似球粒构成,另有少量鲕粒和豆粒,以及骨屑(钙藻、有孔虫、腹足类、双壳类),为广阔浅水潟湖和环潮坪环境的产物。这些沉积物具有干旱到半干旱气候条件下早期(同沉积)潮上带白云岩化的特征(Mehdi 等,2003;Wilmsen 和 Neuweiler,2008;Lachkar 等,2009)。

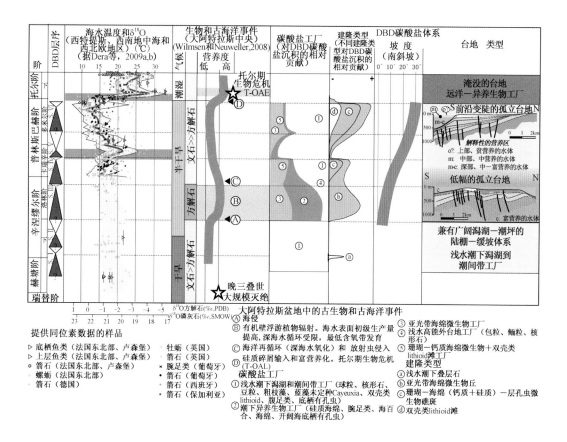

图 5-18 在年代地层图上的综合对比内容包括:下侏罗统 $\delta^{18}O$ 的变化和推测的海水温度(据 Dera 等,2009a,b),大阿特拉斯中央盆地的古气候、碳酸盐沉淀、营养水平以及主要生物和古海洋事件(据 Wilmsen 和 Neuweiler,2008),布达哈山(DBD)的古碳酸盐工厂、建隆类型、碳酸盐台地类型(本次研究)

继层序Ⅰ和Ⅱ特征的潟湖—环潮坪碳酸盐工厂消失之后,较深水潮下带(中和外缓坡)的底栖悬食生物和海绵微生物丘成为层序Ⅲ碳酸盐岩沉积的主角。层序Ⅲ具有的特征包括:地层厚度不大、布达哈山(DBD)台地斜坡缓倾而呈低幅碳酸盐台地的几何形态、缺乏潟湖—环潮坪带碳酸盐沉积、以较深水潮下带沉积物为主。这些特点表明:(1)碳酸盐生产量相对偏低;(2)碳酸盐体系追赶可容空间增加的能力降低;(3)形成陡倾海底地形的能力下降。碳酸盐工厂的这种大转变与拉张构造重新活动同时同步。这次拉张构造活动把先前的广阔环潮坪台地切割成许多旋转的断块。由于断块下盘陆上暴露、而断块上盘下沉到更大的水深,结果,浅水区(例如,在浪基面以上)的面积大幅减少。此外,据 Wilmsen 和 Neuweiler(2008)在大洋盆地同时代的记录表明:(1)伴随着有机壁浮游藻类的辐射海水营养水平增高,(2)初级生产量增加。这两方面的因素可能都驱使洛林期阶底的布达哈山(DBD)碳酸盐工厂产生了上述改变。

在普林斯巴赫期(图 5-18),越来越多的造架生物(像群体珊瑚和层孔虫)、共生的原地泥晶(微生物的)粘结作用,有利于复杂礁群落的发育。这些生物礁群落生长在上斜坡水浅的部位(水深<70m 至 100m),而硅质海绵微生物的生产立足于上斜坡水深的部位(水深 70~140m),后者就是层序Ⅲ的中缓坡生物丘的缩影。原地生物礁的生产迅速改变了碳酸盐体系的格局、大幅提高了浅水碳酸盐的生产,具体表现包括:(1)允许构筑越来越陡的斜坡(坡度由

层序Ⅳ沉积期的8°~19°,在层序Ⅴ和层序Ⅵ持续变陡达到32°);(2)有利于发育具有浅水台地顶部环境的平顶体系,并在其中重新建立起了潟湖—环潮坪碳酸盐工厂。

普林斯巴赫期,珊瑚—海绵—层孔虫生物礁群落在台地边缘发育,大量双壳类lithiotid出现在位于浅水潮下带的外台地和潟湖沉积当中。这些被认为是三叠纪大灭绝结束后物种多样化的复苏(Wilmsen和Neuweiler,2008)。在布达哈山(DBD)这些生物群落也标志着浅水水体营养水平的降低。Leinfelder等(2005)认为,侏罗纪礁体中层孔虫和珊瑚共生是贫营养正常海水的标志。然而,在斜坡带出现硅质海绵微生物丘或许表明中养到富养条件在较深的水体当中依然持续,这可能与水体分层有关。

在若干已发表的断块碳酸盐研究中,上盘是缓坡式沉积的场所并不会演变成高幅体系(Ebdon等,1990;Burchette和Wright,1992;Wilson等,2000;Ruiz-Ortiz等,2004)。对于布达哈山(DBD)观察到的陡斜坡,还有中新世断块台地的陡斜坡(Cross等,1998;Brachert等,2002),可以认为是由以下原因造成的:(1)浅水高产的碳酸盐工厂;(2)与构造沉降无关的相对海平面上升或是受构造沉降而增强的相对海平面上升;(3)粘结生物或微生物岩的斜坡加固作用。布达哈山(DBD)在层序Ⅳ、Ⅴ和Ⅵ沉积期,具备这些所要求的条件。上盘下降和沉积剖面旋转迅速形成可容纳空间,迫使碳酸盐台地主要进行加积和退积,致使台地起伏加大并抑制了台地的进积。

关于拉张构造环境中碳酸盐岩台地的淹没已提出了不同的成因模式,有构造沉降的单因素作用成因、构造活动与区域沉降的复合作用成因,还有因环境变化而降低碳酸盐生产量的多因素作用成因(Santantonio,1993,1994;Bosence等,1998;Blomeier和Reijmer,1999;Ruiz-Ortiz等,2004;Wilmsen和Neuweiler,2008;Lachkar等,2009)。后一种模式可能也适用于布达哈山(DBD)碳酸盐台地。因为布达哈山(DBD)的淹没主要是由于碳酸盐产率降低造成的。事实上,布达哈山(DBD)在层序Ⅵ顶出现陆上暴露事件以后,浅水碳酸盐工厂并没有能够获得重建;虽说也有古地貌高地,即布达哈山(DBD)的中心岛,在随后可容空间迅速增加的进程中,它本来可以作为碳酸盐生产的成核区。然而,有以下一系列因素可能对布达哈山(DBD)的各种碳酸盐工厂产生了负面影响:(1)早托尔期爆发的大洋缺氧事件(Jenkyns,1988),及其相关的灭绝和生物钙化危机(Little和Benton,1995;Wignall等,2005,2006);(2)海水温度的迅速升高(Dera等,2009a,b;Lachkar等,2009);(3)区域性的营养富集(Blomeier和Reijmer,1999)。

5.7 结论

布达哈山(DBD)是早侏罗世在大阿特拉斯海相盆地的半地堑中发育的孤立碳酸盐台地。其中识别出六个具有不同形态、地层结构、层序界面、岩相组成和叠置形式的层序。这些层序是在可容空间不断变化的旋回中形成的地层单元;它们的特点是:数百米厚、大多数由加积到退积的旋回、具有先板后楔的构造控制型几何形态。

布达哈山(DBD)层序中所观察到的可容空间旋回具有复杂的、多重作用的成因,是大地构造作用干扰全球海平面升降信号的结果。此外,伴随沉积中心迁移而改变的碳酸盐生产主体和相对堆积潜力也影响了由构造和海平面变化混合作用产生的可容空间的充填。整个布达哈山(DBD)断块沉降脉冲的累积效应(可能是与大阿特拉斯裂谷盆地南缘的区域演化有关)和全球海平面升降对布达哈山(DBD)碳酸盐台地可容空间变化的影响也许要大于断块的构造旋转。尽管如此,由断块旋转同时引起的上盘下降和上盘上升,造成了层序Ⅲ到Ⅵ界面特征的横向变化和楔状几何形态。

与二叠纪—三叠纪之交生物大灭绝之后的环境变化和生物进化趋势相关联，碳酸盐工厂发生着变化。这种变化也对布达哈山（DBD）碳酸盐台地施加了重要的控制作用，包括：（1）调整台地边缘的剖面结构和台地的构型，比如：斜坡上部的原地碳酸盐生产改变了斜坡的构成和坡度，并且在层序Ⅳ到Ⅵ沉积期发育成前沿变陡的碳酸盐台地；（2）碳酸盐产率和沉积中心位置的变化改变了体系追赶可容空间变化的能力，可能导致了台地淹没。

本次对布达哈山（DBD）的研究表明，在裂谷盆地碳酸盐台地层序发育和淹没的过程中，全球海平面升降和碳酸盐工厂变化所起的作用可能与构造作用同等重要；这就意味着，在其他相似的碳酸盐台地当中，由于人们惯倾向用构造来解读，全球海平面升降和碳酸盐工厂所起的作用可能并没有得到正确识别。为了更好地预测和再现断块碳酸盐台地的地层形态和层序发育，需要更复杂的正演地层学模型，这些模型除了要考虑脉冲式构造旋转，还应考虑持续型断块旋转，考虑由全球海平面升降和整个旋转断块背景式沉降所产生的基准面变化。

参 考 文 献

Agard, J. and Du Dresnay, R. (1965) La région minéralisée du Jbel Bou-Dahar, près de Beni Tajjit (Haut Atlas oriental). Étude géologique et métallogénique. Notes Mém. Serv. Géol. Maroc. ,181,135 – 166.

Arboleya, M. L. , Teixel, A. , Charroud, M. and Julivert, M. (2004) A structural transect through the High and Middle Atlas of Morocco. J. Afr. Earth Sci. ,39,319 – 327.

Aurell, M. , Robles, S. , Bádenas, B. , Rosales, I. , Quesada, S. , Meléndez, G. and García-Ramos, J. C. (2003) Transgressive – regressive cycles and Jurassic palaeogeography of northeast Iberia. Sed. Geol. ,162,239 – 271.

Barattolo, F. and Romano, R. (2005) Some bioevents at the Triassi – Liassic boundary in the shallow water environment. Boll. Soc. Geol. Ital. ,124,123 – 142.

Beauchamp, W. , Barazangi, M. , Demnati, A. and El Alji, M. (1996) Intracontinental rifting and inversion: Missour Basin and Atlas Mountains, Morocco. AAPG Bull. ,80,1459 – 1482.

Bellian, J. A. , Kerans, C. and Jenette, D. C. (2005) Digital outcrop models: applications of terrestrial scanning Lidar technology in stratigraphic modeling. J. Sed. Res. ,75,166 – 176.

Blomeier, D. P. G. and Reijmer, J. J. G. (1999) Drowning of a Lower Jurassic carbonate platform: Jbel Bou Dahar, High Atlas, Morocco. Facies,41,81 – 110.

Blomeier, D. P. G. and Reijmer, J. J. G. (2002) Facies architecture of an Early Jurassic carbonate platform slope (Jbel Bou Dahar, High Atlas, Morocco). J. Sed. Res. ,72,462 – 475.

Bosellini, A. (1989) Dynamics of Tethyan carbonate platforms. In: Controls on Carbonate Platform and Basin Development (Eds P. D. Crevello, J. J. Wilson, J. F. Sarg and J. F. Read), SEPM Spec. Publ. ,44,3 – 13.

Bosence, D. (2005) A genetic classification of carbonate platforms based on their basinal and tectonic settings in the Cenozoic. Sed. Geol. ,175,49 – 72.

Bosence, D. , Cross, N. and Hardy, S. (1998) Architecture and depositional sequences of Tertiary fault – block carbonate platforms; an analysis from outcrop (Miocene, Gulf of Suez) and computer modeling. Mar. Petrol. Geol. ,15, 2003 – 2221.

BouDagher-Fadel, M. K. and Bosence, D. W. J. (2007) Early Jurassic benthic foraminiferal diversification and biozones in shallow-marine carbonates of western Tethys. Senckenb. Lethaea,87,1 – 39.

Brachert, T. C. , Krautworst, U. M. R. and Stueckrad, O. M. (2002) Tectono-climatic evolution of a Neogene intramontane basin (Late Miocene Carboneras subbasin, southeast Spain): revelations from basin mapping and biofacies analysis. Basin Res. ,14,503 – 521.

Burchette, T. P. and Wright, V. P. (1992) Carbonate ramp depositional systems. Sed. Geol. ,79,3 – 57.

Campbell, A. E. and Stafleu, J. (1992) Seismic modeling of an Early Jurassic, drowned carbonate platform: Djebel

Bou Dahar, High Atlas, Morocco. AAPG Bull. ,76,1760 – 1777.

Catuneanu, O. , Abreu, V. , Bhattacharya, J. P. , Blum, M. D. , Dalrymple, R. W. , Eriksson, P. G. , Fielding, C. R. , Fisher, W. L. , Galloway, W. E. , Gibling, M. R. , Giles, K. A. , Holbrook, J. M. , Jordan, R. , Kendall, C. G. St. C. , Macurda, B. , Martinsen, O. J. , Miall, A. D. , Neal, J. E. , Nummedal, D. , Pomar, L. , Posamentier, H. W. , Pratt, B. R. , Sarg, J. F. , Shanley, K. W. , Steel, R. J. , Strasser, A. , Tucker, M. E. and Winker, C. (2009) Towards the standardization of sequence stratigraphy. Earth-Sci. Rev. ,92,1 – 33.

Cobianchi, M. and Picotti, V. (2001) Sedimentary and biological response to sea-level and palaeoceanographic changes of a Lower-Middle Jurassic Tethyan platform margin (Southern Alps, Italy). Palaeogeogr. Palaeoclimatol. Palaeoecol. ,169,219 – 244.

Crevello, P. (1991) High frequency carbonate cycles and stacking patterns: interplay of orbital forcing and subsidence on Lower Jurassic rift platforms, High Atlas, Morocco. In: Sedimentary Modelling: Computer Simulations and Methods for Improved Parameter Definition (Eds E. K. Franseen, W. L. Watney, C. S. C. Kendall and W. Ross), Bull. Kansas State Geol. Surv. ,233,207 – 230.

Cross, N. E. and Bosence, D. W. J. (2008) Tectono-sedimentary models for rift-basin carbonate systems. In: Controls on Carbonate Platform and Reef Development (Eds J. Lukasik and T. Simo), SEPM Spec. Publ. ,89,83 – 105.

Cross, N. E. , Purser, R. H. and Bosence, D. W. J. (1998) The tectono-sedimentary evolution of a rift margin fault – block carbonate platform: Abu Shaar, Gulf of Suez, Egypt. In: Sedimentary and Tectonic Evolution of Rift Basins: The Red Sea – Gulf of Aden (Eds R. H. Purser and D. W. J. Bosence), pp. 271 – 295. Chapman Hall, London.

Davison, I. (2005) Central Atlantic margin basins of North West Africa: geology and hydrocarbon potential (Morocco to Guinea). J. Afr. Earth Sci. ,43,254 – 274.

Dera, G. , Pellenard, P. , Neige, P. , Deconinck, J. -F. , Pucéat, E. and Dommergues, J. -L. (2009a) Distribution of clay minerals in Early Jurassic Peritethyan seas: palaeoclimatic significance inferred from multiproxy comparisons. Palaeogeogr. Palaeoclimatol. Palaeoecol. ,271,39 – 51.

Dera, G. , Pucéat, E. , Pellenard, P. , Neige, P. , Delsate, D. , Joachimski, M. M. , Reisberg, L. and Martínez, M. (2009b) Water mass exchange and variations in seawater temperature in the NW Tethys during the Early Jurassic: evidence from neodymium and oxygen isotopes of fish teeth and belemnites. Earth Planet. Sci. Lett. , 286, 198 – 207.

Dorobek, S. L. (2008) Tectonic and depositional controls on syn-rift carbonate platform sedimentation. In: Controls on Carbonate Platform and Reef Development (Eds J. Lukasik and T. Simo), SEPM Spec. Publ. ,89,57 – 82.

Du Dresnay, R. (1976) Carte Géologique du Haut Atlas D'Anoual-Bou Anane (Haut Atlas Oriental). Ser. Géol. Maroc, Notes et Mem. , No. 246.

Du Dresnay, R. (1977) Le milieu récifal fossile du Jurassique inférieur (Lias) dans le domaine des chaînes atlasiques du Maroc. Bur. Rech. Géol. Min. Mém. ,89,296 – 312.

Ebdon, C. C. , Fraser, A. J. , Higgins, A. C. , Mitchener, B. C. and Strank, A. R. E. (1990) The Dinantian stratigraphy of the East Midlands: a seismostratigraphic approach. J. Geol. Soc. London, 147,519 – 536.

El Hariri, K. , Dommergues, J. -L. , Meister, C. , Souhel, A. and Chafiki, D. (1996) Les ammonites du Lias inférieur et moyen du Haut-Atlas central de Béni Mellal (Maroc): taxinomie et biostratigraphieá haute-résolution. Geobios, 29,537 – 576.

Ellouz, N. , Patriat, M. , Gaulier, J. -M. , Bouatmani, R. and Sabounji, S. (2003) From rifting to Alpine inversion: Mesozoic and Cenozoic subsidence history of some Moroccan basins. Sed. Geol. ,156,185 – 212.

Elmi, S. , Amhoud, H. , Boutakiout, M. and Benshili, K. (1999) Cadre biostratigraphique et environmental de l'évolution du paléorielief du Jebel Bou Dahar (Haut-Atlas oriental, Maroc) au course du Jurassique inférieur et moyen. Bull. Soc. Géol. Fr. ,170,619 – 628.

Gawthorpe, R. L. , Sharp, I. R. , Underhill, J. R. and Gupta, S. (1997) Linked sequence stratigraphic and structural evolution of propagating normal faults. Geology,25,795 – 798.

Gómez,J. J. and Goy,A. (2005) Late Triassic and Early Jurassic palaeogeographic evolution and depositional cycles of the Western Tethys Iberian platform system. Palaeogeogr. Palaeoclimatol. Palaeoecol. ,222,77 – 94.

Graciansky,P. -C. ,de Dardeau,G. ,Dumont,T. ,Jacquin,T. ,Marchand,D. ,Mouterde,R. and Vail,P. R. (1993) Depositional sequence cycles,transgressive/regressive facies cycles and extensional tectonics: example from the Southern Sub-Alpine Jurassic Basin,France. Bull. Soc. Géol. Fr. ,164,709 – 718.

Graciansky,P. -C. ,de Dardeau,G. ,Dommergues,J. L. ,Durlet,C. ,Marchand,D. ,Dumont,T. ,Hesselbo,S. P. ,Jacquin,T. ,Goggin,V. ,Meister,C. ,Mouterde,R. ,Rey,J. and Vail,P. R. (1998) Ammonite biostratigraphic correlation and Early Jurassic sequence stratigraphy in France: comparisons with some U. K. sections. In: Mesozoic and Cenozoic Sequence Stratigraphy of European Basins (Eds P. -C. de Graciansky, J. Hardenbol, T. Jaquinand and P. R. Vail) ,SEPM Spec. Publ. ,60,583 – 622.

Gradstein,F. M. ,Ogg,J. G. ,Smith,A. G. ,Bleeker,W. and Lourens,L. J. (2004) A new geologic time scale,with special reference to Precambrian and Neogene. Episodes,2,83 – 100.

Hallam,A. (1988) A re-evaluation of Jurassic eustasy in the light of new data and the revised Exxon curve. In: Sea-Level Changes – An Integrated Approach (Eds C. K. Wilgus, B. S. Hastings, C. G. St. C. Kendall, H. W. Posamatir, C. A. Ron and J. C. van Wagner) ,SEPM Spec. Publ. ,42,261 – 273.

Hallam,A. (2001) A review of the broad pattern of Jurassic sealevel changes and their possible causes in the light of current knowledge. Palaeogeogr. Palaeoclimatol. Palaeoecol. ,167,23 – 37.

Handford,C. R. and Loucks,R. G. (1993) Carbonate depositional sequences and systems tracts – responses of carbonate platforms to relative sea-level changes. In: Carbonate Sequence Stratigraphy (Eds R. G. Loucks and J. F. Sarg) ,AAPG Mem. ,57,3 – 42.

Haq,B. U. ,Hardenbol,J. and Vail,P. L. (1988) Mesozoic and Cenozoic chronostratigraphy and cycles of sea – level change. In: Sea-Level Changes – An Integrated Approach (Eds C. K. Wilgus, B. S. Hastings, C. G. St. C. Kendall, H. W. Posamatir, C. A. Ron and J. C. van Wagner) ,SEPM Spec. Publ. ,42,71 – 108.

Hardenbol,J. ,Thierry,J. ,Farley,M. B. ,Jaquin,T. ,de Graciansky,P. -C. and Vail,P. R. (1998) Mesozoic and Cenozoic sequence chronostratigraphic framework of European basins – chart 6: Jurassic sequence chronostratigraphy. In: Mesozoic and Cenozoic Sequence Stratigraphy of European Basins (Eds P. -C. de Graciansky, J. Hardenbol, T. Jaquinand and P. R. Vail) ,SEPM Spec. Publ. ,60,3 – 13.

Hesselbo,S. B. (2008) Sequence stratigraphy and inferred relative sea-level change from the onshore British Jurassic. Proc. Geol. Assoc. ,119,19 – 34.

Hesselbo,S. P. and Coe,A. L. (2000) Jurassic sequences of the Hebrides Basin,Isle of Skye,Scotland. In: Field Trip Guidebook (Eds J. R. Graham and A. Ryan) ,pp. 41 – 58. International Sedimentologists Association Meeting,Dublin.

Hesselbo,S. B. and Jenkyns,H. C. (1998) British Lower Jurassic sequence stratigraphy. In: Mesozoic and Cenozoic Sequence Stratigraphy of European Basins (Eds P. -C. de Graciansky, J. Hardenbol, T. Jaquinand and P. R. Vail) , SEPM Spec. Publ. ,60,561 – 581.

Hesselbo,S. P. , Robinson, S. A. , Surlyk, F. and Piasecki, S. (2002) Terrestrial and marine extinction at the Triassic – Jurassic boundary synchronized with major carbon-cycle perturbation: a link to initiation of massive volcanism? Geology,30,251 – 254.

Jacquin,T. ,Dardeau,G. ,Durlet,C. ,de Graciansky,P. -C. and Hantzpergue,P. (1998) The North Sea cycle: an overview of 2nd-order transgressive/regressive facies cycles in Western Europe. In: Mesozoic and Cenozoic Sequence Stratigraphy of European Basins (Eds P. -C. de Graciansky, J. Hardenbol, T. Jaquinand and P. R. Vail) ,SEPM Spec. Publ. ,60,445 – 466.

Jadoul,F. and Galli,M. T. (2008) The Hettangian shallow water carbonates after the Triassic/Jurassic biocalcificationcrisis: the Albenza Formation in the Western Southern Alps. Riv. Ital. Paleontol. Stratigr. ,114,453 – 470.

Jenkyns, H. C. (1988) The Early Toarcian (Jurassic) anoxic event: stratigraphic, sedimentary, and geochemical evidence. Am. J. Sci. ,288,101 – 151.

Kenter, J. A. M. and Campbell, A. E. (1991) Sedimentation on a Lower Jurassic carbonate platform flank: geometry, sediment fabric and related depositional structures (Djebel Bou Dahar, High Atlas, Morocco). Sed. Geol. , 72, 1 – 34.

Kenter, J. A. M. , Harris, P. M. and Della Porta, G. (2005) Steep microbial boundstone-dominated platform margins – examples and implications. Sed. Geol. ,178,5 – 30.

Lachkar, N. (2000) Dynamique sedimentaire d'un basin extensive sur la marge sud-tethysienne: le Lias du Haut Atlas de Rich (Maroc). Thèse de Doctorat, Centre des Sciences de la Terre, Université de Bourgogne, Dijon,275 pp.

Lachkar, N. , Guiraud, M. , El Harfi, A. , Dommergues, J. -L. , Dera, G. and Durlet, C. (2009) Early Jurassic normal faulting in a carboante extensional basin: characterization of tectonically driven platform drowning (High Atlas rift, Morocco). J. Geol. Soc. London,166,413 – 430.

Leeder, M. R. and Gawthorpe, R. L. (1987) Sedimentary models for extensional tilt-block/half-graben basins. In: Continental Extensional Tectonics (Eds M. P. Coward, J. F. Dewey and P. L. Hancock), Geol. Soc. London, Spec. Publ. ,28,139 – 152.

Legarreta, L. and Uliana, M. A. (1996) The Jurassic succession in west-central Argentina: stratal patterns, sequences and paleogeographic evolution. Palaeogeogr. Palaeoclimatol. Palaeoecol. ,120,303 – 330.

Leinfelder, R. R. , Schlagintweit, F. , Werner, W. , Ebli, O. , Nose, M. , Schmid, D. U. and Hughes, G. W. (2005) Significance of stromatoporoids in Jurassic reefs and carbonate platforms – concepts and implications. Facies, 51, 288 – 326.

Li, X. and Grant-Mackie, J. A. (1993) Jurassic sedimentary cycles and eustatic sea-level changes in southern Tibet. Palaeogeogr. Palaeoclimatol. Palaeoecol. ,101,27 – 48.

Little, C. T. S. and Benton, M. J. (1995) Early Jurassic mass extinction: a global long-term event. Geology, 23, 495 – 498.

Markello, J. R. , Koepnick, R. B. , Waite, L. E. and Collins, J. F. (2008) The Carbonate Analogs Through Time (CATT) hypothesis and the global atlas of carbonate fields – a systematic and predictive look at the Phanerozoic carbonate systems. In: Controls on Carbonate Platform and Reef Development (Eds J. Lukasik and T. Simo), SEPM Spec. Publ. ,89,15 – 45.

Marzoli, A. , Bertrand, H. , Knight, K. B. , Cirilli, S. , Buratti, N. , Vérati, C. l. , Nomade, S. b. , Renne, P. R. , Youbi, N. , Martini, R. , Allenbach, K. , Neuwerth, R. , Rapaille, C. d. , Zaninetti, L. and Bellieni, G. (2004) Synchrony of the Central Atlantic magmatic province and the Triassic – Jurassic boundary climatic and biotic crisis. Geology,32, 973 – 976.

Mehdi, M. , Neuweiler, F. and Wilmsen, M. (2003) Les formations du Lias inférieiur du Haut Atlas central de Rich (Maroc): precisions lithostratigraphiques et étapes de l'évolution du basin. Bull. Soc. Géol. Fr. ,174,227 – 242.

Mitchum Jr, R. M. (1977) Seismic stratigraphy and global changes of sea level, part 11: glossary of terms used inseismic stratigraphy. In: Seismic Stratigraphy – Applications to Hydrocarbon Exploration (Ed. C. E. Payton), AAPG Mem. ,26,205 – 212.

O'Dogherty, L. , Sandoval, J. and Vera, J. A. (2000) Ammonite faunal turnover tracing sea level changes during the Jurassic (Betic Cordillera, southern Spain). J. Geol. Soc. London,157,281 – 319.

Olsen, P. E. , Koeberl, C. , Huber, H. , Montanari, A. , Fowell, S. J. , Et Touhami, M. and Kent, D. V. (2002) The continental Triassic – Jurassic boundary in Central Pangea: recent progress and discussion of an Ir anomaly. Geol. Soc. Am. Spec. Pap. ,356,505 – 522.

Olsen, P. A. , Kent, D. V. , Et-Touhami, M. and Puffer, J. H. (2003) Cyclo-, magneto-, and bio-stratigraphic constrains on the duration of the CAMP event and its relationship to the Triassic – Jurassic boundary. In: The Central Atlantic Magmatic Province: Insights from Fragments of Pangea (Eds W. E. Hames, J. G. McHome, P. R. Renne and C. Ruppel), Geophys. Monogr. Ser. ,136,7 – 32.

Pálfy,J. and Smith,P. L. (2000) Synchrony between Early Jurassic extinction,oceanic anoxic event,and the Karoo – Ferrar flood basalt volcanism. Geology,28,747 – 750.

Pomar,L. (2001) Types of carbonate platforms: a genetic approach. Basin Res. ,13,313 – 344.

Pomar,L. and Hallock,P. (2008) Carbonate factories: a conundrum in sedimentary geology. Earth-Sci. Rev. ,87,134 – 169.

Prosser,S. 1993. Rift-related linked depositional systems and their seismic expression. In: Tectonics and Seismic Sequence Stratigraphy (Eds G. D. Williams and A. Dobb),Geol. Soc. London,Spec. Publ. ,71,35 – 66.

Quesada,S. ,Robles,S. and Rosales,I. (2005) Depositional architecture and transgressive – regressive cycles within Liassic backstepping carbonate ramps in the Basque – Cantabrian basin,northern Spain. J. Geol. Soc. London,162,531 – 548.

Rosales,I. ,Quesada,S. and Robles,S. (2006) Geochemical arguments for identifying second-order sea-level changes in hemipelagic carbonate ramp deposits. Terra Nova,18,233 – 240.

Ruiz-Ortiz,P. A. ,Bosence,D. ,Rey,J. ,Nieto,L. M. ,Castro,J. M. and Molina,J. M. (2004) Tectonic control of facies architecture, sequence stratigraphy and drowning of a Liassic carbonate platform (Betic Cordillera, southern Spain). Basin Res. ,16,235 – 258.

Santantonio,M. (1993) Facies associations and evolution of pelagic carbonate platform/basin systems: examples from the Italian Jurassic. Sedimentology,40,1039 – 1067.

Santantonio,M. (1994) Pelagic carbonate platforms in the geologic record: their classification and sedimentary and paleotectonic evolution. AAPG Bull. ,78,122 – 141.

Scheibner,A. H. and Reijmer,J. J. G. (1999) Facies patterns within a Lower Jurassic upper slope to inner platform transect (Jbel Bou Dahar,Morocco). Facies,41,55 – 80.

Schlager,W. (2003) Benthic carbonate factories of the Phanerozoic. J. Earth Sci. ,92,445 – 464.

Schlager,W. (2005) Carbonate Sedimentology and Sequence Stratigraphy. SEPM Concepts in Sedimentology and Paleontology,8. SEPM,Tulsa,Oklahoma,200 pp.

Soussi,M. and Ben Ismaïl,H. (2000) Platform collapse and pelagic seamount facies: Jurassic development of Central Tunisia. Sed. Geol. ,133,93 – 113.

Steiner,C. ,Hobson, A. ,Faure, P. ,Stampfli, D. and Henandez, J. (1998) Mesozoic sequence of Fuerteventura (Canary islands):witness of Early Jurassic sea-floor spreading in the central Atlantic. Geol. Soc. Am. Bull. ,110,1304 – 1317.

Stephen,K. J. and Davies,R. L. (1998) Documentation of Jurassic sedimentary cycles from the Moray Firth Basin,United Kingdom,North Sea. In: Mesozoic and Cenozoic Sequence Stratigraphy of European Basins (Eds P. -C. de Graciansky,J. Hardenbol,T. Jaquinand and P. R. Vail),SEPM Spec. Publ. ,60,481 – 506.

Surlyk,F. (1991) Sequence stratigraphy of the Jurassic-lowermost Cretaceous in East Greenland. AAPG Bull. ,75,1468 – 1488.

Tanner,L. H. ,Lucas,S. G. and Chapman,M. G. (2004) Assessing the record and causes of Late Triassic extinctions. Earth Sci. Rev. ,65,103 – 139.

Teixel,A. ,Arboleya,M. L. ,Julivert,M. and Charroud,M. (2003) Tectonic shortening and topography in the central High Atlas (Morocco). Tectonics,22,1051. doi: 10. 1029/2002TC001460.

Teixel,A. ,Ayarza,P. ,Tesón,E. ,Babault,J. ,Alvarez-Lobato,F. ,Charroud,M. ,Julivert,M. ,Barbero,L. ,Amrhar,M. and Arboleya, M. L. (2007) Geodinámica de las cordilleras del Alto y Medio Atlas: una síntesis. Rev. Soc. Geol. Esp. ,20,333 – 350.

Tesón, E. and Teixel, A. (2008) Sequence of thrusting and syntectonic sedimentation in the eastern Sub-Atlas thrust belt (Dadès and Mgoun valleys, Morocco). Earth Sci. Rev., 97, 103–113.

Thierry, J., Barrier, E., Abbate, E., Ait-Ouali, R., Ait-Salem, H., Bouaziz, S., Canerot, J., Elmi, S., Gelk, M., Georgiev, G., Guiraud, R., Hirsch, F., Ivanik, M., Le metour, J., Le Hyndre, Y. M., Nikishin, A. M., Page, K., Panov, D. L., Pique, A., Poisson, A., Sandulescu, M., Sapunov, I. G., Seghedi, A., Soussi, M., Tarkowski, R. A., Tchoumatchenko, P. V., Vaslet, D., Volozh, Y. A., Voznezenski, A., Walley, C. D., Ziegler, M., Ait-Brahim, L., Bergerat, F., Bracene, R., Brunet, M. F., Cadet, J. P., Guezou, J. C., Jabaloy, A., Lepvrier, C. and Rimmele, G. (2000) Middle Toarcian (180–178 Ma). In: Atlas Peri-Tethys Paleogeographical Maps (Eds J. Dercourt, M. Gaetani, B. Vrielynck, E. B. Barrier, B. Biju-Duval, M.-F. Brunet, J. P. Cadet, S. Crasquin and M. Sandulescu), Vol. I–XX. CCGM/CGMW, Paris, map 8 (40 co-authors).

Van Buchen, F. S. P. and Knox, R. W. O'. B. (1998) Lower and Middle Liassic depositional sequences of Yorkshire (U. K.). In: Mesozoic and Cenozoic Sequence Stratigraphy of European Basins (Eds P.-C. de Graciansky, J. Hardenbol, T. Jaquinand and P. R. Vail), SEPM Spec. Publ., 60, 545–599.

Verwer, K., Della Porta, G., Merino-Tomé, O. and Kenter, J. A. M. (2009a) Controls and predictability of carbonate facies architecture in a Lower Jurassic three-dimensional barrier-shoal complex (Djebel Bou Dahar, High Atlas, Morocco). Sedimentology, 56, 1801–1831.

Verwer, K., Merino-Tomé, O., Kenter, J. A. M. and Della Porta, G. (2009b) Evolution of a high-relief carbonate platform slope using 3D digital outcrop models: Lower Jurassic Djebel Bou Dahar, High Atlas Morocco. J. Sed. Res., 79, 416–439.

Walker, J. D., Geissman, J. W. (2009) 2009 GSA Geologic time scale. Geol. Soc. Am. Today, 19, 60–61. doi: 10.1130/2009.CTS004R2C.

Warme, J. E. (1988) Jurassic carbonate facies of the central and eastern High Atlas Rift, Morocco. In: The Atlas System of Morocco – Studies on its Geodynamic Evolution (Ed. V. H. Jacobshagen). Lect. Notes Earth Sci., 15, 169–199.

Whitesite, J. H., Olsen, P. E., Kent, D. V., Fowell, S. J. and Et-Touhami, M. (2007) Synchrony between the Central Atlantic magmatic province and the Triassic-Jurassic massextinction event? Palaeogeogr. Paleoclimatol. Palaeoecol., 244, 345–367.

Wignall, P. B., Newton, R. J. and Little, C. T. S. (2005) The timing of paleoenvironmental change and cause-and-effect relationships during the Early Jurassic mass extinction in Europe. Am. J. Sci., 305, 1014–1032.

Wignall, P. B., Hallam, A., Newton, R. J., Sha, J. G., Reeves, E., Mattioli, E. and Crowley, S. (2006) An eastern Tethyan (Tibetan) record of the Early Jurassic (Toarcian) mass extinction event. Geobiology, 4, 179–190.

Wilmsen, M. and Neuweiler, F. (2008) Biosedimentology of the Early Jurassic post-extinction carbonate depositional system, central High Atlas rift basin, Morocco. Sedimentology, 55, 773–807.

Wilson, M. E. J., Bosence, D. W. J. and Limbong, A. (2000) Tertiary syntectonic carbonate platform development in Indonesia. Sedimentology, 47, 395–419.

Winterer, E. L. and Bosellini, A. (1981) Subsidence and sedimentation on Jurassic passive continental margin, southern Alps, Italy. AAPG Bull., 65, 394–421.

Wright, V. P. and Burgess, P. M. (2005) The carbonate factory continuum, facies mosaics and microfacies: an appraisal of some of the key concepts underpinning carbonate sedimentology. Facies, 51, 17–23.

第6章 碳酸盐台地的淹没与全球缺氧事件

PHILIPPE LÉONIDE, MARC FLOQUET, CHRISTOPHE DURLET,
FRANCIS BAUDIN BERNARD PITTET, CHRISTOPHE LÉCUYER 著

岳　婷译，金春爽校

摘　要　托尔阶大洋缺氧事件因为伴随着碳酸盐危机、有机质积累和碳同位素追踪显示的碳循环紊乱而闻名于世。在这个古环境下，现有的研究试图更好地抑制造成晚普林斯巴奇期至早托尔期碳酸盐台地淹没的古环境条件。本次研究对南普罗旺斯次级盆地（法国东南部）的几处地层序列做了完整的沉积学、成岩作用和地球化学（稳定同位素和岩石热解）分析。普林斯巴奇阶向托尔阶过渡时，位于两硬底地面下的早成岩作用形成的铁方解石胶结物记录了幕式富营养事件。这些事件是导致碳酸盐台地消失的全球古环境扰动（即全球碳循环紊乱）的前兆，发生在托尔阶大洋缺氧事件之前。随后，托尔阶大洋缺氧事件发生的同时，碳同位素正偏移（2.87‰）跟随着的一个显著负偏移（－3‰），有机质保存以及由软骨鱼类牙釉的氧同位素值推知的海水表面高温（大约25℃）均显示出古环境恶化达到了极点。对南普罗旺斯和东多菲内次级盆地早托尔期沉积序列进行的沉积学对比及地层化学和生物相关性分析，证实了早期发生在南普罗旺斯次级盆地的构造倾斜。这造成了前一个次级盆地中有机质富集层的保存和后一个次级盆地中的沉积间断。因此，当地条件如差异性构造背景，能够显著地更改全球气候事件的沉积特征。

关键词　碳同位素　碳酸盐胶结物　淹没　法国东南部　南普罗旺斯碳酸盐台地　托尔阶大洋缺氧事件

6.1　概况

生物碳酸盐产物快速消失的危机标志了中生代碳酸盐体系沉积历史的间断（Jenkyns，1985；Arthur 等，1988；Jenkyns 等，1994；Weissert 等，1998；Price，1999；Deraet 等，2011）。在这方面，普林斯巴奇阶向托尔阶的过渡是特提斯沉积盆地演化中的关键阶段：碳酸盐危机（Dromart 等，1996；Blomeier 和 Reijmer，1999）不断增加的土壤有机质积累（Jenkyns，1988；Baudin 等，1990）和 $\delta^{13}C$ 追踪显示的碳循环紊乱（Hesselbo 等，2000；Beerling 等，2002）都记录了这一演化阶段的存在。这些特征符合托尔阶大洋缺氧事件（以下简称 T-OAE）（Jenkyns，1988）。这一事件与气候变化发生在同一时期（Bailey 等，2003；Rosales 等，2004a；van de Schootbrugge 等，2005），包括晚普林斯巴奇期气候变冷及随后的早、中托尔期气候变暖（Suan 等，2008b；Dera 等，2009a）。普林斯巴奇阶向托尔阶的过渡也被碳酸盐台地生产的停止（Dromart 等，1996；Blomeier 和 Reijmer，1999）和微化石生物钙化危机（Mattioli 和 Pittet，2002；Erba，2004；Mattioli 等，2004a，b；Tremolada 等，2005）所标记。这一区间以普林斯巴奇末期（Cecca 和 Macchioni，2004）和早托尔期（Hallam，1987；Aberhan 和 Baumiller，2003；van de Schootbrugge 等，2005）生物多样性的恢复为特征。这些事件的起源及因果关系仍存在很多难点和争议（Wignall 等，2005；Newton 等，2006）。南普罗旺斯次级盆地的沉积演化，作为普林斯巴奇阶向托尔阶过渡时环境变化对碳酸盐生成及保存影响的一个实例，具有相当大的信息量。在南普罗旺斯次级盆地这

一过渡的标记是:(1)岩性从透光的浅海沉积环境中纯碳酸盐生成与沉积(晚普林斯巴奇期)到较深的微透光或不透光海洋沉积环境中半深海沉积(早托尔期)(Léonide 等,2007)的变化;(2)贯穿南普罗旺斯次级盆地的较大不整合面的存在加剧了这一过渡。

本次研究集中探讨了以下几个方面:(1)在托尔阶大洋缺氧事件之前和期间,沉积在浅海碳酸盐台地顶部的不整合面,以及相关的含低铁的特殊亮晶方解石胶结物;(2)对浅海相碳酸盐台地沉积事件和泥灰质盆地向海边缘较深处的沉积过程进行对比研究。本文基于野外实地考察,进行了胶结物地层学和有机质含量分析,以及元素和稳定同位素地球化学分析,最后展示了这次综合研究得到的结果。本次研究选取了 15 个连续地区,并对其中 3 个地区进行了地球化学分析:(1)南普罗旺斯次级盆地 Cuers 地区;(2)和(3)是东多菲内次级盆地 La Robine 地区和 Marcoux 地区(图 6-1)。

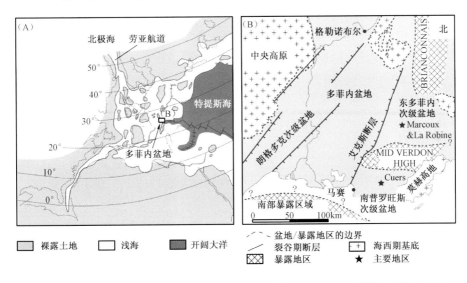

图 6-1 (A)早侏罗世古地理图中东多菲内盆地的位置(Bassoulet 等修改完善,1993);
(B)法国东南部古地理及三大地区(Cuers,La Robine 和 Marcoux)位置图

6.2 方法

碳酸盐成岩作用研究是在胶结物地层学(Meyers,1991)的基础上,建立薄片观察到的成岩作用事件的序列(例如胶结、重结晶、溶解、钻孔、裂缝等)。在南普罗旺斯次级盆地和东多菲内次级盆地普林斯巴奇阶—托尔阶过渡区的 15 处露头(Léonide,2007)采样,并用这些样品磨制了 154 张抛光片。胶结物地层学分析通过使用光学显微镜来观察染在薄片上的茜草色素—钾铁氰化物(Dickson,1966)。本次研究观察完全采用扫描电镜(XL 30 ESEM;FEI/Philips,埃因霍温,荷兰)和带有 Olympus 显微镜(奥林巴斯光学公司,东京,日本)、数码 MRc5 摄像头(Zeiss,Göttingen,德国)的 8200 MKII Technosyn 阴极发光岩相学观测仪(20 kV,600 mA;Technosyn,剑桥,英国)。从普林斯巴奇阶灰岩顶部钻孔硬底面获取的薄片观察到的横切关系让这些不整合面记录的成岩阶段被准确识别出来。连续的成岩阶段可用成岩作用测井曲线图(Durlet 等,1992)表示出来,并可用来确定不整合面的成岩序列(Durlet 和 Loreau,1996;Léonide,2007)。硬底面的成岩序列是一个与沉积缺失同时期发生的所有成岩阶段的有序列表。它起始于现在正位于表面之下的沉积物的沉积之后,结束于下一个沉积

单元沉积之前发生的削截、钻孔或铁氧化物包壳（Loreau 和 Durlet,1999）。这里使用小型粉碎机对方解石胶结物进行微量采样，来分析碳和氧同位素并限制流体成分性质，但是胶结物的细度（小于 150μm）造成很难从这些胶结物所在的单个小层中把样品准确地收集出来。

在里昂第一大学，通过使用连接质谱仪（GV Isoprime 质谱仪；GV 仪器有限公司，曼彻斯特，英国）的自动样品制备外围装置来获得碳同位素数据。根据 NBS19（国家标准局），重复分析方解石中的 $\delta^{13}C$ 同位素，得到其误差小于 ±0.04‰。数值单位为"‰,PDB"（维也纳皮迪箭石）。从南普罗旺斯次级盆地 Cuers 地区的普林斯巴奇阶到中托尔阶系列中共采集了 57 个样品（图 6-1）。为了对两个次级盆地中的碳同位素进行对比，又另外分析了东多菲内次级盆地 La Robine 地区和 Marcoux 地区相同时间段的 56 个样品（图 6-1）。

海水古温度可以通过软骨鱼类牙釉的氧同位素分析估计出来。由于磷灰石对成岩变化不敏感并且具有一个适用于所有鱼类氧元素的特殊分馏方程，所以鱼类牙齿的 $\delta^{18}O$ 同位素是一个较为粗略的温度替代物（Kolodny 等,1983;Lécuyer 等,1999）。从 Cuers 地区早托尔阶的两个泥灰质夹层（tenuicostatum 层和 serpentinum 层）中采集了 29 个牙齿样品。牙齿的尺寸小（从 0.8mm 到 2mm）且具有一定程度的破碎，这些都妨碍了对鲨鱼种类的精确识别，不过可以肯定的是所有牙齿均属于同一个鲨鱼类群。含牙齿的泥灰质夹层反映了近岸浅水沉积环境（Léonide 等,2007）。

在里昂第一大学，通过使用连接质谱仪的自动样品制备外围装置获得了这 29 个样品的 $\delta^{18}O$ 同位素数据，其遵循的方法是由 Lécuyer 等（1996）及 Lécuyer（2004）创建的。磷灰石中 $\delta^{18}O$ 同位素的值以‰V-SMOW（Vienna 标准平均海洋水）为参照，具有 ±0.2‰的精确性和可重复性。海水古温度可以用由 Longinelli 和 Nutti（1973）提出、Kolodny 等（1983）确认的方程 $T℃ = 111.4 - 4.3(\delta^{18}O_p - \delta^{18}O_w)$ 计算得到。在这个方程中，磷酸盐的测量中 $\delta^{18}O_p = \delta^{18}O$，$\delta^{18}O_w = -1‰$[温室时期的海洋水值，据 Marshall（1992）以及 Price 和 Sellwood（1997）]。

根据 Espitalié 等（1985—1986）描述的方法，在巴黎第六大学对从南普罗旺斯次级盆地 Cuers 地区普林斯巴奇阶最上层到中托尔阶采集的 46 个样品，使用原油显示分析仪（OSA）进行岩石热解分析来研究其有机物含量。岩石热解分析可以确定有机物的数量（总有机碳含量，TOC）和类型（陆相还是海相），及其成熟度和保存状态。

6.3 地质背景

6.3.1 古地理学

泛古陆由于受到长期伸展作用而在早、中侏罗世期间发生了分裂（Stampfli 和 Borel,2002）。特提斯海的裂谷作用导致了利古里亚海域的扩张，随后又促进了其西部边缘古北纬 25°地区多菲内盆地的形成（Lemoine 和 de Graciansky,1988;Ziegler,1992;de Graciansky 等,1993;Thierry,2000）（图 6-1A）。这个盆地近似三角形，其封闭的地垒、地堑或者次级盆地由 NNE—SSW 主断层控制（图 6-1B）。这些是朗格多克、东多菲内和南普罗旺斯次级盆地。

南普罗旺斯次级盆地向北受限于"Mid Verdon High"暴露地区（Tempier,1972），向东受限于结晶的"莫赫高地"，向南受限于假想的"地中海或者南部暴露土地"（图 6-1B）（Baudrimont 和 Dubois,1977;Thierry,2000）。向西，"普罗旺斯地区艾克斯"断层从深部的多菲内盆地中划定出了南普罗旺斯次级盆地（Léonide 等,2007;图 6-1B）。

据 de Graciansky 等(1998),多菲内盆地早、中侏罗世序列由四个二阶层序(特提斯地层旋回中用 T4/R4 到 T7/R7 标识)组成,它们记录了与利古里亚特提斯海域扩张有关的裂谷作用旋回。因此,利古里亚裂谷作用动力学及相关的波动沉降率,可以通过碳酸盐岩沉积物的变化来追踪显示,这些沉积物累积在南普罗旺斯次级盆地从滨岸到近岸的开放环境中。在那时,南普罗旺斯次级盆地的古地貌相当于碳酸盐台地(Léonide 等,2007;图 6-2)。

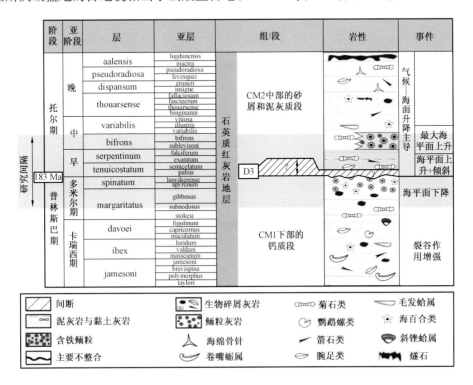

图 6-2　南普罗旺斯次级盆地下侏罗统沉积的等时
—岩相地层学图表(Léonide 等修改完善,2007;Gradstein 等,2004))

6.3.2　岩石地层学和生物地层学

6.3.2.1　南普罗旺斯次级盆地的上普林斯巴奇阶到中托尔阶

从晚普林斯巴奇期到中托尔期,南普罗旺斯次级盆地的石英质红灰岩地层被分成两段(Léonide 等,2007):底部的钙质段(CM1)以及中部的砂屑和泥灰质段(CM2)。晚辛涅缪尔期(raricostatum 层)到晚普林斯巴奇期(spinatum 层)的 CM1 段,从下到上,主要由细粒的含硅质生物碎屑泥粒灰岩、泥灰质灰岩、粗碎屑岩和海胆状粒灰岩组成(Léonide 等,2007)。早托尔期(tenuicostatum 层)到中阿连期(murchisonae 层)的 CM2 段,从下至上大致从泥灰岩向泥灰质灰岩变化,岩性为细粒含硅质生物碎屑泥粒灰岩(图 6-2)、核状粗粒灰岩(Léonide 等,2007)。

D3 不整合沉积间断标志着石英质红灰岩地层 CM1 与 CM2 之间的岩性变化(图 6-2 至图 6-4)。该不整合贯穿整个南普罗旺斯次级盆地,且便于识别和描述。(Garcin,2002;Léonide 等,2007)。

详细地说,D3 是由 3 个独立的界面组成的(图 6-3C 中 D3a、D3b 和 D3c;Léonide,2007)。从剖面图看,这些界面是含有铁质穿孔的硬底面(图 6-4A 到 E)。在 Cuers 地区(图 6-4A 和 B),

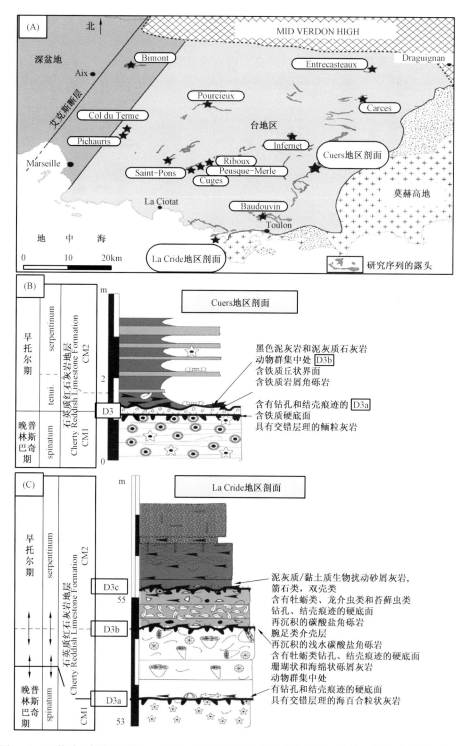

图 6-3 石英质红灰岩地层(Cuers 和 La Cride 地区剖面)内的普林斯巴奇阶—托尔阶过渡期岩相:
(A)南普罗旺斯次级盆地主要剖面的位置;(B)Cuers 地区普林斯巴奇阶—托尔阶过渡期岩相;
(C)La Cride 地区普林斯巴奇阶—托尔阶过渡期岩相

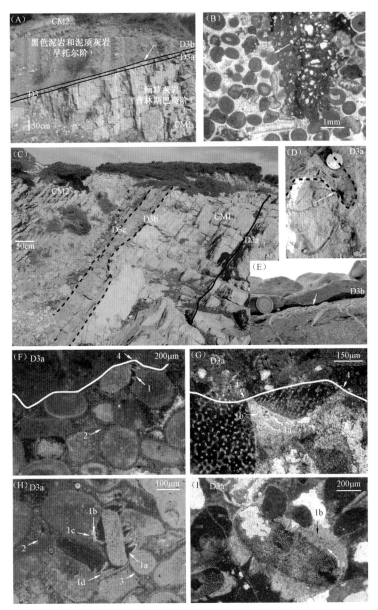

图6-4 石英质红灰岩地层(Cuers 和 La Cride 地区剖面)D3 不整合和D3a 界面处的胶结物(使用阴极发光和茜草色素—钾铁氰化物染色来观察):(A)D3 不整合由两个钻孔面组成("D3a"和"D3b",Cuers 地区剖面,白色箭头)。(B)有结壳和穿孔的含铁质硬底面:带有穿孔(白色箭头处)的鲕粒状灰岩(普林斯巴奇阶)内部充填陆源沉积物(托尔阶初期到 tenuicostatum 层,Cuers 地区剖面);(C)D3 不整合由三个钻孔面构成("D3a","D3b"和"D3c",La Cride 地区剖面;"CM1'"和"CM2":石英质红色石灰岩地层的两段,见图6-2)。(D)D3a 钻孔面(虚线和白色箭头)。钻孔被珊瑚和海绵状砾屑石灰岩相充填(La Cride 地区剖面)。硬币直径2.3cm。(E)珊瑚和海绵状砾屑石灰岩相顶部的含有结壳和钻孔痕迹的硬底面 D3b(白色箭头,La Cride 地区剖面)。硬币直径 1.9cm。(F)和(G)Cuers 和 Saint-Pons 地区剖面中 D3a 侵蚀面和钻孔横切面的取向连生胶结物[(F)中的阴极射线,(G)中的茜草色素—钾铁氰化物染色]。(H)Cuers 地区 D3a 界面中海胆类碎片上的亮晶方解石增生显示出两个非冷光层("1a"和"1c"),它们被"1b"处的几个黄色薄冷光层分隔。(I)Saint-Pons 地区 D3a 界面中海胆类碎片上的亮晶方解石增生染色后在"1a"处显示蓝色,"1b"处显示粉色。标签"1a","1b","1c","1d","2","3"和"4"对应胶结物区域和成岩阶段(详细见文中)。

D3a 截断 CM1 的鲕粒灰岩，并被沉积角砾岩覆盖，包括再沉积的 CM1 的鲕粒和海胆状粒灰岩，以及 CM2 底部的生物碎屑泥粒灰岩(图 6-3B)。D3b 是角砾岩顶部的坚硬面，被 CM2 的泥质灰岩和泥灰岩覆盖(图 6-4A)。菊石、箭石鞘和软骨鱼类牙齿及磷酸盐颗粒都集中在 D3b 之下和 D3b 以上 2~5cm 的地方。D3c 界面是有结介和穿孔的硬底面，只在 La Cride 地区出现(图 6-4C 到 E)。

动物群可根据西欧古特提斯海省菊石判断时间(1997)。D3a 之下，根据腕足类的 Zeilleria (Z.) subovalis(Cuers 地区)、Gibbirhynchia northamptonensis 和 Aulacothyris resputina(Saint-Pons 地区;图 6-3)可以确定 CM1 上部的时代为普林斯巴奇末期(spinatum 层)(Léonide 等，2007)。D3a 和 D3c 之间，菊石类 Neolioceratoides sp.、Protogrammoceras aff. paltum(Cuers 地区)、Harpoceras sp.、Dactylioceras sp.(Carcès 地区,图 6-3A)和腕足类 Gibbirhynchia reyi(Cuers 地区)说明是早托尔期(tenuicostatum 层)(Léonide 等，2007)。在人部分地区，tenuicostatum 层上 D3a 到 D3b 的厚度从几厘米到 20cm 之间变化，但在 La Cride 地区该层厚度达到 2m。这个界面可根据腕足类 Lobothyris arcta 和菊石类 Dactylioceras sp. 进行较精确的定年(图 6-3)。serpentinum 层的上覆沉积物(La Cride 地区剖面 D3c 之上及其他地区剖面 D3b 之上的部分)可以通过腕足类 Soaresirhynchia rustica 和 Homeorhynchia batalleri(Cuers 和 La Cride 地区剖面，图 6-3)定年。因此，D3a 在普林斯巴奇阶向托尔阶过渡的边界(图 6-2)，包括托尔阶初期 paltum 亚层的地层间断，估计为 200~300ka(Guex 等，2001)。

6.3.2.2 东多菲内次级盆地的上普林斯巴奇阶到中托尔阶序列

Mestre(2001)和 Floquet(2003)详细描述了东多菲内次级盆地的 La Robine 和 Marcoux 地区的沉积学，生物地层学和地球化学特征。他们定义了以下岩石地层单位：晚普林斯巴奇期(spinatum 层)的"Calcaires Boudinés"，早托尔期(serpentinum 层)的"Marnes Noires Inférieures"，以及中托尔期(bifrons 层)的"Calcaires Roux Noduleux"。在这两个地区的剖面中，普林斯巴奇阶—托尔阶的过渡被很好地描述，并与具有丰富黄铁矿、木质残骸以及箭石鞘的沉积物硬底面相对应。在局部地区，它覆盖有鱼龙类残骸。在不整合面之下，碳酸盐沉积物为分选好的半深海岩相，与早期风化没有宏观联系。Floquet 等(2003)的生物地层综合研究认为对应于此不整合面的时间间断，包括托尔阶初期，如 tenuicostatum 层，serpentinum 层底部 exaratum 亚层的下半部分。

6.4 结果

6.4.1 南普罗旺斯次级盆地普林斯巴奇阶向托尔阶的过渡

6.4.1.1 D3a 界面的成岩序列

在 D3a 削截面之下的 2~10cm 厚的相中包含多个成岩阶段(图 6-4F 至 I 和图 6-5)。其中一部分，尤其是棘皮类碎片上取向附生的方解石增生，是被钻孔面自身所横切(图 6-4B，F 和 G)，并被其上覆沉积单元所密封。D3a 面的成岩序列包括 5 个连续的成岩阶段(图 6-4 和图 6-5)：

(1)阶段 1：主要表现为沿无壳海胆类碎片的 C 轴发生方解石增生。这个取向连生胶结物，厚可达到 400μm，分层清晰，杂质很少(透明的)、具有非云雾状的带状冷光。分层模式与所有南普罗旺斯次级盆地的露头相似，并可以用茜草色素—钾铁氰化物和阴极射线发光很好地展示出来。它包括 4 个连续的胶结物分区。第一个和第三个没有冷光，被染蓝色(图 6-4G 至 I，图 6-5 和图 6-6A 至 D 中的"1a"和"1c")。第二个和第四个具有橘色—黄色交替冷光，被

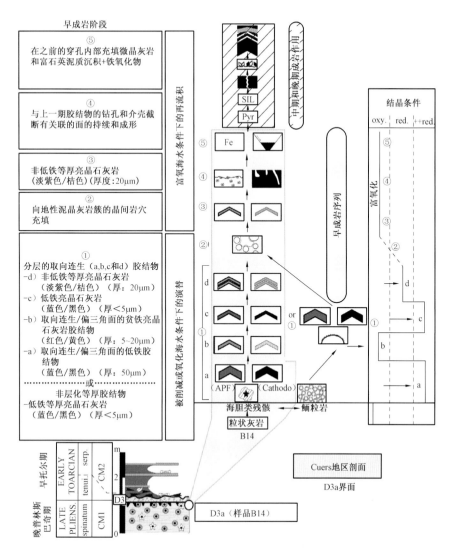

图6-5 D3a界面(Cuers地区剖面)成岩序列:早成岩阶段的细划及
早期方解石胶结物的重结晶演化解释("APF",茜草色素—钾铁氰化物染色)

染粉色(图6-4G至I,图6-5和图6-6A至D中的"1b"和"1d")。在非棘皮类外源化学沉积周围,方解石胶结阶段缺失或者被小的、发微弱冷光的方解石晶体的薄层等厚胶结所替代。

(2)阶段2:粒间孔隙内的向地性的泥晶灰岩刚好位于D3a界面之下。这类泥晶质沉积物包括小的生物碎屑和细碎屑石英。它很明显地覆盖了1d方解石胶结层(图6-4F、H和图6-5中的"2")。

(3)阶段3:等厚的方解石胶结与成岩序列中最后的胶结阶段相对应。这种胶结物(大约20μm厚)杂质含量高(暗淡),发出褐色到橘色的云雾状冷光(图6-4F、H和图6-5中的"3")。

(4)阶段4:环节动物和双壳类钻孔从D3a磨蚀面向下生长;它们横切之前所有的成岩阶段(图6-4F、G和图6-5)。

(5)阶段5:下覆沉积单元之下的D3a界面被铁氧化物包壳(图6-6)。

6.4.1.2 D3b界面的成岩序列

D3b界面的成岩序列与D3a界面相似,也包括5个成岩阶段(图6-6E至H和图6-7):

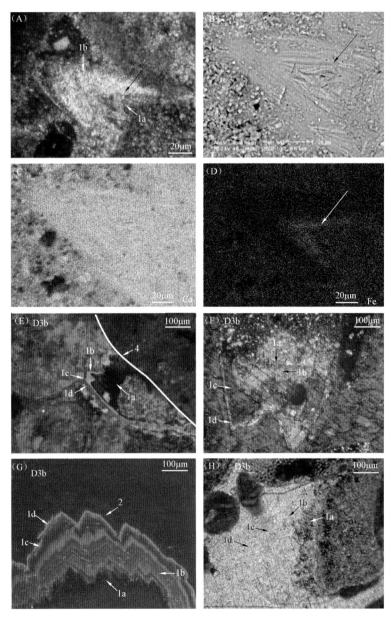

图 6-6 Saint-Pons 地区 D3a 界面的海胆类碎片上的亮晶方解石增生和 D3b 界面的胶结物(使用阴极射线和茜草色素—钾铁氰化物染色来观察):(A)染蓝色的"1a"(黑色箭头)和染粉色的"1b"区域;(B)扫描电镜观察到的亮晶方解石增生(黑色箭头;图 6-4 和图 6-6 的 1a 区域);(C)和(D)使用电子探针分析观察到的低铁方解石 1a 处亮晶方解石中钙、铁的映象[(D)中白色箭头];(E)Cuers 地区 D3b 侵蚀面和钻孔横切面的取向连生胶结物及阶段 1 中的冷光区和非冷光区(详细分区见文中);(F)Saint-Pons 地区 D3b 界面处海胆类碎片上的亮晶方解石增生染色后,呈现出蓝色的"1a"和"1c"区和粉的"1b"和"1d"区;(G)Saint-Pons 地区 D3b 界面处,海胆类残骸上的取向连生胶结物显示出两个非冷光区("1a"和"1c"),两个冷光区("1b"和"1d")和一个云雾状冷光区"2";(H)La Cride 地区 D3b 界面处,海胆类残骸上的亮晶方解石增生染色后呈现出两个蓝色的"1a"和"1c"区,两个粉色的"1b"和"1d"区。标签"1a","1b","1c","1d"和"4"对应于胶结物区域和成岩阶段(详细见文中)

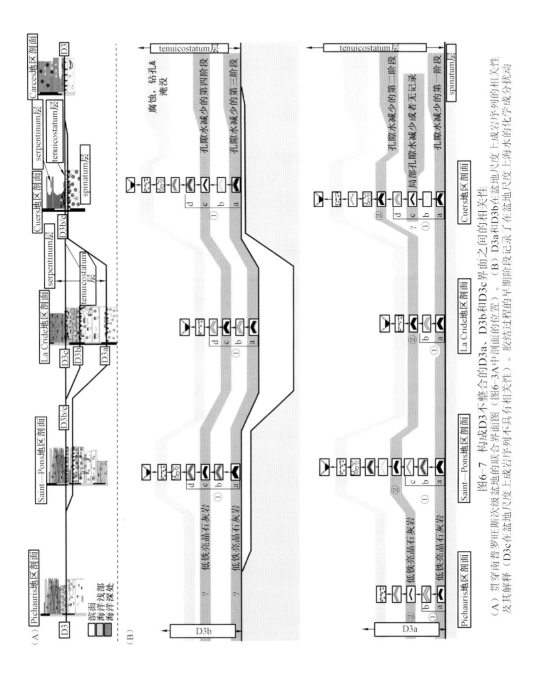

图6-7 构成D3不整合的D3a, D3b和D3c界面之间的相关性

(A) 贯穿南普罗旺斯次级盆地的联合界面剖面图 (图6-3A中剖面的位置)。(B) D3a和D3b在盆地尺度上成岩序列的相关性及其解释 (D3c在盆地尺度上成岩序列不具有相关性)。胶结过程的早期阶段记录了在盆地尺度上海水的化学成分扰动

— 167 —

(1)阶段1:透明的取向附生方解石增生(在棘皮类碎片上)可分为4个连续区带。其中1a,1c和1d被染成蓝色(几乎没有冷光),1b被染成粉色(有较多冷光;图6-6E和H)。

(2)阶段2:等厚胶结物(大约20μm厚)由发出云雾状冷光的暗淡的方解石晶体组成正好位于界面之下(图6-6G中的"2")。

(3)阶段3:洞内泥晶质沉积物正好位于界面之下。这类沉积物具有向地性,通常包含细碎屑石英颗粒。

(4)阶段4和5:硬地面先是被双壳类动物钻孔(图6-6E中的"4"),后来又被铁氧化物包壳。

6.4.1.3　D3c面的成岩序列

D3c面的成岩序列只发生在La Cride地区,包含两个阶段:

(1)阶段1:由暗淡的方解石晶体组成等厚胶结物(10~20μm厚)。

(2)阶段2:粒间内泥晶质沉积物正好位于该界面之下。

6.4.1.4　南普罗旺斯次级盆地的D3硬底面相关性

生物地层学和沉积学数据是研究组成D3不整合沉积间断的三个界面之间地层相关性的基础(Léonide,2007;Léonide等,2007)。就单一界面的保存来说(Carcès地区剖面;图6-7A),在界面之下没有观察到早期层状方解石胶结物,这可能与普林斯巴奇阶大面积的侵蚀有关。在其他地方,D3a和D3b界面有着完整而相似的成岩序列,且每一序列包含两个或者三个染蓝色的或发非云雾状冷光的亮晶方解石胶结物(图6-7B)。南普罗旺斯次级盆地的所有剖面都有着相同的成岩序列。这种条理化现象的出现说明早成岩初期阶段之间相关性的存在(图6-7B)。

6.4.1.5　磷酸盐牙釉质中的碳酸盐、碳同位素和氧同位素

大量的$\delta^{13}C$同位素记录(图6-8和表6-1)反映了南普罗旺斯次级盆地Cuers地区在晚普林斯巴奇期(spinatum层)到中托尔期(variabilis层)内发生的变化。$\delta^{13}C$同位素的值大多在−3‰~2.9‰之间波动(图6-8)。

表6-1　Cuers,Marcoux和La Robine地区晚普林斯巴奇期—晚托尔期序列中碳氧同位素值和碳酸钙含量之间的关系

Cuers地区(南普罗旺斯次级盆地)					
样品	层	$\delta^{13}C$(‰,PDB)	SD $\delta^{13}C$	$CaCO_3$(%)	海拔(m)/D3
CU96	Variabilis	0.77	0.01	86.89	18.13
CU95	Variabilis	1.28	0.002	90.15	17.32
CU94	Variabilis	1.30	0.01	87.3	16.39
CU93	Bifrons	1.45	0.005	80.85	15.61
CU92	Bifrons	1.57	0.09	66.66	15.33
CU91	Bifrons	1.49	0.07	81.9	14.54
CU89	Bifrons	1.57	0.06	92.8	12.84
CU88	Bifrons	1.78	0.05	87.21	12.12
CU87	Bifrons	1.83	0.01	82.8	11.44
CU86	Bifrons	1.26	0.09	80.64	11.03
CU85	Bifrons	1.72	0.19	87	10.56
CU84	Bifrons	1.24	0.004	95.9	10.26

续表

样品	层	$\delta^{13}C(‰, PDB)$	SD $\delta^{13}C$	$CaCO_3(\%)$	海拔(m)/D3
CU83	Bifrons	1.66	0.03	88.72	10.06
CU82	Bifrons	1.80	0.01	87.2	9.86
CU81	Bifrons	2.13	0.10	72.72	9.32
CU80	Bifrons	1.97	0.8	45.45	8.81
CU79	Serpentinum	2.45	0.02	53.9	8.47
CU78	Serpentinum	2.58	0.03	79.54	7.86
CU77	Serpentinum	2.87	0.08	66.39	7.24
CU76	Serpentinum	2.01	0.03	90.2	6.92
CU75	Serpentinum	2.65	0.009	73.68	6.34
CU74	Serpentinum	1.95	0.02	40.8	6.08
CU73	Serpentinum	2.02	0.05	80.46	5.93
CU72	Serpentinum	1.93	0.09	84.09	5.73
CU71	Serpentinum	1.69	0.01	89.51	5.64
CU70	Serpentinum	0.58	—	68.46	5.44
CU68	Serpentinum	0.24	0.04	66.4	4.95
CU66	Serpentinum	−0.01	0.06	77.6	4.46
CU64	Serpentinum	−0.36	0.05	72.72	3.83
CU62	Serpentinum	−0.12	0.03	84.98	3.46
CU60	Serpentinum	−0.38	0.02	45.84	2.95
CU57	Serpentinum	−0.80	0.02	78.26	2.39
CU55	Serpentinum	−0.73	0.06	56.85	2.00
CU53	Serpentinum	−0.93	0.04	69.56	1.63
CU52	Serpentinum	−2.23	0.05	21.34	1.48
CU51	Serpentinum	−2.68	0.001	68.8	1.33
CU50	Serpentinum	−2.83	0.03	25.29	1.21
CU49	Serpentinum	−2.12	0.05	40.4	1.07
CU48	Serpentinum	−3.04	—	22.52	0.95
CU47	Serpentinum	−2.95	0.008	19.6	0.83
CU46	Serpentinum	−1.48	0.038	42.35	0.68
CU45	Tenuicostatum	−0.71	0.06	16.6	0.56
CU44	Tenuicostatum	−0.33	0.007	17.57	0.46
CU43	Tenuicostatum	−0.40	0.02	26.6	0.36
CU43m	Tenuicostatum	0.14	0.02	31.2	0.26
CU42m	Tenuicostatum	0.28	0.03	37.4	0.19

续表

样品	层	$\delta^{13}C(‰,PDB)$	SD $\delta^{13}C$	$CaCO_3(\%)$	海拔(m)/D3
CU42	Tenuicostatum	-0.01	0.02	80	0.14
CU41	Spinatum	1.00	0.10	99	0.00
CU40	Spinatum	0.90	0.02	93.75	-0.15
CU39	Spinatum	1.06	0.01	97.7	-0.47
CU38	Spinatum	0.82	0.06	96.1	-0.74
CU37	Spinatum	1.17	0.11	92.48	-1.00
CU36	Spinatum	0.69	0.02	92.13	-1.84
CU35	Spinatum	0.70	0.09	94	-2.74
CU34	Spinatum	1.34	0.07	96.9	-3.79
CU33	Spinatum	1.07	0.02	99.5	-4.77
CU32	Spinatum	1.53	0.004	97.5	-5.71

La Robine 和 Marcoux 地区(东多菲内次级盆地)

样品	层	$\delta^{13}C(‰,PDB)$	SD $\delta^{13}C$	$CaCO_3(\%)$	海拔(m)/D3
RB-b-3b	Bifrons	1.05	—	74.9	19.30
RB-b-3a	Bifrons	1.26	—	80.78	19.10
RB-b-2	Bifrons	1.10	—	57.91	18.90
RB-b-1	Bifrons	0.68	—	80	18.70
RB-s-20	Serpentinum	1.36	—	43.24	18.50
RB-s-70	Serpentinum	1.33	—	44.53	18.00
MC-1350	Serpentinum	-0.81	—	34.37	17.50
MC-1300	Serpentinum	-0.00	—	26.53	17.00
MC-1200	duplic Serpentinum	-0.70	—	42.18	16.00
MC-1150	Serpentinum	-0.24	—	48.24	15.50
MC-1100	Serpentinum	-0.34	—	52.3	15.00
MC-1050	Serpentinum	0.18	—	41.31	14.50
MC-1000	Serpentinum	0.29	—	42.18	14.00
MC-950	Serpentinum	0.90	—	47.61	13.50
MC-900-h	Serpentinum	0.91	—	46.42	13.00
MC-900	duplic Serpentinum	0.55	—	42.35	12.85
MC-850	Serpentinum	0.74	—	46.82	12.35
MC-800	Serpentinum	0.41	—	39.68	11.85
MC-750	Serpentinum	0.23	—	42.47	11.35
MC-700	Serpentinum	0.02	—	39.38	10.85
MC-610	Serpentinum	0.19	—	43.25	9.95
MC-560	Serpentinum	0.25	—	35.76	9.45
MC-510	Serpentinum	0.37	—	30.98	8.95

续表

样品	层	$\delta^{13}C$(‰, PDB)	SD $\delta^{13}C$	$CaCO_3$(%)	海拔(m)/D3
MC-460	Serpentinum	0.23	—	32.03	8.45
MC-400	Serpentinum	0.31	—	44.01	7.85
MC-350	Serpentinum	0.27	—	38.99	7.35
MC-300	Serpentinum	0.16	—	42.18	6.85
MC-250	Serpentinum	-0.08	—	85.31	6.35
MC-230	Serpentinum	0.74	—	45.38	6.155
MC-210	Serpentinum	1.04	—	35.93	5.955
MC-190	Serpentinum	1.52	—	36.71	5.75
MC-170	Serpentinum	1.47	—	33.33	5.55
MC-150	Serpentinum	1.74	—	39.28	5.35
MC-130	Serpentinum	1.38	—	46.92	5.15
MC-110	Serpentinum	1.17	—	45.63	4.95
MC-90	Serpentinum	1.22	—	41.66	4.75
MC-80	Serpentinum	1.51	—	38.61	4.65
MC-60	Serpentinum	0.89	—	37.3	4.45
MC-30	Serpentinum	1.28	—	35.31	4.15
MC-0	Serpentinum	1.03	—	36.11	3.85
MA-13	Spinatum	0.78	0.03	59.23	3.80
MA-12	Spinatum	0.03	0.02	89.23	3.60
MA-11	Spinatum	-1.23	0.02	84.7	3.30
MA-10	Spinatum	0.33	0.01	52.73	3.05
MA-9	Spinatum	0.24	0.01	49.42	2.60
MA-8	Spinatum	-3.43	0.04	85.15	2.40
MA-7	Spinatum	0.57	0.02	32.54	2.05
MA-6	Spinatum	0.38	0.01	39.21	1.90
MA-5	Spinatum	-2.19	0.003	85.49	1.60
MA-4	Spinatum	0.06	0.03	46.15	1.45
MA-3	Spinatum	-1.89	0.03		1.00
MA-2	Spinatum	0.51	0.02	32.15	0.65
MA-1b	Spinatum	0.48	0.04	41.15	0.40
MA-1a	Spinatum	-1.40	0.03	33.2	0.00

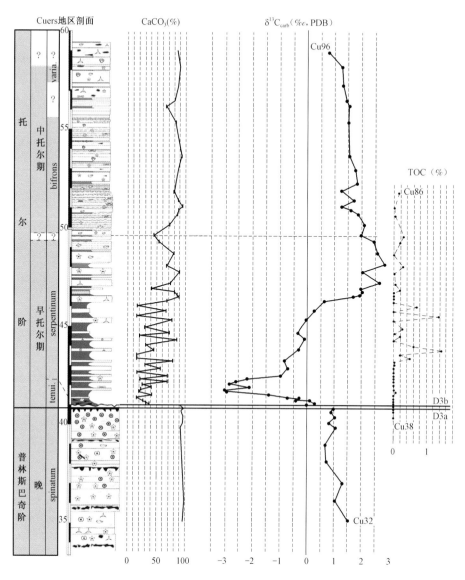

图 6-8 Cuers 地区普林斯巴奇阶—托尔阶过渡时期碳同位素地层、CaCO$_3$ 和 TOC 分布剖面图

生物地层学数据来自于 Léonide 等（2007），碳同位素、CaCO$_3$ 和 TOC 数据列于表 6-1

δ^{13}C 曲线展示了 4 个连续趋势：

（1）在 D3 不整合面下，普林斯巴奇末期（spinatum 层）具有相对稳定的 δ^{13}C 值，约为 1‰；

（2）在 tenuicostatum 层和 serpentinum 层底部，D3 处开始从 1‰ 快速下降到 0，随后，D3 面以上出现从 0 至 -3‰ 的急剧的负变化。在 Cuers 地区，D3 不整合面内记录了这种急剧负变化的开始，并在接近 tenuicostatum/serpentinum 层的边界处达到最大；

（3）在 serpentinum 层上部逐渐增长返回到正值，随后，在 serpentinum 层最上部又是一个约 2.5‰ ~ 2.87‰ 的正偏移；

（4）在 bifrons 和 variabilis 层，同位素值又反弹回 1‰ 的稳定值。从 Cuers 地区早托尔期（tenuicostatum 和 serpentinum 层）沉积物收集到的 29 个牙齿样品中获得的 δ^{18}O 平均值大约在 19.2 到 19.7‰ 之间（图 6-9）。

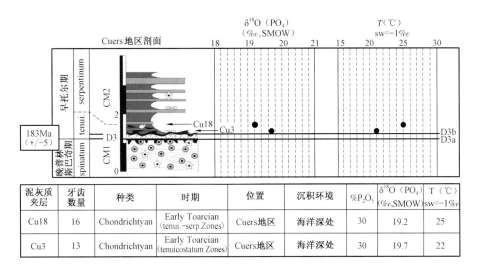

图6-9 Cuers地区软骨鱼类牙釉质中的氧同位素值和由D3不整合面之上的Cu3和Cu18样品计算出的温度。绝对年代引自Gradstein等人(2004)

6.4.1.6 有机物含量

有机物只在serpentinum层(exaratum亚层上部)的泥灰质相中十分丰富,达到了1.5%。在该泥灰质相中TOC含量(46个样品)从0到1.5%变化,且只有13个样品的TOC值高于0.15%(图6-8和图6-10;表6-2)。解释得到的T_{max}均值较低(大约423℃;表6-2)说明有机质含量并没有受到热成岩作用的显著影响,因此,岩石热解分析仪给出的参数可能记录了原始的数据。

表6-2 Cuers地区晚普林斯巴奇期—中托尔期有机地球化学参数:
T_{max},HI和TOC值(＊表示重复测量)

样品	T_{max}(℃)	TOC(%)	HI(mgHC/gTOC)	海拔(m)/D3
CU86	—	0.15	—	10.56
CU84	—	0.00	—	10.06
CU82	—	0.06	—	9.32
CU80＊	422	0.36	111	8.47
CU80	418	0.34	117	8.47
CU79	429	0.18	116	7.86
CU78	—	0.00	—	7.24
CU77＊	421	0.22	—	6.92
CU77	420	0.28	—	6.92
CU76	—	0.00	—	6.34
CU75	—	0.06	—	6.08
CU74	424	0.19	121	5.93
CU73	—	0.00	—	5.73
CU72	—	0.00	—	5.64
CU71	—	0.00	—	5.44

续表

样品	T_{max}(℃)	TOC(%)	HI(mgHC/gTOC)	海拔(m)/D3
CU70	—	0.00	—	5.20
CU69*	426	0.67	364	4.95
CU69	425	0.66	354	4.95
CU68	—	0.00	—	4.71
CU67	417	1.33	421	4.46
CU66	—	0.00	—	4.12
CU65	422	0.31	200	3.83
CU64	412	0.03	—	3.68
CU63*	423	0.19	242	3.46
CU63	422	0.22	209	3.46
CU62	—	0.05	—	3.19
CU61*	421	0.64	331	2.95
CU61	419	0.59	315	2.95
CU60	421	1.43	512	2.73
CU60	421	1.47	479	2.73
CU59*	435	0.15	146	2.53
CU59	433	0.17	117	2.53
CU58	430	0.20	235	2.39
CU57	—	0.00	—	2.22
CU56*	—	0.01	—	2.00
CU56	—	0.03	—	2.00
CU55	—	0.00	—	1.83
CU54	—	0.00	—	1.63
CU53	431	0.02	—	1.48
CU52	—	0.02	—	1.33
CU50	—	0.00	—	1.07
CU49	—	0.03	—	0.95
CU48	—	0.02	—	0.83
CU47	—	0.07	—	0.68
CU46	—	0.02	—	0.56
CU45	—	0.00	—	0.46
CU44	—	0.00	—	0.36
CU43M	—	0.00	—	0.26
CU43	—	0.00	—	0.19
CU42M	—	0.01	—	0.14
CU42	—	0.00	—	0.00
CU40*	—	0.00	—	-0.47

续表

样品	T_{max}(℃)	TOC(%)	HI(mgHC/gTOC)	海拔(m)/D3
CU40	—	0.00	—	-0.47
CU39	—	0.00	—	-0.74
CU38*	—	0.00	—	-1.00
CU38	—	0.00	—	-1.00
平均值	423	0.18	258	
最小值	412	0.00	111	
最大值	435	1.47	512	

在普林斯巴奇阶和托尔阶最下部的碳酸盐岩层中有机质几乎消失，这是由于氧化条件不利于有机质的保存。相反，Cuers地区的泥灰质相利于有机质的保存，尤其是最厚的层段，其TOC值在0.6%到1.5%之间变化(图6-8)。这些层段中有机质的来源可以用含氢指数(HI)判断出来。12个富有机质样品的含氢指数值在111~512mg HC/g TOC范围内。来自serpentinum层的Cu60, Cu61和Cu67样品具有较高的HI值(分别为400mg HC/g TOC和500mg HC/g TOC)，这说明有机质主要来自于水下物源(图6-10)。不稳定有机物的良好保存一定程度说明其为次氧化—缺氧条件的产物。其他的样品则显示出较低的HI值，

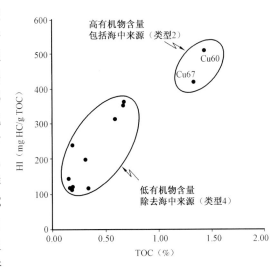

图6-10 含氢指数与TOC值类型2和3反映出

其变化趋势可以从图6-10中清晰地看出。可见，这些样品中的有机质可能有着相同的浮游植物源，但是当TOC减少时这种情况会发生很大的变化。

尽管注意到Cuers地区属于暴露风化，这也可以用来解释其TOC和HI值的减少，但是出现这样的趋势还是很可能与沉积物沉积时氧含量的波动有关。不管怎样，这两组富有机质样品都清楚地显示出托尔阶缺氧环境的存在。

6.4.2 东多菲内次级盆地普林斯巴奇阶向托尔阶的过渡

碳同位素：大量 $\delta^{13}C_{carb}$ 同位素的记录(图6-11和表6-1)与晚普林斯巴奇期(spinatum 层)到中托尔期(variabilis 层)之间的沉积情况相吻合，这一现象被 La Robine 和 Marcoux 地区的菊石动物群的聚集所证明(Floquet 等，2003)。La Robine 和 Marcoux 组合序列中(图6-11右) $\delta^{13}C_{carb}$ 曲线的演化显示了三个连续趋势：

(1) 不整合面之下的普林斯巴奇末期(spinatum 层)出现的急剧而强烈的变化(包括负偏移)；

(2) 不整合面之上和 serpentinum 层底部(exaratum 亚层上部)之间出现向正值转变，并在接近 exaratum 和 falciferum 亚层的边界处达到峰值，约1.7‰；

(3) 在 falciferum 亚层回返到稳定值0‰(或在其上部出现轻微的负值)。

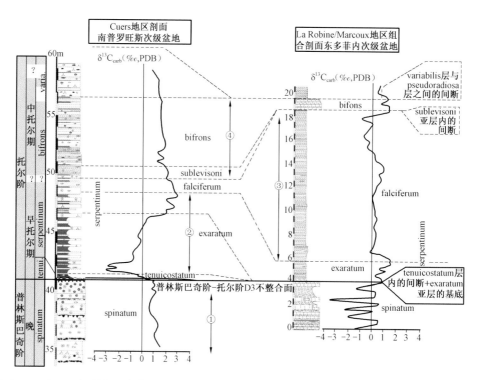

图 6-11 Cuers 地区剖面(南普罗旺斯次级盆地)与 La Robine/Marcoux 地区组合剖面(东多菲内次级盆地)中生物地层学和化学地层学相关性对比。生物地层学数据来自于 Floquet 等(2003),碳同位素数据列于表 6-2

6.4.3 南普罗旺斯次级盆地与东多菲内次级盆地地球化学指标的关联性

生物地层学和化学地层学在南普罗旺斯次级盆地与东多菲内次级盆地沉积序列之间的相关性如下(图 6-11):

(1)对普林斯巴奇阶—托尔阶不整合面的不同表现。在南普罗旺斯次级盆地,这个不整合面被两个钻孔横切面间的早期低铁亮晶胶结物所突现,而在东多菲内次级盆地,它却是一个记录了海洋侵蚀作用的简单硬底面(Floquet 等,2003)。

(2)两个次级盆地的沉积速率差异导致了早托尔期 $\delta^{13}C_{carb}$ 记录之间的差异。在南普罗旺斯次级盆地,tenuicostatum 层记录了急剧的碳同位素负迁移,它开始于 D3 不整合间断,穿过 D3b 或 D3c 上方数厘米处的第一个泥灰质层,并继续到达 serpentinum 层底部(exaratum 亚层的第一部分)(图 6-8 和图 6-11)。随后在 exaratum 亚层,$\delta^{13}C_{carb}$ 开始增加,直到接近 serpentinum 层内 exaratum 和 falciferum 亚层的边界处达到最大。而东多菲内次级盆地,tenuicostatum 层和 exaratum 亚层下部在 D3 不整合面处缺失。$\delta^{13}C_{carb}$ 在 exaratum 亚层的上部开始向正值变化,并在接近 exaratum 和 falciferum 亚层时达到极值,这在 D3 面上非常明显(图 6-11)。

(3)$\delta^{13}C_{carb}$ 曲线在 falciferum 亚层(图 6-11)有着相同的趋势,包括在 falciferum 亚层顶部出现的轻微减小(注意东多菲内次级盆地 sublevisoni 亚层的缺失与沉积间断有关;Floquet 等,2003)。

(4)在 bifrons 层期间 $\delta^{13}C_{carb}$ 值发生了变化,这是由于东多菲内次级盆地当时的沉积压实与间断(最大厚度 1m, apertum 和 bifrons 层面缺失; Floquet 等,2003),然而在南普罗旺斯次级盆地同时期的沉积却是相对扩张的(Cuers 地区达到 7m;图 6-11)。

6.5 讨论

6.5.1 普林斯巴奇阶与托尔阶之间不整合处成岩作用的解释

南普罗旺斯次级盆地的 D3 不整合面序列显示出了一个复杂的、多相的沉积和成岩历史，并有三个紧密间隔的硬底面。D3a 和 D3b 界面的成岩序列中观察到的取向连生亮晶方解石增生物和叶片状胶结物，含杂质少，发出自形、非云雾状的带状冷光。这些岩石学特征说明此处存在保存较好的（非重结晶）从低镁钙比的水中沉淀出来的低镁方解石胶结物（Wilkinson 和 Algeo，1989；Meyers，1991；Moore，2001；Dickson，2002；Holland，2004）。

在下侏罗统，在霰石海时期与方解石海时期的过渡时期（Sandberg，1983；Wilkinson 和 Algeo，1989；Dickson，2002），这样的水往往存在于大气潜水领域地区或者海洋潜水领域地区。但是，在研究区外部台地的背景下，普林斯巴奇阶碳酸盐岩下部不显示地表特征（潮间带到潮上带储层，古喀斯特结构，沼泽相，未成熟结构或者渗流胶结物等等）。相反，D3a 和 D3b 界面的形态从本质上记录了碳酸盐台地淹没期间的水下磨损和生物侵蚀。基于这些原因推测在沉积物—海水分界面下循环的海洋潜水域是最有可能造成这种早期取向连生增长物沉淀的环境。

与这些早期亮晶胶结物相关的另一个关键点是它们被茜草色素—钾铁氰化物染色的带状面（图 6-6A）。这些染蓝的区域反映初期低铁方解石结晶，而红色反映贫铁方解石结晶（Evamy，1963，1969）。在方解石晶体中，铁的结合是直接与氧化还原反应相关的。在酸性电解水中，铁和氧形成铁氧化合物，不能与方解石晶格相组合。在孔隙水减少的条件下，Fe^{2+} 是游离的，并可以与方解石晶格结合（图 6-6D）。D3a 界面中两个广泛分布的富铁层，加上 D3b 界面两个后期形成的富铁层，都表明孔隙水减少的情况发生在与界面相关的沉积间断的四个不同时期。一个显著的事实是南普罗旺斯次级盆地的 D3a 和 D3b 界面中的早成岩作用特征具有很好的相关性（图 6-7），这反映出了整个南普罗旺斯次级盆都普遍发生了孔隙水减少的情况。

一种假说可以解释这些胶结物结晶的机理，那就是在整个南普罗旺斯次级盆地尺度上，结晶的物理化学条件发生了变化。由于营养物输入（营养物和有机质）造成的富营养化导致了碳酸盐台地幕式低氧或者甚至是缺氧事件的发生。很多作者（Hallock 和 Schlager，1986；Zempolich，1993；Cobianchi 和 Picotti，2001；Galluzzo 和 Santantonio，2002）对此有争议，他们认为富营养化是海流上涌，海平面下降，陆源输入增加和气候变化导致的。

Mallarino 等（2002）描述说这种胶结物和 Monte Kumeta（西西里岛）地区的普林斯巴奇阶—托尔阶边界处碳酸盐硬底面上的 D3 早期含低铁方解石增生相似的。基于胶结物的成岩作用和流体包裹体分析，Mallarino 等（2002）解释说它们的形成反映一种浅层（大约 20m）海水富营养化事件。这些人得出结论，认为富营养化是营养物质过剩的结果，从而导致了海水加深过程中出现缺氧或者次氧化环境。西西里岛盆地和南普罗旺斯次级盆地石灰质储层中相似胶结物的同时性似乎说明这些碳酸盐工厂扰动机理在古特提斯洋西部领域内很广泛，并不是局部的。在晚普林斯巴奇期发生的相似的碳酸盐生产扰动在亚平宁山脉中北部（Galluzzo 和 Santantonio，2002）和高阿特拉斯山脉（摩洛哥）（Blomeier 和 Reijmer，1999；Ettaki 和 Chellai，2005；Wilmsen 和 Neuweiler，2008）的储层中也有记录。

6.5.2 碳同位素异常与化学地层学的相关性

来自于南普罗旺斯次级盆地 Cuers 地区和其他盆地，如 North Yorkshine（英国）和 Peniche（卢西塔尼亚盆地）的碳同位素曲线展示于图 6-12。North Yorkshine 地区剖面由保存完好，

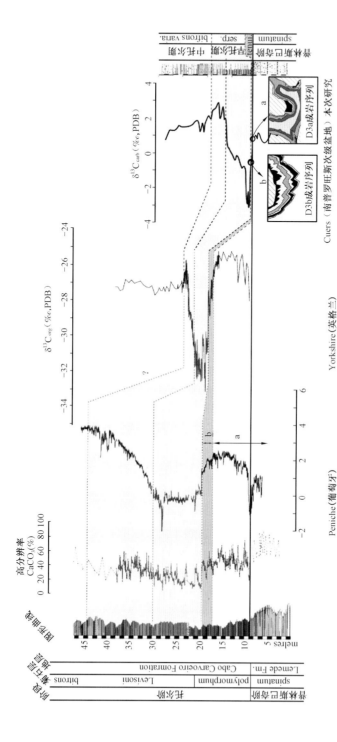

图6-12 Cuers、Peniche 和Yorkshire 地区碳同位素剖面图之间的相关性(部分改编自Suan 等, 2008a)

注意沉积物的压实记录了Yorkshire地区polymorphum-levisoni层碳同位素的迁移和Cuers地区polymorphum层的极限压实状态(ultenuicostattum层)(D3不整合部分记录了该段压实)。Peniche和Yorkshire地区polymorphum层"a"段普林斯巴奇阶-托尔斯阶边界处第一次负和正的同位素偏移与南普罗旺斯次级盆地D3a面的早期低铁方解石期同时代的。南普罗旺斯次级盆地D3b早期低铁方解石的保存和碳同位素b段的急剧负变化都与polymorphum层上部的第二次碳同位素负迁移有关。Peniche地区碳同位素数据来自于Hesselbo 等(2007),碳酸盐含量数据来自于Suan (2008a)。Yorkshire地区碳同位素数据来自于Cohen等(2004)和Kemp 等(2005)

不受生物扰动的、层状富有机物泥岩组成,沉积在约 40°N 的大陆架处(Kemp 等,2005),但是 Peniche 地区剖面反映了在约 20°N 处晚普林斯巴奇期到早托尔期的连续沉积(Elmi,2006)。因此,这两个地区剖面是不同纬度大洋缺氧事件的典型代表,提供了地层学上完整而准确的全球事件的记录。这三个盆地碳同位素曲线之间的相关性:① 普林斯巴奇阶—托尔阶过渡时期不同的变化;以及② 早托尔期相同的变化。

(1)Hesselbo(2007)、Suan(2008a)等发现的第一个由负到正的碳同位素变化被记录在 Peniche 地区剖面中,显示了在普林斯巴奇阶—托尔阶边界处的负变化以及随后的 polymorphum 层的连续正变化,例如 tenuicostatum 层。Cohen(2004)、Kemp 等(2005)也提供了证据证明在 North Yorkshire 地区剖面内部那时存在正向的变化(图 6-12 中厚、白段"a")。在南普罗旺斯次级盆地这时期的 D3a 面,第一个碳同位素变化没有被记录可能是因为剥蚀作用或其他的非沉积作用(图 6-12 中的矩形 a)。在东多菲内次级盆地,普林斯巴奇阶地层碳同位素曲线从 0 到负值的剧烈变化(图 6-11)可以解释为第一个碳同位素变化时两者具有不同的胶结物以及泥质和钙质层的矿物学稳定性不同造成。

(2)第二次由负到正的碳同位素变化也被很好地记录下来,并且和 Peniche 及 North Yorshire 地区剖面有相关性(Cohen 等,2004;Kemp 等,2005;Hesselbo,2007;Suan 等,2008a)。最初的负变化开始于 polymorphum 层的上部(即 tenuicostatum 层),随后在 levisoni 层(即 serpentinum 层)的上部达到负向的最大值。在南普罗旺斯次级盆地,tenuicostatum 层(即 polymorphum 层)顶部的第二个碳同位素变化的初始负变化记录在 D3a 和 D3b 之间,并且刚好在 D3b 正上方(图 6-12 右手边的窄淡红色条带"b")。早托尔阶(exaratum 亚层)时期的 Cuers 地区 $\delta^{13}C$ 曲线的负极值在 -3‰(图 6-12 右手边的黄色条带)处,是和 North Yorshire 地区(Cohen 等,2004;Kemp 等,2005)中的有机物 $\delta^{13}C$ 负极值 -33‰ 以及 Peniche 地区(Hesselbo 等,2007)中的碳酸盐(-2‰)和树木化石(大约 30‰)记录的 $\delta^{13}C$ 负极值有好的相关性。这些负向的变化也是和其他一些地区是类似的:例如,威尔士(Mochras;Jenkyns 等,2002),英格兰东北部(约克郡;Jenkyns 和 Clayton,1997)和西北德国(Dotternhaussen;Schmid-Röhl 等,2002)。随后在南普罗旺斯次级盆地 Cuers 地区 serpentinum 层时期发生了正变化(图 6-12 右手边的上部淡蓝色条带),此时的峰值为 2.87‰,接近 exaratum 或 falciferum 亚层边界处,以此为标记可以发现 North Yorshire 地区有机物 $\delta^{13}C$ 极值为 -25‰,Peniche 地区的为 4‰(图 6-12 左手边的上部亮蓝色带)。正极值也是和其他西欧古特提斯盆地的数据报告有很好的相关性(Küspert,1982;Jenkyns 和 Clayton,1986,1997;Hesselbo 等,2000,2007;Schouten 等,2000;Beerling 等,2002;Jenkyns 等,2002;Kemp 等,2005;Emmanuel 等,2006)。

Suan 等人(2008b)建议两个碳循环扰动是和两幕古环境应力是相对应的:普林斯巴奇阶—托尔阶边界处到早托尔阶(tenuicostatum 层)是第一幕;exaratum 到 falciferum 亚层是第二幕(包括 Jenkyns 于 1988 最初定义的托尔阶大洋缺氧事件期间的最剧烈的应力)。这两幕也和海底动物群及菊石类快速灭绝有相关性(Alméras 等,1995;Little 和 Benton,1995;Macchioni 和 Cecca,2002;Cecca 和 Macchioni,2004)。D3a 面的早期胶结物(例如低铁方解石增生层),暗示出南普罗旺斯次级盆地碳酸盐的幕式削减,这很有可能和其他盆地古环境应力的第一幕相吻合(Hesselbo 等,2007;Suan 等,2008b)。类似地,据一些同时期地区剖面记载,D3b 面的早期低铁胶结物和南普罗旺斯次级盆地碳同位素突然的负变化都与第二幕应力具有等时关系(Suan 等,2008b),例如,对剧烈扰动的古环境和托尔阶大洋缺氧事件(Hesselbo 等,2000,2007;McArthur 等,2000;Jenkyns 等,2001;Rosales 等,2004a;Kemp 等,2005)。

6.5.3 碳酸盐台地对托尔阶大洋缺氧事件中海洋古地理变化的响应

6.5.3.1 托尔阶大洋缺氧事件前的扰动

普罗旺斯次级盆地内部,晚普林斯巴奇阶沉积序列的上部(margaritatus 和 spinatum 层)(图6-13A)记录的向上进积变浅模式被认为是有效可容空间被碳酸盐沉积物充填的结果(Léonide,2007;Léonide 等,2007)。早期的台地在晚普林斯巴奇期的剥蚀导致整个南普罗旺斯次级盆地和东多菲内次级盆地具有相对海平面下降的特征(图6-13A),并对应于 Léonide 等(2007)定义的二阶沉积序列中的低位体系域。

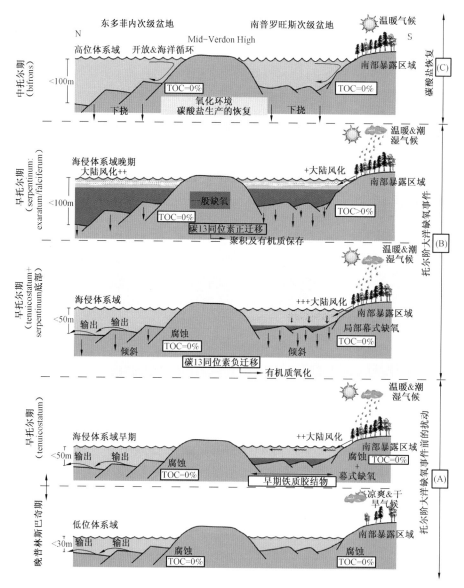

图6-13 图解古海洋地质模型中,法国东南部沿南北向横切面的南普罗旺斯次级盆地和东多菲内次级盆地在晚普林斯巴奇期到中托尔期之间的气候变化与差异沉降:(A)托尔阶大洋缺氧事件前的扰动;(B)托尔阶大洋缺氧事件;(C)碳酸盐恢复(详细见文中)(LST,低位体系域;TST,海侵体系;HST,高位体系域)

全球气候变冷可能发生在晚普林斯巴奇期。那时高纬度地区出现的钟乳状方解石说明那时冰盖发育,至少具有幕式冰期条件的特征。(Price,1999;Guex 等,2001;Morard 等,2003)。另外,由箭石鞘(Rosales 等,2004a,b)和腕足类化石(Suan 等,2008b)的氧同位素分析可以推断出当时温度下降和 spinatum 层的冰点。通过氧同位素分析和 Mg/Ca 比研究全球海洋水高盐度尚存在争议(McArthur 等,2000;Bailey 等,2003;van de Schootbrugge 等,2005)。但是,长期的降温并没有导致海平面显著的下降,也不会导致整个南普罗旺斯次级盆地内部台地的暴露(图6-13A)。

之后的阶段,全球变暖(Dera 等,2011)和相应的扰动,如大陆风化作用的增强和随后的海水富营养化,都可能对早托尔期的碳酸盐形成产生消极影响(Blomeier 和 Reijmer,1999;Cobianchi 和 Picotti,2001;Mattioli 和 Pittet,2004;Woodfine 等,2008;Dera 等,2009a;Mattioli 等,2009)。西班牙(Cantabrian 盆地;Rosales 等,2004b)、英国和德国(Bailey 等,2003)以及葡萄牙(Lusitanian 盆地;Suan 等,2008b)等地的沉积序列中都记录了这一时期海水温度的上升。

为了根据含磷酸盐的牙釉质中 $\delta^{18}O$ 值来估算古温度,判断早侏罗世海水的同位素组成($\delta^{18}O$ 海水)就变得十分重要。现代表层海水 $\delta^{18}O$ 海水的平均值大约在 0‰(SMOW)。现代极地冰盖的全部融化将使 $\delta^{18}O$ 海水值降到 -1‰(SMOW),考虑到当时并无冰盖,所以通常用它来计算侏罗纪时期的古温度。由南普罗旺斯次级盆地(Cuers 地区)中软骨鱼类牙釉质的氧同位素比率计算出的海水温度表明早托尔期的温度应该在 22~25℃ 之间(图6-9 和图6-13)。这个值与西特提斯盆地其他地区(21~27℃ 之间)是相似的(McArthur 等,2000;Jenkyns 等,2002;Bailey 等,2003;Rosales 等,2004a;van de Schootbrugge 等,2005;Gómez 等,2008;Metodiev 和 Koleva-Rekalova,2008;Suan 等,2008b;Dera 等,2009b)。

海平面和温度的上升导致了碳酸盐台地的淹没,但是南普罗旺斯次级盆地 D3 硬底面上的含低铁晶石的出现(图6-13A)却是和幕式缺氧与侵蚀事件相关联的。此次淹没事件的开始对应于二级沉积序列中的早期海侵体系域,据 Léonide 等(2007)。

6.5.3.2 托尔阶大洋缺氧事件

早托尔期海面变化速率上升(Thierry,2000)导致了南普罗旺斯次级盆地和东多菲内次级盆地(图6-13B)以及世界各地的沉积盆地黑色页岩的沉积。这些黑色页岩沉积物(图6-13B)对应于 Léonide 等(2007)定义的二级沉积序列中海侵和晚期海侵体系域。这些页岩具有托尔阶大洋缺氧事件的特征,它们首先被 Jenkyns(1988)描述,并在整个西欧特提斯海域边缘都有发现(Jenkyns 和 Clayton,1997;Jenkyns 等,2001,2002;McArthur 等,2000;Rosales 等,2004a,b);早托尔期富有机物沉积说明这些边缘的海水当时发生了强烈的层化(Küspert,1982;Schouten 等,2000;Röhl 等,2001;Bucefalo Palliani 等,2002,2003;Schmid-Röhl 等,2002)。在水下有机物,水下碳酸盐和大陆矿物中的正、负碳同位素迁移记录了托尔阶大洋缺氧事件(Hesselbo 等,2000,2007;Schouten 等,2000;Kemp 等,2005;Woodfine 等,2008)。托尔阶大洋缺氧事件期间,碳同位素变化的起源目前还是一个激烈的争论点(Küspert,1982;Hesselbo 等,2000,2007;Kemp 等,2005;van de Schootbrugge 等,2005;Wignall 等,2005)。

早托尔期的古气候重建表明季风性大气环流控制了特提斯海域中纬度边缘地区的气候,并导致了高的季节性降雨量(Chandler 等,1992;Bjerrum 等,2001)。全球变暖引起的水文机制变化假设会引起大陆架海水盐度的降低,那么这将导致海水层化的加剧(Saelen 等,1996;Bailey 等,2003)。锇和锶同位素比值的数据也能为水文机制的增强和大陆风化的加剧提供佐证(Cohen 等,1999,2004;Cohen 和 Coe,2007)。因此,早托尔期全球变暖和高降雨量,将增强大陆风化和碳酸盐台地的营养物质输入,从而导致幕式海水层化和相关的缺氧或贫氧发生(图6-13B,第一个框图)。此外,由于大量 CO_2 输入进海洋—大气系统,引起海水酸度的增加,这也将导致碳酸盐岩危机的发生(Kleypas 等,1999;Gattuso 和 Buddemeier,2000;Woodfine 等,2008)。

总之,南普罗旺斯次级盆地碳酸盐台地生产的终止可能是对全球气候变化引起的海水营养变动的一种响应,正如同时期的其他碳酸盐台地(Blomeier 和 Reijmer,1999;Mallarino 等,2002;Woodfine 等,2008)。低铁方解石胶结物(D3a/D3b)和硅质碎屑沉积物包括有机物(黑色泥灰泥和页岩)及相关的碳同位素示踪都记录了这一幕式或短期海水条件的变化(可能由于营养供给导致底层水幕式缺氧或贫氧;图6-13A 和 B)。

6.5.3.3 托尔阶大洋缺氧事件后的碳酸盐生产的恢复

南普罗旺斯次级盆地中托尔期(bifrons 层)相对海平面上升的最大值和海岸超覆具有陆源储层消失,碳酸盐沉积速率低的特点(图6-13C)。总之,根据 Gradstein 等(2004)的时间标度,在800ka 的时间跨度里,bifrons 层沉积了50cm 至2~5m 厚的石灰岩。这个过程导致了动物群的聚集和后生矿物的汇聚(如磷酸盐和海绿石)。大量底栖动物群及其扰动作用(Léonide 等,2007),还有稳定碳同位素值恢复到约1~2‰(图6-11 和图6-13C)都可以证明海平面达到高位时水域开放和海水富氧。高位水平面被认为是古特提斯西部及其周围地区边缘的海平面升降和区域下挠联合造成的(Thierry,2000)。

6.5.4 局部和/或区域沉降对托尔阶大洋缺氧事件沉积记录的影响

古特提斯洋西部边缘的区域构造沉降(Lallam 等,1997;de Gracianskyet 等,1998;Thierry,2000;El Arabi 等,2001)可能曾强烈的影响发生在托尔阶大洋缺氧事件之前的全球气候扰动的沉积记录。构造起始和发展于受限的次级盆地,导致了局部稳定缺氧条件的形成。在 van de Schootbrugge 等(2005)和 Wignall 等(2005)等提出的缺氧模型中被重建出来的古气候变化和大洋循环重组的联合作用,似乎与那些导致大陆板块重排和气候变化加剧的构造活动密不可分。

南普罗旺斯次级盆地托尔阶沉积序列底部的碳酸盐台地淹没(tenuicostatum 层期间,D3上)和随后的加深(serpentinum 层期间)不仅与营养改变(如富营养化)导致的碳酸盐危机有关,同时也和局部或区域的构造有关(图6-13)。构造倾斜引起了不同的沉降,从而导致了局部地区黑色泥页岩赋存在半氧化到缺氧环境的倾斜块体底部(图6-13B)。在普罗旺斯南部次级盆地和多菲内东部次级盆地之间的 exaratum 和 falciferum 亚层沉积厚度的显著差异可以证明这个构造倾斜的存在,这可以用倾斜动力学中不同的时序和剧烈程度来解释(Floquet 等,2003;Léonide,2007;Léonide 等,2007)。这个构造倾斜与当时的伸展构造的主要阶段及影响整个古特提斯洋边缘的显著沉降是相符的[例如:巴黎的凯尔西和多菲内盆地(de Graciansky 等,1998;Floquet 等,2003);摩洛哥盆地(Lallam 等,1997;El Arabi 等,2001)]。

6.6 结论

对南普罗旺斯和东多菲内次级盆地(普罗旺斯东南部)的普林斯巴奇末期和早托尔期沉积序列进行详尽的成岩作用和化学地层研究可以为托尔阶大洋缺氧事件(T-OAE)的沉积记录对比提供佐证:

(1)南普罗旺斯次级盆地托尔阶初期(tenuicostatum 层)多相不整合内的早期铁方解石胶

结物记录了海底这一幕式减少环境(是否为缺氧环境有待研究),它是由营养供给和随后的营养变动所诱发的。狭义地解释,这些事件与托尔阶大洋缺氧事件发生前的大环境扰动相一致。这种胶结物是全球缺氧事件前兆阶段的纪录。

(2)南普罗旺斯次级盆地中碳同位素显著的负变化起始于 tenuicostatum 层,并在 serpentinum 层(exaratum 亚层的底部)达到极值(大约 -3‰),这些都已经被记录下来。这个负偏移与普罗旺斯南部次级盆地泥灰岩—灰岩互层中的有机物富集是同一时代的,且紧随其后的是 serpentinum 层(exaratum/falciferum 亚层的边界)的正偏移。狭义地解释,这些变化是与托尔阶大洋缺氧事件相对应的,反映出了全球古环境的扰动。

(3)托尔阶初期含磷酸盐的软骨鱼类牙齿中氧同位素测定表明表层海水温度大约在 22~25℃之间。这个值与世界其他几大地区早托尔期全球变暖的记录是一致的。

(4)因此,普林斯巴奇阶末期位于南普罗旺斯次级盆地的浅水碳酸盐岩台地的淹没可能是因为全球气候改变造成的海水富营养化导致的,也可能是在托尔阶初期由普林斯巴奇阶—托尔阶地层边界处的强烈下沉导致的。

(5)碳同位素地层与生物地层在南普罗旺斯次级盆地和东多菲内次级盆地的关联性表明托尔阶最初期的构造倾斜存在差异时序。板块倾斜导致了南普罗旺斯次级盆地倾斜块体底部有机质富集层的保存,并且造就了东多菲内次级盆地的沉积间断。

(6)中托尔期(bifrons 层)碳酸盐沉积的恢复是与全球变暖驱动的海平面抬升最大值和古特提斯洋西部边缘区域沉降活动的综合作用密切相关的。

参 考 文 献

Aberhan, M. and Baumiller, T. K. (2003) Selective extinction among Early Jurassic bivalves: a consequence of anoxia. Geology, 31, 1077 – 1080.

Alméras, Y., Mouterde, R., Elmi, S. and Rocha, R. B. (1995) Legenre Nannirhynchia (Brachiopoda, Rhynchonellacea, Norellidae) dans le Toarcien portugais. Palaeontogr. Abt. A, 237, 1 – 38.

Arthur, M. A., Dean, W. E. and Pratt, L. M. (1988) Geochemical and climatic effects of increased marine carbon burial at the Cenomanian/Turonian boundary. Nature, 335, 714 – 717.

Bailey, T. R., Rosental, Y., Mc Arthur, J. M., Van de Schootbrugge, B. and Thirlwall, M. F. (2003) Paleoceanographic changes of the Late Pliensbachian-Early Toarcian interval: a possible link to the genesis of an Oceanic Anoxic Event. Earth Planet. Sci. Lett., 212, 307 – 320.

Bassoulet, J. P., Elmi, S., Poisson, A., Cecca, F., Bellion, Y., Guiraud, R. and Baudin, F. (1993) Mid Toarcian. In: Atlas Tethys Paleoenvironmental Maps (Eds J. Dercourt, L. E. Ricou and B. Vrielynck), pp. 63 – 84. BEICIP-FRANLAB, Rueil-Malmaison.

Baudin, F., Herbin, J. -P. and Vandenbroucke, M. (1990) Mapping and geochemical characterization of Toarcian organic matter in the Mediterranean Tethys. Org. Geochem., 16, 677 – 687.

Baudrimont, A. F. and Dubois, P. (1977) Un basin Mésogéen du domaine péri-alpin: le Sud-Est de la France. Bull. Centres Rech. Explor. -Prod. Elf Aquitaine, 1, 261 – 308.

Beerling, D. J., Lomas, M. R. and Gröcke, D. R. (2002) On the nature of gas-hydrate dissociation during the Toarcian and Aptian Oceanic Anoxic Event. Am. J. Sci., 302, 28 – 49.

Bjerrum, C. J., Surlyk, F., Callomon, J. H. and Slingerland, R. L. (2001) Numerical paleoceanographic study of the Early Jurassic transcontinental Laurasian Seaway. Paleoceanography, 16, 390 – 404.

Blomeier, D. P. G. and Reijmer, J. J. G. (1999) Drowning of a Lower Jurassic carbonate platform: Jbel Bou Dahar, High Atlas, Morocco. Facies, 41, 81 – 110.

Bucefalo Palliani, R., Mattioli, E. and Riding, J. B. (2002) The response of marine phytoplankton and sedimentary organic matter to the early Toarcian (Lower Jurassic) oceanic anoxic event in northern England. Mar. Micropaleontol., 46, 223–245.

Bucefalo Palliani, R., Mattioli, E. and Riding, J. B. (2003) Erratum to "The response of marine phytoplankton and sedimentary organic matter to the early Toarcian (Lower Jurassic) oceanic anoxic event in northern England". Mar. Micropaleontol., 48, 167–168.

Cecca, F. and Macchioni, F. (2004) The two early Toarcian (Early Jurassic) extinctions events in ammonoids. Lethaia, 37, 35–56.

Chandler, M. A., Rind, D. and Ruedy, R. (1992) Pangean climate during Early Jurassic: GCM simulation and the sedimentary record of paleoclimate. Geol. Soc. Am. Bull., 104, 543–559.

Cobianchi, M. and Picotti, V. (2001) Sedimentary and biological response to sea-level and palaeoceanographic changes of a Lower-Middle Jurassic Tethyan platform margin (Southern Alps, Italy). Palaeogeogr. Palaeoclimatol. Palaeoecol., 169, 219–244.

Cohen, A. S. and Coe, A. L. (2007) The impact of the Central Atlantic Magmatic Province on climate and on the Sr- and Os-isotope evolution of seawater. Palaeogeogr. Palaeoclimatol. Palaeoecol., 244, 1–4.

Cohen, A. S., Coe, A. L., Bartlett, J. M. and Hawkesworth, C. J. (1999) Precise Re-Os ages of organic-rich mudrocks and the Os isotope composition of Jurassic seawater. Earth Planet. Sci. Lett., 167, 159–173.

Cohen, A. S., Coe, A. L., Harding, S. M. and Schwark, L. (2004) Osmium isotope evidence for the regulation of atmospheric CO_2 by continental weathering. Geology, 32, 157–160.

Dera, G., Pellenard, P., Neige, P., Deconinck, J.-F., Pucéat, E. and Dommergues, J.-L. (2009a) Distribution of clay minerals in Early Jurassic Peritethyan seas: Palaeoclimatic significance inferred from multiproxy comparisons. Palaeogeogr. Palaeoclimatol. Palaeoecol., 271, 39–51.

Dera, G., Pucéat, E., Pellenard, P., Neige, P., Delsate, D., Joachimski, M. M., Reisberg, L. and Martinez, M. (2009b) Water mass exchange and variations in seawater temperature in the NW Tethys during the Early Jurassic: evidence from neodymium and oxygen isotopes of fish teeth and belemnites. Earth Planet. Sci. Lett., 286, 198–207.

Dera, G., Brigaud, B., Monna, F., Laffont, R., Pucéat, E., Deconinck, J.-F., Pellenard, P., Joachimski, M. M. and Durlet, C. (2011) Climatic ups and downs in a disturbed Jurassic world. Geology, in press.

Dickson, J. A. D. (1966) Carbonate identification and genesis as revealed by staining. J. Sed. Petrol., 36, 491–505.

Dickson, J. A. D. (2002) Fossil echinoderms as monitor of the Mg/Ca ratio of Phanerozoic oceans. Science, 298, 1222–1224.

Dromart, G., Allemand, P., Garcia, J. P. and Robin, C. (1996) Variation cyclique de la production carbonatée au Jurassique le long d'un transect Bourgogne-Ardèche, Est-France. Bull. Soc. Géol. France, 167, 423–433.

Durlet, C. and Loreau, J. P. (1996) Séquence diagénétique intrinsèque des surfaces durcies: mise en évidence de surface d'émersion et de leur ablation marine. Exemple de la plateforme bourguignonne, Bajocien (France). CR Acad. Sci. Paris, 323, 389–396.

Durlet, C., Loreau, J. P. and Pascal, A. (1992) Signature diagénétique des discontinuités et nouvelles représentation graphique de la diagenèse. CR Acad. Sci. Paris, 314, 1507–1514.

El Arabi, H., Ouahhabi, B. and Charrière, A. (2001) Les séries du Toarcien-Aalénien du SW du Moyen-Atlas (Maroc): précisions stratigraphiques et signification paléogéographique. Bull. Soc. Géol. France, 172, 723–736.

Elmi, S. (2006) Pliensbachian/Toarcian boundary: the proposed GSSP of Peniche (Portugal), Vol. Jurassica, IV, 5–16.

Emmanuel, L., Renard, M., Cubaynes, R., De Rafelis, M., Hermoso, M., Lecallonec, L., Le Solleuz, A. and Rey, J. (2006) The "Schistes Carton" of Quercy (Tarn, France): a lithological signature of a methane hydrate dissociation event in the early Toarcian. Implications for correlations between Boreal and Tethyan realms. Bull. Soc. Géol. France,177,237 – 247.

Erba,E. (2004)Calcareous nannofossils and Mesozoic oceanic anoxic events. Mar. Micropaleontol. ,52,85 – 106.

Espitalié, E. J., Deroo, G. and Marquis, F. (1985 – 1986) La pyrolyse Rock Eval et ses applications. Rev. Inst. Fr. Pet. ,40,563 – 579,755 – 784;41,73 – 89.

Ettaki,M. and Chellai,E. H. (2005)Le Toarcien inferieur du Haut Atlas de Todrha-Dades(Maroc): sedimentologie et lithostratigraphie. CR Geosci. ,337,814 – 823.

Evamy,B. D. (1963)The application of a chemical staining technique to a study of dedolomitization. Sedimentology, 2,164 – 170.

Evamy,B. D. (1969)The precipitational environment and correlation of some calcite cements deduced from artificial staining. J. Sed. Petrol. ,39,787 – 792.

Floquet,M. ,Cecca,F. ,Mestre,M. ,Macchioni,F. ,Guiomar,M. ,Baudin,F. ,Durlet,C. and Almeras,Y. (2003) Mortalité en masse ou fossilisation exceptionnelle: le cas des gisements d'âge Toarcien inférieur et moyen de la région de Digne-Les-Bains(Sud-Est de la France). Bull. Soc. Géol. France,174,159 – 176.

Galluzzo,F. and Santantonio,M. (2002)The Sabina Plateau: a new element in the Mesozoic paleogeography of the Central Apennines. Boll. Soc. Geol. Italiana Special Paper,21,561 – 588.

Garcin,Y. (2002)Evolution de la Plate-forme Provençale d'âge éliensbachien à Aalénien: crises tectoniques,climatiques et de la production carbonatée. Mémoire de D. E. A. ,Université de Provence,Marseille,58 pp.

Gattuso,J. P. and Buddemeier,R. W. (2000)Ocean biogeochemistry: calcification and CO_2. Nature,407,311 – 313.

Gómez,J. J. ,Goy,A. and Canales,M. L. (2008)Seawater temperature and carbon isotope variations in belemnites linked to mass extinction during the Toarcian(Early Jurassic)in Central and Northern Spain. Comparison with other European sections. Palaeogeogr. Palaeoclimatol. Palaeoecol. ,258,28 – 58.

de Graciansky,P. C. ,Dardeau,G. ,Dumont,T. ,Jacquin,T. ,Marchand,D. ,Mouterde,R. and Vail,P. R. (1993) Depositional sequence cycles,transgressive-regressive facies cycles,and extensional tectonics: example from the southern Subalpine Jurassic basin. Bull. Soc. Géol. France,164,709 – 718.

de Graciansky,P. C. ,Jacquin,T. and Hesselbo,S. P. (1998)The Ligurian cycle: an overview of lower Jurassic 2nd-order transgressive/regressive facies cycles in Western Europe. In:Mesozoic and Cenozoic Sequence Stratigraphy of European Basins(Eds P. C. de Graciansky,J. Hardenbol,T. Jacquin and P. R. Vail) ,SEPM Spec. Publ. ,60,467 – 479.

Gradstein,F. M. ,Ogg,J. G. ,Smith,A. G. ,Bleeker,W. and Lourens,L. J. (2004)A new geologic time scale,with special reference to Precambrian and Neogene. Episodes,27,83 – 100.

Groupe Français d'Etude du Jurassique(1997)Biostratigraphie du Jurassique ouest-européen et méditerranéen:zonations parallèles et distribution des invertébrés et microfossiles(Eds E. Cariou and P. Hantzpergue) ,Bull. Centres Rech. Explor. -Prod. Elf Aquitaine,17,440.

Guex,J. ,Morard,A. ,Bartolini,A. and Moretinni,E. (2001)Découverte d'une importante lacune stratigraphique à la limite Domérien-Toarcien: implications paléo-océanographiques. Bull. Soc. Vaud. Sci. Nat. ,87,277 – 284.

Hallam,A. (1987)Radiations and extinctions in relation to environmental change in the marine Jurassic of northwest Europe. Paleobiology,13,152 – 168.

Hallock,P. and Schlager,W. (1986)Nutrient excess and the demise of coral reefs and carbonate platforms. Palaios,1, 389 – 398.

Hesselbo,S. P. , Gröcke, D. R. ,Jenkyns, H. C. ,Bjerrum, C. J. ,Farrimond, P. L. ,Morgans-Bell, H. S. and Green, O. R. (2000)Massive dissociation of gas hydrates during a Jurassic oceanic anoxic event. Nature,406,392 – 395.

Hesselbo,S. P. ,Jenkyns,H. C. ,Duarte,L. V. and Oliveira,L. C. V. (2007)Carbon-isotope record of the Early Jurassic(Toarcian) Oceanic Anoxic Event from fossil wood and marine carbonate(Lusitanian Basin,Portugal). Earth

Planet. Sci. Lett. ,253,455 – 470.

Holland, H. D. (2004) The geologic history of seawater. In: Treatise on Geochemistry, Vol. 6. The Oceans and Marine Geochemistry (Ed. H. Elderfield; Exec. Eds H. D. Holland and K. K. Turekian, pp. 583 – 625. Elsevier Pergamon, Kidlington, Oxford.

Jenkyns, H. C. (1985) The Early Toarcian and Cenomanian-Turonian anoxic events in Europe: comparisons and contrasts. Geol. Rundsch. ,74/3,505 – 518.

Jenkyns, H. C. (1988) The early Toarcian (Jurassic) anoxic event: stratigraphic, sedimentary and geochemical evidence. Am. J. Sci. ,288,101 – 151.

Jenkyns, H. C. and Clayton, C. J. (1986) Black shales and carbon isotopes in pelagic sediments from the Tethyan Lower Jurassic. Sedimentology,33,87 – 106.

Jenkyns, H. C. and Clayton, C. J. (1997) Lower Jurassic epicontinental carbonates and mudstones from England and Wales: chemostratigraphic signals and the early Toarcian anoxic event. Sedimentology,44,687 – 706.

Jenkyns, H. C. ,Gale, A. S. and Corfield, R. M. (1994) Carbonand oxygen-isotope stratigraphy of the English Chalk and Italian Scaglia and its palaeoclimatic significance. Geol. Mag. ,131,1 – 34.

Jenkyns, H. C. ,Gröcke, D. and Hesselbo, S. P. (2001) Nitrogen isotope evidence for water mass denitrification during the early Toarcian Oceanic Anoxic Event. Paleoceanography,16,593 – 603.

Jenkyns, H. C. ,Jones, C. E. ,Gröcke, D. R. ,Hesselbo, S. P. and Parkinson, D. N. (2002) Chemostratigraphy of the Jurassic system: applications, limitations and implications for palaeoceanography. J. Geol. Soc. London, 159, 351 – 378.

Kemp, D. B. ,Coe, A. L. ,Cohen, A. S. and Schwark, L. (2005) Astronomical pacing of methane release in the Early Jurassic period. Nature,437,396 – 399.

Kleypas, J. A. ,Buddemmeier, R. W. ,Archer, D. ,Gattuso, J. -P. ,Langdon, C. and Opdyke, B. N. (1999) Geochemical consequences of increased atmospheric carbon dioxide on coral reefs. Science,284,118 – 120.

Kolodny, Y. ,Luz, B. and Navon, O. (1983) Oxygen isotope variation in phosphate of biogenic apatites. 1. Fish bone apatite-rechecking the rules of the games. Earth Planet. Sci. Lett. ,64,298 – 404.

Küspert, W. (1982) Environmental change during oil shale deposition as deduced from stable isotopic ratio. In: Cyclic and Event Stratification (Eds S. Einsele and A. Seilacher), pp. 482 – 501. Springer, Berlin.

Lallam, S. ,Sahnoun, E. ,El Hatimi Hervouet, Y. and Tejera de Leon, J. (1997) Mise enévidence de la dynamique de la marge téthysienne de l'Hettangien à l'Aalénien dans la Dorsale calcaire (Tétouan, Rif, Maroc). CR Acad. Sci. Paris,324,IIa,923 – 930.

Lécuyer, C. (2004) Oxygen isotope analysis of phosphate. In: Handbook of Stable Isotope Analytical Technics (Ed. P. de Groot), Elsevier B. V. ,22, I,482 – 496.

Lécuyer, C. ,Grandjean, P. and Emig, C. C. (1996) Determination of oxygen isotope fractionation between water and phosphate from living lingulids: potential application to palaeoenvironmental studies. Palaeogeogr. Palaeoclimatol. Palaeoecol. ,126,101 – 108.

Lécuyer, C. ,Grandjean, P. and Sheppard, S. M. F. (1999) Oxygen isotope exchange between dissolved phosphate and water at temperatures ≤35℃: inorganic versus biological fractionations. Geochim. Cosmochim. Acta,63,855 – 862.

Lemoine, M. and de Graciansky, P. C. (1988) History of a passive continental margin: the western Alps in the Mesozoic, Introduction. Bull. Soc. Géol. France,8,4.

Léonide, P. (2007) Réponses des plates-formes carbonatées aux changements paléo-océanographiques, paléo-climatiques et tectoniques: le bassin sud-provençal au Jurassique inférieur à moyen. Thèse de Doctorat, Université de Provence, Marseille,385 pp.

Léonide, P. ,Floquet, M. and Villier, L. (2007) Interaction of tectonics, eustasy, climate and carbonate production on the sedimentary evolution of an early-middle Jurassic extensional basin (Southern Provence Sub-basin, SE France).

Basin Res. ,19,125-152.

Little, C. T. S. and Benton, M. J. (1995) Early Jurassic mass extinction: a global long term event. Geology, 23, 495-498.

Longinelli, A. and Nutti, S. (1973) Oxygen isotope measurements of phosphate from fish teeth and bones. Earth Planet. Sci. Lett. ,20,337-340.

Loreau, J. P. and Durlet, C. (1999) Diagenetic stratigraphy of discontinuity surfaces: an application to paleoenvironments and sequence stratigraphy. N. Jb. Geol. Pal. ,1,381-407.

Macchioni, F. and Cecca, F. (2002) Biodiversity and biogeography of Middle-Late Liassic ammonoids: implications for the Early Toarcian mass extinction. Geobios,35,150-164.

Mallarino, G. , Goldstein, R. H. and Di Stefano, P. (2002) New approach for quantifying water depth applied to the enigma of drowning of carbonate platforms. Geology,30,783-786.

Marshall, J. D. (1992) Climatic and oceanographic isotopic signals from the carbonate rock record and their preservation. Geol. Mag. ,129,143-160.

Mattioli, E. and Pittet, B. (2002) Contribution of calcareous nannoplankton to carbonate deposition: a new approach applied to the Early Jurassic of central Italy. Marine Micropalaeontology,45,175-190.

Mattioli, E. and Pittet, B. (2004) Spatial and temporal distribution of calcareous nannofossils along a proximAl-distal transect in the Umbria-Marche basin (Lower Toarcian; Italy). Palaeogeogr. Palaeoclimatol. Palaeoecol. , 205, 295-316.

Mattioli, E. , Pittet, B. , Bucefalo Palliani, R. , Röhl, H. J. , Schmid-Röhl, A. and Morettini, E. (2004a) Phytoplankton evidence for the timing and correlation of palaeoceanographical changes during the early Toarcian Oceanic Anoxic Event (Early Jurassic). J. Geol. Soc. London,161,685-693.

Mattioli, E. , Pittet, B. , Young, J. R. and Bown, P. R. (2004b) Biometric analysis of Pliensbachian-Toarcian (Lower Jurassic) coccoliths of the family Biscutaceae: intra-and interspecific variability versus palaeoenvironmental influence. Mar. Micropaleontol. ,51,5-27.

Mattioli, E. , Pittet, B. , Petitpierre, L. and Maillot, S. (2009) Dramatic decrease of pelagic carbonate production by nannoplankton across the Early Toarcian anoxic event (T-OAE). Global Planet. Change,65,134-145.

McArthur, J. M. , Donovan, D. T. , Thirlwall, M. F. , Fouke, B. W. and Mattey, D. (2000) Strontium isotope profile of the early Toarcian (Jurassic) oceanic anoxic event, the duration of ammonite biozones, and belemnite palaeotemperatures. Earth Planet. Sci. Lett. ,179,269-285.

Mestre, M. (2001) Evénements biosédimentaires et tectoniques au Domérien supérieur-Toarcien enregistrés dans la série réduite de La Robine-Marcoux (Nappe de Digne, Bassin du Sud-Est, Réserve Géologique de Haute-Provence). Mémoire de D. E. A. Université de Provence, Marseille,38 pp.

Metodiev, L. and Koleva-Rekalova, E. (2008) Stable isotope records ($\delta^{18}O$ and $\delta^{13}C$) of Lower-Middle Jurassic belemnites from the Western Balkan mountains (Bulgaria): palaeoenvironmental application. Appl. Geochem. , 23, 2845-2856.

Meyers, W. J. (1991) Calcite cement stratigraphy: an overview. In: Luminescence Microscopy: Quantitative and Qualitative Aspects (Eds C. E. Backer and O. C. Kopp), SEPM Short Course,25,133-147.

Moore, C. (2001) Carbonate Reservoirs: Porosity Evolution and Diagenesis in a Sequence Stratigraphic Framework. Elsevier, Amsterdam.

Morard, A. , Guex, J. , Bartolini, E. and De Wever, P. (2003) A new scenario for the Domerian-Toarcian transition. Bull. Soc. Géol. France,174,351-356.

Newton, R. J. , Reeves, E. , Kafousia, N. , Wignall, P. B. and Bottrell, S. H. (2006) Questioning the global nature of the Toarcian carbon isotope excursions. Geochim. Cosmochim. Acta Suppl. ,70,A443.

Price, G. D. (1999) The evidence and implications of polar ice during the Mesozoic. Earth-Sci. Rev. ,48,183-210.

Price, G. D. and Sellwood, B. W. (1997) "Warm" palaeotemperatures from high late Jurassic palaeolatitudes (Falkland Plateau). Ecological, environmental or diagenetic controls? Palaeogeogr. Palaeoclimatol. Palaeoecol., 129, 315 – 327.

Röhl, H. J., Schmid-Röhl, A., Oschmann, W., Frimmel, A. and Schwark, L. (2001) The Posidonia Shale (Lower Toarcien) of SW-Germany: an oxygen depleted ecosystem controlled by sea level and palaeoclimate. Palaeogeogr. Palaeoclimatol. Palaeoecol., 165, 27 – 52.

Rosales, I., Quesada, S. and Robles, S. (2004a) Paleotemperature variations of Early Jurassic seawater recorded in geochemical trends of belemnites from Basque-Cantabrian basin, northern Spain. Palaeogeogr. Palaeoclimatol. Palaeoecol., 203, 253 – 275.

Rosales, I., Robles, S. and Queseda, S. (2004b) Elemental and oxygen isotope composition of early Jurassic belemnites: salinity vs temperature signals. J. Sed. Res., 74, 342 – 354.

Saelen, G., Doyle, P. and Talbot, M. R. (1996) Stable isotope analyses of belemnite rostra from the Whitby Mudstone Fm., England: surface water conditions during deposition of a marine black shale. Palaios, 11, 97 – 117.

Sandberg, P. A. (1983) An oscillating trend in Phanerozoic non-skeletal carbonate mineralogy. Nature, 305, 19 – 22.

Schmid-Röhl, A., Röhl, H. J., Oschmann, W., Frimmel, A. and Schwark, L. (2002) Palaeoenvironmental reconstruction of the Lower Toarcian epicontinental black shales (Posidonia Shale, SW Germany): global versus regional control. Geobios, 35, 13 – 20.

van de Schootbrugge, B., McArthur, J. M., Baley, T. R., Rosental, Y., Wright, J. D. and Miller, K. G. (2005) Toarcian Anoxic Event: an assessment of global using belemnite C isotope records. Paleoceanography, 20, PA3008.

Schouten, S., Kaam-Peters, M. E., Rijpastra, I., Schoell, M. and Sinnighe Damste, J. S. (2000) Effects of an oceanic anoxic event on the stable carbon isotopic composition of an early Toarcian carbon. Am. J. Sci., 300, 1 – 22.

Stampfli, G. M. and Borel, G. D. (2002) A plate tectonic model for the Paleozoic and Mesozoic constrained by dynamic plate boundaries and resorted synthetic oceanic isochrones. Earth Planet. Sci. Lett., 196, 17 – 33.

Suan, G., Pittet, B., Bour, Y., Mattioli, E., Duarte, L. V. and Mailliot, S. (2008a) Duration of the Early Toarcian carbon isotope excursion deduced from spectral analysis: consequence for its possible causes. Earth Planet. Sci. Lett., 267, 666 – 679.

Suan, G., Mattioli, E., Pittet, B., Mailliot, S. and Lécuyer, C. (2008b) Evidence for major environmental perturbation prior to and during the Toarcian (Early Jurassic) oceanic anoxic event from the Lusitanian Basin, Portugal. Paleoceanography, 23, PA1202.

Tempier, C. (1972) Les faciès calcaires du Jurassique provençal, Vol. 4. Travaux du Laboratoire des Sciences de la Terre, Saint-Jérôme, Marseille, 361 pp.

Thierry, J. (2000) Middle Toarcian. In: Atlas Peri-tethys, Paleogeographical Maps (Eds J. Dercourt, M. Gaetani, B. Vrielynck, E. Barrier, B. Biju-Duval, M. F. Brunet, J. P. Cadet, S. Crasquin and M. Sandulescu), Map 8, pp. 61 – 70. CCGM/CGMW, Paris.

Tremolada, F., van de Schootbrugge, B. and Erba, E. (2005) Early Jurassic schizosphaerellid crisis in Cantabria, Spain: implications for calcification rates and phytoplankton evolution across the Toarcian Oceanic Anoxic Event. Paleoceanography, 20, PA2011.

Weissert, H., Lini, A., Fölmi, K. B. and Kuhn, O. (1998) Correlation of Early Cretaceous carbon isotope stratigraphy and platform drowning events: a possible link? Palaeogeogr. Palaeoclimatol. Palaeoecol., 137, 189 – 203.

Wignall, P. B., Newton, R. J. and Little, C. T. S. (2005) The timing of paleoenvironmental change and cause-and-effect relationship during the early Jurassic mass extinction in Europe. Am. J. Sci., 305, 1014 – 1032.

Wilkinson, B. H. and Algeo, T. J. (1989) Sedimentary carbonate record of calcium magnesium cycling. Am. J. Sci., 289, 1158 – 1194.

Wilmsen, M. and Neuweiler, F. (2008) Biosedimentology of the Early Jurassic post-extinction carbonate depositional system, central High Atlas rift basin, Morocco. Sedimentology, 55, 773 – 807.

Woodfine, R. G. , Jenkyns, H. C. , Sarti, M. , Baroncini, F. and Violante, C. (2008) The response of two Tethyan carbonate platforms to the early Toarcian (Jurassic) oceanic event: environmental change and differential subsidence. Sedimentology, 55, 1011 – 1028.

Zempolich, W. G. (1993) The drowning succession in the Jurassic Carbonates of the Venetian Alps, Italy: a record of supercontinental breakup, gradual eustatic rise, and eutrophication of shallow-waters environments In: Carbonate Sequence Stratigraphy: Recent Developments and Applications (Eds R. G. Loucksand J. F. Sarg), Am. Assoc. Petrol. Geol. Mem. , 57, 63 – 105.

Ziegler, P. A. (1992) Plate tectonics, plate moving mechanisms and rifting. Tectonophysics, 215, 9 – 34.

第7章 浅水碳酸盐台地的环境变化速率和等时性

ANDRÉ STRASSER, STÉPHANIE VÉDRINE, NOÉMIE STIENNE 著

金春爽 译, 张天舒 校

摘 要 最近几十年的全球变化在浅水碳酸盐台地上留下了明显的印迹。对过去环境变化的速率和等时性进行的评估达到什么程度,才能对比今天环境变化的过程和产物?在瑞士 Jura 山脉,晚牛津期的碳酸盐台地剖面已被详细记录下来。在菊石生物地层学和沉积层序分级堆积的基础上建立的时间框架说明这些剖面由低振幅、高频率的海平面变化形成,与轨道周期一致。这个最小单元,基本层序可以归因于 20ka 的岁差周期。通过识别层序地层元素,例如一个基本层序内的最大洪泛面,可以获得甚至更高的时间分辨率,并且可以重建水深和海平面变化。沉积速率极不规则,常见不连续面。将一个剖面的不连续面与其他剖面进行对比,其结果是不确定的,因为这些界面也可能是在局部发育,并且不一定具有等时性。在基本层序尺度,沉积相的垂向演化可以由水深变化来解释,水深变化可能由海平面变化(异旋回)引起,或是自旋回引起的。黏土矿物分布表明海平面的变化,以及内陆地区的降雨量。生物的变化受水深和水体质量控制,同时也受养分和黏土输入控制。碳、氧同位素显示不出基本层序的显著变化。原始生态、矿物学和地球化学信息部分被同质化,因为按时间取平均数是通过生物扰动作用和风暴再造作用影响,以及成岩改造获取的。因此,解释碳酸盐台地的环境变化是有可能的,但是时间分辨率并不会好于几千年。导致环境变化的沉积过程的速率和等时性只可以被估计在这个时间框架内。

关键词 碳酸盐台地 旋回地层 环境变化 牛津阶 瑞士 Jura

7.1 概况

过去一个世纪的全球气候以很高的速率变化着:大气中的二氧化碳增加约 80pp,二氧化碳和其他温室气体的增加使得全球平均温度上升 1℃,海水和内陆冰融化的热膨胀使得全球平均海平面上升 17cm(IPCC,2007)。除了温室气体,太阳辐射和火山悬浮颗粒的变化也会影响在十年至百年尺度的气候系统(Crowley,2000;Servonnat 等,2010)。这些快速的变化会对陆地和海洋生态系统产生重大影响,当然也会影响到浅水碳酸盐台地。珊瑚作为重要的碳酸盐生产者特别敏感,增加水的温度会导致颜色变淡(Penaflor 等,2009);海水酸化会导致碳酸盐生产减慢(Jokiel 等,2008);风暴增强会损害珊瑚礁生长(Emanuel,2005);大雨导致的内陆营养增加会有利于海藻生长,而妨碍珊瑚生长(McCook,1999)。百年尺度的变化趋势会被一些极端事件,如厄尔尼诺—南方涛动现象所打断,这可能会导致短期温度值达到高峰,导致大量珊瑚被漂白。一些生物体可以恢复,其他生物体则会死亡(Smith 等,2008;Brandt,2009)。释放的生态位将很快被新的适应环境的生物体所占据。

现今的碳酸盐台地是全球气候变化的许多方面的完美记录,并且现在的沉积记录可以达到一年至一天的时间分辨率。而且,这种记录可以与仪器测量的气候和海洋学参数进行对比。例如,珊瑚生长范围的分析可以重建水温变化和新陈代谢(Bessat 和 Buigues,2001),在监测风暴基础上,也可用风暴岩来记录和对比这些变化(Goff 等,2010)。

研究地质历史过程中的碳酸盐台地,进行这种高精度定年及对比是不可能的。结合珊瑚化石、潮汐束或纹泥可能记录短期变化,但仍有局限性,在全球尺度上精确对比是不可能的。至今,分析古代碳酸盐台地的最佳时间分辨率是由轨道(米兰科维奇)旋回给出的。如果沉积物忠实地记录了轨道周期,时间分辨率可能达到20ka(与春分、秋分的平分点一致;Berger 等,1989;Strasser 等,2006)。然而,即使可以建立旋回地层时间框架,监测到的十年到百年尺度的现代的全球变化与研究中的千年尺度的古代的全球变化仍存在很大差异。

本文以170ka的时间间隔,利用瑞士和法国的Jura山脉晚牛津期(晚侏罗世)碳酸盐台地沉积记录,分析环境变化。为解释Jura台地环境变化的影响因素,本文讨论了20ka层序内的海平面变化,以及陆源物质输入、碳、氧同位素和生物变化。由于可用的年代地层框架太粗而不能与全球范围的快速变化进行对比,这种解释是地区性的。本文试图找到以下问题的答案:

(1)从古碳酸盐台地所获的最高时间分辨率是多少?

(2)沉积记录反映的环境变化在区域上可以对比到多大范围?

(3)对过去环境变化的速率和等时性进行的评估达到什么程度,可以作为对比今天环境变化的产物的实例?

尽管本次研究仅涉及了地球漫长历史中很短的时间窗,也仅限于浅水碳酸盐台地,但强调了过去及今后一直对沉积环境相互作用的复杂性。

7.2 古地理、地层、古气候背景

晚侏罗世,现今的Jura山脉是碳酸盐台地的一部分,西面是巴黎盆地,东南是特提斯洋,北部是起源海西期的伦敦—布拉邦特地块,莱茵河和波西米亚地块(图7-1)。古纬度大约在26°~27°N(Dercourt 等,1993)。台地形态由基底断裂的同沉积活化作用和台地南部陆缘盆地的演化所确定(Allenbach,2001)。

瑞士Jura牛津阶的岩石地层由Gygi(1995,2000)建立。地层组和段的年代地层标定是由菊石生物地层学(在分区级别)、矿物学和地层对比给出的(Gygi 和 Persoz,1986;Gygi,1995,2000)。另外,Gygi 等(1998)识别出大型层序边界,可以很好地与Hardenbol 等(1998)在几个欧洲盆地识别出的层序边界进行对比(图7-2)。Strasser(2007)收集了19个台地剖面的信息,其中包括4个位于法国东南部的盆地内的剖面,这4个剖面由菊石提供良好的生物地层标定(图7-3),进而提出了牛津阶和科姆地阶旋回地层的时间尺度。

潟湖相富含生产碳酸盐的有机质,点礁和鲕粒滩表示温暖的亚热带水域环境。水深在开放性潟湖中的几十米到低洼岛屿滨岸的潮间带和潮上带之间。来自海西地块的陆缘碎屑定期被输送到台地和盆地中(Gygi 和 Persoz,1986;Pittet,1996),揭示了内地的潮湿环境。图7-3给出的瑞士Jura山气候演化趋势(Pittet,1996;Hug,2003)可以与Abbink(2001)根据孢粉分析提出的北海南部气候进行对比。研究区间处于Jura台地由温暖、季节性湿润到弱潮湿气候的转换期,而北海的记录表明是由温暖到变冷的时期。瑞士和法国台地测得的碳同位素曲线和半深海剖面表明:早—中牛津阶重要的负向变化由甲烷释放引起,正向变化则与板块构造活动有关(Louis-Schmid 等,2007)。然而,研究的这段时间区间,意大利翁布里亚—马尔凯盆地的气候演化曲线是稳定的(Bartolini 等,1996)。法国亚高山盆地半深海沉积物中测得的黏土矿物组合显示出在整个牛津阶的分布相对平稳(Deconinck 等,1985),但在Jura盆地呈现为更复杂的模式(Gygi 和 Persoz,1986)。

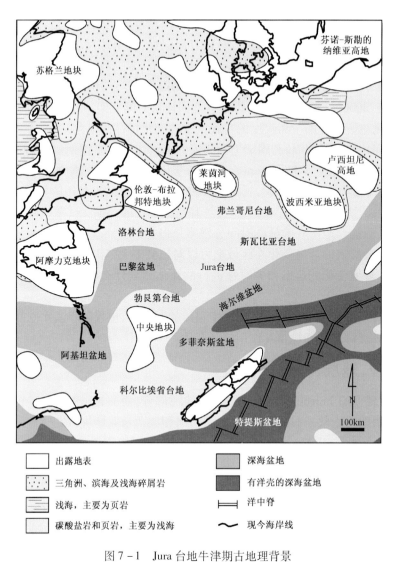

图 7-1 Jura 台地牛津期古地理背景

据 Carpentier 等(2006)修改,据 Enay 等(1980)、Ziegler(1990)和 Thierry(2000)

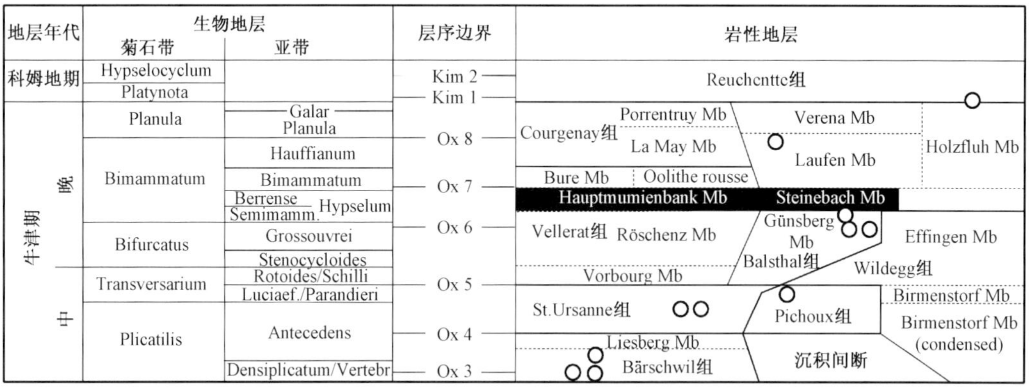

图 7-2 中—晚牛津期地层表

Hauptmumienbank 和 Steinebach 段被突出显示。岩性地层和生物地层据 Gygi(1995,2000),圆圈表示发现菊石生物地层标志的地方。层序界面据 Haedenbol 等(1998)和 Gygi 等(1998)。Fm:地层;Mb:段

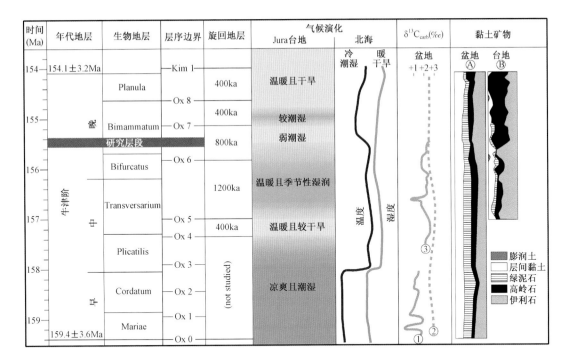

图 7-3 Jura 台地的气候演化与出版资料的对比

气候演化据北海南部(Abbink 等,2001);碳同位素曲线据:(1)Louis-Schmic(2006);(2)Bartolini 等(1996);(3)Pardden 等(2001)。黏土矿物组合是根据:(A)Deconinck 等(1985);及(B)Gygi 和 Persoz(1986)。年代地层、生物地层和层序地层据 Hardenbol 等(1998),旋回地层据 Strasser(2007)。详见正文

7.3 资料和方法

本次研究选取了 Pertuis, Savagnières, Gorges de Court, Hautes—Roches, Vorbourg 和 Voyeboeuf 六个剖面(图 7-4、表 7-1),Vedrine(2007)对其进行了厘米级的记录和密集取样。而且,Stienne(2010)对 Hautes—Roches 和 Vorbourg 剖面做了 100% 的样品覆盖分析。分析中利用了薄片、抛光片和灰泥冲洗,在没有采样处,通过放大镜观察野外露头来确定沉积相。在偏光显微镜或双目放大镜下,使用邓纳姆(1962)分类法进行了微相分析,并对岩石组成进行了半定量估计。要特别注意沉积结构和沉积间断面(Clari 等,1995;Hillgartner,1998)。所有这些沉积学上的信息用来解释沉积环境。利用 Vail 法(1991)命名法解释层序地层。高分辨率层序地层和旋回地层解释采用 Strasser 等(1999)的概念:利用层序地层方法来解释那些形成于与轨道旋回一致的高频、低幅度海平面变化条件下的沉积层序。

表 7-1 本次研究中剖面位置、地层范围和作者(坐标参考瑞士国家地形图 1:25000)

代码	名称	位置	剖面基底坐标	地层	作者
Pe	Pertuis	沿 Dombresson 到 Pertuis 公路边	561'850/216'050	中—晚奥陶世	Pittet(1996) Védrine(2007)
Sa	Savagnières	在 St-Imier 和 VAl-de-Ruz 之间的公路边	566'750/219'750	中—晚奥陶世	Pittet(1996) Védrine(2007)

续表

代码	名称	位置	剖面基底坐标	地层	作者
GC	Gorges de Court	在 Moutier 和 Court 之间的公路边	593′200/234′300	中—晚奥陶世	Pittet(1996) Hug(2003) Védrine(2007)
HR	Hautes-Roches	在 Hautes-Roches 村庄南部树林小路边	594′950/238′250	中—晚奥陶世	Pittet(1996) Dupraz(1999) Védrine(2007) Stienne(2010)
Vo	Vorbourg	在 Vorbourg 教堂上方的路上	593′850/247′625	中—晚奥陶世	Pittet(1996) Védrine(2007) Stienne(2010)
Vb	Voyeboeuf	在 Porrentruy 和 Courgenay 之间的公路边	574′160/251′120	晚奥陶世	Védrine(2007)

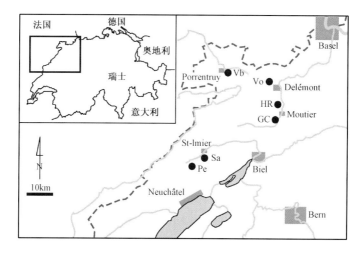

图 7-4 位于瑞士西北部的研究剖面地理位置图

Pe:Pertuis;Sa:Savagniè-res;GC:Gorges de Court;HR:Hautes-Roches;Vo:Vorbourg;Vb:Voyeboeuf

对 Savagnières 剖面分析了黏土矿物(Védrine,2007)。在纳沙泰尔(瑞士)大学地质学院,用 Scintag XRD 2000 衍射仪(由美国加利福尼亚州的库比提诺 Scintag 公司生产)进行了样品制备和 X 射线衍射。样品制备采用了 Kubler(1990)描述的方法。在半定量估计黏土矿物组成时,用每秒计数量(cps)来测量标志每种黏土矿物(蒙脱石,绿泥石,伊利石,层间高岭石)的峰值强度,这样给出的是没有校正的相对百分含量。变化超过 10% 时就认为有效。石英含量是利用薄片和 X 射线衍射分析来估计体积百分含量,当石英体积含量不足够多时,利用 2~16 碎片的每秒计数量来估计石英含量。伊利石结晶度指数表示为:在小于 2μm 馏分衍射指数背景值测量上,中等高度基峰值和峰值高度的宽度之间的比值。

Vorbourg 和 Savagnières 剖面全部样品都进行了氧($\delta^{18}O$)和碳($\delta^{13}C$)同位素分析,优先选择微晶灰岩区,代表了有机和无机来源均质化的碳酸盐泥(Védrine,2007)。测量是在洛桑大学(瑞士)使用 Finnigan Delta Plus XL 质谱仪(Thermo Fisher Scientific,Waltham,MA,美国)开展的,测量中使用了附带的 GasBench II 和 PAL 自动取样器。岩石粉末和 100% 磷酸在 90℃ 下反应。在连续的 CO_2 气流下,每个样品进行了 10 种测量。原始测量结果用由国家标准局制定的国内标准(卡拉大理岩)进行了校正。用 10 个校正过的原始结果均值计算了代表每个样品的碳和氧同位素值。误差包括测量误差和再现性误差,小于 ±0.1‰。

7.4 沉积层序

为了达到本文的研究目的,选择层序界面 Ox6 附近的层段(图 7-5)。这个边界位于 Bifurcatus 菊石带的上部(Gygi 等,1998 年,图 7-2)。在研究的剖面中,植物碎片、鸟眼状微生物席或低能泻湖沉积物表明了相对的浅水沉积相特征,包括珊瑚礁,鲕粒滩和富含似核形石的潟湖。根据 Hardenbol 等(1998)建立的欧洲盆地层序—时间地层表,有证据表明在 Ox6 之上的 Bimammatum 菊石带最下部存在以百万年海侵—海退旋回为尺度的最大海退事件。在瑞士 Jura 山区,在 Röschenz 和 Günsberg 段(图 7-2)顶部能识别出该层段,植物碎片和轮藻表明存在土壤和淡水湖泊。上覆的 Hauptmumienbank 和 Steinebach 段含有珊瑚骨架灰岩、鲕粒和富含似核形石的粒泥灰岩和泥粒灰岩,因此表明整个 Jura 台地都发生了海侵。Ox6 层序最大洪泛面位于 Semimammatun 和 Berrrense 分区的边界(Hardenbol 等,1998),这与含有棘皮动物,腕足类和珊瑚的研究剖面相对应(图 7-5,小尺度层序 10)。瑞士 Jura 记录的沉积体系演化,至少部分与在整个欧洲可以识别出的海平面变化相关。

高分辨率层序地层和旋回地层学分析 Jura 山中—晚牛津期地层表明,沉积相的分布在空间和时间上都是非常不均匀的(Pittet,1996;Dupraz,1999;Hug,2003;Védrine,2007;Stienne,2010)。分析的控制因素有:差异沉降形成的盆地形态,高频率海平面波动引起的水深变化,输送沉积物的流体模式。沉积相变化和界面解释出分级叠加的沉积层序(通常体现为变深—变浅趋势)。在由生物地层学和年代地层学定义的时间框架内,(图 7-3),上面提到的作者提到了很多这些层序与岁差和偏心率的轨道周期相符。图 7-5 中显示的小尺度层序是在 Strasser(2007)研究基础上定义的。建议在 Röschenz 和 Günsberg 段顶部的 Ox6 层序边界和最大海退面间的层段划分为四个小尺度的层序。因为小层序对应于 100ka 的短偏心率周期,这个时间间隔代表 400ka 的长偏心率周期(Berger 等,1989;Strasser 等,2006)。400ka 层序的最大洪泛面位于 Pertuis,Savagnières Hautes—Roches 鲕粒滩的顶部,其上的沉积相指示了由于高位沉积的进积作用产生的较低能量和较多黏土注入的环境。小尺度层序 10 内的最大洪泛面(也是 400 千年层序的最大洪泛面)在 Hardenbol 等(1998)的 Semimammatun—Berrrense 分区边界图中提及。

小尺度层序由单层或层组组成,在大多情况下可以描述成沉积层序(图 7-6 给出了野外的例子)。这些层序叫做基本层序,也是最小的单元,在这些单元中,沉积相演化意味变深—变浅趋势,并且/或者所划分的界面指示了海平面下降的界面(Strasser 等,1999)。但是,解释这些层序通常很难,因为沉积体系固有的自旋回过程也会起作用。这些过程包括沉积体的进积或侧向迁移,如潮坪、浅滩和三角洲朵体,这会形成向上变浅的沉积相变化趋势(Ginsburg,1971;Pratt 和 James,1986;Satterley,1996)。Burgess 和 Wright(2003)以及 Burgess(2006)通过正演模拟得出碳酸盐生产率和沉积物运移方向随时间产生的变化,这些变化能形成复杂和重复的地层模式。

这些过程不依赖于全球海平面升降或地台的沉降,但依赖于区域生态参数、水流模式、水道迁移、朵体转换和海底形态。因此,由自旋回过程引起的地层叠加将显示出无序模式,或者是侧向上连续性受限的有序叠加模式(Drummond 和 Wilkinson,1993;Burgess,2006)。

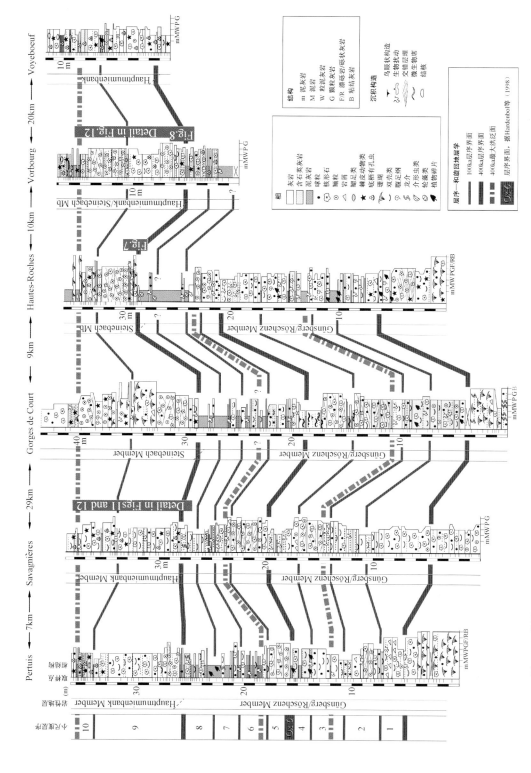

图7-5 基于Pittet(1996)、Hug(2003)、Vedrine(2007)、Strasser(2007)的层序地层学和旋回地层学的剖面对比研究
编号的小尺度层序解释对应于100ka的短偏心率旋回。讨论见正文

图 7-6 研究层段野外露头照片

(A) Voyeboeuf 剖面：小尺度层序 9 的上部和小尺度层序 10 的底部。标出了基本层序单元的层序界面(SB)和最大海泛面(MFS)。锤子(白色箭头)作为比例尺(33cm)。(B) Vorbourg 剖面：小尺度层序 8 的顶部和小尺度层序 9 的下部。标出了基本层序单元的层序界面和最大海泛面。小尺度层序 9 的第 3 个基本层序海侵部分颗粒灰岩显示自生的再活化面(与图 7-8 和图 7-12 中的记录对比，锤子作为比例尺)

然而，如果五个基本层序可以组成一个好界定的小尺度层序(100ka)，如果他们表现为侧向上连续，他们可以暂时归因于持续时间为 20ka 的岁差周期(伯杰等,1989)。这样，研究 Jura 台地环境变化就有了一个 20ka 的时间分辨率。

接下来集中讨论小尺度层序 8、小尺度层序 9 最顶部的基本层序和小尺度层序 10 的海侵部分，也就是持续了 170ka 的层段(Védrine,2007)。Stienne(2010)获得了一个甚至更高的时间分辨率，集中在小尺度层序 8 和小尺度层序 9 间的界限。选择这个层段是因为它代表了长期海侵的开始。由于台地上可容空间的产生使得沉积记录保存良好。

7.5 海平面变化

垂向上定义沉积层序的相演化反映了水深的变化。全球海平面升降形成可容空间,可容空间的充填依靠沉积物的供给、水动力条件和造架生物的生长潜力。在170ka 的研究区间内很难估计水深。仅在 Geoges de Court 剖面 35m 处观察到了指示潮坪环境的鸟眼构造(图 7-5)。另外,没有观察到任何显示潮间带和潮上带的特征(如根迹或轮藻)。造礁珊瑚在水深几厘米到几十米区域中生长繁盛(取决于波浪的能量和透光带的深度),但珊瑚碎屑可以冲到海滩上或滚落到深水区。鲕粒状灰岩的交错层理指示了潮流的活动,从潮间带至数米水深(Immenhauser, 2009)处最为活跃。富含碳酸盐岩泥的泥灰岩和沉积物在浪基面以下堆积,也就是在簸选的颗粒灰岩更深的水体环境,或者堆积在障壁岛保护下的不受高能影响的浅水区。狭盐性生物,如腕足类和棘皮类动物代表正常的海洋条件,但不能反映水深。基于这种模糊的指示特征,不可能重建水深变化曲线,甚至很难重建相对海平面变化曲线。然而,对 Vorbourg 剖面小尺度层序 8 中的基本单元 5 进行了尝试性的水深、相对海平面曲线恢复(图 7-20)。

层序的详细分析表明沉积记录绝不是连续的(Stienne, 2010, 图 7-9)。隐约的不连续面表现为指示低能环境的薄层泥岩沉积,由水动力加强造成的贝壳凸面向上堆积,或指示较低沉积速率的生物扰动富集现象。伞状构造(遮蔽孔隙)表明碳酸盐岩泥在贝壳凸面向上堆积之后就已经形成了。化石的方向通常是水平的(可能受压实的影响),但局部有倾斜或垂直的;这可以解释为穴居动物活动引起的化石移位。含有示顶底腔孔充填(腹足纲、珊瑚、介形类)的化石通常看起来被旋转了。样品 VO9 解释为:至少两种风暴沉积的混合,两种风暴沉积之间由一薄层泥岩隔开。局部通过潜穴和包裹沉积物之间的孔隙度和渗透率对比可以知道,白云岩化和去白云岩化意味着流体在沉积物中循环。样品 VO8 和 VO12 中的灰岩结核,是在早成岩阶段沉积物中围绕碳酸盐岩颗粒发育的碳酸盐岩次生加大形成的(Savrda 和 Bottjer,1988),这种次生加大可能在硫酸盐递减带内文石溶解被加强(Cherns 等,2008)。当沉积物堆积被扰乱时,初期的硬灰岩层可以在海底快速形成(几个月或几年内),首先是微生物黏合,然后是胶结作用(Dravis, 1979; Hillgärtner 等, 2001)。化石碎片的破碎、磨损、微晶化、钻孔、结壳作用显示出中到高的石化级别(定义见 Flessa 等, 1993),表明在相对较长时期内他们一直处于石化活动带(Davies 等, 1989)。

为了重建一个假设的海平面曲线,首先要对沉积物进行去压实作用,以便估算沉积物堆积时的最小可容空间。在最初埋藏 100m 后,碳酸盐岩泥中颗粒的机械重组和脱水作用会导致孔隙度损失的 10% 到 30%(Moore, 1989)。Shinn 和 Robbin(1983)认为机械和去水化压实作用会使孔隙度减少 20% 到 70%。如果很早就发生碳酸盐胶结作用,孔隙度会更低(Halley 和 Harris,1979)。埋藏越深,化学压实作用越重要。Jura 山区可能没有埋藏超过 2km 的情况(Trümpy,1980),但颗粒接触处的压溶作用证实了研究区沉积物的溶解作用。Goldhammer(1997)提出埋藏达到 1000m 时,压实作用使得碳酸盐岩泥孔隙度减少约 50%,使得碳酸盐岩砂孔隙度减少约 15%。根据 Enos(1991)的研究,泥质陆源沉积物和泥质碳酸盐岩沉积物没有明显不同的压实曲线。但是,黏土的压溶作用看起来加强了碳酸盐岩的化学压实作用(Bathurst,1987)。在这些发表的数据基础上,采用了下列的去压实系数:颗粒灰岩 1.2,泥岩 2.5。对于泥粒灰岩和粒泥灰岩,采用中等系数,1.5 和 2,泥灰岩压实系数取 3(Strasser 和 Samankassou,2003)。

在晚侏罗世,高纬度地区和山区可能存在大陆冰川,但体积会很小(Fairbridge,1976;

Frakes 等,1992;Eyles,1993;Price,1999)。因此轨道控制的气候变化只会导致小的冰川性相对海平面波动。但是,通过热扩散和最顶层海水的收缩,日照变化一定会影响低纬度海平面波动(Gornitz 等,1982)。现今海平面的上升一半是由于冰川融化,一半是由于热扩散(IPCC,2007)。另外,热量引起的深层水循环体积的变化(Schulz 和 Schäfer - Neth,1998),和/或水在湖泊和地下蓄水层的保留和释放也可以导致海平面变化(Jacobs 和 Sahagian,1993)。因为极地冰盖的缓慢堆积和迅速融化,如在第四纪,并不是主体过程,推测受轨道控制的日照变化或多或少直接转化为对称的海平面升降(Read,1995)。根据基本层序的去压实厚度(分别根据岁差和短偏心率周期),估计在侏罗纪末期和早期白垩纪时期形成的幅度只有几米(Strasser 和 Samankassou,2003;Strasser 等,2004)。相比,Crevello(1991)提出了一个记录在摩洛哥早期的侏罗纪碳酸盐岩台地上幅度为 2~3m 的岁差控制的高频率海平面波动,并且 Aurell 和 Bádenas(2004)估计有 5~10m 的海平面变化幅度可能由西班牙启莫里晚期碳酸盐斜坡的偏心率周期引起的。

在图 7-10 的重建中显示差异性去压实的基本层序。相对于一个岁差周期,假定持续时间是 20ka。海平面上升的周期假定对称,但是要将其叠加在一个长期上升趋势上,使得沉积层序没有受到地表剥蚀。幅度随意绘在 110cm 处,但在高频率海平面变化的幅度范围内,这来自对侏罗纪的贝里亚斯组的重建,在那里根据潮间带的特征可以得出更好的估算(strasser 等,2004)。海底变化随时间的变化可以通过沉积记录得到很好的解释。一些潮下侵蚀和在层序基底处无沉积作用要考虑到下面沉积物的瘤化作用。在相应的时间间隔里,层序界面形成了。上覆的石灰岩床被解释为相应的海侵矿床,那是因为不断上升的海平面使得陆源碎屑推向更向大陆的位置。通过水流作用和无沉积作用,沉积物的堆积重复地被再活化作用阶段打断,产生了不连续以及初期的硬灰岩层。富含生物扰动的波状起伏的界面,被看作为退积和进积之间的转换面,也就是最大洪泛面。持续下降的相对海平面导致了陆源碎屑的进积作用,为高位期沉积的主体。这些进积沉积被几乎瞬间堆积的暴风沉积所打断。考虑到泥灰岩的固结作用,一个不连续界面指示了较低的沉积速率。碳酸盐岩结核保留的原始结构可能会被风暴侵蚀。第二个层序界面位于下一个灰岩层底部。

海平面曲线与两个层序界面(最快速的下降)和最大洪泛面(最快速的上升)相联系。水的深度选择在最小值(几十厘米)但有可能是几米。然而,如果以 3cm/ka 速度沉降,根据 Wildi 等(1989)提出的晚侏罗纪瑞士 Jura 平均速率,总的趋势意味着长期上升了 5.5cm/ka。相比 Ramajo 和 Aurell(2008)所估计的,在 Bimammatum 区长期海平面上升速率为 2cm/ka,这基于他们在西班牙一个卡洛期—牛津期剖面的研究。图 7-10 分析获得更高值似乎是现实的,因为它形成于海侵阶段界面 Ox6 之后(图 7-2;Strasser,2007)。根据图 7-10 中的图例,最快的海平面上升发生在以 30cm/ka 为速率的 20ka 周期开始之后的 8600 年(相比之下,在 1993~2003 年期间,全球平均海平面上升的速率为 310cm/ka 左右;IPCC,2007 年)。

7.6 陆源注入

在研究的剖面中,硅质碎屑在横向和纵向上所占比例是变化的(图 7-5);它们的物源区为围绕 Jura 台地出露的地区(图 7-1;Gygi 和 Persoz,1986)。石英颗粒次棱角状,直径范围从 0.02mm 到 0.1mm;它们局部约占 40%,但是,一般来说,低于 10%。长石和重矿物含量较低。

高石英含量更多出现在小尺度基本层序界面附近(图7-11,Védrine,2007)。

黏土矿物分析根据来自Savagnières剖面的22个样品(图7-11)。伊利石、绿泥石的趋势是类似的,都与高岭石趋势相反。伊利石、蒙脱石混合层倾向与高岭石的趋势相关。缺少蒙脱石以及混合层相对丰富是牛津期瑞士Jura地层的典型特征(Gygi和Persoz,1986)。要么缺乏蒙脱石是一种原始特性,要么是由于在埋藏成岩或者成土阶段,蒙脱石转换成伊利石蒙脱石混合层所形成的(Gygi和Persoz,1986;Chamley,1989)。绿泥石含量一般在小尺度基本层序界面较高,在最大洪泛面附近较低。伊利石和蒙脱石混合层在小尺度层序界面附近以及偶尔在基本层序附近较高。另外,伊利石和蒙脱石混合层含量高值也出现在小尺度最大洪泛面附近,而最低值出现在基本最大洪泛面。这个高岭石趋势显示了在剖面底部为低值,然后向上大幅增加,但没有发现与层序地层解释的关联。伊利石大量存在于一些层序界面,但在其他层序界面则没有,并没有识别出有规律的关系。高岭石、伊利石的比率在小尺度层序界面附近以及小尺度最大洪泛面附近较高。通常伊利石低结晶度指示由岩屑形成的伊利石,这与瑞士Jura低埋藏深度一致。相对最高值(对应于结晶较差的伊利石)出现在小尺度和基本层序界面附近。

从内陆的结晶山丘搬运到台地上,硅质碎屑的侵蚀作用受控于出露地区的海平面和降雨量。低海平面增加了河流下切侵蚀潜力,并且有利于海岸三角洲的进积作用。增加的陆地表面也增加了黏土向土壤转化的潜力(取决于基质的类型和气候条件;Retallack,2001)。在海侵阶段,这些黏土被再次搬运分散在台地上。降雨量的增加有助于基岩改造,并且增强了侵蚀潜力。河流向台地更活跃地搬运硅质碎屑。不能排除风力搬运,但是假设Jura台地上发现的硅质碎屑多数由河流搬运(Gygi和Persoz,1986)。

在台地上,碳酸盐岛屿局部偶尔发育,正如在Hautes—Roches的轮藻门和在Pertuis、Gorges de Court、Hautes—Roches和Voyeboeuf的植物碎屑所指示的那样(图7-5)。然而,它们的表面可能不足够大,存活时间不足以经历重要的成土过程。

高岭石—伊利石(K/I)比率一般用于古气候重建和反映气候和地势起伏变化在化学风化和成土过程的影响。高K/I比通常指示一个更潮湿、温暖气候(Curtis,1990;Thiry,2000)。Deconinck(1993)发现,高岭石优先聚集在近端的台地。在Savagnières剖面,较高的K/I比率出现在层序界面,因此指示了在海平面的下降期间更潮湿气候,伴随较高石英含量所指示的侵蚀力增强。然而,K/I比率在小尺度层序9和10的最大洪泛面附近也为高值,虽然没有泥灰土或较高的石英含量(图7-11)。这种情况可以解释为在那些时期由于潮湿气候、海平面快速上升侵蚀力小,或者由在台地广泛洪泛时期高岭石的再次搬运引起的。

在瑞士Jura牛津阶,硅质碎屑(黏土和石英)主要集中在层序界面,即它们出现在海平面下降过程中。对瑞士Jura和西班牙的索里亚地区的等时露头的小规模(100ka)层序进行比较(Pittet和Strasser,1998)。在那里,硅质碎屑都集中在最大洪泛层。这个对比解释为,在Jura,海平面下降阶段为潮湿气候,在西班牙则是干旱气候,并且,在Jura海平面上升的时候是更为干旱的气候,而西班牙此时正在下雨。这样看来,侏罗纪Jura(26°至27°N)和西班牙古纬度的差异(23°到24°N;Dercourt等,1993)足以导致他们的气候相反。大气环流调整与轨道周期相一致被视为引起这些气候变化的原因(Matthews和Perlmutter,1994)。

和硅质碎屑一起,养分也被水冲进台地水域;这里主要发育珊瑚点礁,普遍发育微生物结壳。从洁净的珊瑚表面到微结壳(比如,有孔虫、红藻类、苔藓虫和海绵),然后到微生物的序列,指示了从贫养到中等营养,最后到富营养环境的演化过程(Dupraz,1999;Dupraz 2002)。

侵蚀地块到研究区数百千米(图7-1)。因此,硅质碎屑很可能首先沉积在靠近物源区的三角洲河口地区,然后由水流搬运至 Jura 台地。另外一个因素是硅质碎屑在台地上经过地形凹陷而生成河道,如 Pittet(1996)和 Huge(2003)的描述,这导致了硅质碎屑不均匀的侧向分布。被高能事件再次移动搬运到其他地区之前,石英和黏土可能被阻塞在这样的凹陷部位一段时间。这些过程可能发生显著的时间滞后,这解释了为什么大量的硅质碎屑和海平面变化之间的相关性并不明确。此外,风暴再作用和生物扰动作用产生的沉积混合,根据沉积速率,可以按时间平均为几十厘米深度对应几百到几千年(Flessa等,1993;Kidwell等,2005)。

7.7 碳氧同位素

Vorbourg 剖面20个样品、Savagnières 剖面24个样品测量了碳($^{13/12}$C)、氧($^{18/16}$O)同位素比值。(图7-12Védrine,2007)。在两个地区中 δ^{13}C 的值在 -0.5‰ ~ +2‰ 之间变化,然而在 Savagnières,δ^{18}O 的值在 -5‰ ~ -4‰ 之间变化,在 Vorbourg,在 -5.5‰ ~ -3.5‰ 之间变化。δ^{18}O 和 δ^{13}C 之间没有相关性,这暗示着原始同位素信息没有因为成岩作用发生明显的改变(艾伦和马修斯,1982)。层序界面或最大洪泛面没有明显的、系统的关系,并在两剖面之间没有相关的趋势。

在瑞士和法国 Jura 贝利亚斯组潮缘带碳酸盐,Joachimski(1994)指出在出露界面之下的 δ^{13}C 值通常耗尽,暗示着源自土壤的二氧化碳的影响。同时,δ^{18}O 可能增加,因为在蒸发过程中孔隙水中的^{16}O 被除去。这里分析的层序没有显示任何长期出露的沉积学标志,这也被同位素记录中缺乏特征性变化所证实。

有机质氧化作用,光合作用和呼吸作用会在碳酸盐台地上形成大量水体进而提高 δ^{13}C 值,使其明显不同于开阔海(Patterson 和 Walter,,1994;Immenhauser 等,2003;Colombie 等,2010)。δ^{18}O 的值取决于温度和盐度,两者在台地浅水中在时间和空间上都有很大的变化。Plunkett(1997)从瑞士 Jura 台地中牛津阶样本中基于 -3.5‰的 δ^{18}O 平均值计算的古温度在 26℃ ~ 27℃。根据测量的浮游有孔虫和箭石中的氧同位素,Frakes 等(1992)指出晚牛津期海洋表面温度达到27℃。

两个研究剖面的 C 和 O 同位素值没有显示出表明显著的环境变化的波动,然而沉积层序中的相趋势表明海平面变化,黏土矿物的集合体意味着内陆气候的变化。两个剖面随时间的轻微变化可能是因为碳酸盐岩的原始组成不同,提供 C 和 O 的碳酸盐生产物的水体不同,以及引起碳酸盐早期成岩稳定性的孔隙流体的成分不同(Van der Kooij 等,2009)。这些差异在一个结构化的、普遍发育横向和纵向相变化的碳酸盐台地上是可以发现的(facies mosaics;Strasser 和 Védrine,2009),那里的海平面、气候变化影响了海水的温度和盐度,并且早期成岩作用由于局部的、事件性的侵蚀和淡水晶状体的出现而被划分出来(正如 Plunkett 通过阴极发光研究所证明的那样,1997)。此外,通过生物扰动和风暴再作用所做的按时间平均可能使原始信息均质化。

7.8 生物变化

在 Gunsberg 或 Roschenz 和 Hauptmum-ienbank 或 Steinebach 段之间的边界上,岩性由泥灰质快速变成灰岩是由于一次主要的海侵事件。在此期间,向 Jura 台地输入的硅质碎屑减少,这为珊瑚礁的形成和鲕粒沉积提供了空间。然而,在之前受海洋的影响,显示棘皮动物(如海百合类和海胆类)相对更加繁盛,并且在 Gunsberg 或 Roschenz 段的上部发现有腕足动物。这类生物和腹足动物、双壳类、介形虫类、苔藓虫类以及龙介虫一起繁盛于泥灰质为主的,受保护的潟湖里。底栖有孔虫(尤其 Pseudocyclammina 和 Ammobaculites)通常在 Vorbourg 剖面的泥灰质和灰质部分中都有富集。在 Hautes—Roches 这两个种属集中出现在靠近研究剖面底部的层段(图 7-7)。在这些泥灰质段中发现有低丰度 Dasycladacean 藻和珊瑚,而且主要在灰岩层和结核中。

鲕粒发育的 Hauptmumienbank 段珊瑚丰度低,但是在 Gorges de Court 和 Hautes—Roches 的 Steinebach 段的珊瑚水平层中珊瑚占主体。在后者的露头中,这些水平层被解释成珊瑚席,它的生长被定期扰动(参见下文)。珊瑚丰度低,且主要是 Isastraea 和 Microsolenida。生物侵蚀作用产生的胃形钻孔很重要。占绝大多数的微结壳有 Placopsilina, Bullopora 和 Nubecularids,此外柱状的微生物岩也有发现。从灰质占主导的沉积物到珊瑚席的迅速转变至少部分因为环境从中等营养到营养匮乏的转变:伴随着硅质碎屑物而来的营养变少了,所以发育低多样性的珊瑚成为可能。那些在 Hauptmumienbank 段发现的鲕粒在泥灰质的 Roschenz 段中也有出现(Mumie 在德语中译为妈咪,描述了在一个核周围形成不规律的包裹层)。在 Hautes—Roches,Vorbourg 和 Voyeboeuf 研究层段的鲕粒属于 Védrine 的 2 型:他们在直径上从几毫米到 1cm,有着光滑的表面,在泥灰主导的外层中有可能包裹有机物(如 serpulid 虫或者 Bullopora)。在 Savagnières 和 Pertuis 剖面中 3 型和 4 型也有出现。3 型鲕粒直径达到 5cm,4 型直径达到 10cm。3 型和 4 型都有朵叶状表层,且含有 Bacinella 和 Lithocodium(绿藻名)。3 型有不规则的叠层皮层,4 型鲕粒没有叠层。3 型和 4 型鲕粒通常跟有孔虫 Mohlerina basiliensis 一起出现,这暗示了它们相同的生态需求。1 型鲕粒在 Gorges de Court 有发现:它们直径有几毫米,光滑的表层,不含微结壳。3 型和 4 型鲕粒指示了很低的能量条件,而 1 型和 2 型的光滑表层暗示了它们在海底的滚动。1 型和 2 型可以在动荡的海水中形成,然而 3 型和 4 型(有需要光照的 Lithocodium 和 Bacinella 的出现)显示了清澈的水体条件。Lithocodium 和 Bacinella 的生长归因于营养水平的提升(Immenhauser 等,2005;Schlagintweit 等,2010)。然而,在瑞士 Jura 的牛津阶,这些生物经常与珊瑚的高度多样化一起出现,这又暗示了相当匮乏的营养条件(Dupraz,1999;Dupraz 和 Strasser,2002)。

观察到的动物群和鲕粒的演化(图 7-7 和图 7-8)不能直接反映海平面的变化,但是可以反映营养情况的变化。在 Guns-berg 或 Roschenz 段的异养生物包括腹足类,Bourguetia striata,以藻类和微生物席为食,以及滤食性动物,例如腕足类和海百合类(Dupraz,1999)。在 Hauptmumienbank 和 Steinebach 段的形成过程中,硅质碎屑物的输入及与之伴随的营养物质通常降低,但是不规则的台地形态导致了环境条件随后重要的横向差异。在 Hautes—Roches,虽然珊瑚的生长被风暴事件产生的碎石,被生物侵蚀和微生物结壳,以及最后被鲕粒沙波的推进所中断多次,但是珊瑚席可以在地形高点固着生长。同时,在凹陷部位形成鲕粒潟湖。在整个研究的时间段中,盐度可能是常量。因为,发现狭盐性的动物群没有变化。

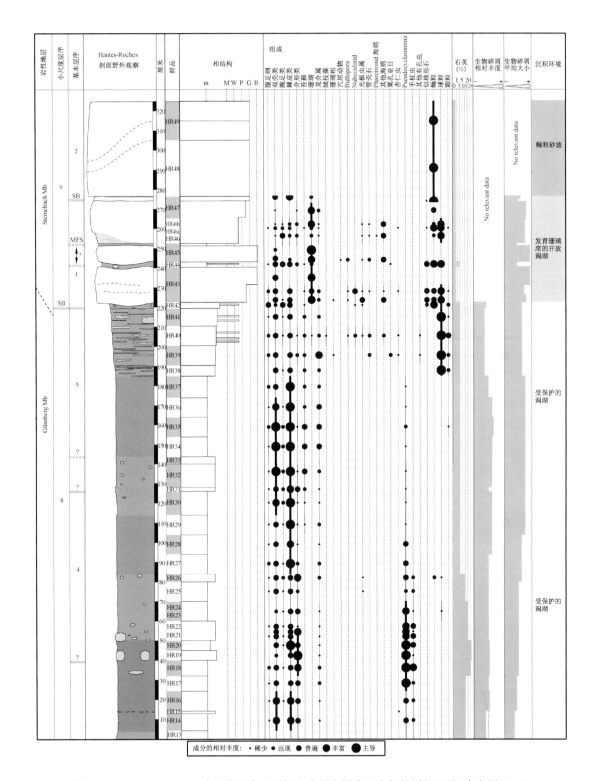

图7-7 Hautes-Roches 小尺度层序8顶部和小尺度层序9底部的详细记录(参考图7-5) 小尺度层序内的基本层序标号为1-5。样品覆盖露头的100%。环境解释根据组成的相对丰度(由薄片估计)和沉积构造而定。符号解释见图7-5。SB:层序界面;MSF:最大海泛面。据Stienne(2010)有改动

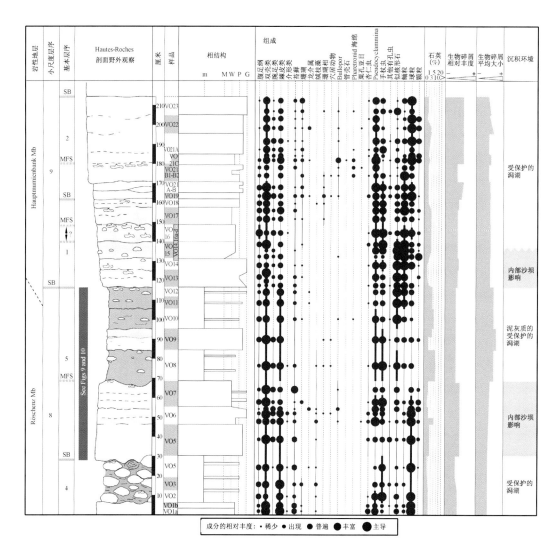

图 7-8 部分 Vorbourg 剖面的详细记录

详细解释见图 7-7 说明,符号见图 7-5。据 Stienne(2010) 有改动

在绝大多数所研究的样品中可以看到生物扰动(主要为 Thalassinoides 海生迹),这暗示了沉积物在沉积之后的改造,按时间平均有几十厘米(Stienne,2010)。此外,风暴事件造成了生物碎屑的侵蚀和再沉积。因此,所观察到的化石的分布不能反映生物的确切位置,这也解释了在此潟湖环境中的生物序列是渐变的。

7.9 高分辨率对比

在对 Jura 台地的研究中,旋回地层学是获得各剖面之间高分辨率对比的最佳方法(图 7-5)。基于给定的生物地层和年代地层框架(图 7-2 和图 7-3),可作出假设:观察到的基本层序相对应于20ka 的轨道岁差周期。在图 7-13 中,尝试对比了 Vorbourg 和 Hautes—Roches 剖面的一个基本层序,以估计环境变化的等时性。这两个剖面位于瑞士 Jura 不同背斜上,这意味着他们之间的古地理距离大于今天的 10km。Vorbourg 的层序主要有鲕粒、似核形石颗粒灰岩、泥粒岩和粒泥灰岩,而在 Hautes—Roches 珊瑚占主体(图 7-7 和图 7-8)。Vorbourg 剖面,海侵面直接覆于层序界面之上,层序界面位于灰岩层底部,而 Hautes—Roches 剖面,层序

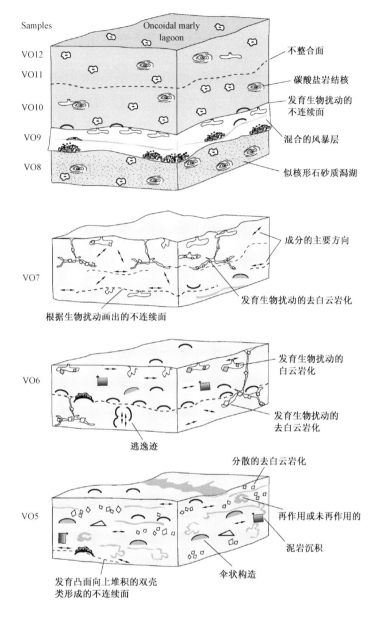

图 7-9 Vorbourg 剖面选定样品的三维素描图

显示沉积、成岩和不连续特征(无比例尺)。对比见图 7-8;符号见图 7-5。据 Stienne(2010)有改动

界面仍然是泥灰质,但在珊瑚碎屑下是含有鲕粒薄层。

　　Vorbourg 的第一个不连续以泥晶灰岩碎屑为标志,表明了粘结的碳酸盐泥被水流和波浪所改造。既然包裹着沉积物的是颗粒灰岩,必须假设沉积这种碳酸盐泥的是一个安静的阶段。在 Hautes—Roches,层序下部包含珊瑚碎片,指示静水环境中珊瑚席生长和随后被破坏的过程。因此,Hautes—Roches 的珊瑚碎砾和 Vorbourg 的泥晶灰岩碎屑可能同时形成于同一高能阶段。Vorbourg 的第二个不连续以铁浸渍为标志,表明了沉积速率减小。增强的孔隙流体的循环也导致了微亮晶化(microsparitization)。在 Hautes—Roches,生长在碎砾上的珊瑚席表现出了死亡的迹象(生物侵蚀和微生物结壳),这可能是由于环境的变化,从贫养至中营养型或富营养所造成的。相同的环境变化会引起 Vorbourg 中沉积作用的减少。在 Hautes—Roches,改造过的珊瑚和双壳类碎片,以及 Vorbourg 的贝壳堆积指示了风暴活动。

Vorbourg 的第三个不连续存在大量生物扰动,意味着沉积速率的减小,使很多洞穴集中存在。在 Hautes—Roches,这个界面是与第二个珊瑚骨架岩顶部界面相关,在此界面上发育的硬灰岩层,现在发现是上覆泥灰岩的再沉积砾石。第二、第三个不连续都体现了最大洪泛面的特征,海平面的上升速率增加导致沉积饥饿以及黏土和养分注入从而导致珊瑚的死亡。沉积然后再次抬升,但在 Vorbourg 的第四个不连续的铁浸渍意味着环境再次变化。这可能与珊瑚骨架上厚微生物结壳所指示的 Hautes—Roches 富营养化作用是同期的。在 Vorbourg 下一个基本层序的层序界面和海侵界面位于含有再沉积的双壳类和铁浸渍的碎屑地层底部。在 Hautes—Roches,层序界面和海侵界面位于在沙波以下薄鲕粒层的底部。

当然,这些不连续的对比是可推测的,因为局部因素的变化可能会导致沉积饥饿,并且不同的位置沉积饥饿的时间变化也很大。水体的富营养化可以影响一大片水域,但是它对生物的影响也许也被局部环境所改变。因此,如果他们没有发生在同一个基本层序中时,即他们形成在同一个 20ka 周期内,我们目前还无法说清形成这些界面所需要的时间或者他们的等时性。如果层序地层单元(如最大洪泛面)可以在一个基本层序中被识别出来,那么时间分辨率可以提高到区分一个 20ka 周期内较早和较晚的阶段。

7.10 讨论:速率和等时性估算

为了估算地质历史时期中环境变化的速率和等时性,一个精确的地质年代格架是必要的。在侏罗纪晚期,通过菊石地层识别出的最有效的以几十万年计的生物地层年代分辨率(Hardenbol 等,1998)。然而,该时间分辨率依赖于辐射日期而不是精确测量的生物地层学。比如,Schoene 等(2010)将三叠纪—侏罗纪界限处的灭绝时间间隔限定为 29×10^4a 以内。旋回地层可以提供 2×10^4a 的时间分辨率(与岁差周期时间相符,据 Strasser 等,2006)。δ^{13}C 值的变化可以发生在几千年之内,但是由于碳在海洋中的保留时间以十万年计(Weissert 等,2008),δ^{13}C 值相应的变化幅度会略有降低。地球磁场的极性转换发生在几千年内(Langreis 等,2010)。在任何情况下,只有在所研究的沉积物允许使用年代测定方法的情况下,就是说生物、沉积、地球化学或磁性的变化被准确持续地记录下来,高时间分辨率才能够被保证。

深水沉积物中,在潮汐流、波浪作用面以下,并且无底流侵蚀作用,可以认为沉积记录是完整的,在两个时间连接点之间平均沉积速率可以计算出来。贫氧或厌氧的底部水体会进一步通过生物扰动降低平均时间。例如,碳循环中主要的扰动导致海洋缺氧事件(OAE)1a,,Van Breugel 等(2007)估算 δ^{13}C 值的负偏移在缺氧事件形成的黑色页岩以下,而该海洋缺氧事件在 0.6~1.4Ma 间的意大利西斯蒙岩石露头中可见,在 1.9Ma 间的中太平洋深海钻探计划 463 站也可见(假设沉积速率连续、剖面连续)。该事件可归因于 Ontong—Java 大型火成岩群初期 CO_2 的释放(梅内格蒂 等,1998),或由于天然气水合物分解造成的甲烷释放(Beerling 等,2002)。Méhay 等(2009)详细分析了西斯蒙海洋缺氧事件时间间隔,得出结论是快速(0.6~0.7Ma)的 CO_2 释放引起了碳同位素负偏移和海洋酸化,从而导致钙质超微化石钙化程度降低(Weissert 和 Erba,,2004)。

揭示深海环境变化的另一个实例是地中海东部腐泥质 S1。根据 De Lange 等(2008)的研究,腐泥质的形成开始于大约 $9770 \pm 350^{14}C$ a BP,终止于大约 $5710 \pm 440^{14}C$ a BP。然而,分析的 18 个岩心,依据测量的替代物(G 曲霉上的 $\delta^{18}O$, Ba/Al 和 Mn/Al)显示,腐泥质形成从开始到最后差别高达 2000a;这里提到的年龄是很有意义的。腐泥质形成归因于高流量的有机物质和水下缺氧环境的保存条件。两种条件都与地中海由于尼罗河提供淡水注入而形成的修复性循环模式有关,其自身依靠地球公转控制北东非气候变化(Cramp 和 O'Sullivan,1999)。

湖泊沉积物由于时间分辨率高,甚至可达到每年,可以很好地记录环境变化。例如,在纽瓦克盆地三叠系 Lockatong 组中,Olsen(1984)发现湖平面随着岁差周期变化一致,缺氧环境保存下来的纹泥平均每年沉积 0.24mm。

在浅水碳酸盐台地,沉积作用几乎是连续的(Goldhammer 等,1993;Sadler,1994;Strasser 等,1999)。相对海平面下降而出露会导致无沉积和侵蚀(图 7-14),生态因素会影响有机碳酸盐岩生产力,水流和波浪会改造沉积碎屑物质的分布。水深和海底地貌导致横向上强烈相变,使得基于岩性的等时对比具有一定难度。获得沉积环境均质化的最好机会是在最大海泛期间,位于退积作用和进积作用的转换位置,此时整个台地的水深最大。然而,即使在这个时候沉积物高生产速度也会超过海平面上升速度,使得海底深度很浅(Kendall 和 Schlager,1981;Colombié 和 Strasser,2005),这种情况再次引起相变。精确的测年和对比方法比深水沉积物更难获得,在生物地层学中有价值的浮游生物化石非常稀少,分层的水体会获得不同的地球化学组分,早期淡水成岩作用会改变原始的地球化学特征。同样,最大海泛时期也是浮游生物被冲到台地上、以及与广海相连通的台地以上水体的最好机会。如果暴露,层序界面代表明显的时间中断,如果由于出露而没有沉积作用,层序界面通常是穿时的。随着时间推移,海侵面逐渐上侵到古地貌之上则被剥蚀:因此,海泛面是不等时的,除非他们由越过障壁岛的瞬时海泛形成(图 7-14)。

如果有证据可以表明所观察到的沉积层序与地球运转轨道变化控制下的海平面升降一致,那么在最理想的情况下,时间分辨率可以达到 2×10^4a。在意大利北部的三叠纪 Latemar 台地中,Zuhlke(2004)发现了频率更高的时间单位(亚米兰科维奇旋回),但他们的成因和持续时间都值得怀疑。如果假设在地球运转轨道控制下的海平面升降在晚侏罗世是对称的(参见上文)并且沉降速率保持恒定,那么可以推测在为期 2×10^4a 的海平面升降旋回中,会出现最大海泛面(图 7-10)。如果可以定义一个最大海泛面或相对深水的或开阔海沉积相的地层,那么得到的时间分辨率甚至可以高于 2×10^4a。然而,这种时间分辨率很大程度上取决于最大海泛面如何被记录下来。它的物理表达随着时间的迁移可能发生变化,主要是因为在此期间沉积速率和台地的地貌可能发生变化(Catuneanu 等,1998)。图 7-15 总结了三种不同的情形:(1)沉积相表现为水深逐渐加深然后变浅,但没有发育界面(最大海泛区);(2)沉积作用连续,但是可以发现界面或快速相变(最大海泛面);(3)由于没有或强烈减小的沉积作用,发育凝缩段(凝缩的地层,有或者没有界面发育)。当在台地范围内对比同一个基本层序的时候,等时性被定义为 2×10^4a 的间隔或者在此间隔基础上继续细分,但是无法确定这些细分的准确的持续时间(见图 7-13)。

海平面的升降速率取决于 2×10^4a 周期海平面升降的幅度与盆地沉降速率,仅可以做出

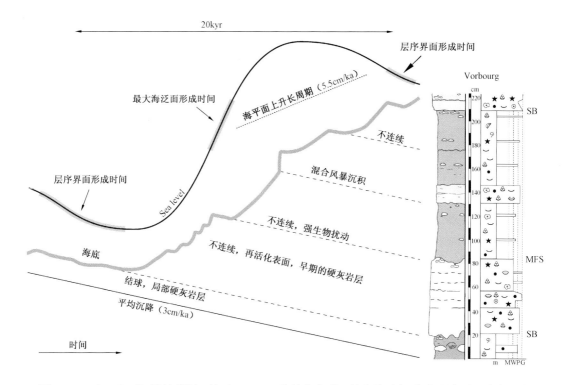

图7-10 在20ka的时间间隔内,基于Vorbourg的基本序列5的差异压实,海底和海平面升降初步重建和演化。讨论在参考文中,符号定义见图7-5

具大误差界限的估算。然而,从图7-10中我们可以得知,现今的海平面上升速率约为地质历史时期的10倍(从1993到2002的平均值310cm/ka相比30cm/ka)。当然,这种对比并不合理,因为温室效应下高频率海平面变化幅度,比如晚侏罗世,比寒室效应下的变化幅度小得多,比如第四纪和今天。然而,通过对比一个海平面升降周期中一小部分,偏差会减小(Strasser和Samankassou,2003)。例如,Toscano与Macintyre2003年做出的佛罗里达、巴哈马以及加勒比的海平面恢复曲线,说明海平面变化幅度在10.6~7.7ka之间为520cm/ka,而从7.7~2ka之间为147cm/ka,从2~0.4ka为93cm/ka。

对于侏罗纪碳酸盐台地时间分辨率最好的研究是解释高频率海平面波动(图7-16)。自旋回形成的相变化没有明显的模式,虽然它们可能间接受控于海平面,例如:在早期海侵和晚期高位条件下,潮坪的进积作用或者是浅滩的侧向迁移会更加显著。风暴沉积在20ka周期内没有明显的分布。可以想象,正如今天这样,风暴强度增加伴随海洋表面温度的增加(Emanuel,2005),但是在瑞士Jura剖面证明这点是不可能的,因为风暴层保存不完整,无法进行强度的重建。由于搬运过程的原因,黏土矿物和海平面有一定关系。但是黏土矿物从内陆到研究区要经历很长的搬运过程,其携带的气候信号会延迟。生物的变化取决于多种因素(例如,水能、水温、水化学、浊度、营养水平、基质和物种间的关系),因此其与海平面变化有复杂的关系。此外,将沉积物按时间平均很难确定准确的时间。最后,由于沉积物的按时间平均和成岩改造,$\delta^{13}C$和$\delta^{18}O$记录的变化不够准确到20ka尺度的层序。

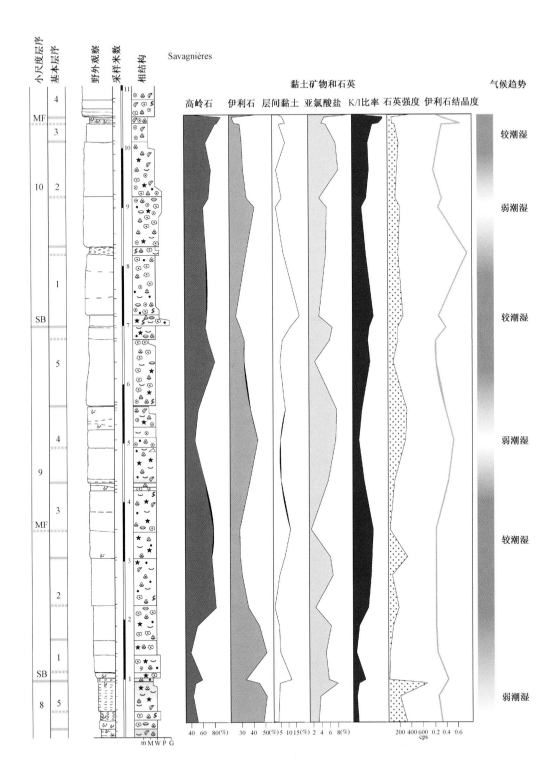

图7-11 在Savagnières剖面黏土矿物和石英的分布（据Védrine，2007，修改）

讨论见正文，IS：层间黏土；SB：层序界面；MF：最大洪泛。符号定义见图7-5

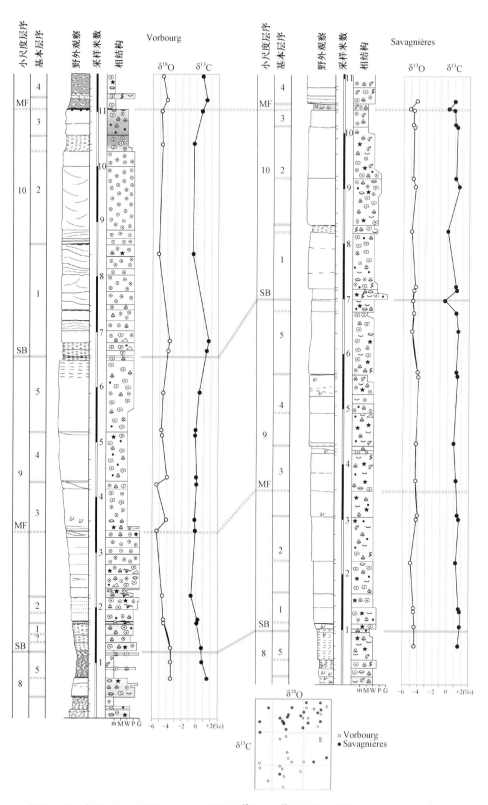

图 7-12 在 Vorbourg 和 Savagnières 剖面 $\delta^{13}C$ 和 $\delta^{18}O$ 的分布（据 Védrine,2007,修改）
讨论见正文。SB:层序界面;MF:最大洪泛。符号定义见图 7-5

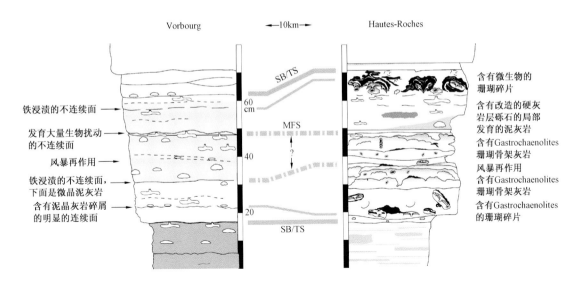

图 7 – 13 Vorbourg 和 Hautes – Roches 基本层序 1 中（在小尺度层序 9 的基础上）
不连续面的初步对比（与图 7 – 7 和图 7 – 8 比较）

讨论见正文。SB：层序界面；TS：海侵面；MFS：最大洪泛面

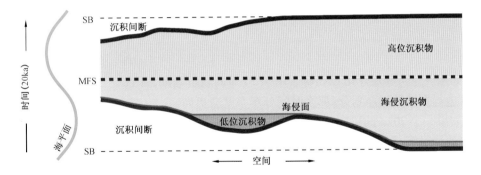

图 7 – 14 一个 20ka 海平面旋回中台地地貌对层序界面、海侵面和最大海泛面等时性的影响

在浅水台地，低位沉积一般缺失或限定在凹陷区。虚线指示层序界面和最大洪泛面的理论位置

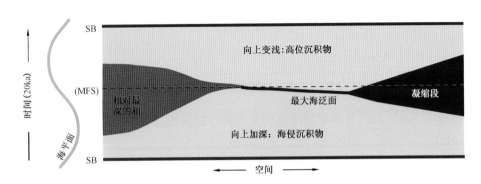

图 7 – 15 一个 20ka 层序中最大海泛情况的不同表达

虚线指示最大洪泛面的理论位置。假定层序界面（SB）上没有沉积间断。讨论见正文

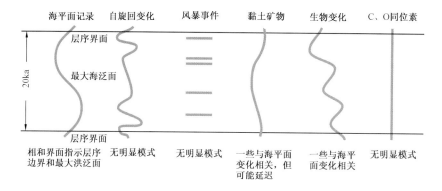

图 7-16 记录一个浅水的、碳酸盐为主的、相当于 20ka 岁差周期的基本层序中环境变化的不同潜力
讨论见正文。SB：层序界面；MFS：最大洪泛面

7.11 结论

这项研究表明埋藏较浅的碳酸盐台地对环境变化非常敏感，但是很难用足够高的时间分辨率解释古沉积记录，从而实现它与现今全球变化的直接定量的比较。主要的挑战是建立一个狭窄的时间框架，在这个时间框架内可以解释环境变化的记录和估算它们的变化率。剖面对比与研究是为了估算区域或者全球范围有多大程度的等时性。

7.11.1 前景

(1) 如果一个旋回地层格架可以建立，一个高达 20ka 的时间分辨率就可以获得。

(2) 如果一个沉积层序可以归属于一个 20ka 的轨道周期，并且一个高分辨率的层序地层分析可以进行更细的层序划分，那么时间分辨率就会更高。

(3) 如果一个 20ka 的层序及其层序地层元素可以在一个碳酸盐岩台地范围内可对比，那么在最好的情况下，等时性可以在几千年时间范围内被评估出来。

(4) 这个时间分辨率可比得上更新世和全新世的时间分辨率。更新世和全新世的气候、海洋、生物和沉积过程已被很好的认知，可以更好的量化。

7.11.2 局限性

(1) 通常在碳酸盐台地剖面上很难建立明确的旋回地层格架，因为碳酸盐台地自旋回过程活跃，并且差异沉降可能歪曲了高频、轨道控制的海平面变化信息。

(2) 在潮下沉积的 20ka 均质相层序内，很难识别特征的层序地层元素。

(3) 由于横向沉积相变及多变的沉积速率，很难进行剖面之间的层序地层和界面对比。

(4) 经历生物扰动作用、风暴重建和成岩作用进行的按时间平均，通常改变了原始的环境变化记录。

(5) 通过旋回地层和高分辨率层序地层分析获得的沉积过程的速度仅可以被估算为时间段内的平均值。

7.11.3 从局部到区域到全球

(1) 对 20ka 层序明确的对比有时很难，即使是短距离对比。

(2) 即使可以追索出界面，但是很难评价一个岩性界面是否等时。

(3) 从一个碳酸盐台地体系到另一个体系，进行区域或全球范围对比需由生物地层学、化学地层学及磁性地层学来确定。因此,高时间分辨率变得更难获得。

(4)在全球范围内对比同期的20ka层序受限于足够精确的地层学方法,因此不能确定瑞士Jura观察的环境变化是否是局部的、区域的还是全球范围的。

(5)如果Jura台地环境变化记录确实是全球范围的,就不能说它与其他地区的记录在多大程度上是等时的。

7.11.4 从过去到现在

(1)在Jura山脉的牛津阶台地,可用20ka的时间分辨率,但反映环境变化的界面和沉积相变的等时性只能依据20ka层序的时间估算来大概估计。

(2)今天在几十年内,全球环境发生了显著变化。

(3)从研究的沉积记录来看,只能说环境变化发生在几千年的间隔期内,但是如果越过了门槛,比如洪泛越过了障壁岛,就可以在局部获得对于环境变化所做出快速反应的证据。

(4)在20ka时间间隔里,牛津期珊瑚礁经历几个阶段的生长和死亡,但无法建立与短期海平面和气候变化的直接对比。

(5)今天,观察到的浅水珊瑚礁普遍死亡。

(6)虽然过去的环境变化的确切速率和幅度仍然难以了解,但是在浅水碳酸盐台地上的沉积和古生态记录,可以作为实例来了解现今环境快速演化的在未来的结果。

参 考 文 献

Abbink, O., Targarona, J., Brinkhuis, H. and Visscher, H. (2001) Late Jurassic to earliest Cretaceous palaeoclimatic evolution of the southern North Sea. Global Planet. Change, 30, 231–256.

Allan, J. R. and Matthews, R. K. (1982) Isotope signatures associated with early meteoric diagenesis. Sedimentology, 29, 797–817.

Allenbach, R. P. (2001) Synsedimentary tectonics in an epicontinental sea: a new interpretation of the Oxfordian basins of northern Switzerland. Eclogae Geol. Helv., 94, 265–287.

Aurell, M. and Bádenas, B. (2004) Facies and depositional sequence evolution controlled by high-frequency sea-level changes in a shallow-water carbonate ramp (late Kimmeridgian, NE Spain). Geol. Mag., 141, 717–733.

Bartolini, A., Baumgartner, P. O. and Hunziker, J. (1996) Middle and Late Jurassic carbon stable-isotope stratigraphy and radiolarite sedimentation of the Umbria-Marche Basin (Central Italy). Eclogae Geol. Helv., 89, 811–844.

Bathurst, R. G. C. (1987) Diagenetically enhanced bedding in argillaceous platform limestones: stratified cementation and selective compaction. Sedimentology, 34, 749–778.

Beerling, D. J., Lomas, M. R. and Gröcke, D. R. (2002) On the nature of methane gas-hydrate dissociation during the Toarcian and Aptian oceanic anoxic events. Am. J. Sci., 302, 28–49.

Berger, A., Loutre, M. F. and Dehant, V. (1989) Astronomical frequencies for pre-Quaternary palaeoclimate studies. Terra Nova, 1, 474–479.

Bessat, F. and Buigues, D. (2001) Two centuries of variation in coral growth in a massive Porites colony from Moorea (French Polynesia): a response of ocean-atmosphere variability from south central Pacific. Palaeogeogr. Palaeoclimatol. Palaeoecol., 175, 381–392.

Brandt, M. E. (2009) The effect of species and colony size on the bleaching response of reef-building corals in the Florida Keys during the 2005 mass bleaching event. Coral Reefs, 28, 911–924.

Burgess, P. M. (2006) The signal and the noise: forward modeling of allocyclic and autocyclic processes influencing peritidal carbonate stacking patterns. J. Sed. Res., 76, 962–977.

Burgess, P. M. and Wright, V. P. (2003) Numerical forward modeling of carbonate platform dynamics: an evaluation of complexity and completeness in carbonate strata. J. Sed. Res., 73, 637–652.

Carpentier, C., Martin-Garin, B., Lathuilière, B. and Ferry, S. (2006) Correlation of reefal Oxfordian episodes and climatic implications in the eastern Paris Basin (France). Terra Nova, 18, 191–201.

Catuneanu, O., Willis, A. J. and Miall, A. D. (1998) Temporal significance of sequence boundaries. Sed. Geol., 121, 157–178.

Chamley, H. (1989) Clay Sedimentology. Springer-Verlag, Berlin, 623 pp.

Cherns, L., Wheeley, J. R. and Wright, V. P. (2008) Taphonomic windows and molluscan preservation. Palaeogeogr. Palaeoclimatol. Palaeoecol., 270, 220–229.

Clari, P. A., Dela Pierre, F. and Martire, L. (1995) Discontinuities in carbonate successions: identification, interpretation and classification of some Italian examples. Sed. Geol., 100, 97–121.

Colombié C. and Strasser, A. (2005) Facies, cycles, and controls on the evolution of a keep-up carbonate platform (Kimmeridgian, Swiss Jura). Sedimentology, 52, 1207–1227.

Colombié, C., Lécuyer, C. and Strasser, A. (2010) Carbon-and oxygen-isotope records of palaeoenvironmental and carbonate production changes in shallow-marine carbonates (Kimmeridgian, Swiss Jura). Geol. Mag., first view, 1–21. doi: 10.1017/S0016756810000518.

Cramp, A. and O'Sullivan, G. (1999) Neogene sapropels in the Mediterranean: a review. Marine Geol., 153, 11–28.

Crevello, P. D. (1991) High-frequency carbonate cycles and stacking patterns: interplay of orbital forcing and subsidence on Lower Jurassic rift platforms, High Atlas, Morocco. In: Sedimentary Modeling: Computer Simulations and Methods for Improved Parameter Definition (Eds. E. K. Franseen, W. L. Watney, C. G. St. C. Kendall and W. Ross), Mem. Kansas Geol. Surv., 233, 207–230.

Crowley, T. J. (2000) Causes of climate change over the past 1000 years. Science, 289, 270–277.

Curtis, C. D. (1990) Aspects of climate influence on the clay mineralogy and geochemistry of soils, paleosols and clastic sedimentary rocks. J. Geol. Soc. London, 147, 351–357.

Davies, D. J., Powell, E. N. and Stanton, R. J. J. (1989) Relative rates of shell dissolution and net sediment accumulation-a commentary: can shell beds form by the gradual accumulation of biogenic debris on the sea floor? Lethaia, 22, 207–212.

De Lange, G. J., Thomson, J., Reitz, A., Slomp, C. P., Principato, M. S., Erba, E. and Corselli, C. (2008) Synchronous basinwide formation and redox-controlled preservation of a Mediterranean sapropel. Nat. Geosci., 1, 606–610.

Deconinck, J.-F. (1993) Clay mineralogy of the Late Tithonian-Berriasian deep-sea carbonates of the Vocontian Trough (SE France): relationships with sequence stratigraphy. Bull. Centres Rech. Explor.-Prod. Elf Aquitaine, 17, 223–234.

Deconinck, J.-F., Beaudoin, B., Chamley, H., Joseph, P. and Raoult, J.-F. (1985) Contrôles tectonique, eustatique et climatique de la sédimentation argileuse du domaine subalpin français au Malm-Crétacé. Rev. Géol. Dyn. Géogr. Phys., 26, 311–320.

Dercourt, J., Ricou, L. E. and Vrielynck, B. (1993) Atlas: Tethys Palaeogeographic Maps. CCGM, Paris.

Dravis, J. (1979) Rapid and widespread generation of recent oolitic hardgrounds on a high energy Bahamian platform, Eleuthera Bank, Bahamas. J. Sed. Petrol., 49, 195–208.

Drummond, C. N. and Wilkinson, B. H. (1993) Carbonate cycle stacking patterns and hierarchies of orbitally forced eustatic sea level change. J. Sed. Petrol., 63, 369–377.

Dunham, R. J. (1962) Classification of carbonate rocks according to depositional texture. AAPG Mem., 1, 108–121.

Dupraz, C. (1999) Paléontologie, paléoecologie et évolution des faciés récifaux de l'Oxfordien Moyen-Supérieur (Jura suisse et français). GeoFocus, 2, 247 pp.

Dupraz, C. and Strasser, A. (2002) Nutritional modes in coralmicrobialite reefs (Jurassic, Oxfordian, Switzerland): evolution of trophic structure as a response to environmental change. Palaios, 17, 449–471.

Emanuel, K. (2005) Increasing destructiveness of tropical cyclones over the past 30 years. Nature, 436, 686–688.

Enay, R., Cariou, E., Debrand Passard, S., Menot, J.-C. and Rioult, M. (1980) Middle-Oxfordian. In: Synthèse paléogé-ographique du Jurassique français (Eds. R. Enay and C. Mangold). Doc. Lab. Géol. Fac. Sci. Lyon,5,181 – 184.

Enos, P. (1991) Sedimentary parameters for computer modeling. In: Sedimentary Modeling: Computer Simulations and Methods for Improved Parameter Definition (Eds. E. K. Franseen, W. L. Watney, C. G. St. C. Kendall and W. Ross). Mem. Kansas Geol. Surv. ,233,63 – 99.

Eyles, N. (1993) Earth's glacial record and its tectonic setting. Earth-Sci. Rev. ,35,1 – 248.

Fairbridge, R. W. (1976) Convergence of evidence on climatic change and ice ages. Ann. NY Acad. Sci. , 91, 542 – 579.

Flessa, K. W. ,Cutler, A. H. and Meldahl, K. H. (1993) Time and taphonomy: quantitative estimates of time-averaging and stratigraphic disorder in a shallow marine habitat. Paleobiology ,19,266 – 286.

Frakes, L. A. ,Francis, J. E. and Syktus, J. I. (1992) Climate Modes of the Phanerozoic. Cambridge Univ Press, Cambridge, 274 pp.

Ginsburg, R. N. (1971) Landward movement of carbonate mud: new model for regressive cycles in carbonates (abstract). AAPG Bull. ,55,340.

Goff, J. A. ,Allison, M. A. and Gulick, S. P. S. (2010) Offshore transport of sediment during cyclonic storms: Hurricane Ike (2008), Texas Gulf Coast, USA. Geology ,38,351 – 354.

Goldhammer, R. K. (1997) Compaction and decompaction algorithms for sedimentary carbonates. J. Sed. Petrol. ,67, 26 – 35.

Goldhammer, R. K. , Lehmann, P. J. and Dunn, P. A. (1993) The origin of high-frequency platform carbonate cycles and third-order sequences (Lower Ordovician El Paso Gp, west Texas): constraints from outcrop data and stratigraphic modeling. J. Sed. Petrol. ,63,318 – 359.

Gornitz, V. ,Lebedeff, S. and Hansen, J. (1982) Global sea-level trend in the past century. Science,215,1611 – 1614.

Gygi, R. A. (1995) Datierung von Seichtwassersedimenten des Späten Jura in der Nordwestschweiz mit Ammoniten. Eclogae Geol. Helv. ,88,1 – 58.

Gygi, R. A. (2000) Integrated stratigraphy of the Oxfordian and Kimmeridgian (Late Jurassic) in northern Switzerland and adjacent southern Germany. Mem. Swiss Acad. Sci. ,104,152.

Gygi, R. A. and Persoz, F. (1986) Mineralostratigraphy, litho-and biostratigraphy combined in correlation of the Oxfordian (Late Jurassic) formations of the Swiss Jura range. Eclogae Geol. Helv. ,79,385 – 454.

Gygi, R. A. ,Coe, A. L. and Vail, P. R. (1998) Sequence stratigraphy of the Oxfordian and Kimmeridgian stages (Late Jurassic) in northern Switzerland. SEPM Spec. Publ. ,60,527 – 544.

Halley, R. B. and Harris, P. M. (1979) Fresh-water cementation of a 1,000-year-old oolite. J. Sed. Petrol. , 49, 969 – 988.

Hardenbol, J. ,Thierry, J. ,Farley, M. B. ,Jacquin, T. ,De Graciansky, P. -C. and Vail, P. R. (1998) Jurassic sequence chronostratigraphy. SEPM Spec. Publ. ,60. Chart.

Hillgätner, H. (1998) Discontinuity surfaces on a shallowmarine carbonate platform (Berriasian-Valanginian, France and Switzerland). J. Sed. Res. ,68,1093 – 1108.

Hillgätner, H. ,Dupraz, C. and Hug, W. (2001) Microbially induced cementation of carbonate sands: are micritic meniscus cements good indicators of vadose diagenesis? Sedimentology,48,117 – 131.

Hug, W. (2003) Sequenzielle Faziesentwicklung der Karbonatplattform des Schweizer Jura im Späten Oxford und frühesten Kimmeridge. GeoFocus,7,156 pp.

Immenhauser, A. (2009) Estimating palaeo-water depth from the physical rock record. Earth-Sci. Rev. ,96,107 – 139.

Immenhauser, A. , Della Porta, G. and Kenter, J. A. M. (2003) An alternative model for positive shifts in shallowmarine carbonate $\delta^{13}C$ and $\delta^{18}O$. Sedimentology,50,953 – 959.

Immenhauser, A. ,Hillgätner, H. and van Betum, E. (2005) Microbial-foraminiferal episodes in the Early Aptian of the southern Tethyan margin: ecological significance and possible relation to oceanic anoxic event 1a. Sedimentology, 52,77 – 99.

IPCC(2007)Summary for policymakers. In: Climate Change 2007: The Physical Science Basis. Contribution of Working Group I to the Fourth Assessment Report of the Intergovernmental Panel on Climate Change(Eds. S. Solomon, D. Qin, M. Manning, Z. Chen, M. Marquis, K. B. Averyt, M. Tignor and H. L. Miller). 18pp. Cambridge University Press, Cambridge.

Jacobs, D. K. and Sahagian, D. L. (1993) Climate-induced fluctuations in sea level during non-glacial times. Nature, 361, 710 – 712.

Joachimski, M. M. (1994) Subaerial exposure and deposition of shallowing upward sequences: evidence from stable isotopes of Purbeckian peritidal carbonates (basal Cretaceous), Swiss and French Jura Mountains. Sedimentology, 41, 805 – 824.

Jokiel, P. L., Rodgers, K. S., Kuffner, I. B., Andersson, A. J., Cox, E. F. and Mackenzie, F. T. (2008) Ocean acidification and calcifying reef organisms: a mesocosm investigation. Coral Reefs, 27, 473 – 483.

Kendall, C. G. St. C. and Schlager, W. (1981) Carbonates and relative changes in sea level. Marine Geol., 44, 181 – 212.

Kidwell, S. M., Best, M. M. R. and Kaufman, D. S. (2005) Taphonomic trade-offs in tropical marine death assemblages: differential time-averaging, shell loss, and probable bias in siliciclastic vs. carbonate facies. Geology, 33, 729 – 732.

Kübler, B. (1990) "Cristallinité" de l'Ilite et mixed-layers: brève révision. Schweiz. Mineral. Petrogr. Mitt., 70, 89 – 93.

Langreis, C. G., Krijgsman, W., Muttoni, G. and Menning, M. (2010) Magnetostratigraphy-concepts, definitions, and applications. Newsl. Stratigr., 43, 207 – 234.

Louis-Schmid, B. (2006) Feedback Mechanisms between Carbon Cycling, Climate and Oceanography. Unpublished PhD thesis, ETH Zurich, Switzerland, 132 pp.

Louis-Schmid, B., Rais, P., Bernasconi, S. M., Pellenard, P., Collin, P. – Y. and Weissert, H. (2007) Detailed record of the mid-Oxfordian (Late Jurassic) positive carbon-isotope excursion in two hemipelagic sections (France and Switzerland): a plate-tectonic trigger? Palaeogeogr. Palaeoclimatol. Palaeoecol., 248, 459 – 472.

Matthews, M. D. and Perlmutter, M. A. (1994) Global cyclostratigraphy: an application to the Eocene Green River Basin. IAS Spec. Publ., 19, 459 – 481.

McCook, L. J. (1999) Macroalgae, nutrients and phase shifts on coral reefs: scientific issues and management consequences for the Great Barrier Reef. Coral Reefs, 18, 357 – 367.

Méhay, S., Keller, C., Bernasconi, S. M., Weissert, H., Erba, E., Bottini, C. and Hochuli, P. A. (2009) A volcanic CO_2 pulse triggered the Cretaceous Oceanic Anoxic Event 1a and a biocalcification crisis. Geology, 37, 819 – 822.

Menegatti, A. P., Weissert, H., Brown, S., Tyson, R. V., Farrimond, P., Strasser, A. and Caron, M. (1998) High-resolution $\delta^{13}C$ stratigraphy through the early Aptian "Livello Selli" of the Alpine Tethys. Paleoceanography, 13, 530 – 545.

Moore, C. H. (1989) Carbonate Diagenesis and Porosity. Dev. Sedimentol., 46, 338 pp.

Olsen, P. E. (1984) Periodicity of lake-level cycles in the Late Triassic Lockatong Formation of the Newark basin (Newark Supergroup, New Jersey and Pennsylvania). In: Milankovitch and Climate, Understanding the Response to Orbital Forcing (Eds. A. L. Berger, J. Imbrie, J. Hays, G. Kukla and B. Saltzman). NATO ASI Series C, 126, 129 – 146.

Padden, M., Weissert, H. and de Rafélis, M. (2001) Evidence for Late Jurassic release of methane from gas hydrate. Geology, 29, 223 – 226.

Patterson, W. P. and Walter, L. M. (1994) Depletion of ^{13}C in seawater ΣCO_2 on modern carbonate platforms: significance for the carbon isotopic record of carbonates. Geology, 22, 885 – 888.

Peñaflor, E. L., Skirving, W. J., Strong, A. E., Heron, S. F. and David, L. T. (2009) Sea-surface temperature and thermal stress in the Coral Triangle over the past two decades. Coral Reefs, 28, 841 – 850.

Pittet, B. (1996) Contrôles climatiques, eustatiques et tectoniques sur des systèmes mixtes carbonates-siliciclastiques de plate-forme: exemples de l' Oxfordien (Jura suisse, Normandie, Espagne). Unpublished PhD thesis, University of Fribourg, Fribourg, Switzerland, 258 pp.

Pittet, B. and Strasser, A. (1998) Long-distance correlations by sequence stratigraphy and cyclostratigraphy: examples and implications (Oxfordian from the Swiss Jura, Spain, and Normandy). Geol. Rundschau, 86, 852 – 874.

Plunkett, J. M. (1997) Early Diagenesis of Shallow Platform Carbonates in the Oxfordian of the Swiss Jura Mountains. Unpublished PhD thesis, University of Fribourg, Fribourg, Switzerland, 155 pp.

Pratt, B. R. and James, N. P. (1986) The St George Group (Lower Ordovician) of western Newfoundland: tidal flat island model for carbonate sedimentation in shallow epeiric seas. Sedimentology, 33, 313 – 343.

Price, G. D. (1999) The evidence and implications of polar ice during the Mesozoic. Earth-Sci. Rev., 48, 183 – 210.

Ramajo, J. and Aurell, M. (2008) Long-term Callovian-Oxfordian sea-level changes and sedimentation in the Iberian carbonate platform (Jurassic, Spain): possible eustatic implications. Basin Res., 20, 163 – 184.

Read, J. F. (1995) Overview of carbonate platform sequences, cycle stratigraphy and reservoirs in greenhouse and icehouse worlds. SEPM Short Course, 35, 1 – 102.

Retallack, G. J. (2001) Soils of the Past-An Introduction to Paleopedology, 2nd edn. Blackwell, Oxford, 404 pp.

Sadler, P. M. (1994) The expected duration of upward-shallowing peritidal carbonate cycles and their terminal hiatuses. Geol. Soc. Am. Bull., 106, 791 – 802.

Satterley, A. K. (1996) The interpretation of cyclic successions of the Middle and Upper Triassic of the Northern and Southern Alps. Earth-Sci. Rev., 40, 181 – 207.

Savrda, C. E. and Bottjer, D. J. (1988) Limestone concretion growth documented by trace-fossil relations. Geology, 16, 908 – 911.

Schlager, W. (1993) Accommodation and supply-a dual control on stratigraphic sequences. Sed. Geol., 86, 111 – 136.

Schlagintweit, F., Bover-Arnal, T. and Salas, R. (2010) New insights into Lithocodium aggregatum Elliott 1956 and Bacinella irregularis Radoíćć 1959 (Late Jurassic-Lower Cretaceous): two ulvophycean grreen algae (? Order Ulotrichales) with a heteromorphic life cycle (epilithic/euendolithic). Facies, 56, 509 – 547.

Schoene, B., Guex, J., Bartolini, A., Schaltegger, U. and Blackburn, T. J. (2010) Correlating the end-Triassic mass extinction and flood basalt volcanism at the 100 ka level. Geology, 38, 387 – 390.

Schulz, M. and Schäfer-Neth, C. (1998) Translating Milankovitch climate forcing into eustatic fluctuations via thermal deep water expansion: a conceptual link. Terra Nova, 9, 228 – 231.

Servonnat, J., Yiou, P., Khodri, M., Swingedouw, D. and Denvil, S. (2010) Influence of solar variability, CO_2 and orbital forcing between 1000 and 1850 AD in the IPSLCM4 model. Clim. Past, 6, 445 – 460.

Shinn, E. A. and Robbin, D. M. (1983) Mechanical and chemical compaction in fine-grained shallow-water limestones. J. Sed. Petrol., 53, 595 – 618.

Smith, L. D., Gilmour, J. P. and Heyward, A. J. (2008) Resilience of coral communities on an isolated system of reefs following catastrophic mass bleaching. Coral Reefs, 27, 197 – 205.

Stienne, N. (2010) Paléoécologie et taphonomie comparative en milieux carbonatés peu profonds (Oxfordien du Jura Suisse et Holocène du Belize). GeoFocus, 22, 248 pp.

Strasser, A. (2007) Astronomical time scale for the Middle Oxfordian to Late Kimmeridgian in the Swiss and French Jura Mountains. Swiss J. Geosci., 100, 407 – 429.

Strasser, A. and Samankassou, E. (2003) Carbonate sedimentation rates today and in the past: Holocene of Florida Bay, Bahamas, and Bermuda vs. Upper Jurassic and Lower Cretaceous of the Jura Mountains (Switzerland and France). Geologica Croatica, 56, 1 – 18.

Strasser, A. and Védrine, S. (2009) Controls on facies mosaics of carbonate platforms: a case study from the Oxfordian of the Swiss Jura. IAS Spec. Publ., 41, 199 – 213.

Strasser, A., Pittet, B., Hillgärtner, H. and Pasquier, J.-B. (1999) Depositional sequences in shallow carbonate-dominated sedimentary systems: concepts for a high-resolution analysis. Sed. Geol., 128, 201 – 221.

Strasser, A., Hillgärtner, H. and Pasquier, J.-B. (2004) Cyclostratigraphic timing of sedimentary processes: an example from the Berriasian of the Swiss and French Jura Mountains. SEPM Spec. Publ. ,81,135 – 151.

Strasser, A. , Hilgen, F. J. and Heckel, P. H. (2006) Cyclostratigraphy-concepts, definitions, and applications. Newsl. Stratigr. ,42,75 – 114.

Thierry, J. (2000) Early Kimmeridgian (146 – 144 Ma). In: Atlas Peri-Tethys (Eds. J. Dercourt, M. Gaetani, B. Vrielynck, E. Barrier, B. Biju-Duval, M. F. Brunet, J. P. Cadet, S. Crasquin and M. Sandulescu), Map 10. CCGM/CGMW, Paris.

Thiry, M. (2000) Palaeoclimatic interpretation of clay minerals in marine deposits: an outlook from the continental origin. Earth-Sci. Rev. ,49,201 – 221.

Toscano, M. A. and Macintyre, I. G. (2003) Corrected western Atlantic sea-level curve for the last 11,000 years based on calibrated ^{14}C dates from Acropora palmata framework and intertidal mangrove peat. Coral Reefs,22,257 – 270.

Trümpy, R. (1980) Geology of Switzerland, a Guide-Book. Part A: An Outline of the Geology of Switzerland. Wepf & Co. , Basel,104 pp.

Vail, P. R. , Audemard, F. , Bowman, S. A. , Eisner, P. N. and Perez-Cruz, C. (1991) The stratigraphic signatures of tectonics, eustasy and sedimentology-an overview. In: Cycles and Events in Stratigraphy (Eds. G. Einsele, W. Ricken and A. Seilacher), pp. 617 – 659. Springer-Verlag, Heidelberg.

Van Breugel, Y. , Schouten, S. , Tsikos, H. , Erba, E. , Price, G. D. and Sinninghe Damsté, J. S. (2007) Synchronous marine and terrestrial biomarkers at the onset of the early Aptian oceanic anoxic event 1a: evidence for the release of ^{13}C-depleted carbon into the atmosphere. Paleoceanography,22,PA1210.

Van der Kooij, B. , Immenhauser, A. , Csoma, A. , Bahamonde, J. and Steuber, T. (2009) Spatial geochemistry of a Carboniferous platform-margin-to-basin transect: balancing environmental and diagenetic factors. Sed. Geol. , 219, 136 – 150.

Védrine, S. (2007) High-frequency palaeoenvironmental changes in mixed carbonate-siliciclastic sedimentary systems (Late Oxfordian, Switzerland, France, and southern Germany). GeoFocus,19,216.

Védrine, S. (2008) Co-occurrence of the foraminifer Mohlerina basiliensis with Bacinella-Lithocodium oncoids: palaeoenvironmental and palaeoecological implications (Late Oxfordian, Swiss Jura). J. Micropal. ,27,35 – 44.

Védrine, S. , Strasser, A. and Hug, W. (2007) Oncoid growth and distribution controlled by sea-level fluctuations and climate (Late Oxfordian, Swiss Jura Mountains). Facies,53,535 – 552.

Weissert, H. and Erba, E. (2004) Volcanism, CO_2 and palaeoclimate: a Late Jurassic-Early Cretaceous carbon and oxygen isotope record. J. Geol. Soc. London,161,695 – 702.

Weissert, H. ,Joachimski, M. and Sarnthein, M. (2008) Chemostratigraphy. Newsl. Stratigr. ,42,145 – 179.

Wildi, W. , Funk, H. , Loup, B. , Amato, E. and Huggenberger, P. (1989) Mesozoic subsidence history of the European marginal shelves of the Alpine Tethys (Helvetic realm, Swiss Plateau and Jura). Eclogae Geol. Helv. , 82, 817 – 840.

Ziegler, P. A. (1990) Geological Atlas of Western and Central Europe. Shell International Petroleum Maatschappij, The Hague,233 pp.

Zühlke, R. (2004) Integrated cyclostratigraphy of a model Mesozoic carbonate platform-the Latemar (Middle Triassic, Italy). SEPM Spec. Publ. ,81,183 – 211.

第8章 地层不整合的沉积演化特征

NIELS RAMEIL, ADRIAN IMMENHAUSER, ANITA É. CSOMA,
GEORG WARRLICH 著

郭彬程 译, 张天舒 校

摘 要 浅海碳酸盐序列的不连续界面可能代表了在地质记录中的古大陆边缘—浅海海域重要的时间间断。了解隐藏于主要的不连续界面中的地质信息, 在古环境分析、层序地层学、重建海平面变化及盆地演化研究中具有重要意义。本文中描述的位于阿曼苏丹国阿普弟阶下 Shu'aiba 组顶部, 是在区域上延伸的($>100000km^2$)且具有长期(10Ma)和复杂的地质历史的一个显著实例。Shu'aiba 组顶部的不连续界面形成于 Oman 台地的高部位, 并代表了晚阿普弟阶时间间隔。陆棚内巴布盆地与海洋边缘的同时代的碳酸盐沉积表明强制海退、且层序逐步下降的过程。通过总结研究区域内的关于阿曼露头和岩心方面的沉积学、层序地层学、岩相学、古生物学和地球化学证据, 补充部分新的资料, 恢复了这一过程背景。在野外, 这种不连续表示为一个低地势、受剥蚀的界面, 表明海相硬灰岩层在这一阶段占据主导地位。关于瞬时大气(成岩作用)初期阶段(同位素的变化、大气胶结物、颗粒周围的裂缝等)模糊特征虽然已经显现出来, 但是对于它们的解释需要认真详细的工作。此特征是显著的, 一系列的相对海平面下降幅度高达数十米, 从早期到后期的阿普弟阶边界到阿普弟阶的结束在中东和其他地方均被记录。尽管研究区古地理位置在热带气候区, 但具有明显的深切喀斯特地貌特征, 许多长期全球性暴露界面极少发育。事实证明, 现代世界没有提供阿曼阿普弟阶碳酸盐系统真正的类似物, 现实中的、受海浪剥蚀的海岸阶地和 Shu'aiba 顶部不连续之间的形态相似性在受到批判性的讨论。这可能意味着, 在暴露期的几个百万年间, Shu'aiba 顶部间断反复经历了浅水洪泛和暴露, 并且在下伏岩石记录中发育有每次海侵的迁移部分。这里展示的资料证明了陆缘浅海碳酸盐的尖灭界面的复杂性, 并且可以作为其他主要不连续的例证。

关键词 碳酸盐岩 成岩作用 不连续面 海相硬灰岩层 相对海平面变化 地表暴露

8.1 概述

浅海相碳酸盐不连续面在层序地层分析和重建过去的海平面变化中发挥关键作用(Van Wagoner 等, 1988; Clari 等, 1995; Hillgärtner, 1998; Immenhauser 等, 1999; Sattler 等, 2005; Schlager, 2005)。此外, 不连续面对于油气藏也有重大意义, 它们经常控制或改变碳酸盐岩储层岩石物理性质。不连续性可划分储层并有助于产生密封的地层圈闭(Budd 等, 1995; Immenhauser 等, 2000a)。另外, 这个关系到地层不连续的过程也可能导致次生孔隙的形成(Moore, 2001; Warrlich 等, 2010)。因此, 利用环境中包含关于上覆和下伏岩石的相关信息和解开这些表面的复杂起源史具有相同的重要性。尤其在浅水部位和高振幅期间, 高频率海平面变化, 存在许多不连续的界面, 这些不连续的界面是广泛的间断的积累过程的标志。在本文中, 阿曼苏丹国阿普弟阶 Shu'aiba 组顶部不整合作为一个典型实例, 代表着区域上的这个重要的不连续界面有一个复杂的、多方面的历史。本文主要研究对象是阿普弟阶台地环境下的 Shu'aiba 组的岩心、露头(图 8-1), 而上 Shu'aiba 组(Bab 段), 即 Bab 台地内部盆地的沉积充填, 没有做研究。因此,术语"Shu'aiba 组"被用来作为下 Shu'aiba 组台地碳酸盐岩的同义词使用,术语"Shu'aiba 顶部不连续面"是指从最上面的阿普弟阶到阿尔布阶巴尔乌姆尔组黏土质陆棚内盆地相中区分出下阿普弟阶 Shu'aiba 组台地相碳酸盐岩的不连续面(图 8-1)。鉴于 Shu'aiba 组灰岩是中东地区一个主要储层,有相当数量的应用和基础文章都在研究这些碳酸盐岩。这些文章包括:Owen 和 Nasr(1958);Simmons(1994);Vahrenkamp(1996);Masse 等(1997);Al-Awar 和 Humphrey(2000);Sharland 等(2001);Borgomano 等(2002);Van Buchem 等(2002,2010a);Pittet 等(2002);Immenhauser 等(2004);Yose 等(2006,2010);Droste(2010);以及 Forbes 等(2010)。令人惊讶的是,这些研究中只有少数几篇文章提供了对下 Shu'aiba 组顶部不连续面浅显的描述。

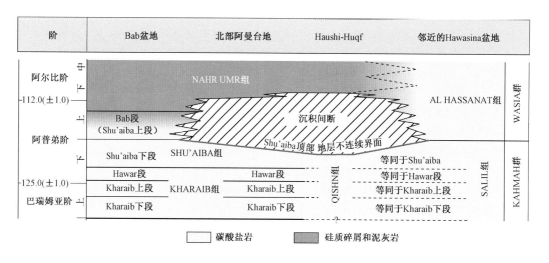

图 8-1 阿曼中白垩世地层构架

根据 Simmons 和 Hart(1987)、Hughes Clark(1988)、Simmons(1994)、Witt 和 Gokdag(1994)、Le Metour 等(1995)、Vahrenkamp(1996)、Masse 等(1997,1998)、Immenhauser 等人(1999)、Skelton 和 Masse(2000)、Immenhauser 等(2004)和 Greselle 与 Pittet(2005)。据 Gradstein 等(2004)的年代数值

在之前对于陆缘浅海碳酸盐岩的不连续处理的研究中发现了一个类似的情况。研究往往处理具体方面例如古代不连续,包括古生态学(Goldring 和 Kazmierczak,1974;Reitner 等,1995;Taylor 和 Wilson,2002;Bromley 和 Heinberg,2006;Carmona 等,2007)或沉积学、层序地层学(Ruffell 和 Wach,1998a)、沉积学及岩相学(Purser,1969;Wright,1982;Brett 和 Brookfield,1984;Carson 和 Crowley,1991;Reolid 等,2007)或地球化学(Joachimski,1994;Yang,2001;Theiling 等,2007)。论文采用内容更全面、区域更广泛的办法,这些在公开文献中是不多见的(Kennedy 和 Garrison,1975;Fürsich,1979;Wright,1988;Clari 等,1995;Hillgärtner,1998;Di Stefano 和 Mindszenty,2000;Immenhauser 等,2000a;Sattler 等,2005)。

下文中所采取的方法是利用最新的来自露头和地下岩心资料来对先前公布的 Shu'aiba 顶部的不连续相关资料的进行补充。这些信息包括利用区域地层学、沉积学、层序地层学、古生物学、岩相—成岩和地球化学对露头到薄片的观测结果。通过观察储层的岩心得到的 Shu'aiba 顶部不连续界面特征,与发现的类似露头特征做比较,来评估出不同成岩作用过程对于主要不连续的影响。

本论文的目的有三个:首先,审查和汇编现有的露头和地下数据,以先前未公布的沉积岩石学证据为基础,按要求补充这些研究结果;其次,严格定位海相硬灰岩层如暴露于地表和多次埋藏叠加印迹等沉积过程环境的研究结果；最后，在更大水深、构造和气候环境中评估这些研究结果以适应不断变化的中白垩世环境。本文的研究深入分析了有趣的、复杂的不连续界面，这里使用的方法可以适用于对其他在类似环境中的不连续界面的研究。

8.2 地层和地质构造背景

阿普弟阶Shu'aiba组位于阿拉伯台地东南部原生大陆架沉积,跨越了晚二叠世到早白垩世土仑阶,在整个海湾地区是一个重要的油气藏(Murris,1980;Vahrenkamp,1996;Alsharhan 和 Nairn,1997;Sharland 等,2001;Borgomano 等,2002;Van Buchem 等,2002,2010b;Droste 和 Van Steenwinkel,2004;Immenhauser 等,2004;Yose 等,2006,2010;Droste,2010)。阿拉伯台地由碳酸盐台地复杂体组成,该区域可细分出一个台地内部盆地(Bab 盆地)直到(南—)西部的研究区,其位于碳酸盐斜坡边缘,连接东部和东北部(图 8 - 2B)。Bab 盆地下 Shu'aiba 段(下阿普弟阶)加上"上 Shu'aiba 段"或者"Bab 段"(上阿普弟阶,图 8 - 1)的 Shu'aiba 组地层厚度大约为100m,局部最大厚度130m。在研究区,只有下 Shu'aiba 段存在台地相,同时上 Shu'aiba 段的同期地层缺失。在北部露头地层厚度达到58m(Wadi Bani Kharus)至85m(Wadi Mu'aydin and Jabal Madar;Pittet 等,2002;Sattler 等,2005),而通过测量,Haushi-Huqf高地最东南部的下 Shu'aiba 段同期地层沉积露头,其地层厚度仅有13m(Wadi Baw;Immenhauser 等,2004)。Bab 盆地中的厚度岩心记录的厚度是在35~108m 之间,分别位于近端和盆地边缘最东南部井区。

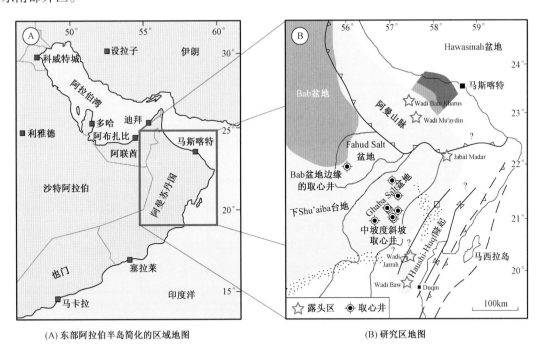

(A) 东部阿拉伯半岛简化的区域地图　　(B) 研究区地图

图 8 - 2　研究区古地理/古构造图

下阿普弟阶的区域用蓝色表示并添加了白色符号。深蓝色表示较深的区域,浅蓝色表示较浅的区域。主要的构造区域用灰色线勾画轮廓(据 Masse 等,1997,1998;Pollastro,1999,经修改)

Shu'aiba组与主要为 Albian Nahr Umr 组的上覆地层由区域可对比的不整合分隔开来。这一不整合代表了一个几百万年时间的沉积中断,在这里被称为 Shu'aiba 顶部不连续界面

(Immenhause 等,2000a;图 8-1),而且可能与 Rub Al Khali 盆地(阿曼南部)西部和南部边缘的构造抬升相关,在那里,这个不连续界面切入下古生界碎屑岩(Montenat 等,2003;Immenhauser 等,2004)。北部阿曼山脉沉积间断的持续时间,近似在 3.5~4Ma 之间(Immenhauser 和 Scott,1999),但向岸方向,向南、西南以及向 Bab 盆地方向减少(Scott,1990;图 8-2)。在研究区南部(岩心与 Haushi-Huqf 露头带),基于 Gradstein 等(1995)的时间模式估算间断持续时间超过 5Ma(Immenhauser et al.,2004)。根据 Gradstein 等(2004)最新的时间尺度,晚阿普弟期持续时间被延长,因此,间断的持续时间可能更长。Van Buchem 等(2010b)认为,间断最长的持续时间大约是 10Ma,这取决于阿曼台地的相对位置。这一不连续的意义非常关键,因为它直接位于阿曼和其他海湾地区 Shu'aiba 油气藏的盖层(上覆 Nahr Umr 组)下面。在下伏的储层中,在成岩作用的不同阶段产生的次生孔隙具有重要意义,这些次生孔隙与这一不连续紧密相关(Warrlich 等,2010)。

8.3 术语、研究区和研究方法

在本文中,"不连续界面"这个专用名词被 Clari 等用来描述地层记录中与沉积间断有关的任何界面(例如地表暴露界面、海相硬灰岩层、沉积间断面、坚固基底和岩石基底)。后来,Sattler 等在 2005 年的参考文献提到,这种不连续记录了海相硬灰岩层和地面暴露阶段,在这里被称为"复合界面"或"多基因界面"。

从北到南,目前的研究介绍了阿曼山脉中的两个主要露头带(Wadi Bani Kharus 和 Wadi Mu'aydin)以及位于 Jabal Madar 的露头。露头资料分别与来自阿曼中心北部 Fahud 和 Ghaba 含盐盆地各构造域的 9 口钻井岩心进行对比(图 8-2B)。本文也展示了从 Haushi-Huqf 高地到阿曼山脉南部到阿曼内陆盆地东部/东南部的露头观察结果(图 8-2B;Wadi Jarrah 和 Wadi Baw;Immenhauser 等,2004)。在所有露头地点,Shu'aiba 顶部界面暴露达几百至几万平方米。

Shu'aiba 组顶部的岩心和露头剖面通过厘米级到分米级的测量记录下来,比例尺分别是 1:25 或 1:50。在记录的时候,特别注意识别了成岩作用的变化,这种变化可以在野外工作中清晰辨认出来。这些特征包括,界面的成矿作用,在不连续面之下石灰岩的漂白作用,溶解角砾岩,以及开放的溶蚀孔和铸模孔。

磨制 42 个岩心、露头样品岩石薄片用来做显微镜热台阴极发光观察。染色的岩心样品岩石薄片(茜素红 S 和铁氰化钾;Dickson,1965),由阿曼石油开发部提供,这些薄片被收集整理以用作目前的研究。被打磨的岩石光面是利用来自 Shu'aiba 顶部界面露头样品中选出的 5 个样品制作而成的。利用德国波鸿—鲁尔大学标准的透射光显微镜(平面偏振光)和阴极发光显微镜技术观察岩石薄片。德国波鸿—鲁尔大学阴极发光试验通过显微镜热台阴极发光试验得以实现(型号 HC1-LM;Neuser,1995),在磨光的薄片两侧使用了镀金技术。在样品的表面,电子束加速电压为 14keV,电子束流密度为大约 $9\mu A/mm^{-2}$。请参阅 Immenhauser 等人(2000b,2004)和 Vahrenkamp(2010)的碳氧同位素的分析方法和本文中的微量元素数据。

8.4 数据集

这里引用的数据主要来源于作者们公布的数据以及先前企业工作人员发表的证据,这些证据来源于一些未公布的生产报告。本文中相关数据来源表述如下:当汇编 Shu'aiba 顶部地层间断的沉积学、古生态学、岩相学和(层序)地层学方面的野外和岩心数据时,现有数据(包括本文和已发表的)的空白,被作者们收集的补充信息所填补。这些数据包括露头的重新观察研究、岩心的重新观察研究,收集补充样品,重新观察岩石薄片或分析新的薄片。在某些情况下,现有的地层剖面被更加详细地描述记录,或集中某些相关研究(表 8-1)。

表 8-1 Shu'aiba 顶部不连续界面的海相硬灰岩层判别特征(引自参考文献)

阿曼露头带	说明	阿曼地下	说明
钻孔 存在,在不同露头出现频率变化(0~15个/m²)。在局部,食石双壳类的形状被保存下来(图 8-3E)。孔洞被部分充填,充填物为 Nahr Umr 页岩或次生氧化铁的浸透	钻孔的数量相对来说并不可观。观测值考虑了相当短暂的缺失阶段的证据。由于不连续面顶部的侵蚀作用(海侵?),钻孔在局部的发育情况是变化的	存在但不经常观测到。大部分生物钻孔的迹象来自岩心 A	观测到的生物钻孔很少,可能是由岩心观测窗口有限而导致的
海洋的浸渍作用 不连续面铁—锰的浸渍是所有露头中的普遍特征。浸渍的范围从薄层状(图 8-5A)到团块状(图 8-5D)。在 Wadi Mu'aydin,次生钻孔和矿化面处于 Shu'aiba 顶部不连续面下方 25cm 处(图 8-8A)。在 Jabal Madar,7~8 层的铁/锰团块出现在不连续面下面的石灰岩中(图 8-8B)	铁—锰的浸渍作用也可能是因为在埋藏成岩至最近的抬升作用期间,由于再活化作用和次生沉淀作用而产生的。磷酸盐的浸染比较罕见	在岩心 B 和岩心 C 中缺失,岩心 A 中存在。通常浸渍呈现出 0.5mm 厚的黑色斑点或微红色结壳,覆盖在钻孔表面	露头带的浸渍作用虽然比较薄,但特征相似
早期的海相潜水胶结物 体积上并不重要? 多数明显现象是在较薄的(大约 50μm)等厚线边缘,叶片状方解石环绕颗粒岩中的颗粒发育。这一阶段为无阴极发光到斑状阴极发光	对于微钻探和地球化学分析来说,海相胶结实在是太小了	较薄的等厚线边缘的胶结物存在于原生粒间孔	

这里展示的对于同位素和微量元素数据的汇编都是来自公开的文献。碳和氧同位素数据来自大量泥晶灰岩基质,其中指出,这些泥晶灰岩基质来自海洋、大气或埋藏碳酸盐岩胶结物(Wagner 等,1995;Vahrenkamp,1996,2010;Immenhauser 等,2000b,2004;Sattler 等,2005;Warrlich 等,2010)。同位素和微量元素数据见表 8-2。

表 8-2 Shu'aiba 顶部不连续中地表暴露阶段的相关特征(引自参考文献)

阿曼露头带	说明	阿曼地下	说明
次生孔隙 露头带中的开放孔隙,发育在不连续下面完全岩化的碳酸盐岩中,但不普遍存在。与暴露有关的孔隙目前普遍被埋藏胶结物所封盖。根迹的开放孔隙空间除外	与大气流体有关的孔隙必须和埋藏淋滤作用有关的孔隙区分开来	存在大量的开放孔洞和铸模(毫米到厘米级,铸模主要由厚壳蛤外壳形成;图 8-6C 和 D)。次生孔隙十分丰富,存在于 Shu'aiba 顶部不连续面以下,厚度在 3~25m 之间	上、下角砾化沉积间断界限可能平行或倾斜于地层层面。至少部分次生孔隙生成于地下
根迹(Rhizoturbation) 根迹十分稀少,但存在于 Shu'aiba 顶部界面富铁结壳(图 8-7)和下伏碳酸盐岩最上面的几厘米中(图 8-9)。根及根状物部分被针铁矿或等轴方解石胶结物(图 8-9)或蜂窝状构造所封闭(图 8-7 和图 8-9)	新近的根迹不能与阿普弟阶根迹相混淆	岩心样品无法证明根迹存在	缺少根迹可能由于岩心有限的观测窗口尺度偏小

阿曼露头带	说明	阿曼地下	说明
浸渍作用 Shu'aiba 顶部不连续面（图 8-4B）的阿普弟阶硅结砾岩封闭了钻孔洞并且被阿尔比阶 Nahr Umr 页岩覆盖（图 8-4B）。许多在 Shu'aiba 顶部界面的铁—锰的矿化作用都以核状出现	与硅结砾岩有关的岩化石英砂的出现和成土成因一致。颗粒周围裂隙与根迹同时出现（图 8-7A）以及浸渍作用，证明了在阿普弟阶存在地表暴露阶段	露头带与铁—锰有关的浸渍作用虽然比较薄，但特征相似。并没有发现成土成因的证据	至于根迹，岩心观测窗口太小以至于无法描述土壤相关的铁—锰浸渍作用
漂白作用 在 Wadi Mu'aydin 和 Wadi Bani Kharus，黑灰色 Shu'aiba 石灰岩位于不整合下方，自界面向下局部被漂白。漂白作用向下延伸大约 10cm，局部可达 30~40cm	即使在新鲜的露头中，漂白作用也仅发育在不整合下的石灰岩中，在其他地方并不发育	岩心中没有观测到漂白作用	由于存在石油浸染，漂白作用可能很难识别
表层岩溶 局部观测切割岩样（蚀刻板块）后认为其构造类似，这些样品来自 Shu'aiba 组露头区最上面几厘米处。在 Wadi Bani Kharus，充填棱角状溶蚀凹槽的是细粒层内角砾岩，角砾岩碎屑的大小在几毫米范围内（图 8-4E）	阿普弟阶新近表层岩溶的不同之处可能在于该处充填了阿尔比阶页岩	在岩心 B 和 C 中，不连续界面并没有呈现出一个无规律的、具棱角的地形。毫米到厘米级的沉降被上覆的 Nahr Umr 泥质相填充（图 8-6B）	岩心尺度小限制了表层岩溶特征的观察
大气胶结 等轴的、孔洞充填的、不发光的亮晶上覆于海相等厚的环边胶结物，它存在于不整合下面几米厚的沉积间隔中（图 8-9D 和 E）。在局部，这种胶结物封闭了根状结核和富铁结壳中成土成因形成的孔洞（图 8-7）。悬挂的/微型钟乳石胶结物，作为阿普弟阶渗流成岩作用环境中的岩相残余物，存在但十分稀少（图 8-9B）	流体中的夹杂物证明了这些胶结物为大气成因	大气成因的、等轴的、孔洞充填的、不发光的亮晶，受到晶洞的横切，晶洞反过来被块状埋藏物、橘色发光的分区带的方解石填充	图 8-9B 所示为悬挂的胶结物的实例
来自大气胶结物的碳、氧同位素数据 从 Shu'aiba 顶部地层基质碳酸盐岩中获取碳、氧同位素数据（图 8-10）。变化范围从不足 2‰ 到大于 6‰。氧同位素数据耗减大约 2‰-5‰。受影响的沉积间隔可达下伏岩石向下延伸 1~2m 的范围。镶嵌的亮晶胶结物充填了根系铸模：$\delta^{18}O$ 为 -6.7‰，$\delta^{13}C$ 为 +1.0‰	充填根系铸模的镶嵌亮晶的具有相当高的 $\delta^{13}C$ 数值可能指示了发育不良的、薄层、主要为无机土壤。关于 $\delta^{13}C$ 数值变低的记录，可能部分是由流体上升过程中的圈闭埋藏导致	同位素的变化（C）通过几口井得到了证实。沉积间隔在 1~7m 厚度之间。$\delta^{13}C$ 变化幅度约 <2‰。没有氧同位素数据	更高密度取样可能得到更显著的同位素变化
大气胶结物微量元素基本丰度（ppm/%） 充填根系铸模的镶嵌亮晶胶结物：MgO = 300±80；MnO = 110±200；FeO = 1950±220；SrO = 探测下限以下	SrO 丰度耗减与缺乏 Sr 的大气水相符合。升高的 FeO 数值与含有铝红土的土壤相一致	无可用数据	

孔隙度方面的数据,无论是露头带还是地下,都是来自 Immenhauser 等(2000b,2004)和 Warrlich 等(2010)。由于埋藏浅,位于阿曼东部 Haushi-Huqf 高地(图 8-2B)的 Qishn 组的露头带(图 8-1)横穿 Shu'aiba 不连续界面的顶部,部分与 Shu'aiba 组在时间和沉积相方面是对等的——保存了储集性能,并且可以详细描述出孔隙度和渗透率的趋势(Immenhauser 等,2004)(表 8-2)。

8.5 Shu'aiba 顶部不连续界面在露头和地下的特征

在露头方面,Shu'aiba 顶部的不连续界面显示了多种显著的特征,这些特征可以通常用来表征海相硬灰岩层。然而,通过进一步观察,发现在岩相学、地球化学和生物学方面的证据,证明这一界面经过地表暴露(Immenhauser 等,2000)。因此,Shu'aiba 顶部不连续有可以作为一个"多基因界面"(Sattler 等,2005)。

该地区中—晚阿普弟期 Shu'aiba 台地顶部重要的、持续时间长的地表暴露,在横穿整个 Bab 台内盆地边缘的地震剖面上特征明显(Yose 等,2006,2010;Droste,2010),经阿曼山脉北部露头证实,存在低位楔(Hillgärtner 等,2003;Gréselle 和 Pittet,2005),并且在 Qatar 的 Shu'aiba 台地边缘发育下切谷(Raven 等,2010)。有人认为,相对海平面下降导致其暴露地表,这与海面升降事件有关,并且可能被构造因素强化(Montenat,2003;Immenhauser 等,2004)。然而,几百万年时间的沉积间断,包括阿拉伯东北部的 Shu'aiba 碳酸盐岩台地长期的地面暴露,只在台地顶部的岩石记录中保存下来极少的信息。还没有发现来自露头和岩心的关于深切喀斯特地形的特征或发育良好的古土壤层这些确凿证据。在露头研究方面,Shu'aiba 顶部不连续界面似乎平行于下伏灰岩层,尽管在阿曼南部,它深深切入中生代和古生代的岩层。因此,在区域尺度上,Shu'aiba 顶部的不连续面被视为一个角度不整合(Wagner,1990)。阿曼地下的地震剖面显示了一个类似于喀斯特构造的大型几何形态。然而,把这些几何形态解释为喀斯特是有争议的,而且相关数据主要来自未发表的生产报告。在阿拉伯联合酋长国 Shu'aiba 顶部界面的类似结构,被 Yose 等人有说服力地解释为潟湖水塘(2010)。海相硬灰岩层特征和地表暴露的特征相结合,共同发育在单一的不连续上,这个不连续为一个数百万年的沉积间断,这使得对于 Shu'aiba 顶部不连续的研究具有普遍的意义。

在所有的三处露头点中(阿曼山脉、Jabal Madar 和 Haushi-Huqf 高地,图 8-2B),不连续特征为一个明显的、形态稳定的、局部呈现从锈色到黑色的铁、锰氧化物和局部硅的矿化作用形成的界面(图 8-3)。在 Wadi Bani Kharus 和 Wadi Mu'aydin(图 8-2 至图 8-4),界面存在明显的上限,起伏平缓。在 Jabal Madar,界面呈现不规则的波状,明显呈碗状或被拉长的凹陷形态,其地势起伏的幅度有 20cm(图 8-5)。在打磨的岩石板片和薄片中,发现界面与上覆的 Nahr Umr 组呈明显的圆齿状接触(图 8-6A 和图 8-7A)。此外,(生物)碎屑的截面表明其为侵蚀组分(图 8-4E;Immenhauser 等,2000a,2004)。深切侵蚀到地台截面的时间窗可能或者与持续时间很长的地表暴露有关,或者与浪基面相关的,Shu'aiba 台地暴露时前 Nahr Umr 沉积重新注入期间的海侵剥蚀有关。

图 8-3 Shu'aiba 顶部不连续界面露头特征

(A)在 Wadi Mu'aydin 这个不连续界面在地形上被标识出来(红色标识)。Wadi 基底和 Natih 洋脊可见的最高点之间的海拔差距大约是 300m。(B)Haushi-Huqf 高地南部,Shu'aiba 顶部不连续界面广阔的露头。比例尺人高约 1.8m。(C)在 Wadi Mu'aydin,漂白的 Shu'aiba 顶部不连续界面(白色标识)。注意到界面的漂白作用不是通体都有的,而是呈现斑点状。左下角作为比例尺的地质学家约 1.8m 高。(D)在 Wadi Mu'aydin Shu'aiba 顶部不连续界面的棱角状内碎屑的地图(11cm 宽)。(E)Haushi-Huqf 高地南部,Shu'aiba 顶部不连续界面钻孔和氧化铁斑点。作为比例尺的钢笔宽度为 7mm

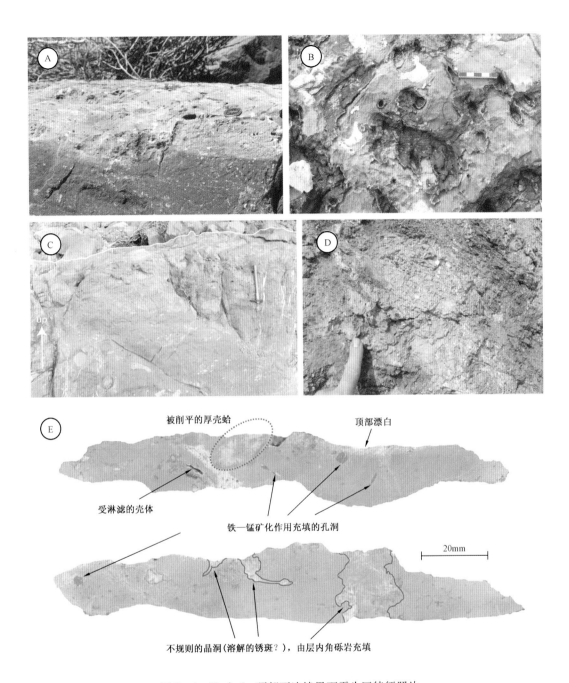

图 8-4 Shu'aiba 顶部不连续界面露头区特征照片

(A) Wadi Bani Kharus 中漂白的 Shu'aiba 顶部不连续界面。以硬币(直径 2cm)作为比例尺。(B) Wadi Bani Kharus 西部 Shu'aiba 顶部不连续界面的硅结砾岩。硅结砾岩覆盖了钻孔和厚壳蛤孔洞,上覆 Nahr Umr 页岩。直尺细分间隔值为 1cm。(C) 在 Jabal Madar,Shu'aiba 最顶部的灰岩的浸染斑点。Shu'aiba 顶部不连续界面用白色标记,地质锤长度为 50cm。(D) 在 Jabal Madar 的 Shu'aiba 顶部不连续面之上几分米处 Nahr Umr 细粒沉积中的浸染斑点。(E) 来自 Wadi Bani Kharus Shu'aiba 顶部界面岩石样品,具侵蚀作用、出露、海相硬灰岩层阶段特征。上覆的铁—锰结壳在取样过程中断开

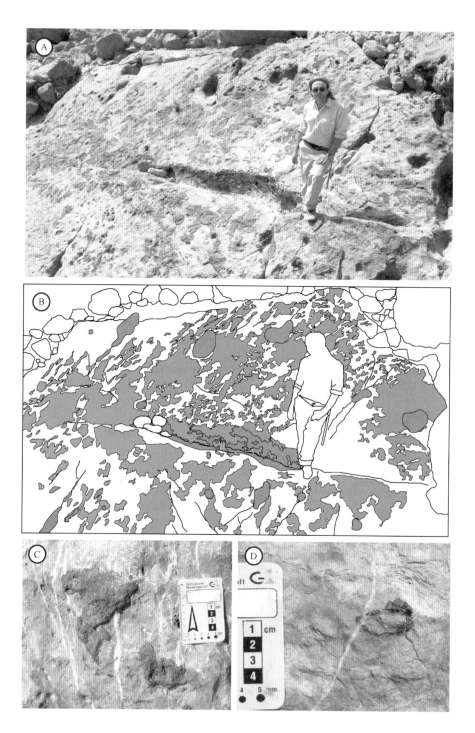

图 8-5 Jabal Madar 的 Shu'aiba 顶部不整合面(图 8-2B)

(A)Shu'aiba 顶部不整合面上 Fe—Mn 元素的注入,部分被剥蚀。注意长条形和碗状的溶解凹点("香蕉孔")。人作比例尺,大约 1.8m 高。(B)图 A 的素描。Fe—Mn 壳标为棕色,溶解凹点为灰色。(C)Fe—Mn 壳盖在 Shu'aiba 石灰岩的晚白垩世方解石纹理之上。(D)Fe/Mn 充填的厚壳蛤模,注意晚白垩世方解石纹理切过该模,地质时代可能属于阿普弟期

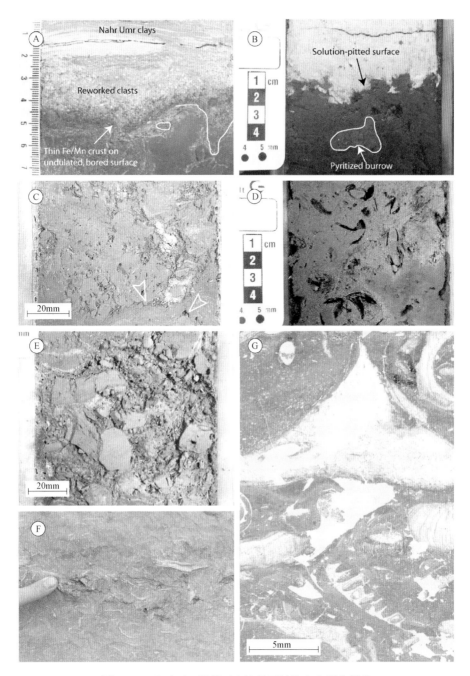

图 8-6 Shu'aiba 顶部不连续界面的取心和露头照片

(A)Shu'aiba 顶部界面的中心。界面呈现波状起伏,发育钻孔,并被毫米级厚的铁—锰结壳覆盖。上覆的 Nahr Umr 页岩基底呈现出粒度向上变细的趋势。(B)Shu'aiba 顶部界面中心不规则的溶解凹痕。注意缺乏铁—锰结壳。(C)向下的 Nahr Umr 页岩充填垂直延伸的晶洞,直接位于下 Shu'aiba 顶部不连续面之下。淋滤作用影响了缝合作用(白色箭头)。棕色的灰岩基质是由于石油浸染。(D)受石油浸染的下 Shu'aiba 地层顶部的厚壳蛤铸模。(E)位于下 Shu'aiba 顶部界面之下的溶解角砾岩。注意大多数角砾为厚壳蛤碎屑。棕色的灰岩基质是由于石油的沾染。(F)小容积的次生孔隙位于 Jabal Madar 的 Shu'aiba 顶部露头之下,与地下灰岩强烈的淋滤作用形成对比。(G)地下的 Shu'aiba 灰岩中的次生大孔隙,淋滤作用下的结构性选择和非结构性选择形成这种次生大孔隙。蓝色树脂指示开放性孔隙空间

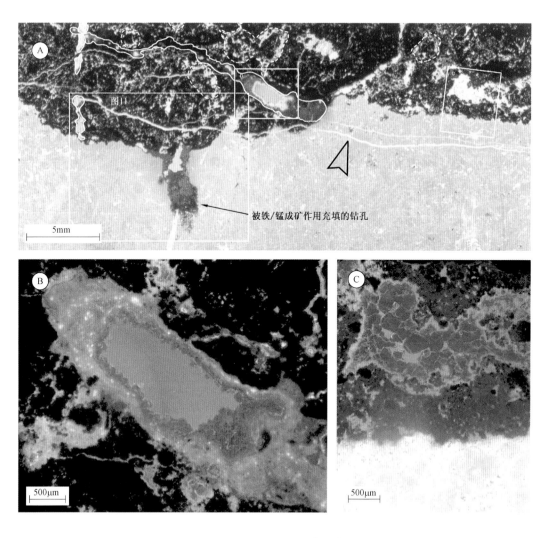

图 8-7 Jabal Madar 地区古土壤特征

(A)Shu'aiba 顶部的铁氧化物结壳被解释为与阿普弟晚期成土过程有关。注意结壳基质的凝块结构、根迹(白线)、颗粒周围的裂缝(虚线)。Shu'aiba 顶部界面的钻孔(中心靠左)充填晚期结壳[图8-11(6)]。中白垩世记载的晚期方解石脉,切割 Shu'aiba 灰岩和上覆结壳(黑色箭头)。成土孔隙中的(B)和(C)胶结物显示了大气胶结物典型的黑蓝色、固有的、狭小的橙色区域阴极发光(Bruckchen 和 Richter,1994)。过度曝光的 CL 图像中,明亮橙色的 Shu'aiba 中成岩改造灰岩[下方的(C)]。注意铁—锰结壳凝结状内部结构

在 Wadi Bani Kharus(图8-2B),不连续界面之下的岩相是一套细粒的、递变的、似球粒颗粒灰岩。在 Wadi Mu'aydin(图8-2),这是一个粗粒的骨骼砾屑灰岩,包括厚壳蛤类碎屑、珊瑚碎屑、似核形石、棘皮动物和大型的未被磨圆的内碎屑(图8-3D)。在 Jabal Madar(图8-2),发育似球粒和生物碎屑颗粒灰岩,局部含有厚壳蛤类碎屑(图8-8)。在所有的剖面中,颗粒灰岩沉积间断在地层中向下穿过了若干过渡相,变成了浅灰色生物碎屑粒泥灰岩到泥粒灰岩,这一特征是大部分 Shu'aiba 组的特征。

不连续界面上覆地层最上面的阿普弟阶/阿尔比阶(Albian)Nahr Umr 组,发育有与下面的 Shu'aiba 组碳酸盐岩截然相反的沉积相(Immenhauser 等,1999)。Nahr Umr 沉积底部是黄褐色到浅绿色的黏土质泥粒灰岩,富含小圆片虫属 Orbitolina texana。小圆片虫类有孔虫类在局

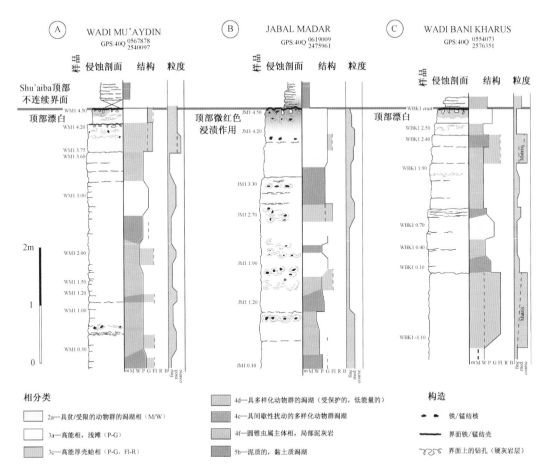

图 8-8 阿曼北部 Shu'aiba 顶部不整合露头剖面描述图(位置见图 8-2B)

部部位占岩石体积的 70%。在 Jabal Madar,不连续界面上覆的是贫瘠的浅绿色到黄褐色黏土。不连续界面下伏的类似于 Shu'aiba 组碳酸盐岩,在 Jabal Madar 底部发育几十厘米的红色的 Nahr Umr 组地层(图 8-4C)。在所有的露头中,Shu'aiba/Qishn 和 Nahr Umr 地层沉积相之间的对比体现在剥蚀的差异上,较软的 Nahr Umr 沉积物更容易被侵蚀,只保留较坚硬的 Shu'aiba 组碳酸盐岩(图 8-3A 到 C)。

在地下露头中,Shu'aiba/Nahr Umr 的过渡体现为不同岩心特征的差异性。但在所有的被检测的岩心中,仅有一个岩心是厚壳蛤类漂砾岩,或含厚壳蛤粒泥灰岩/泥粒灰岩,代表了在不整合之下的原生厚壳蛤类层状生物礁或经搬运的厚壳蛤类碎屑沉积物。在岩心 A 中,界面呈波状起伏(图 8-6A)。界面发育几毫米厚的含钻孔的黑色铁/锰结壳层。2~3cm 厚的沉积间断,直接发育在界面上面,富含直径达 5mm 的再沉积岩屑。这种岩屑间断表明在递变至贫瘠的、成层的浅灰色—浅蓝色的 Nahr Umr 黏土之前,呈明显向上变细的趋势。大部分钻探的 Shu'aiba 组油气藏都发育油斑,这一现象在界面处戛然而止(图 8-6A 和 B)。

8.5.1 海相硬灰岩层沉积阶段的特征

在这里简单描述一下表 8-1 所展示的海相硬灰岩层沉积事件的指相特征。值得注意的方面包括食石的双壳类和海绵产生的生物钻孔,海相胶结阶段的出现,以及次生铁和锰氧化物的渗入(图 8-7A)。这些特征指出了碳酸盐海底早期成岩水下岩化作用,其中确凿的证据是含有钻孔。请参考详细描述这些特征的文献。

在所有研究的露头中,Shu'aiba 顶部界面发育的钻孔穿过下伏岩石结构(图 8 - 3E)。在局部地区,钻孔保存了侵蚀他们的食石的双壳类形状。然而,通常接下来的风化作用使得钻孔变得模糊,往往很难找到具体钻孔所属的钻孔有机生物体(双壳类、海绵等)。一般来说,在与其他长期沉积缺失界面相比较的时候,钻孔程度的意义相对来说微不足道,如波兰的启莫里阶不连续(Goldring 和 Kazmierczak,1974)或法国和瑞士 Jura 山脉的下白垩统不连续(Hillgärtner,1998)。钻孔之间的横截现象并不普遍。相对稀少的钻孔和缺乏横截现象被解释为硬灰岩层沉积时间有限的证据。考虑到 Shu'aiba 顶部不整合持续时间总共有几个百万年,这可能表明,这些在野外非常显著的特征仅仅体现了复杂得多的演化过程中最后那个相对短期的阶段。

在早期海底岩化作用阶段发生沉淀的胶结阶段,在很多情况下,由于地表暴露阶段和埋藏成岩作用的叠加,很难精准确定。根据先前对早期海洋胶结物的研究(Tucker 和 Wright,1990;Bruckschen 和 Richter,1994;Flügel,2004),海相最显著的特征是等厚的薄边缘,叶片状的方解石晶体包裹在颗粒灰岩的组分上(图 8 - 9A;Immenhauser 等,2000a;Al-Awar 和 Humphrey,2000;Sattler 等,2005)。后来作者的观点,一个具有普遍意义的问题是,很多海相硬灰岩层显然缺乏明确的岩相学证据来证明发育广泛的早期海洋胶结。对这一有争议观点的一种可能的解释是,通常发育在浅埋藏阶段的胶结作用,实际上是早期海洋孔隙水的沉淀所导致的早期水下胶结作用。举一个例子,在 Wadi Mu'ayd 的 Shu'aiba 组顶部不连续面之下的碳酸盐岩(图 8 - 8B),在这里似球粒和骨骼颗粒灰岩普遍被不均匀发光的块状的方解石胶结。根据未压实的碳酸盐岩结构判断,这些胶结物可能发生同沉积/早成岩沉积作用,尽管它们具有不均匀发光的模式。如果这个假设是正确的,这个早期海相胶结阶段可能在物质来源方面是暂时稳定的(即高镁方解石),并且在埋藏过程中至少已经有部分被稳定的低镁方解石取代,这样导致观察到不均匀的发光模式。然而,由于这些晶体粒度小,直接地球化学取样是不可能的,并且目前这些假设仍然是推测。另外(或补充说明),由于粒度小,早期海相方解石胶结阶段并不明显,最好用扫描电子显微镜来识别。虽然很多硬灰岩层界面都有共同点(Fürsich,1979;Brett 和 Brookfield,1984;Pomoni-Papaioannou,1994;Clari 等,1995;Hillgärtner,1998;Reolid 等,2007),但是次生的铁和锰氧化物渗入不连续界面,以及局部二氧化硅结壳并不是与海相硬灰岩层相关的先验证据。来自上覆 Nahr Umr 页岩的铁和锰氧化物的埋藏成岩重聚集,以及下伏 Shu'aiba/Nahr Umr 接触面的渗入,这些因素不应该不予考虑进去。在 Jabal Madar 得到了这一过程的证据,与区域隆起和抬升有关的后阿普弟阶方解石脉(Gréludd 等,2006;Immenhauser 等,2007),被厘米厚的 FeO 结壳覆盖封闭。相反,也可见渗入的氧化物被晚成岩作用的方解石脉横截的实例(图 8 - 5D);后者被认为代表了海洋硬灰岩层特征。这些实例被观察海洋微型钻孔虫和其他未鉴定的岩石内生物体所证实,这些生物体局部穿入到铁—锰结壳以及侧向的一些部位,在那里不连续不直接渗入到灰岩界面(Immenhauser 等,2004)。

8.5.2 地表暴露阶段特征

在这里简单描述一下表 8 - 2 列出的指示地表暴露阶段的特征。值得注意的方面包括根迹、表层喀斯特特征、漂白作用、与土壤有关的界面的矿物渗入、大气成岩作用有关的产生的次生孔隙、地球化学模式和大气胶结物。这些特征体现了一个地表暴露的阶段,根迹、表层喀斯特与大气胶结物是最具指相性的特征。请参考引用的详细描述这些特征的参考文献。

Shu'aiba 顶部不连续界面发育的根迹和齿槽状结构是先前发育上覆古土壤的直接证据。

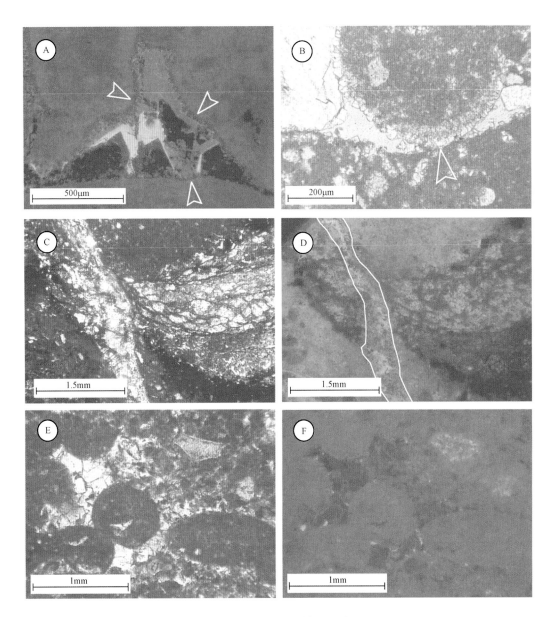

图8-9 成岩特征镜下照片

(A)阴极发光图,沿孔隙壁发育的早期海相等厚的、叶片状胶结物(白色箭头)。位于孔隙中心的发光块状胶结物反映了埋藏序列(Bruckschen和Richter,1994)。Wadi Mu'aydin,样品编号WM1,4.2m。(B)地下悬挂状的胶结物(箭头)指示了Shu'aiba顶部不连续面的大气潜水成岩作用。(C)至(F)配对的透射光和阴极发光显微照相显示,(C)和(D)根迹铸模充填蜂窝状结构,其位置在Wadi Bani Khraus,Shu'aiba顶部界面之下大约10cm处。根迹铸模被晚期埋藏方解石脉切割(白色线条),这证明根迹铸模的年代为中白垩世。样品编号WBK1,2.75m。(E)和(F),块状亮晶在极薄的等厚的边缘胶结物基础上长满了原生粒间孔隙。暗淡无光的CL和其共生位置揭示了大气成因。这类的胶结物只发现于Shu'aiba顶部不连续界面的地层间隔中

根据Klappa(1980)研究,具有明显边界的根状物为细根穿过硬化的岩石形成的钻孔。在研究的沉积间断中缺失发育良好的钙质层,并且整体上根状物稀少,然而,这表明土壤发育没有达到高级阶段,或者界面下包含了大量岩石记录的古土壤已经在随后的海侵过程中被侵蚀掉。此外,必须留意根迹特征的区别,这些特征与白垩纪的地表暴露有关,并且它们是不久前形成

的高地上的植物和岩石内部苔藓活动的结果。有时,可以根据岩相学的关系区分开两种情况。根迹横截与埋藏相关的缝合线或阿曼山脉在晚白垩纪/新近纪抬升产生的方解石脉,从来源上说,这些根迹形成年代较新。相反,新鲜暴露的露头上的根迹被晚成岩的方解石脉或者缝合线横截(图 8-9C 和 D),这可能是阿普弟阶之后经埋藏以及后白垩纪抬升和剥蚀的结果(Hanna,1990;Grélaud 等,2006)。

露头带和岩心的表层喀斯特特征被解释为溶解点蚀(solution pitting)的结果。溶解点蚀是一种(不成熟的)表层喀斯特的典型特征(Klimchouk,2004)。暴露的 Shu'aiba 顶部不整合上的新近形成的喀斯特特征不应该与阿普弟阶的喀斯特相混淆。在新鲜露头中给出了阿普弟阶的喀斯特特征的证据,在这里阿尔比阶 Nahr Umr 页岩填实了溶解特征。区域地震图在界面上显示出的大型几何形状,的确类似于喀斯特的结构;然而把它们解释为喀斯特是一个有争议的问题。Shu'aiba 顶部不连续新鲜暴露的碳酸盐漂白作用(图 8-4),可能与红色石灰土地层有关(Wright,1994),以及阿普弟阶地表暴露阶段的风化作用有关。不连续下面储备流体的化学性质改变和阿尔比阶上覆的盖层相可能是另外的因素,可能导致碳酸盐被漂白和染色的具体程度。典型的漂白发育在台地顶部的暴露带,其特点为低孔隙度、低渗透率。正是在这些低孔隙度露头环境下,漂白作用最明显。相反,岩心数据指出了在地下环境中产生可观的次生孔隙度和渗透率。漂白作用不能在岩心研究中得以表述;笔者认为,这主要与岩心普遍含油斑有关,这些油斑掩盖了岩石原来的颜色。

Shu'aiba 顶部不整合阿普弟阶界面的渗入可能与土壤发育状态有关,包括含水的铁氧化物(针铁矿、赤铁矿和褐铁矿),局部的锰氧化物(黑锰矿)结壳作用和硅质壳层。这些沉积物与很多形成于温暖的、半干旱条件的现代风化层具有共同特征(Lee 等,2003;Reolid 等,2007)但铁—锰外壳也形成于海洋硬灰岩层之上(Distefano 和 Mindszenty,2000)。Shu'aiba 顶部不连续界面上的硅质壳层(图 8-4B)在新近暴露的某些部位封闭了钻孔,被阿尔比阶黏土质沉积物覆盖,这证明了硅质壳层形成于阿普弟期(Immenhauser 等,2000a)。硅质壳层可以形成于海底硬灰岩层早期成岩作用阶段以及当大陆性半干旱到干旱气候条件下土壤的沉淀(Webb 和 Golding,1998)。例如 Jura 山脉和瑞士阿尔卑斯山早白垩世时期水下的受二氧化硅渗入的硬灰岩层,这些实例都已经被 Godet 等人描述过(2011)。

在 Shu'aiba 顶部不连续界面之下,骨架颗粒的次生孔隙(例如,厚壳蛤类)和基质(图 8-6C 至 G)存在从出现(露头)到富含(岩心)的现象,但是在上覆的阿尔比阶 Nahr Umr 组地层中不发育。孔隙空间是由异化粒和基质的结构选择性和结构非选择性的浸出产生的。次生孔隙空间体积和孔隙直径(毫米至厘米级)在岩心和露头中变化很大。根据研究的样品,孔隙空间或者保持开放或者被 Nahr Umr 地层细粒沉积封闭(图 8-6C),黄铁矿(图 8-6B)、针铁矿、等径的大气方解石胶结物分别以固有的发蓝色光(图 8-7B 和 C)或埋藏方解石阶段(发暗橙色到亮橙色光)为特征。然而大多数的孔隙空间可能形成较晚,孔隙被大气胶结物充填被认为是:(1)形成于阿普弟期;(2)与阿普弟阶地表暴露阶段有关。

对于碳和氧同位素数据的具体形式(Allan 和 Matthews,1982;Joachimski,1994;Wagner 等,1995;Tobin 等,1999;Immenhauser 等,2008),以及微量元素的含量(Wagner 等,1995;Immenhauser 等,2000b)被认为是早先碳酸盐岩海底地表暴露的标志。不连续界面下的 ^{13}C 的消耗通常是包含同位素轻质土壤 CO_2 的大气胶结物中的碳酸盐的溶解与再析出的结果。大量的低氧同位素比例的碳酸盐通常表明记录在大气胶结物中的缺少 ^{18}O 的大气水的影响(Lohaman,1988)。向更低的碳和氧同位素值的变化在 Shu'aiba 顶部不连续之下的露头和地下都有出现(图8－10;Wagner 等,1995;Immenhauser 等,2000b;Vahrenkamp,1996,2010)。我们可能感兴趣的是,记录特征同位素样式的石灰岩沉积间断的厚度在露头带比 Oman 地下岩心薄得多(图8－10)。具体来说,在 Jabal Madar、Wadi Mu'aydin 或 Wadi Bani Kharus(图8－2 和图8－10),向更低的碳氧同位素比值的变化过程,被记录在厚度小于1m 的沉积间断中,然而,阿曼 D 油田的地下岩心显示这个沉积间断的厚度大约为5m(图8－10A;Vahrenkamp,2010)。

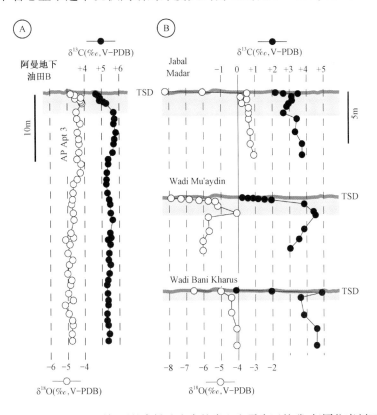

图8－10 Shu'aiba 顶部不整合挑选出来的岩心和露头区的碳、氧同位素剖面

(A)来自阿曼 B 井地下的 $\delta^{13}C$ 值记录(Vahrenkamp,2010,做修改)。灰色的阴影区指示了一种以碳同位素递减变化为特征的岩石间隔,这种变化为可论证为与地表暴露时期土壤带 CO_2 有关。(B)来自 Jabal Madar、Wadi Mu'aydim 和 Wadi Bani Kharus 的露头区同位素剖面。$\delta^{13}C$ 和 $\delta^{18}O$ 记录都呈现了朝不整合方向数值减小的趋势。总的来说,同位素的这种变化幅度与阿曼地下相比在露头剖面的变化幅度相当大,但是受影响的岩石间隔(灰色阴影)其地层较薄(1~2m)。这些差异可能与海侵期间成土间隔中的差异性侵蚀削平作用有关(图8－12C)。注意(A)和(B)之间垂直比例的不同。TSD = Shu'aiba 顶部不连续界面。AP Apt XY =阿拉伯板块层序 Apt XY(Van Buchem 等,2010a)

然而,除了大气解释,对 $\delta^{18}O$ 和 $\delta^{13}C$ 值降低这种变化的其他解释是必须要考虑的。沿着缝合线、断裂通道向上流动的流体(Immenhauser 等,2007),和由于沉积相或与不连续有关的孔隙度差异导致的流体的阻碍是埋藏带的普遍特征。同位素轻质埋藏流体优先沿流动障碍迁移,可能会叠加在海相同位素标志上。至于 Shu'aiba/Nahr Umr 地层边界,Nahr Umr 黏土成为主要的流动障碍,因此,缺少 ^{13}C 的上升流体在界面下积蓄,可能在地层边界下面产生相对薄的沉积间断。然而,考虑到固体岩石、流体包裹体和沉积学证据(Immenhauser 等,2000;Sattler 等,2005),土壤带二氧化碳和出露阶段大气水渗透的影响不能被忽视,大部分观测到的同位素变化可能归因于地表暴露。

8.5.3 不明显的晚成岩特征

地质学上对不连续的合理解释,对于从那些晚成岩阶段特征,尤其是那些早期大气成岩或者早期水下成矿作用产生的相似特征中,区分出与海相硬灰岩层和地表暴露阶段相关的早期大气淡水岩特征,是十分必要的。表明埋葬带的典型特征见表8-3,在这里简要叙述一下,详细内容参考引用的文献。

表8-3 与埋藏成岩作用过程相关的Shu'aiba顶部不连续界面识别特征(引自参考文献)

阿曼露头带	说明	阿曼地下	说明
次生孔隙 孔隙通常被铁—锰结壳、Nahr Umr细粒沉积(图8-6C)、大气或埋藏胶结物所封闭(图8-5D),或(很少!)保持开放。在露头,没有溶解角砾岩,既没有像岩心中那样的开放孔隙(图8-6E),也没有发现被后期胶结作用"愈合"的原地角砾岩	溶解扩张的裂缝和深层(地下)岩溶的特征引自Jabal Madar	对于异化颗粒与基质的结构选择性和非结构选择性的淋滤作用导致大量的毫米到厘米级开放性晶洞、铸模和溶解角砾岩的形成(图8-6C至G)。在角砾间隔带的次生孔隙度可达最大值约40%,渗透率达到几百毫达西	原生特征,例如地层层面或缝合面(图8-6C),都受到淋滤作用影响。溶解扩张的缝合面指示晚期埋藏淋滤作用。一些次生孔隙充填来自上覆阿尔比阶Nahr Umr页岩(图8-6D)
浸渍作用 在Jabal Madar,部分晚于晚白垩世的铁—锰结壳中与隆起相关的岩脉(图8-5C),指示了地下环境富氧化铁流体的再活化作用。不整合下面存在灰岩中黄铁矿成矿作用,但不普遍	黄铁矿可能由于剥露作用过程中存在的雨水而被氧化	Shu'aiba顶部不连续界面之下,黄铁矿的矿化作用普遍发育在几米厚的地层间隔中(图8-6B)	岩心中的浸渍作用和露头带比较起来没有那么显著
埋藏胶结物 浅埋藏:无光—发光、粗粒、铁浸染的块状方解石。埋藏:自形白云石长度大约200μm。白云石相除外缘发亮光外,都不发光	晚期方解石脉显示了复杂的带状构造并发明亮的光	次生孔隙空间部分被等轴的、半自形到自形的第一分区埋藏方解石胶结物充填,呈现单一的暗橙色光。在某些油田中,大量的晶粒间孔隙空间与微斜方晶的次生方解石相关	关于阿曼地区阿普弟阶地下埋藏成岩作用的数据是相当多的,但是主要是未发表的保密的工业报告
碳、氧同位素数据 平均浅埋藏块状方解石$\delta^{18}O$值为$-7.6‰$,$\delta^{13}C$值为$+1.0‰$。富微晶基质的白云石的数据为$\delta^{18}O$在$-8.3‰\sim-4.4‰$;$\delta^{13}C$在$-0.13‰\sim1.5‰$	由于这些晶体尺寸过小而无法获得来自单个白云石晶体的同位素数据		根据Vahrenkamp(2010)研究,阿曼地区地下内部碳同位素数据没有受到成岩作用改造。与此相反,氧同位素数据报告反映出了这一改变
微量元素丰度(ppm/%) 平均浅埋藏块状方解石微量元素值为: MgO = 1240 ± 80; MnO = 1180 ± 200; FeO = 6710 ± 200。 埋藏的全形白云石微量元素值为: Mg = 21%; MnO = 625 ± 190; FeO = 低于测量下限; SrO = 低于测量下限	无数据		

铁—锰氧化物注入的差异形成了顶部海相硬灰岩层,并且那些与地表土壤带相关的沉积过程已被描述和讨论(Di Stefano和Mindszenty,2000)。另一种形成或者至少影响和改变现存的铁—锰结壳的方式,是由从上覆Nahr Umr组页岩(埋藏压实作用所排出的)流下来的流体或从下覆地层单元上涌的流体形成的。其他地区同位素漂移和铁—锰成矿作用的类似的特征

已被描述(Winterer 和 Sarti,1994)。这种流体可以在埋藏带逐步压实过程中被排出(Karsten 等,2003;Immenhauser 等,2007)。很有可能,但很难确定的是,露头中的这些次生外壳的形成和改变很多都持续到现在。对比阿普弟阶的渗入作用,一些 Jabal Madar 表面裂隙中的铁氧化物和喀斯特孔洞,在中新世地层抬升和倾斜期间(或者之后)已经形成(图 8-5C;Gréloud 等,2006;Immenhauser 等,2007)。这些铁氧化物和喀斯特孔洞与地层层面的倾斜关系指示了这个过程,并明确指出地下环境中富含铁—锰流体的再活化作用。

  岩相学证据表明,所研究的角砾质沉积间断的岩心(图 8-6C、E 和 G)中的次生孔隙空间与埋藏淋滤作用有关,最大可获得 40%左右的次生孔隙度,渗透率可达几百个毫达西。一些次生孔隙接下来被晚期的、发光的方解石充填,或者被上覆 Nahr Umr 组阿尔比阶页岩(图 8-9A)封闭。然而,岩心中大部分孔隙空间仍然保持开放(图 8-6G)。一个可能的解释是,对储集岩起到埋藏淋滤作用的侵蚀流体与烃类成熟度和先前的负载有关。这种观点可以解释为什么在那些油田钻探中取得的岩心中可以观察到广泛发育的次生孔隙。脉冲式的侵蚀流体可能与北部晚白垩世阿曼蛇绿石的富集有关,在相邻的 Fahud 和 Ghaba Salt 盆地触发了流体流动的变化(图 8-2;Immenhauser 等,2007)。与地下储集岩研究截然不同,在本次研究观察的大量露头或者公开出版的研究成果中,没有发现在 Shu'aiba 顶界面之下存在渗透淋滤的证据。所以,作为先决条件发育在岩心中的埋藏淋滤阶段并不存在。露头中发现的少量的开放孔隙和孔洞被解释为标志了一个早期(全新世)岩溶作用阶段(后期形成的变化)。与获取岩心的油气藏不同,阿曼山脉和 Jabal Madar 露头区由于(halo)构造抬升剥蚀暴露出来。开放孔洞的相对年龄因它们与,比如,Neogene 抬升阶段有关的方解石脉的交叉切割关系而被确认。然而,不能排除一些白垩纪的孔隙直到今天仍然保持开放状态,并且那些形成于剥蚀暴露作用早期岩溶作用的孔隙已经被富含铁的外壳所封闭,并且由于活化的铁和锰氧化物这些外壳仍会继续形成。总之,地下油气藏中大多数有效的次生孔隙空间是在埋藏(中期)阶段由上覆 Nahr Umr 盖层之下的 Shu'aiba 不整合面顶部封闭的不饱和流体形成的。然而,在阿曼山脉和 Jabal Madar 的露头研究中,并没有发现这一特殊的成岩特征。在这些地区,只有次生孔隙的边缘才是形成于阿普弟阶和剥蚀暴露阶段(后期)后来的大气成岩作用。

8.6 讨论

8.6.1 海相硬灰岩层和暴露阶段的相对年代学

  Shu'aiba 下段顶部不连续面显示的岩相学、古生物学和地球化学特征指示了海洋硬灰岩层和地表的暴露阶段。因此,Shu'aiba 顶部不连续面被归类为一个多基因的界面。值得讨论的一个关键问题是,地表暴露出现在海相硬灰岩层形成之前的还是海相硬灰岩层形成于地表暴露阶段之前。先前对于阿曼地区(Immenhauser 等,2000a)白垩纪多基因不连续界面的研究工作提出两个不同的现象:(1)两个海相硬灰岩层形成阶段被一个地表暴露阶段隔开;(2)一个海相硬灰岩层形成先于地表暴露阶段。在阿尔比阶 Nahr Umr 地层复合界面发现的海相硬灰岩层形成在地表暴露之后的间接证据(Immenhauser 等,1999)包括大量发育在硬灰岩层表面保存完好的薄壳牡蛎类,硬灰岩层被泥质碳酸盐岩覆盖。在地表暴露和再次洪泛条件下的这种微小特征的保存潜力被认为可以忽略不计。由此,在这个例子中,结壳以及未知比例的钻孔似乎形成于地表暴露之后。

  Shu'aiba 顶部不连续面,野外和地下的证据(表 8-1 和表 8-2)表明,普遍重要的海洋硬灰岩层阶段之后是一个短期地表暴露阶段,最后可能是一个先于黏土质沉积物埋藏的短期海洋硬灰岩层阶段。显然,这代表了一个非常重要的长期沉积间断,Shu'aiba 顶部不连续可能已经经历了一系列复杂的长期暴露和沉积缺失阶段。然而,如果是这样的话,比如复杂的、叠合的生物穿孔特征或深切喀斯特地形这些证据随后已被破坏。相反的,对于长期的暴露或缺失阶段的海相碳酸盐岩,并不存在记录下来的海洋硬灰岩层还是任何观察到的露头或岩心的地面暴露留下来的重要痕迹,这些都需要进一步探讨。

对于最后记录下来的 Shu'aiba 不连续演化阶段的一个尝试性的逐步解释,在图 8 - 11 这张草图中呈现出来,在这里做一个总结:Shu'aiba 不连续(1)在海洋里被石蛏属双壳类钻孔,(2)大多数的钻孔在随后被针铁矿结壳所封闭,(3)只局部保存新生钻孔,这被认为至少是部分海洋生物注入的证据。在大多数情况下,发生在钻孔之后的早期的矿化作用封闭了钻孔。随后,浸渍的表面受到侵蚀(4)可能发育在短期的地表暴露之下。在一些地区,先前浸渍作用形成的残余物质被保存下来。证据是次生富铁矿化作用的沉淀。(5)清楚地显示了地表暴露的证据,例如根系痕迹和颗粒周围的裂缝。根系痕迹也可能穿入到早期的结壳中(图 8 - 7 和图 8 - 11)或者下覆的 Shu'aiba 灰岩中(图 8 - 9C 和 D)。这一次生结壳可能形成于较薄的,覆盖植物的砖红色土壤盖层之下,植物的根系侵蚀进入下覆的石灰岩中。其他观察到的与古土壤相关的特征,如漂白作用、硅结砾岩、或厘米级的表层喀斯特构造(图 8 - 3C,8 - 4A、B 和 E)可能都是从这一阶段开始形成的。然而,关于古土壤的形成和植物活性的证据零星分布。同样,只有不连续下面的具有^{13}C 消耗特点的薄地层间隔(图 8 - 10),与薄的土壤层以及有限的生物活性相一致。当 Shu'aiba 顶部不连续埋藏于晚阿普弟期/阿尔比期 Nahr Umr 海相碎屑岩下面的时候这一最后的阶段就宣告结束了。晚白垩世的构造活动表现为不连续界面的裂缝和成脉作用(图 8 - 11 中的(6))。

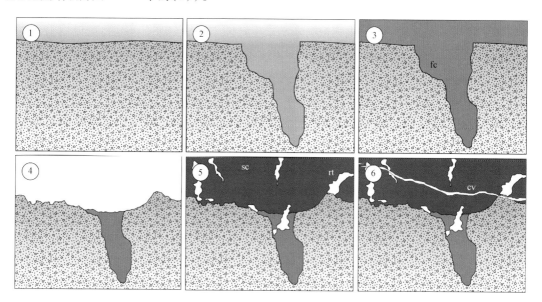

图 8 - 11 Shu'aiba 顶部不连续界面成岩特征形成的简要模式图
(1)和(2)表示海相硬灰岩层和生物穿孔的形成过程。(3)局部海相针铁矿(亮棕色,"fc")封闭的钻孔,也被新的钻孔切割。(4)短暂的地表侵蚀导致地表削平。(5)含有红土的结壳(深棕色,"sc")形成于地表侵蚀阶段;结壳保存了较高陆地植物留下的根迹"rt"以及颗粒周围的裂缝。(6)方解石岩脉:"cv"横切根迹留下的痕迹,以及 FeO 结壳和下伏的 Shu'aiba 灰岩。Shu'aiba 顶部不连续界面快速埋藏于 Nahr Umr 碎屑之下,抑制晚期的硬灰岩层阶段(见图 8 - 12,水深环境和文字讨论)

8.6.2 海洋和大气叠合埋藏特征

除了之前已经描述的复杂的海洋和大气阶段之外,成岩作用过程有可能已经叠合隐藏的或假的海洋或大气特征。Shu'aiba 顶部不连续面的海相硬灰岩层阶段是显著的,在露头中容易识别出来(图 8-3B 和 E)。矿化作用的模式(主要是针铁矿)和形态上的电阻特征变化很小,成为主要的野外识别标志。事实上,界面特征提供了区域尺度的硬灰岩层阶段的证据。这里有岩相学证据,然而,许多(甚至多数)露头带的富铁—锰矿化作用可能来源于晚成岩作用。这方面的证据意味着,看似显著的野外识别标志可能导致错误的结论;这一警告来自 Clari 等(1995),他们认为过高估计铁—锰结壳可能导致对于不连续面的不正确的分类。

可以假设,Shu'aiba 顶面的次生铁和锰氧化物外壳的形成和演化的三个阶段是分开的:(1)一个或几个早阿普弟期海相硬灰岩层阶段,以界面上沾染了薄的铁和锰氧化物为特征,部分封闭了孔洞或是被孔洞穿切(图 8-3E 和 8-6A,图 8-11 中的(3));(2)可能与晚期阿普弟阶出露有关,短期的土壤层形成阶段提供了含水的铁和锰氧化物以及硅结砾岩(图 8-4B,图 8-11 中的(5)),它们来自砖红色土壤和晚阿普弟期海侵形成的 Nahr Umr 碎屑物;(3)一个或几个埋藏或与剥露作用有关的再活化作用和沉淀事件(图 8-5C)。显然,必须要注意,不要混淆较新的,厘米级厚度的矿化作用,这种成矿作用用薄层的或者透明的白垩系渗入作用在不连续界面之上或者之下分别封闭了赛诺曼阶地层以及较新的方解石脉(图 8-5)。并不是在任何地区所有阶段都被保存下来。

另一个可供参考的例子是 Jabal Madar,它包含了后来地下改造作用的多基因界面的复杂演化过程。在这里,Shu'aiba 顶面分米级的溶解凹坑的形成可能与埋藏有关,晚成岩溶解过程(深成的喀斯特构造,Immenhauser 等,2007;假的大气表层喀斯特构造"香蕉洞";图 8-5A)。这些溶解凹坑最初沿着与抬升有关的裂缝系统发育,并且尤其频繁出现在交叉点,这里不同的连接点相互交叉(Immenhauser 和 Rameil,2011)。证明这些特征是由于埋藏形成的证据是在 Jabal Madar 具有溶解—加宽的特征,Immenhauser 等人做了详细的描述(2007)。

8.6.3 Shu'aiba 顶部不连续界面区域演化特征

碳酸盐台地顶部几百万年沉积间断的积累,包括地表暴露的证据,需要一个(系列)重要的海平面下降事件,目的是解释如此长的时间里无沉积或侵蚀。事实上,不能排除 Shu'aiba 台地经历了一系列短暂的晚阿普弟期浅水洪泛事件的可能性。然而,在附近的 Bab 盆地地震测线(图 8-12B)和岩石记录中缺少存在实质性海侵的证据。实际上,阿曼和阿拉伯联合酋长国 Bab 盆地边缘的地震剖面(Droste,2010)似乎表明晚阿普弟期期间强制海退和层序逐渐下降的过程(图 8-12)。与此相反,沿着 Bab 盆地南缘远到西北部的油田的地震横剖面(Strohmenger 等,2010),显示出强制海退,但是在 Shu'aiba 层序 4b 和 5 之间的边界(图 8-12B)只有一个明显的下降事件。这种模式可以选择性的被解释为台地重新泛滥期形成的削截顶超。从地震图像本身来看,很难确定具有次一级下降特征的削截顶超或者海退导致观测到的大尺度几何形态。在这两种情况下,侵蚀和削截某种程度上发生在海平面下降时期进积斜坡的顶部。然而,假设标志性的削截在台地顶部发生,测井应预期显示不完整的向上变浅序列。特别是,斜坡顶部浅海相碳酸盐岩会受到剥蚀,从而更深水的沉积物直接沉积于不连续面之下。根据 Yose 等人的研究观点(2006),这不是油田"A"的情况,因此提出在这一地区强制海退(具有可能的次一级下降过程)被看作是更为重要的过程。此外,图 8-12A 中的地震横剖面显示出斜坡顶部 4a 和 4b 附近夷平的斜坡,这也表明进积斜坡最上面的部分被保留了下来。由于 Haushi-Huqf High 构造抬升,研究区(Fahud Salt 盆地,图 8-2B)最西边地区的情况比较复杂

(Immenhauser 等,2004)。沿着阿曼 Bab 盆地边缘(图 8-2B)在 Nahr Umr 底部之下的大面积 AP Apt 4a 和 4b 对应的斜坡地形顶部,以及 Shu'aiba 顶部向东较远的削截之下(图 8-12A 和 B),一些井的数据显示不完整的向上变浅的旋回。因此,削截顶超几何形态很可能出现在这个地区。关键的信息是,从 Bab 盆地边缘,层序 AP Apt 4a 和 4b 在 Shu'aiba 台地顶部无(可能甚至是负的)可容空间阶段,有清晰的退覆/进积特征。然而,层序 AP Apt 4b 和 5 之间的边界,清晰显示了一个显著的相对海平面下降过程。Van Buchem 等把这一海平面下降描述为最小数量值大约 40m,其过渡期超过 2~3Ma,这与 Maurer 等(2010)、Raven 等(2010)的估计和 Al-Husseini 与 Matthews(2010)的模拟海平面曲线相符合。相对海平面的下降导致 Arabian 台地普遍大面积的地表暴露(Sharland 等,2001)。

这些结果大体上与在阿曼北部海洋古边缘基于野外露头工作的层序地层解释相一致(Hillgärtner 等,2003;Grésell 和 Pittet2005)。Grésell 和 Pittet(2005)提出对于晚阿普弟阶,三个斜坡体系的强制进积过程,每一个都代表了一个三级层序(图 8-12C;分别为层序 AP Apt 4a 到 AP Apt 5 和层序Ⅱ到Ⅳ)。估计早/晚阿普弟期边界周围相对海平面第一次下降的幅度变化在 Haushi-Huqf High 的 ca 30m(Immenhauser 和 Matthews,2004),50m(Van Buchem 等,2002)甚至更多,在 Oman 北部的海洋古边缘,可达 100m(Hillgärtner 等,2003)。在阿曼北部,Hillgärtner 等(2003)解释为相对海平面下降是由于区域构造抬升,从而形成 50~80m 的海平面下降。根据 Hillgärtner 等(2003),第二次相对海平面下降的幅度似乎类似于第一次,而第三次相对海平面下降,发生在中—晚阿普弟期(AP Apt 4b/5;图 8-12),显然相比不显著。这一观测值与较早对于 Bab 盆地周围地震数据的讨论相反(Yose 等,2006,2010;Maurer 等,2010;Strohmenger 等,2010),这些讨论提出第三次和最后一次的海平面下降最为明显。对于这一差异的一个可能的解释是阿曼北部古边缘与阿曼内陆盆地之间构造背景的差异(图 12-2)。总之,横穿东南 Arabian 台地,发现了中到晚阿普弟期三个海平面事件的证据。海平面下降的幅度估计达到了几十米的范围,并且导致了长期的 Shu'aiba 碳酸盐台地地表暴露。这一观点在 Shu'aiba 顶部界面演化过程中有重要意义。

8.6.4 全球范围的 Shu'aiba 顶部不连续界面的演化

回顾在阿曼阿尔比阶 Nahr Umr 地层沉积期间相对海平面变化的驱动机理,Immenhauser 和 Matthews(2004)估算了海平面下降速率和可能的对比速率的振幅以及已知的与比如大洋中脊体积的改变、火山活动、局部或区域的构造运动、海洋沉积物沉积体积的变化过程有关的幅度。笔者认为,那些未知的机理或者大陆冰川体积的变化(冰川性海面升降)可以解释为地区性几十米尺度的迅速的海平面下降。

白垩纪中期可能存在的大陆冰层至关重要,因为这个时期一般是指全球极端温暖的时期,是代表地质记录中全球"温室效应"环境最好的实例(Hay,2011)。然而,实际的白垩纪温室环境的稳定性问题,成为当下争论的主题,长期以来认为,白垩纪气候温和稳定越来越受到质疑(Kemper,1987;Frakes 和 Francis,1988;Weissert 和 Lini,1991;Francis 和 Frakes,1993;Sellwood 等,1994;Pirrie 等,1995;Stoll 和 Schrag,1996;Immenhauser,2005)。关于在巴瑞姆亚期—阿尔比期中白垩世温室环境发生较短暂变冷事件(寒流)的证据,以至少在高纬度地区存在季节性亚冷冻环境为特征,来自冰坠石(冰携碎屑)和钙芒硝状方解石(六水方解石之后的假晶,一个在零摄氏度以上的 $CaCO_3$ 不稳定阶段;Kemper,1987;Frakes 和 Francis,1990),主要来自两半球极点附近地区以及动物记录。汇总全球冰坠石和钙芒硝状方解石的产状(Heimhofer 等,2008;图 8-13A)表明,常年或季节性大陆冰层可能出现在晚阿普弟期到早阿尔比期。

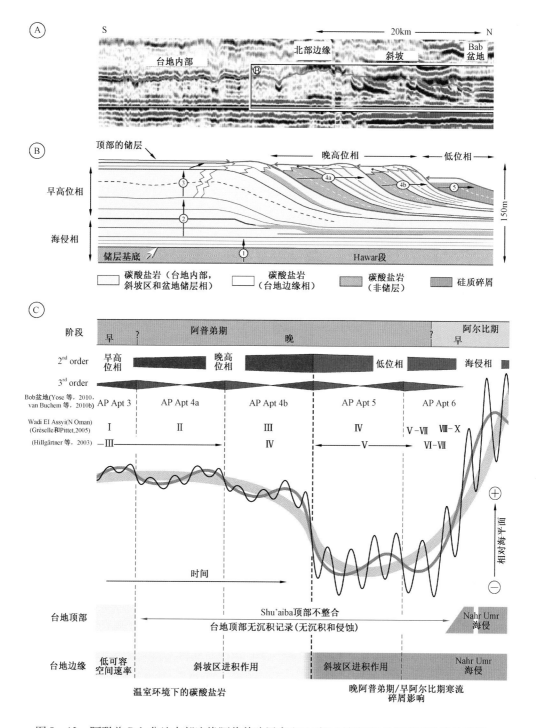

图 8-12 阿联酋 Bab 盆地南部边缘阿普弟阶层序和地震地层概况以及海平面曲线示意图
(A)二维垂直切片上的地震几何形态(振幅数据,沿储层基底拉平),据 Yose 等(2006)做修改。(B)层序地层格架和储层特征。储层被划分为四个主要沉积相,对应长期(二级)海平面旋回,六个沉积层序(Ap1 至 Ap6),对应三级海平面变化,据 Yose 等(2010)做修改。(C)相对海平面曲线示意图,包括对比阿拉伯东南部的不同层序地层图件以及全球/区域气候变化的关系。台地顶部以及台地边缘的沉积和无沉积阶段(见讨论内容)

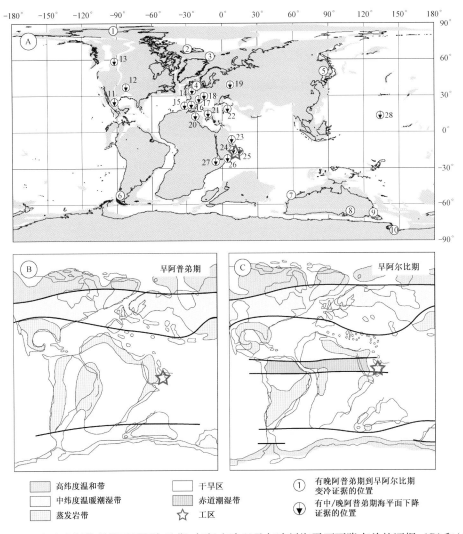

图 8-13 (A)晚阿普弟期到早阿尔比期,气候变冷以及与冰川海平面下降有关的证据,(B)和(C)分别为早阿普弟期和早阿尔比期全球古气候区的重建(据 Chumakov 等 1995 年研究,经修改)。工区用红色五角星表示(见文字讨论)

有关冰点(玄能岩,坠石)的实物证据的位置如下:(1)阿拉斯加东北部(Dettermann 等,1975);(2)Sverdrup 盆地,加拿大北极圈(Kemper,1987);(3)Svalbard(Pickton,1981);(4)北海盆地中部(Ruckheim 等,2006a,2006b);(5)Primorye 南部,俄罗斯(Krassilov,1973);(6)卡拉法特旱谷,巴塔哥尼亚(pirrie 等,2004);(7)埃克斯茅斯高原(Clarke 和 Jenkyns,1999);(8)伊罗曼加盆地,澳大利亚中部(Frakes 和 Francis,1988;De Lurio 和 Frakes,1999);(9)吉普斯兰盆地,澳大利亚东部(Constantine 等,1998);(10)James Ross 岛屿,南极洲(Ditchfield 等,1994)。大约在早/晚阿普弟期边界或在晚阿普弟期明显的海平面下降(地表暴露,从海相到陆地沉积的转变,无沉积,侵蚀)的位置用向下的箭头表示;(11)和(12)圣罗莎峡谷,墨西哥和得克萨斯州中北部(Scott 等,2000);(13)亚伯达南部,加拿大(Ardies 等,2002;Zaitlin 等,2002);(14)英国(Ruffell 和 Wach1998b);(15)卢西坦统盆地,葡萄牙西部(Burla 等,2008,2009);(16)西班牙东北部(Peropadre 等,2008;Bover Arnal 等,2009);(17)Vocontian 盆地,S France(Friedrich 等,2003);(18)赫尔维蒂推覆体,瑞士(Brisibeds,诺兰菊石区;Follmi 和 Gainon,2008);(19)俄罗斯台地(Sahagian 等,1996);(20)突尼斯北部(Arnaud-Vanneau 等,2005;Echihaoui,2005;Chaabani 和 Razgallah,2006);(21)意大利(Csoma 等,2004);(22)土耳其(Yilmaz 和 Altiner,2006);(23)伊朗(Van Buchem 等,2010c);(24)阿拉伯联合酋长国(Yose 等,2006,2010;Maurer 等,2010;Raven 等,2010);(25)阿曼山脉(Hillgärtner 等,2003);(26)也门 Socotran 台地和 Hadhdramaut 地区(Morrison 等,1997;Schroeder 等,2010);(27)埃塞俄比亚(Bosellini 等,1999);(28)太平洋平顶海山(Rohl 和 Ogg,1996)。编辑内容基于 Heimhofer 等(2008)研究。环阿普弟阶/阿尔比阶边界古地理的重建来自 IFM-GEOMAR 地图编制者(http://www.odsn.de),并修改

同样,同位素记录、古植物学和超微化石数据揭示了这一时期在北半球和南半球的一个变冷事件,(晚阿普弟期/早阿尔比期寒流;Hochuli 等,1999;Price,2003;Puceat 等,2003;Pirrie 等,2004;Steuber 等,2005;Mutterlose 等,2009)。这种变冷的趋势,特别是在南半球,也反映在 Chumakov 等(1995)的古气候地图中,地图中设置了一个沿南极海岸的温带湿润气候带(图 8-13B 和 C)。

对于研究区,阿尔比期赤道潮湿的气候带的形成与上述变冷趋势相关,因为它揭示了 Arabian 台地从更干旱的早阿普弟期到阿尔比期完全潮湿的环境的变化(图 8-13B 和 C)。然而,没有来自阿曼台地阿普弟阶蒸发岩的报告,所以完全干旱环境出现在气候变化之前似乎是不可能的。向更潮湿环境的变化也触发了风化模式的变化,这体现在岩性地层中为从下到中阿普弟阶碳酸盐台地体系(下 Shu'aiba 台地,包括进积碳酸盐斜坡,层序 AP Apt 1 到 4,图 8-12)到阿普弟晚期和阿尔比早期有更多的硅质碎屑的注入的转变过程(层序 AP Apt 5 和 Nahr Umr 组,图 8-12,参见 Droste,2010;Yose 等,2010)。

假设导致下 Shu'aiba 平台顶部地面暴露时间延长的重复性的、高幅度晚阿普弟期海平面下降源于至少部分冰川性海面升降,那么在其他地区也应该存在这些事件的记录。实际上,晚阿普弟期主要的海平面下降证据遍及全球(图 8-13A)。Peropadre 等(2008)描述,在西班牙的东北部中阿普弟阶下切深度从 10~73m 变化的下切谷体系的多相演化特征(从 *furcata* 到 *subnodosocostatum* 菊石带)。根据 Bover-Arnal 等(2009)对相同露头的另一种解释认为,海平面下降事实上被位于坡脚的强制海退楔体系域记录下来,这与在阿曼发现的强制海退特征非常相似。Bover-Arnal 等(2009)估计海平面下降导致强制海退楔发育达至少 60m 的幅度。在突尼斯,被铁氧化物和磷酸盐结壳、被钻孔以及局部被喀斯特化的,与 Shu'aiba 顶部不连续在外貌上很相似的不连续界面,它们被发现于靠近并且在阿普弟阶 Serj 组顶部的碳酸盐台地沉积物中(Arnaud-Vanneau 等,2005)。与阿曼类似,突尼斯界面显示了非常复杂的成岩史的证据(Echihaoui,2005),并且已被解释为重复性的相对海平面明显下降的结果(Arnaud-Vanneau 等,2005)。Grèselle 和 Pittet(2005)以及 Immenhauser(2005)提出了对选择的海平面数据的详细对比,但结果并不令人信服;这可能是由于海平面升降复杂的相互作用、基底的抬升和下沉、空间效应、大地水准面的变形以及其他因素的影响,主要的全球海平面事件往往在全球范围不同步的(Immenhauser,2005)。此外,众所周知,白垩纪浅水区时间分辨率差及与生物地层标志的地区性特征相关的问题,以及不同时间尺度的应用,这些因素使得在许多情况下从不同古地理环境获得的海平面变化事件对比变得复杂(Hardenbol 等,1998;Immenhauser 和 Scott,1999)。

8.6.5 缺少深切的大气喀斯特特征

一个(系列)造成 Shu'aiba 台地长期的地表暴露的几十米幅度的海平面下降(图 8-12),出现在白垩纪中期热带地区(图 8-13C),与之形成鲜明对比的是在这些环境下缺少许多长期暴露界面的深切的喀斯特特征(Wright,1988;Saller 等,1999)。以 Shu'aiba 顶部不整合为例,在局部地区微弱的表层岩溶作用穿透几厘米进入石灰岩(图 8-4E),但较大规模的喀斯特界面特征在露头及地震上的证据是孤立的,不太令人信服或存在争议。Raven 等(2010)描述在 Shu'aiba 顶部不连续面之下存在几十米宽、长达 25m 的硅质碎屑填充的裂缝,这是通过对研究区北部 Qatar 地区电缆测井解释得到的。与深切谷体系的描述出自同一作者,这些推测的岩溶特征是地表暴露的大型构造目前唯一有说服力的证明。

有人提出,低起伏的 Arabian 台地可能会通过限制远离界面的溶解的 $CaCO_3$ 的运输率,来减缓碳酸盐的溶解作用的潜力(Immenhauser 等,1999)。Grelaud 等(2006)讨论了一个可比较的例子,位于阿曼赛诺曼阶 Natih 组,同样缺乏地表暴露界面上的主要的喀斯特特征,并得出类似的结论:喀斯特的发育可能由于缺乏来自土壤的有机酸、受限的水力压头以及极平缓坡度的沉积剖面而受到抑制(范围在 20~30m)。

另外一种说明为什么界面本身仅显示了土壤发育和表层岩溶的微弱证据,可能是在下 Shu'aiba 台地从最初的阿普弟期最晚期到阿尔比期的再次洪泛期间被波浪侵蚀的结果(图 8-10)。然而,可以支持这种假设的侵蚀碎屑在所有研究的岩心和露头中明显缺乏。并且,缺乏由于生物侵蚀作用明显去除石灰岩序列的证据。后者的观测值也许并不奇怪,因为浪成阶地环境不适宜于固着和生物侵蚀的生物体的生存(石蛏属双壳类、棘皮类动物、有孔海绵状物等)。利用现代的浪成阶地作为类比,大量的岩溶化的石灰岩,当在岩石基底之上被波浪和水流搬运时,结合了机械波浪作用而具有侵蚀性(Trenhaile,2002),同时石灰岩自身也受侵蚀。阿曼 Shu'aiba 顶部不连续的低起伏地形变得明显,这里露头充分暴露,其延伸距离达数千米,Huqf-Haushi high 露头带就是这种情况(图 8-3B)。关于波浪侵蚀的低起伏地形的一个现代实例,来自阿根廷 Valdes 半岛,虽然规模有很大差异,这个实例显示在图 8-14 中,可以供大家参考。这一现代实例的形态特征与阿曼 Shu'aiba 顶部不连续很相似(图 8-14B 和 C)。

然而,必须强调的是晚阿普弟期的气候,虽然确定的是热带且太潮湿的气候不能形成蒸发岩的沉积,但极其广阔的、平坦的、定期出露和下切的下 Shu'aiba 碳酸盐台地,在现代仍然没有真正的类比物。与许多现代波浪侵蚀阶地围绕着一个地形抬升的内陆分布(图 8-13A)相反的是,下 Shu'aiba 台地在西部被 Bab 盆地围绕,往东是印度洋,往北是 Hawasina 盆地(图 8-2B)。这一发现可能意味着波浪侵蚀从西部、北部和东部向岩溶化的下 Shu'aiba 台地顶部推进,形成一个广阔的浅海水下硬灰岩层准平原。鉴于岩块、砾石和砂级碎屑的搬运需要一个最小水动力,硬灰岩层准平原以上的水深必须使得海底起码在有效的浪基面以内。考虑到阿曼众多露头中相对稀缺的钻孔,与现代实例相似,这个区域性的硬灰岩层环境不适宜固着的、钻孔生物生存。

考虑到波浪侵蚀的准平原假说,一系列的问题仍然存在。(1)许多现代碳酸盐海岸其特征是浪成喀斯特(海岸灰岩沟,图 8-14D)以及海水—淡水交界面的岩溶作用(Mylroie 和 Carew,1995)。为什么在这里提到的阿普弟阶实例没有类似特征的证据?类似的,在许多抬升的海相阶地上缺乏滩岩沉积意味着什么(Ramkumar 等,2000)?(2)现代波浪侵蚀海相阶地往往以向陆地一侧向上坡度变陡的斜坡为界(图 8-14A),该斜坡受到波浪侵蚀作用,并且提供岩块转变为砂级碎屑,结合波浪和水流,从而具有侵蚀性。如果在阿曼晚阿普弟期中存在一个逐渐被侵蚀的低起伏海岸,那么就没有保存在岩石记录中的证据了。(3)由于在浅海、极广阔的水下低起伏台地的顶部受到浪基面—海洋基底相互作用而产生的波浪递降问题一定不能被忽视。在波浪衰减区,有效浪基面横切向陆地方向变浅的基底剖面,波浪的能量逐渐被摩擦过程和基底沉积物的变形吸收(参见 Immenhauser,2009;以供讨论)。产生了这样一个问题,关于如何使得在区域上广阔的水下准平原侵蚀和搬运碳酸盐碎屑所需的水动力条件在逐渐远离周边盆地的过程中保持不变。潮汐起到了怎样的作用?列举这些问题证明了直接比较 Shu'aiba 顶部不连续、基于较小的热带喀斯特岛屿的喀斯特水文学模型以及现代受冰川作用影响的碳酸盐台地(Back 等,1984;Mylroie 和 Carew,1995)是最为关键的。

然而,类比区域上广阔的、波浪侵蚀浅海阶地,作为一个尝试性的研究设想,似乎合情合理

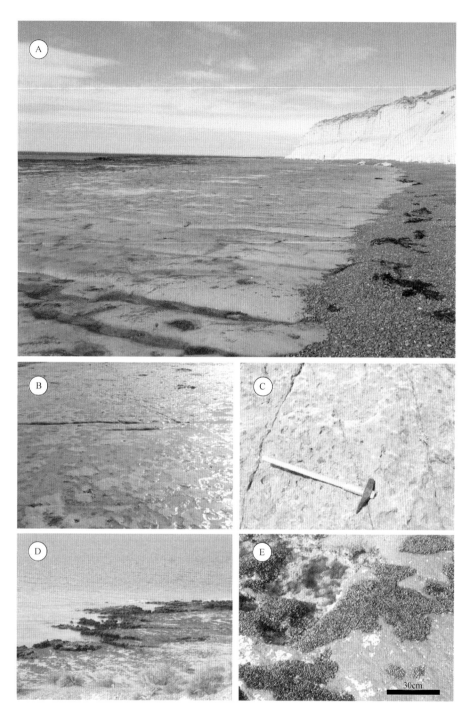

图 8-14 现代海洋阶地潮间带(Valdes 半岛,阿根廷)可作为 Shu'aiba 顶部不连续的类比物
(A)位于低潮线的阿根廷海洋阶地。注意右侧的砾石。悬崖高度 18m。(B)阿根廷硬灰岩层面的特写照片。注意海相固着和石内生物有机体。图片的横宽为 1.5m。注意与(C)地形的相似。下阿普弟阶 Shu'aiba 顶部不连续的特写照片(Wadi Bani Kharus,阿曼,图 8-2B)。比例尺地质锤长度为 50cm。(D)在潮间—潮下边界处海岸喀斯特岩溶沟特征(阿根廷)。前景灌木丛的高度约 1m。(E)阿根廷硬灰岩层界面局部被向海方向的潮间—潮下边界处的小型固着双壳类生物(*Mytilus* sp.)占据

的是，保存在 Shu'aiba 顶部不连续的大气特征只代表了最后的短期暴露以及台地顶部硬灰岩层准平原在最晚期的阿普弟期/阿尔比期再次洪泛事件（图 8-12C）。图 8-11 中对于野外实例的讨论意味着，第（1）阶段代表了经波浪磨蚀的、再次洪泛之后直接削平的碳酸盐岩表面，实际上仍然缺乏更早历史的任何地表记录。低幅度、高频率的海平面变化造成了重复的、短暂的洪泛事件（图 8-12C）。在洪泛事件之间的一个短期间隔与在 Shu'aiba 顶部不连续界面上形成于硬灰岩层阶段的低密度的钻孔具有很好的相符性。通常，低密度的钻孔被认为指示了一个短期的沉积间断阶段（Goldring 和 Kazmierczak，1974）。间断的洪泛事件很可能是被在低幅度、高频率的海平面下降期间发育的短暂的表层喀斯特间隔所中断，所以地表暴露在形态上的和岩相上的大量证据被下一次海侵事件侵蚀掉。

总之，在 Shu'aiba 顶部不连续面上，通常易观察的、暴露良好、空间极其广泛（数平方千米）的露头，至少在形态上与波浪磨损的海岸准平原相似（图 8-14A 到 C）。在研究的露头和岩心中，Shu'aiba 顶部不连续缺乏深切岩溶特征，同时表层岩溶和大气成岩作用的其他特征被保存下来。阿曼以及 Arabian 半岛的其他地区地层间隔的工业地下数据（地震图、电缆测井）延伸了观测窗口，并且指出在阿拉伯联合酋长国 Shu'aiba 顶部不连续面里存在局部大规模的喀斯特特征。然而，缺乏（或无？）保存下来的 Shu'aiba 顶部界面中代表数百万年的沉积间断的深切喀斯特特征，仍然是一个具有挑战性和了解甚少的问题。

8.6.6 深层意义

目前的研究被认为是一个有用的典型范例，用来调查其他中生代岩石记录中的同样具有代表性的区域不连续。在本质上，同时揭示了所涉及到的复杂过程，它强调了作为过去环境变化的存档信息，深入分析不连续面的巨大潜力。然而，获取这些存档信息，需要严谨的和多学科的方法。当处理一个综合露头和钻井岩心数据的时候，考虑使用不同尺度的观测窗口是很重要的。由此产生的观测偏差包括岩心和露头之间的几个数量级（厘米与米，甚至千米），从而可能在考虑明显的构造存在（没有）或者其横向变化的时候容易导致误判。一个显而易见的结果是，基于岩心的研究应该补充露头类比的野外工作。然而，最重要的方法也许是在进一步观察基础上，乐意丢弃明显的野外证据，这样可以揭示比最初预期更复杂的情况。

8.7 结论

Shu'aiba 顶部台地不连续界面代表了一个区域重要的、数百万年的沉积间断（达到 10Ma 时间），具有复杂的地质历史。来自三个露头带和地下岩心资料的区域地层、沉积学、层序地层学、岩相学、古生物学和地球化学的证据显示：（1）证明海相硬灰岩层阶段；（2）证明地表暴露阶段，也许接下来是重复的、短暂的海相缺失；（3）不确定的来源且特征模糊；（4）后期成岩作用特征。

根据野外观察结果，Shu'aiba 顶部不连续主要以海相硬灰岩层为特征。这些特征包括铁、锰氧化物的矿化作用、食石的双壳类和海绵钻孔。进一步详细的工作揭示了地表暴露面的一个不引人注目的特点，如碳、氧同位素在界面以下的偏移、大气胶结、颗粒周围的裂缝或表层岩溶。总的来说，界面形态呈现出明显的低起伏，类似于一个波浪磨蚀的水下准平原。在地震图像上，特别是 Qatar 地下，即阿曼研究区西北部地区，Shu'aiba 顶部不连续下部发育几十米宽的硅质碎屑充填的裂缝长达 25m。在阿曼南部，与 Shu'aiba 顶部不连续对等的界面出现在古生代岩石上面，这证明了存在区域隆起。

来自 Bab 盆地南缘的地震剖面证明了从早/晚阿普弟期边界开始在下 Shu'aiba 碳酸盐台地发生强制海退和潜在的层序下降。Yose 等(2010)定义的层序 AP Apt 4a 和 4b 在 Shu'aiba 台地上一个没有(或者可能甚至是负的)可容空间时期从 Bab 盆地边缘退覆/进积,这导致了地层区域不连续的形成。层序 AP Apt 4b 和层序 AP Apt 5 之间的界限代表了一次标志性的相对海平面下降,估计下降幅度达到几十米。这一事件的驱动机理和在阿普弟晚期海平面下降中的类似实例可能包括冰川性海平面升降和区域构造作用(晚阿普弟期/早阿尔比期寒流)。事实上,来自其他地区,如西班牙和突尼斯,也有对类似的海平面下降的描述。

明显缺乏深切的这种预想出现在长期陆地条件的热带气候下的喀斯特特征,这是该界面主要的判断标志。我们承认现代世界可能不会对这种区域性低起伏的暴露面提供真正的类比物,不连续界面与现代波浪切割的海相阶地有很多共性。这种不连续界面可以或者不可以说明现在暴露的 Shu'aiba 顶部不连续不是数百万年地表暴露阶段的产物,而只不过是最后未知数量的高频率、低幅度的相对海平面变化,每次海侵覆盖了地表暴露的碳酸盐准平原这样一个过程的记录。伴随每次海侵事件,大部分以先前大陆喀斯特阶段为特征的大气成岩作用记录,在台地被完全海侵并埋藏在 Nahr Umr 碎屑物中之前,就已经被去除了。因此,核心意思是,在主要的不连续界面演化过程中,与最后(或最晚期)阶段有关的特征,通常在野外占主体,地表观察导致了对它们地质演化并不完整的认识。

参 考 文 献

Al-Awar, A. A. and Humphrey, J. D. (2000) Diagenesis of the Aptian Shuaiba formation at Ghaba North Field, Oman. In: Middle East Models of Jurassic/Cretaceous Carbonate Systems(Eds A. S. Alsharhan and R. W. Scott), SEPM Spec. Publ., 69, pp. 173 – 184.

Al-Husseini, M. I. and Matthews, R. K. (2010) Tuning Late Barremian-Aptian Arabian Plate and global sequences with orbital periods. In: Barremian-Aptian Stratigraphy and Petroleum Habitat of the Eastern Arabian Plate(Eds F. S. P. van Buchem, M. I. Al-Husseini, F. Maurer and H. J. Droste), GeoArabia Spec. Publ., 4, Vol. 1, pp. 199 – 228.

Allan, J. R. and Matthews, R. K. (1982) Isotope signatures associated with early meteoric diagenesis. Sedimentology, 29, 797 – 817.

Alsharhan, A. S. and Nairn, A. E. M. (1997) Sedimentary Basins and Petroleum Geology of the Middle East. Elsevier, Amsterdam, 843 pp.

Ardies, G. W., Dalrymple, R. W. and Zaitlin, B. A. (2002) Controls on the geometry of incised valleys in the Basal Quartz Unit(Lower Cretaceous), western Canada sedimentary basin. J. Sed. Res., 72, 602 – 618.

Arnaud-Vanneau, A., Zghal, I. and Echihaoui, A. (2005) The three last Upper Aptian cycles of the Serj formation. In: Aptian-Turonian Events in Central Tunisia(Eds A. Arnaud-Vanneau and I. Zghal), Géol. Alp., S. Spéc. Coll. Exc., 5, pp. 82 – 85.

Back, W., Hanshaw, B. B. and van Driel, J. N. (1984) Role of groundwater in shaping the Eastern Coastline of the Yucatan Peninsula, Mexico. In: Groundwater as a Geomorphic Agent(Ed. R. G. LaFleur), pp. 280 – 293. Allen & Unwin, London.

Borgomano, J., Masse, J.-P. and Al Maskiry, S. (2002) The Lower Aptian Shuaiba carbonate outcrops in Jebel Akhdar, northern Oman: Impact on static modelling for Shuaiba petroleum reservoirs. AAPG Bull., 86, 1513 – 1529.

Bosellini, A., Russo, A. and Schroeder, R. (1999) Stratigraphic evidence for an Early Aptian sea-level fluctuation: the Graua Limestone of south-eastern Ethiopia. Cretaceous Res., 20, 783 – 791.

Bover-Arnal, T., Salas, R., Moreno-Bedmar, J. A. and Bitzer, K. (2009) Sequence stratigraphy and architecture of a late Early-Middle Aptian carbonate platform succession from the western Maestrat Basin(Iberian Chain, Spain).

Sed. Geol. ,219,280 – 301.

Brett,C. E. and Brookfield, M. E. (1984) Morphology, faunas and genesis of Ordovician hardgrounds from southern Ontario,Canada. Palaeogeogr. Palaeoclimatol. Palaeoecol. ,46,233 – 290.

Bromley,R. G. and Heinberg, C. (2006) Attachment strategies of organisms on hard substrates: a palaeontological view. Palaeogeogr. Palaeoclimatol. Palaeoecol. 232,429 – 453.

Bruckschen,P. and Richter,D. K. (1994) Zementstratigraphische Grundmuster in marinen Karbonatablagerungen des Phanerozoikums-ein Abbild der normalen Beckenentwicklung. Zbl. Geol. Palä ontol. Teil I,1993,959 – 972.

van Buchem,F. S. P. , Pittet, B. , Hillgärtner, H. , Grötsch, J. , AlMansouri, A. I. , Billing, I. M. , Droste, H. J. , Oterdoom,W. H. and van Steenwinkel,M. (2002) High-resolution sequence stratigraphic architecture of Barremian-Aptian carbonate systems in northern Oman and the United Arab Emirates(Kharaib and Shu'aiba formations) GeoArabia,7,461 – 500.

van Buchem,F. S. P. , Al-Husseini,M. I. , Maurer, F. and Droste, H. J. (Eds) (2010a) Barremian-Aptian Stratigraphy and Petroleum Habitat of the Eastern Arabian Plate. GeoArabia Spec. Publ. ,4,Manama,Bahrain,614 pp.

van Buchem,F. S. P. , Al-Husseini, M. I. , Maurer, F. , Droste, H. J. and Yose, L. A. (2010b) Sequence-stratigraphic synthesis of the Barremian-Aptian of the Eastern Arabian Plate and implications for the petroleum habitat. In: Barremian-Aptian Stratigraphy and Petroleum Habitat of the Eastern Arabian Plate(Eds F. S. P. van Buchem,M. I. Al-Husseini,F. Maurer and H. J. Droste),GeoArabia Spec. Publ. ,4,Vol. 1,pp. 9 – 48.

van Buchem,F. S. P. , Baghbani,D. , Bulot,L. G. , Caron, M. , Gaumet, F. , Hosseini, A. , Keyvani, F. , Schroeder, R. , Swennen, R. , Vedrenne, V. and Vincent, B. (2010c). Barremian-Lower Albian sequence stratigraphy of southwest Iran(Gadvan,Dariyan and Kazhdumi formations) and its comparison with Oman,Qatar and the United Arab Emirates. In: Barremian-Aptian Stratigraphy and Petroleum Habitat of the Eastern Arabian Plate (Eds F. S. P. van Buchem,M. I. Al-Husseini,F. Maurer and H. J. Droste),GeoArabia Spec. Publ. ,4,Vol. 2,pp. 503 – 548.

Budd,D. A. ,Saller,A. H. and Harris,P. M. (eds.) (1995) Unconformities and Porosity in Carbonate Strata: AAPG Memoir,63,Tulsa,313 pp.

Burla,S. ,Heimhofer, U. , Hochuli, P. A. , Weissert, H. and Skelton, P. (2008) Changes in sedimentary patterns of coastal and deep sea successions from the North Atlantic (Portugal) linked to Early Cretaceous environmental change. Palaeogeogr. Palaeoclimatol. Palaeoecol. 257,38 – 57.

Burla,S. , Oberli, F. , Heimhofer, U. , Wiechert, U. and Weissert, H. (2009) Improved time control of Cretaceous coastal deposits: New results from Sr isotope measurements using laser ablation. Terra Nova,21,401 – 409.

Carmona,N. B. , Mángano,M. G. , Buatois, L. A. and Ponce,J. J. (2007) Bivalve trace fossils in an early Miocene discontinuitiy surface in Patagonia, Argentina: burrowing behaviour and implications for ichnotaxonomy at the firmground-hardground divide. Palaeogeogr. Palaeoclimatol. Palaeoecol. ,225,329 – 341.

Carson,G. A. and Crowley,S. F. (1991) The glauconite-phosphate association in hardgrounds: examples from the Cenomanian of Devon,southwest England. Cretaceous Res. ,14,69 – 89.

Chaabani,F. and Razgallah,S. (2006) Aptian sedimentation:an example of interaction between tectonics and eustatics in Central Tunisia. In: Tectonics of the Western Mediterranean and North Africa (Eds G. Moratti and A. Chalouan) ,Geol. Soc. Lond. Spec. Publ. ,262,55 – 74.

Chumakov,N. M. ,Zharkov, M. A. , Herman, A. B. , Doludenko, M. P. , Kalandadze, N. N. , Lebedev, E. L. , Ponomarenko, A. G. and Rautian, A. S. (1995) Climatic belts of the mid-Cretaceous time. Stratigr. Geol. Corr. , 3, 241 – 260.

Clari,P. A. ,Dela Pierre,F. and Martire,L. (1995) Discontinuities in carbonate successions: identification,interpretation and classification of some Italian examples. Sed. Geol. ,100,97 – 121.

Clarke,L. J. and Jenkyns,H. C. (1999) New oxygen isotope evidence for long-term Cretaceous climate change in the Southern Hemisphere. Geology,27,699 – 702.

Constantine,A. , Chinsamy, A. , Vickers-Rich, P. and Rich, T. H. (1998) Periglacial environments and polar dinosaurs. S. Afr. J. Sci. ,94,137 – 141.

Csoma, A. E., Goldstein, R. H., Mindszenty, A. and Simone, L. (2004) Diagenetic salinity cycles and sea level along a major unconformity, Monte Camposauro, Italy. J. Sed. Res., 74, 889–903.

De Lurio, J. L. and Frakes, L. A. (1999) Glendonites as a paleoenvironmental tool: Implications for early Cretaceous high latitude climates in Australia. Geochim. Cosmochim. Acta, 63, 1039–1048.

Dettermann, R. L., Reiser, H. N., Brosgé, W. P. and Dutro, J. T. (1975) Post-Carboniferous stratigraphy, northeastern Alaska. US Geol. Surv. Prof. Pap., 886, 1–46.

Di Stefano, P. and Mindszenty, A. (2000) Fe-Mn-encrusted "Kamenzita" and associated features in the Jurassic of Monte Kumeta(Sicily): subaerial and/or submarine dissolution? Sed. Geol., 132, 37–68.

Dickson, J. A. D. (1965) A modified staining technique for carbonates in thin section. Nature, 205, 587.

Ditchfield, P. W., Marshall, J. D. and Pirrie, D. (1994) High latitude palaeotemperature variation: new data from the Tithonian to Eocene of James Ross island, Antarctica. Palaeogeogr. Palaeoclimatol. Palaeoecol., 107, 79–101.

Droste, H. J. (2010) High-resolution seismic stratigraphy of the Shu'aiba and Natih Formations in Oman: implications for Cretaceous epeiric carbonate platform systems. In: Mesozoic and Cenozoic Carbonate Systems of the Mediterranean and the Middle East: stratigraphic and diagenetic reference models (Eds F. S. P. van Buchem, K. Gerdes and M. M. Esteban), Geol. Soc. Spec. Publ., 329, pp. 143–160.

Droste, H. J. and Van Steenwinkel, M. (2004) Stratal geometries and patterns of platform carbonates: the Cretaceous of Oman. In: Seismic Imaging of Carbonate Reservoirs and Systems (Eds G. P. Eberli, J. L. Massaferro and J. F. R. Sarg), AAPG Mem., 81, pp. 108–206.

Echihaoui, A. (2005) The demise of carbonate platform at Jebel El Hamra. In: Aptian-Turonian Events in Central Tunisia (Eds A. Arnaud-Vanneau and I. Zghal), Géol. Alp., S. Spéc. Coll. Exc., 5, pp. 76–77.

Flügel, E. (2004) Microfacies of Carbonate Rocks: Analysis, Interpretation and Application. Springer, Berlin, Heidelberg, New York, 976 pp.

Föllmi, K. B. and Gainon, F. (2008) Demise of the northern Tethyan Urgonian carbonate platform and subsequent transition towards pelagic conditions: The sedimentary record of the Col de la Plaine Morte area, central Switzerland. Sed. Geol., 205, 142–159.

Forbes, G. A., Jansen, H. S. M. and Schreuers, J. (2010) Lexicon of Oman Subsurface Stratigraphy. GeoArabia Spec. Publ., 5. Gulf PetroLink, 371 pp.

Frakes, L. A. and Francis, J. E. (1988) A guide to Phanerozoic cold polar climates from high-latitude ice-rafting in the Cretaceous. Nature, 333, 547–549.

Frakes, L. A. and Francis, J. E. (1990) Cretaceous palaeoclimates. In: Cretaceous Resources, Events, and Rhythms (Eds R. N. Ginsburg and B. Beaudoin), pp. 287–373. Kluwer Academic Publishers, Dordrecht, The Netherlands.

Francis, J. E. and Frakes, L. A. (1993) Cretaceous climates. Sedimentol. Rev., 1, 17–30.

Friedrich, O., Reichelt, K., Herrle, J. O., Lehmann, J., Pross, J. and Hemleben, C. (2003) Formation of the Late Aptian Niveau Fallot black shales in the Vocontian Basin (SE France): evidence from foraminifera, palynomorphs, and stable isotopes. Mar. Micropaleontol., 49, 65–85.

Fürsich, F. T. (1979) Genesis, environments and ecology of Jurassic hardgrounds. Neues Jb. Geol. Paläontol. Abh., 158, 1–63.

Godet, A., Föllmi, K., Stille, P., Bodin, S., Matera, V. and Adatte, T. (2011) Reconciling strontium-isotope and K-Arages with biostratigraphy: the case of the Urgonian platform, Early Cretaceous of the Jura Mountains, Western Switzerland. Swiss J. Geosci., 104, 147–160.

Goldring, R. and Kazmierczak, J. (1974) Ecological succession in intraformational hardground formation. Paleontology, 17, 949–962.

Gradstein, F. M., Agterberg, F. P., Ogg, J. G., Hardenbol, J., Veen, P. V., Thierry, J. and Huang, Z. (1995) A Triassic, Jurassic and Cretaceous time scale. SEPM Spec. Publ., 54, 95–126.

Gradstein, F., Ogg, J. and Smith, A. (Eds) (2004) A geologic time scale. Cambridge University Press, Cambridge, 589 pp.

Grélaud, C., Razin, P., Homewood, P. W. and Schwab, A. M. (2006) Developement of incisions on a periodically emergent carbonate platform(Natih Formation, Late Cretaceous, Oman). J. Sed. Res., 76, 647 – 669.

Gréselle, B. and Pittet, B. (2005) Fringing carbonate platforms at the Arabian Plate margin in northern Oman during the Late Aptian-Middle Albian: evidence for high-amplitude sea-level changes. Sed. Geol., 175, 367 – 390.

Hanna, S. S. (1990) The Alpine deformation of the central Oman Mountains. In: The Geology and Tectonics of the Oman Region(Eds A. H. F. Robertson, M. P. Searle and A. C. Ries), Geol. Soc. Spec. Publ., 49, pp. 341 – 359.

Hardenbol, J., Thierry, J., Farley, M. B., Jacquin, T., DeGraciansky, P. -C. and Vail, P. R. (1998) Mesozoic-Cenozoic sequence chronostratigraphic framework. In: Sequence Stratigraphy of European Basins(Eds P. -C. DeGraciansky, J. Hardenbol, T. Jacquin, P. R. Vail and M. B. Farley), SEPM Spec. Publ., 60, pp. 3 – 13.

Hay, W. W. (2011) Can humans force a return to a "Cretaceous" climate? Sed. Geol., 235, 5 – 26.

Heimhofer, U., Adatte, T., Hochuli, P. A., Burla, S. and Weissert, H. (2008) Coastal sediments from the Algarve: low-latitude climate archive for the Aptian-Albian. Int. J. Earth Sci. (Geol. Rundsch.), 97, 785 – 797.

Hillgärtner, H. (1998) Discontinuity surfaces on a shallowmarine carbonate platform(Berriasian, Valanginian, France and Switzerland). J. Sed. Res., 68, 1093 – 1108.

Hillgärtner, H., van Buchem, F. S. P., Gaumet, F., Razin, P., Pittet, B., Grötsch, J. and Droste, H. (2003) The Barremian-Aptian evolution of the Eastern Arabian carbonate platform margin(Northern Oman). J. Sed. Res., 73, 756 – 773.

Hochuli, P. A., Menegatti, A. P., Weissert, H., Riva, A., Erba, E. and Silva, I. P. (1999) Episodes of high productivity and cooling in the early Aptian Alpine Tethys. Geology, 27, 657 – 660.

Hughes Clarke, M. W. (1988) Stratigraphy and rock unit nomenclature in the oil-producing area of interior Oman. J. Petrol. Geol., 11, 5 – 60.

Immenhauser, A. (2005) High-rate sea-level change during the Mesozoic: new approaches to an old problem. Sed. Geol., 175, 277 – 296.

Immenhauser, A. (2009) Estimating palaeo-water depth from the physical rock record. Earth-Sci. Rev., 96, 107 – 139.

Immenhauser, A. and Matthews, R. K. (2004) Albian sea-level cycles from Oman: the "Rosetta Stone" approach. GeoArabia, 9, 11 – 46.

Immenhauser, A. and Rameil, N. (2011) Interpretation of ancient epikarst features in carbonate successions — a note of caution. Sed. Geol., 239, 1 – 9.

Immenhauser, A. and Scott, R. W. (1999) Global correlation of middle Cretaceous sea-level events. Geology, 27, 551 – 554.

Immenhauser, A., Schlager, W., Burns, S. J., Scott, R. W., Geel, T., Lehmann, J., Van Der Gaast, S. and Bolder-Schrijver, L. J. A. (1999) Late Aptian to Late Albian sea-level fluctuations constrained by geochemical and biological evidence(Nahr Umr Fm, Oman). J. Sed. Res., 69, 434 – 446.

Immenhauser, A., Creusen, A., Esteban, M. and Vonhof, H. B. (2000a) Recognition and interpretation of polygenic discontinuity surfaces in the Middle Cretaceous Shu'aiba, Nahr Umr, and Natih Formations of northern Oman. GeoArabia, 5, 299 – 322.

Immenhauser, A., Schlager, W., Burns, S. J., Scott, R. W., Geel, T., Lehmann, J., van der Gast, S. and Bolder-Schrijver, L. J. A. (2000b) Origin and correlation of disconformity surfaces and marker beds, Nahr Umr Formation, northern Oman. In: Middle East models of Jurassic/Cretaceous Carbonate Systems (eds A. S. Alsharhan and R. W. Scott), SEPM Spec. Publ., 69, pp. 209 – 225.

Immenhauser, A., Hillgärtner, H., Sattler, U., Bertotti, G., Schoepfer, P., Homewood, P., Vahrenkamp, V., Steuber, T., Masse, J. -P., Droste, H., TaAl-van Koppen, J., van der Kooij, B., van Bentum, E., Verwer, K., Hoogerduijn Strating, E., Swinkels, W., Peters, J., Immenhauser-Potthast, I. and Al Maskery, S. (2004) Barremian-lower Aptian Qishn Formation, Haushi-Huqf area, Oman: a new outcrop analogue for the Kharaib/Shu'aiba reservoirs. GeoArabia, 9, 153 – 194.

Immenhauser, A., Dublyansky, Y. V., Verwer, K., Fleitman, D. and Pashenko, S. (2007) Textural, elemental, and isotopic characteristics of Pleisocene phreatic cave deposits(Jabal Madar,Oman). J. Sed. Res. ,77,68 – 88.

Immenhauser, A. , Holmden, C. and Patterson, W. P. (2008) Interpreting the carbon-isotope record of ancient shallow Epeiric Seas: lessons from the recent. In: Dynamics of Epeiric Seas (Eds B. R. Pratt and C. Holmden), Geol. Assoc. Can. Spec. Publ. 48, pp. 135 – 174.

Joachimski, M. M. (1994) Subaerial exposure and deposition of shallowing upward sequences: evidence from stable isotopes of Purbeckian peritidal carbonates(basal Cretaceous), Swiss and French Jura Mountains. Sedimentology, 41,805 – 824.

Karsten, M. , Machel, H. G. and Bachu, S. (2003) New insights into the origin and migration of brines in deep Devonian aquifers, Alberta, Canada. J. Geochem. Expl. ,80,193 – 219.

Kemper, E. (1987) Das Klima der Kreidezeit. Geol. Jb. , Reihe A,96,5 – 185.

Kennedy, W. J. and Garrison, R. E. (1975) Morphology and genesis of nodular phosphates in the Cenomanian Glauconite Marl of south-east England. Lethaia,8,339 – 360.

Klappa, C. (1980) Rhizoliths in terrestrial carbonates: classification, recognition, genesis and significance. Sedimentology,27,613 – 629.

Klimchouk, A. (2004) Towards defining, delimiting and classifying epikarst: its origin, processes and variants of geomorphic evolution. Speleogen. Evol. Karst Aquif. ,2,13.

Krassilov, V. A. (1973) Climatic changes in Eastern Asia as indicated by fossil floras. I. Early Cretaceous. Palaeogeogr. Palaeoclimatol. Palaeoecol. ,13,261 – 273.

Le Métour, J. , Michel, J. C. , Béchennec, F. , Platel, J. P. and Roger, J. (1995) Geology and Mineral Wealth of the Sultanate of Oman. (Ministry of Petroleum and Minerals, Directorate General of Minerals, Sultanate of Oman and Bureau de Recherches et Minières, France). Muscat, Oman, 285 pp.

Lee, Y. W. , Lee, Y. I. and Hisada, K. -I. (2003) Paleosols in the Cretaceous Goshoura and Mifune groups, SW Japan and their paleoclimate implications. Palaeogeogr. Palaeoclimatol. Palaeoecol. ,199,265 – 282.

Lohman, K. C. (1988) Geochemical patterns of meteoric diagenetic systems and their application to studies of paleokarst. In: Paleokarst(Eds N. P. James and P. W. Choquette), pp. 58 – 80. Springer, Berlin.

Masse, J. -P. , Borgomano, J. and Al Maskiry, S. (1997) Stratigraphy and tectonosedimentary evolution of a late Aptian-Albian carbonate margin: the northeastern Jebel Akhdar(Sultanate of Oman). Sed. Geol. ,113,269 – 280.

Masse, J. -P. , Borgomano, J. and Al Maskiry, S. (1998) A platform-to-basin transition for lower Aptian carbonates (Shuaiba Formation) of the northeastern Jebel Akhdar(Sultanate of Oman). Sed. Geol. ,119,297 – 309.

Maurer, F. , Al-Mehsin, K. , Pierson, B. J. , Eberli, Gregor. P. , Warrlich, G. , Drysdale, D. and Droste, H. J. (2010) Facies characteristics and architecture of Upper Aptian Shu'aiba clinoforms in Abu Dhabi. In: Barremian-Aptian Stratigraphy and Petroleum Habitat of the Eastern Arabian Plate (Eds F. S. P. van Buchem, M. I. Al-Husseini, F. Maurer and H. J. Droste), GeoArabia Spec. Publ. ,4, Vol. 2, pp. 445 – 468.

Montenat, C. , Barrier, P. and Soudet, H. J. (2003) Aptian faulting in the Haushi-Huqf(Oman) and the tectonic evolution of the southeast Arabian platform-margin. GeoArabia,8,643 – 662.

Moore, C. H. (2001) Carbonate reservoirs: porosity evolution and diagenesis in a sequence stratigraphic framework. Dev. Sedimentol. ,55,444.

Morrison, J. , Birse, A. , Samuel, M. A. , Richardson, S. M. , Harbury, N. and Bott, W. F. (1997) The Cretaceous sequence stratigraphy of the Socotran platform, the Republic of Yemen. Mar. Petrol. Geol. ,14,685 – 699.

Murris, R. J. (1980) Middle East: stratigraphic evolution and oil habitat. AAPG Bull. ,64,597 – 618.

Mutterlose, J. , Bornemann, A. and Herrle, J. (2009) The Aptian-Albian cold snap: evidence for "mid" Cretaceous icehouse interludes. N. Jb. Geol. Paläont. Abh. ,252,217 – 225.

Mylroie, J. E. and Carew, J. L. (1995) Karst development on Carbonate Islands. In: Unconformities and Porosity in Carbonate Strata(Eds D. A. Budd, A. H. Saller and P. M. Harris), AAPG Memoir,63, pp. 55 – 76.

Neuser, R. D. (1995) A new high-intensity cathodoluminescence microscope and its application to weakly luminescing minerals. Bochum. Geol. Geotechn. Arb. ,44,116 – 118.

Owen, R. M. S. and Nasr, S. N. (1958) Stratigraphy of the Kuwait-Basra area. In: A Symposium: Habitat of Oil (Ed. L. G. Weeks), AAPG, Tulsa, Oklahoma pp. 1252 – 1278.

Peropadre, C., Meléndez, N. and Liesa, C. L. (2008) Variaciones del nivel del mar registradas como valles incisos en la Formación Villarroya de los Pinares en la subcuenca de Galve (Teruel, Cordillera Ibérica). Geo-Temas, 10, 167 – 170.

Pickton, C. A. G. (1981) Palaeogene and Cretaceous dropstones in Spitsbergen. In: Earth's Prepleistocene Glacial Record (Eds M. J. Hambrey and W. B. Harland), pp. 567 – 569. Cambridge University Press, Cambridge.

Pirrie, D., Doyle, P., Marshall, J. D. and Ellis, G. (1995) Cool Cretaceous climates—new data from the Albian of Western Australia. J. Geol. Soc. London, 152, 739 – 742.

Pirrie, D., Marshall, J. D., Doyle, P. and Riccardi, A. C. (2004) Cool early Albian climates; new data from Argentina. Cretaceous Res. ,25 ,27 – 33.

Pittet, B., van Buchem, F. S. P., Hillgärtner, H., Razin, P., Grötsch, J. and Droste, H. (2002) Ecological succession, paleoenvironmental change, and depositional sequences of Barremian-Aptian shallow-water carbonates in northern Oman. Sedimentology, 49, 555 – 581.

Pollastro, R. M. (1999) Ghaba salt basin province and Fahud salt basin province, Oman-geological overview and total petroleum systems. US Geol. Surv. Bull. ,2167 ,1 – 41.

Pomoni-Papaioannou, F. (1994) Paleoenvironmental reconstruction of a condensed hardground-type depositional sequence at the Cretaceous-Tertiary contact in the Parnassus-Ghiona zone, central Greece. Sed. Geol. ,93 ,7 – 24.

Price, G. D. (2003) New constraints upon isotope variation during the early Cretaceous (Barremian-Cenomanian) from the Pacific Ocean. Geol. Mag. ,140 ,513 – 522.

Pucéat, E., Lécuyer, C., Sheppard, S. M. F., Dromart, G., Reboulet, S. and Grandjean, P. (2003) Thermal evolution of Cretaceous marine waters inferred from oxygen isotope composition of fish tooth enamels. Paleoceanography, 18 ,7 – 1 – 7 – 12.

Purser, B. H. (1969) Syn-sedimentary marine lithification of Middle Jurassic limestones in the Paris Basin. Sedimentology, 12, 205 – 230.

Ramkumar, M., Ramayya, M. P. and Gandhi, M. S. (2000) Beachrock exposures at wave cut terraces of Modern Godavari delta: Their genesis, diagenesis and indications on coastal submergence and sealevel rise. Indian J. Mar. Sci., 29, 219 – 223.

Raven, M. J., van Buchem, F. S. P., Larsen, P. -H., Surlyk, F., Steinhardt, H., Cross, D., Klem, N. and Emang, M. (2010) Late Aptian incised valleys and siliciclastic infill at the top of the Shu'aiba Formation (Block 5, offshore Qatar). In: Barremian-Aptian Stratigraphy and Petroleum Habitat of the Eastern Arabian Plate (Eds F. S. P. van Buchem, M. I. Al-Husseini, F. Maurer and H. J. Droste), GeoArabia Spec. Publ. ,4, Vol. 2, pp. 469 – 502.

Reitner, J., Wilmsen, M. and Neuweiler, F. (1995) Cenomanian/Turonian sponge microbialite deep-water hardground community (Liencres, northern Spain). Facies, 32, 203 – 212.

Reolid, M., Abad, I. and Martin-Garcia, J. M. (2007) Palaeo-environmental implications of ferruginous deposits related to a middle-upper Jurassic discontinuity (prebetic zone, betic cordillera, Southern Spain). Sed. Geol. ,203 ,1 – 16.

Röhl, U. and Ogg, J. G. (1996) Aptian-Albian sea level history from guyots in the western Pacific. Paleoceanography, 11, 595 – 624.

Rückheim, S., Bornemann, A. and Mutterlose, J. (2006a) Integrated stratigraphy of an Early Cretaceous (Barremian-Early Albian) North Sea borehole (BGS 81/40). Cretaceous Res. ,27 ,447 – 463.

Rückheim, S., Bornemann, A. and Mutterlose, J. (2006b) Planktic foraminifera from the mid-Cretaceous (Barremian-Early Albian) of the North Sea Basin: Palaeoecological and palaeocenographic implications. Mar. Micropaleontol. ,58 ,83 – 102.

Ruffell, A. and Wach, G. (1998a) Firmgrounds-key surfaces in the recognition of parasequences in the Aptian Lower Greensand Group, Isle of Wright (southern England). Sedimentology, 45, 91 – 107.

Ruffell, A. and Wach, G. (1998b) Estuarine/offshore depositional sequences of the Cretaceous Aptian/Albian boundary, England. In: Mesozoic and Cenozoic Sequence Stratigraphy of European Basins (Eds P. Ch. de Graciansky, J. Hardenbol, Th. Jacquin and P. R. Vail), SEPM Spec. Publ., 60, 411 – 422.

Sadler, P. M. (1981) Sediment accumulation rates and the completeness of stratigraphic sections. J. Geol., 89, 569 – 584.

Sahagian, D., Pinous, O., Olferiev, A. and Zakharov, V. (1996) Eustatic curve for the Middle Jurassic-Cretaceous based on Russian platform and Siberian stratigraphy: zonal resolution. AAPG Bull., 80, 1433 – 1458.

Saller, A. H., Dickson, J. A. D. and Matsuda, F. (1999) Evolution and distribution of porosity associated with subaerial exposure in Upper Paleozoic platform limestones, west Texas. AAPG Bull., 83, 1835 – 1854.

Sattler, U., Immenhauser, A., Hillgätner, H. and Esteban, M. (2005) Characterization, lateral variability and lateral extent of discontinuity surfaces on a Carbonate Platform (Barremian to Lower Aptian, Oman). Sedimentology, 52, 339 – 361.

Schlager, W. (2005) Carbonate sedimentology and sequence stratigraphy. SEPM Concepts in Sedimentology and Palaeontology. 8, Tulsa, Oklahoma, 200 pp.

Schroeder, R., van Buchem, F. S. P., Cherchi, A., Baghbani, D., Vincent, B., Immenhauser, A. and Granier, B. (2010) Revised orbitolinid biostratigraphic zonation for the Barremian-Aptian of the eastern Arabian Plate and implications for regional stratigraphic correlations. In: Barremian-Aptian Stratigraphy and Petroleum Habitat of the Eastern Arabian Plate (Eds F. S. P. van Buchem, M. I. Al-Husseini, F. Maurer and H. J. Droste), GeoArabia Spec. Publ., 4, Vol. 1, pp. 49 – 96.

Scott, R. W. (1990) Chronostratigraphy of the Cretaceous carbonate shelf, southeastern Arabia. In: The Geology and Tectonics of the Oman Region (Eds. A. H. F. Robertson, M. P. Searle and A. C. Ries), Geol. Soc. Spec. Publ., 49, 89 – 108.

Scott, R. W., Schlager, W., Fouke, B. and Nederbragt, S. A. (2000) Are mid-Cretaceous eustatic events recorded in Middle East carbonate platforms? In: Middle East models of Jurassic/Cretaceous Carbonate Systems (Eds A. S. Alsharhan and R. W. Scott), SEPM Spec. Publ., 69, pp. 77 – 88.

Sellwood, B. W., Price, G. D. and Valdes, P. J. (1994) Cooler estimates of Cretaceous temperatures. Nature, 370, 453 – 455.

Sharland, P. R., Archer, R., Casey, D. M., Davies, R. B., Hall, S. H., Heward, A. P., Horbury, A. D. and Simmons, M. D. 2001. Arabian Plate Sequence Stratigraphy. GeoArabia Spec. Publ.. 2, Gulf PetroLink, Manama, Bahrain, 371 pp.

Simmons, M. D. (1994) Micropalaeontological biozonation of the Kahmah Group (Early Cretaceous), Central Oman Mountains. In: Micropalaeontology and Hydrocarbon Exploration in the Middle East (Ed. M. D. Simmons), pp. 177 – 206. Chapman & Hall, London.

Simmons, M. D. and Hart, M. B. (1987) The biostratigraphy and microfacies of the Early to mid-Cretaceous carbonates of Wadi Mi'aidin, Central Oman Mountains. In: Micropaleontology of Carbonate Environments (Ed. M. B. Hart), pp. 176 – 207. Ellis Horwood, Chichester, UK.

Skelton, P. W. and Masse, J. -P. (2000) Synoptic guide to the Lower Cretaceous rudist bivalves of Arabia. In: Middle East models of Jurassic/Cretaceous Carbonate Systems (Eds A. S. Alsharhan and R. W. Scott), SEPM Spec. Publ., 69, pp. 89 – 99.

Steuber, T., Rauch, M., Masse, J. -P., Graaf, J. and Malkoç M. (2005) Low-latitude seasonality of Cretaceous temperatures in warm and cold episodes. Nature, 437, 1341 – 1344.

Stoll, H. M. and Schrag, D. P. (1996) Evidence for glacial control of rapid sea level changes in the Early Cretaceous. Science, 272, 1771 – 1774.

Strasser, A., Pittet, B., Hillgärtner, H. and Pasquier, J. -B. (1999) Depositional sequences in shallow carbonate-dominated sedimentary systems: concepts for a high-resolution analysis. Sed. Geol., 128, 201 – 221.

Strohmenger, C. J., Steuber, T., Ghani, A., Barwick, D. G., Al-Mazrooei, S. H. A. and Al-Zaabi, N. O. (2010) Sedimentology and chemostratigraphy of the Hawar and Shu'aiba depositional sequences, Abu Dhabi, United Arab Emirates. In: Barremian-Aptian Stratigraphy and Petroleum Habitat of the Eastern Arabian Plate (Eds F. S. P. van Buchem, M. I. Al-Husseini, F. Maurer and H. J. Droste), GeoArabia Spec. Publ. ,4, Vol. 2, pp. 341 – 365.

Taylor, P. D. and Wilson, M. A. (2002) A new terminology for marine organisms inhabiting hard substrates. Palaios, 17, 522 – 525.

Theiling, B. P., Railsback, L. B., Holland, S. M. and Crowe, D. E. (2007) Heterogeneity in geochemical expression of subaerial exposure in limestones, and its implications for sampling to detect exposure surfaces. J. Sed. Res. ,77, 159 – 169.

Tobin, K. J., Steinhauff, D. M. and Walker, K. R. (1999) Ordovician meteoric carbon and oxygen isotopic values: implications for the latitudinal variations of ancient stable isotopic values. Palaeogeogr. Palaeoclimatol. Palaeoecol. , 150, 331 – 342.

Trenhaile, A. S. (2002) Modeling the development of marine terraces on tectonically mobile rock coasts. Mar. Geol. 185, 341 – 361.

Tucker, M. E. and Wright, V. P. (1990) Carbonate Sedimentology. Blackwell Science, Oxford, 482 pp.

Vahrenkamp, V. C. (1996) Carbon isotope stratigraphy of the Upper Kharaib and Shuaiba Formations: implications for the Early Cretaceous evolution of the Arabian Gulf region. AAPG Bull. ,80, 647 – 662.

Vahrenkamp, V. C. (2010) Chemostratigraphy of the Lower Cretaceous Shu'aiba Formation: A $\delta^{13}C$ reference profile for the Aptian Stage from the southern Neo-Tethys Ocean. In: Barremian-Aptian Stratigraphy and Petroleum Habitat of the Eastern Arabian Plate (Eds F. S. P. van Buchem, M. I. Al-Husseini, F. Maurer and H. J. Droste), GeoArabia Spec. Publ. ,4, Vol. 1, pp. 107 – 137.

Van Wagoner, J. C., Posamentier, H. W., Mitchum, R. M., Vail, P. R., Sarg, J. F., Loutit, T. S. and Hardenbol, J. (1988) An overview of the fundamentals of sequence stratigraphy and key definitions. In: Sea-Level Changes-An Integrated Approach (Eds C. E. Wilgus, B. S. Hastings, Ch. G. St. C. Kendall, H. W. Posamentier, C. A. Ross and J. C. Van Wagoner), SEPM Spec. Publ. ,42, pp. 38 – 45.

Wagner, P. D. (1990) Geochemical stratigraphy and porosity controls in Cretaceous carbonates near the Oman Mountains. In: The Geology and Tectonics of the Oman Region (Eds A. H. F. Robertson, M. P. Searle and A. C. Ries), Geol. Soc. Spec. Publ. ,49, 127 – 137.

Wagner, P. D., Tasker, D. R. and Wahlman, G. P. (1995) Reservoir degradation and compartmentalization below subaerial unconformities: limestone examples from West Texas, China, and Oman. In: Unconformities and Porosity in Carbonate Strata(Eds D. A. Budd, A. H. Saller and P. M. Harris), AAPG Mem. ,63, pp. 177 – 195.

Warrlich, G., Hillgärtner, H., Rameil, N., Gittins, J., Mahruqi, I., Johnson, T., Alexander, D., Wassing, B., van Steenwinkel, M. and Droste, H. (2010) Reservoir characterisation of data-poor fields with regional analogues: a case study from the Lower Shu'aiba in the Sultanate of Oman. In: Barremian-Aptian Stratigraphy and Petroleum Habitat of the Eastern Arabian Plate (Eds F. S. P. van Buchem, M. I. Al-Husseini, F. Maurer and H. J. Droste), GeoArabia Spec. Publ. ,4, Vol. 2, pp. 577 – 604.

Webb, J. A. and Golding, S. D. (1998) Geochemical mass-balance and oxygen-isotope constraints on silcrete formation and its paleoclimatic implications in southern Australia. J. Sed. Res. ,68, 981 – 993.

Weissert, H. and Lini, A. (1991) Ice age interludes during the time of Cretaceous greenhouse climate? In: Controversies in Modern Geology(Eds D. W. Müller, J. A. McKenzie and H. Weissert), pp. 173 – 191. Academic Press, London.

Winterer, E. L. and Sarti, M. (1994) Neptunian dykes and associated features in southern Spain: mechanisms of formation and tectonic implications. Sedimentology, 41, 1109 – 1132.

Witt, W. and Gökdag, H. (1994) Orbitolinid biostratigraphy of the Shuaiba Formation(Aptian), Oman-implications for reservoir development. In: Micropalaeontology and Hydro carbon Exploration in the Middle East (Ed. M. D. Simmons), pp. 222 – 234. Chapman & Hall, London.

Wright, V. P. (1982) The recognition and interpretation of paleokarsts: two examples from the Lower Carboniferous of south Wales. J. Sed. Res. ,52, 83 – 94.

Wright, V. P. (1988) Paleokarst and paleosols as indicators of paleoclimate and porosity evolution: a case study from the Carboniferous of South Wales. In: Paleokarst (Eds N. P. James and P. W. Choquette), pp. 329 – 341. Springer, Berlin, Heidelberg, New York.

Wright, V. P. (1994) Paleosols in shallow marine carbonate sequences. Earth-Sci. Rev., 35, 367 – 395.

Yang, W. (2001) Estimation of duration of subaerial exposure in shallow-marine limestones-an isotopic approach. J. Sed. Res., 71, 778 – 789.

Yılmaz, İ. Ö. and Altıner, D. (2006) Cyclic palaeokarst surfaces in Aptian peritidal carbonate successions (Taurides, southwest Turkey): internal structure and response to mid-Aptian sea-level fall. Cretaceous Res., 27, 814 – 827.

Yose, L. A., Ruf, A. S., Strohmenger, C. J., Schuelke, J. S., Gombos, A., Al-Hosani, I., Al-Maskary, S., Bloch, G., AlMehairi, Y. and Johnson, I. G. (2006) Three-dimensional characterization of a heterogeneous carbonate reservoir, Lower Cretaceous, Abu Dhabi (United Arab Emirates). In: Giant Hydrocarbon Reservoirs of the World: From Rocks to Reservoir Characterization and Modelling (Eds P. M. Harris and L. J. Weber), AAPG Mem., 88, pp. 173 – 212.

Yose, L. A., Strohmenger, C. J., Al-Hosani, I., Bloch, G. and Al-Mehairi, Y. (2010) Sequence-stratigraphic evolution of an Aptian carbonate platform (Shu'aiba Formation), eastern Arabian Plate, onshore Abu Dhabi, United Arab Emirates. In: Barremian-Aptian Stratigraphy and Petroleum Habitat of the Eastern Arabian Plate (Eds F. S. P. van Buchem, M. I. Al-Husseini, F. Maurer and H. J. Droste), GeoArabia Spec. Publ., 4, Vol. 2, pp. 309 – 340.

Zaitlin, B. A., Potocki, D., Warren, M. J., Rosenthal, L. and Boyd, R. (2002) Depositional styles in a low accommodation foreland basin setting: an example from the Basal Quartz (Lower Cretaceous), southern Alberta. Bull. Can. Petrol. Geol., 50, 31 – 72.

第9章 侏罗纪斜坡沉积不连续界面的表征

NICOLAS CHRIST,ADRIAN IMMENHAUSER,FRÉDÉRIC AMOUR,MARIA MUTTI,SARA TOMÁ S,SUSAN M. AGAR,ROBERT ALWAY, LAHCEN KABIRI 著

杜业波 译,陈瑞银 校

摘 要 陆缘海碳酸盐沉积体系中广泛地识别了非连续界面,但是其特征却并未被很好地加以认识。在沉积体系中,这种特征代表的是低于生物地层分辨率的沉积间断,但在沉积纪录中占据可观的比例。从应用的角度来看,非连续界面代表了水平流动受阻并形成了储层分隔。本章将摩洛哥高阿特拉斯山的侏罗系(中、上 Bajocian Assoul 组)碳酸盐斜坡沉积体系中的 80 个凝缩段(S1)、固底(S2)和硬底(S3)的地貌特征、铁和锰氧化物及磷酸盐的次生充填情况和古生态学特征做了详细描述。统计表明在 220m 长的碳酸盐岩层序中平均每 10m 有 2 个 S1 界面、1.1 个 S2 界面和 0.4 个 S3 界面。基于两个在地层和空间上相对独立的剖面开展了非连续界面在地层剖面中出现的频率、侧向展布范围及界面附近的沉积相类型变化的定量化研究。研究区的显著特征包括 Assoul 组的巨大厚度以及明显缺少近地表暴露标志。根据本区研究资料,固底和硬底最好解释为最大海退的相关特征体现。相对海平面降低导致了浪基面的降低,波浪和洋流的侵蚀导致了海底的侵蚀和硬化成岩。还需注意的是对非连续界面的解释不能过于简单化,比如区分其与近岸到深海的剖面上由于水深和碳酸盐环境相变引起的非连续特征变化。本区的研究工作对于解释浅水碳酸盐沉积环境和建立更切合实际的碳酸盐岩储层流动模型具有重要意义。

关键词 阿特拉斯山 碳酸盐斜坡 非连续界面 硬底 水动力强度 侏罗系 相对海平面

9.1 概况

非连续界面是沉积物中沉积间断或沉积缺失的常见标志(Bathurst,1975;Clari 等,1995;Immenhauser 等,2000;Sattler 等,2005)。这一沉积间断面通常但并不总是伴随着明显的沉积相变(Hillgartner,1998)。"非连续界面"这一术语由 Heim(1924,1934)引入,Bromley(1975a)将其定义为"小于平行不整合级别的沉积间断"。Clari 等(1995)将其定义为无法解释的(包括海相硬底、陆上暴露面、海侵面等)、具有独立发育的沉积间断和沉积缺失相关特征的界面。

在海相地层中,非连续界面通常被认为是海平面变化在地层剖面对比中的标志层(Kauffman 等,1991;Immenhauser 等,2000)和层序以及旋回地层学中的沉积体系边界的标志(Mitchum 等,1977;Sarg,1988;Handford 和 Loucks,1993;Strasser 等,1999;Sattler 等,2005;Catuneanu 等,2009)。此外,在非连续界面之下通常发育油田开发中的致密层、流动隔层或输导层(Read 和 Horburry,1994;Cander,1995;Wagner 等,1995)。因此,非连续界面的侧向变化(致密或疏松)及其下伏岩层的成岩作用对储层发育有重要意义。尽管具有重要的实用和研究价值,但到目前为止只有关于非连续界面的地层和平面分布量化研究只有有限的几篇文章发表(Hillgärtner,1998;Immenhauser 等,2002;Sattler 等,2005)。

通过对摩洛哥高阿特拉斯山脉侏罗系碳酸盐斜坡沉积体系野外和实验室研究分析,开展非连续性界面在地层及横向展布特征的量化研究,并将研究结果加以记录讨论形成本篇过程导向性的文章。本文有两个目的:一是从在地层中的年代分布、横向展布、界面两侧沉积相的

变化、地貌、古生物特征和其下伏地层的厚度变化等方面详细描述侏罗系碳酸盐斜坡沉积中的非连续界面；为了统计其详细数据，在几百米至上千米的范围内对非连续界面进行横向追踪和描述。二是尝试在古环境背景中讨论这些非连续界面的特征，尤其是解释碳酸盐斜坡体系中控制各类成层或不成层的非连续界面的沉积过程。此外本文还提供了中生代碳酸盐斜坡沉积体系中与非连续界面有关的储层隔层的量化统计分析。

9.2 区域构造背景

高阿特拉斯山中、东部地区的地质动力学演化与中生代联合古陆的解体及原大西洋的裂开（Brede，1987；Beauchamp，1988）密切相关。这些构造运动包括三叠纪初始裂谷的产生、伴生玄武岩的喷发及与西特提斯洋相连的从"西南西"至"东北东"走向的裂谷盆地的形成（图9-1A）。该盆地与撒哈拉地台（冈瓦纳）之间被摩洛哥和Oran Mesetas微板块相隔；这两个微板块在早侏罗世晚期和中侏罗世经历了新一期裂谷运动，被新形成的北东—南西向的中阿特拉斯盆地所隔开（图9-1B；Warme，1988；Hauptmann，1990）。

高阿特拉斯中部地区从三叠纪到早侏罗世持续沉降形成了数千米厚的陆相沉积层序（干裂谷，sensu Warme，1988），其岩性包括砾岩、玄武岩（Roch，1950；Abdeljalil等，1959）、河流相红层和蒸发岩（Mattis，1977；Van Houten，1977）。早侏罗世的海侵导致中、东高阿特拉斯裂谷由陆相沉积转为海相沉积，并发育了一套范围巨大的碳酸盐斜坡沉积体系。

辛涅缪尔期中、东高阿特拉斯裂谷构造开始复活，导致碳酸盐斜坡体系解体（Dubar，1962；Du Dresnay，1979；Crevello，1990），与此相应，在中、东高阿特拉斯南部发育几个小型沉积体系，其中一个即为本文研究区。研究区发育两套继承性的碳酸盐斜坡体系（Pierre，2006），分别是Amellago组（晚土阿辛阶—早巴柔期）和Assoul组（中—晚巴柔期，图9-2），两套地层之间以巴柔阶早期Agoudim海侵泥灰岩隔开（Pierre，2006），目前研究的重点是中、上巴柔阶Assoul组碳酸盐斜坡沉积（图9-2）。

9.3 研究区概况

Amellago峡谷位于Errachidia市西70km，该峡谷可达数百米高，谷内有沿Gheris河延伸约15km的出露良好的Amellago和Assoul组露头剖面。本文重点研究Amellago峡谷Assoul组，工区附近该组总体厚约220m。Pierre（2006）的区域研究表明，在Amellago峡谷东部及东南部Amellago组（下巴柔阶）和Assoul组（上巴柔阶）发育镶边台地沉积。本次研究的野外地质资料表明，至少在Amellago峡谷地区Assoul组在几何形态属低角度碳酸盐斜坡沉积体系。这一碳酸盐岩体系几何形态局部的短期变化是下步区域野外地质研究的焦点，但目前本文应用的统计学方法还无法对其展开预测。

典型露头：为了刻划沉积相带和非连续界面在侏罗系碳酸盐斜坡沉积体系中垂向和平面上的展布，从Amellago峡谷选取了两个典型露头开展研究，这两个露头可以近似代表Assoul组上面三分之一和下面三分之一，其沉积相类型相应的是碳酸盐内斜坡—中斜坡沉积体系（表9-1）。这两个研究露头都位于Assoul组中以石灰岩为主的地层，分别命名为"岛"和"长丘"。为了将这两个露头的地层进行对比，建立和测量了该区的一条地层厚度约110m的主剖面（剖面IUS）。图9-3显示了Amellago峡谷Assoul组地层概况，该剖面不能反映小—中尺度的纵向沉积相变化，但是在时间尺度上一级相变和与之相应的平面延伸数千米的主要非连续界面都能被很好地表现出来。

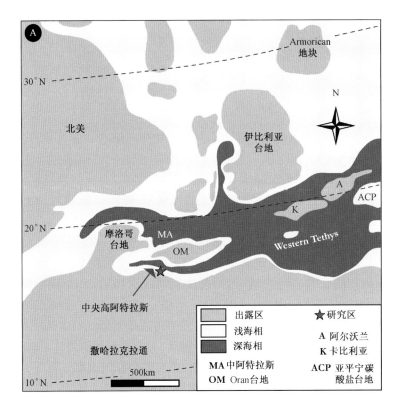

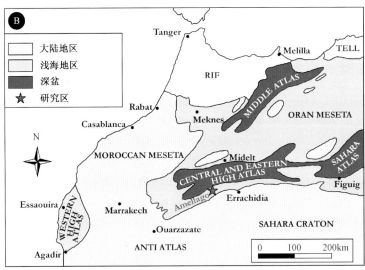

图 9-1 (A)西特提斯中央高阿特拉斯裂谷盆地早
侏罗世古地理,引自 Wilmsen 和 Neuweiler(2008),改自 Bassoullet 等(1993);(B)摩洛哥主要地质单元
分布图,引自 Blomeier 和 Reijmer(1999),改自 DuDresnay(1971)

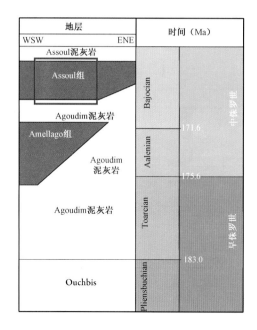

图 9-2 Amellago 地区简化地层图(据 Pierre,2006,有修改)

表 9-1 沉积相组合及解释

沉积相编码	沉积相	结构	骨架和非骨架成分	沉积环境
F1	灰色—青灰色泥灰岩	m	腹足类 2	中部斜坡末端,后浅滩
F2	泥灰岩/生物碎屑粒泥灰岩互层	m,Ms,Ws	海胆类 2-3,介形亚纲类 2-3,双壳类 2	潟湖
F3	生物碎屑—双壳类泥粒灰岩—漂砾灰岩	Ps(Fs)	双壳类 3,海胆类 3,苔藓虫类 2,蓝藻门 1-2,似球粒 2-3,腕足类 1-2	中部斜坡末端
F4	生物碎屑—似球粒泥灰岩	Ms-Ws,Ws	绿藻 1,双壳类 2,核形石 1,似球粒,蓝藻门有孔虫类 0-1,腹足类 2,介形亚纲类 1-2,腕足类 2	似潟湖,中部斜坡末端
F5	似球粒生物碎屑泥粒灰岩	Ws-Ps,Ps	似球粒 3,绿藻 2,双壳类,核形石 1,有孔虫类 1,蓝藻门 2,海胆类 2	潟湖,后浅滩
F6	似核形石漂砾灰岩	Fs	核形石(Cayeuxia)3-4,球粒 3,双壳类 2,细菌和泥晶结核 1,腹足类 2,有孔虫类 1,海胆类 1	潟湖,中部斜坡近端
F7	生物碎屑漂砾灰岩	Fs	腹足类 3,海胆类 2-3,有孔虫类 1,微生物结核 2,珊瑚类 2-3,双壳类 2-3	潟湖,中部斜坡近端
F8	似球状—鲕粒泥粒灰岩—颗粒灰岩	Ps,Ps-Gs,Gs	似球粒 4,鲕粒 2,微生物结核 1,内碎屑 1,珊瑚碎屑 1,有孔虫类 2,海胆类 2,腹足类 2	前部浅滩
F9	鲕粒灰岩	Gs	鲕粒 4,双壳类 3,珊瑚碎屑 3,腹足类 1,有孔虫类 1	浅滩
F10	鲕粒—内碎屑漂砾灰岩	Fs(Rs)	内碎屑(鲕粒 2)3,鲕粒 2-3	后浅滩
F11	微生物(蓝藻)砾状灰岩	Rs	鲕粒 2-3,细菌结核 3,有孔虫类 2,内碎屑 2-3,介形亚纲类 1-2	前部浅滩
F12	珊瑚/微生物/海绵动物/牡蛎礁体	Bs		中部斜坡末端

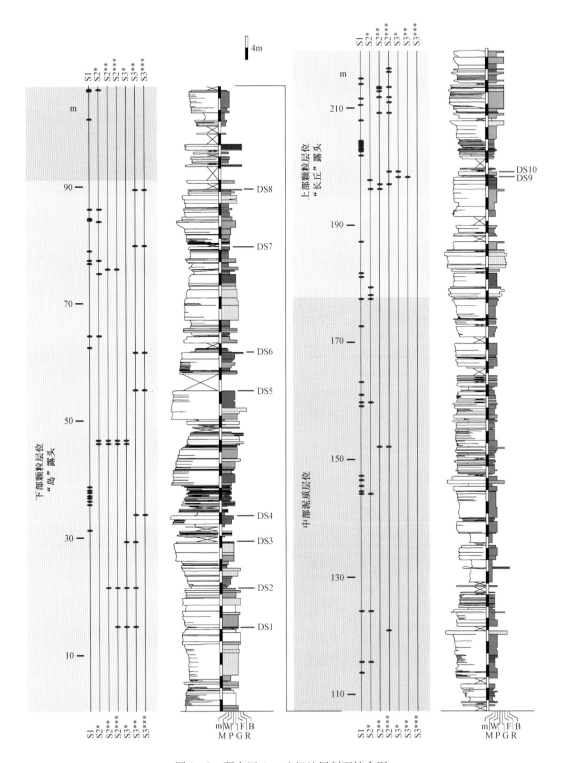

图 9-3 研究区 Assoul 组地层剖面综合图

展示了研究区露头泥质和颗粒质地层及其相对位置。展示了非连续界面(S1 到 S3)的分布和类型。DS1 到 DS10 是主要的 S3 型区域主要非连续界面。沉积相颜色编码参考图 9-8

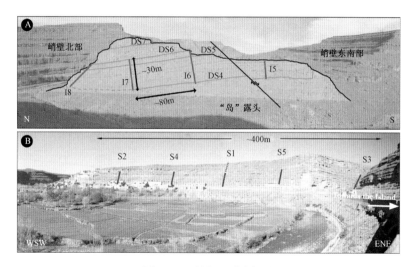

图 9-4 研究区露头概图

(A)"岛"露头。标识了主要区域非连续界面。I5、I6、I7 和 I8 是垂向剖面的标记。(B)"长丘"露头,标识了地层剖面。S2、S3、S4 和 S5 剖面约 20m 厚,剖面 S1 约 45m 厚

(1)"岛"露头。"岛"露头研究区是由 Gheris 河切割的椭圆形剥蚀残余段,位于 Assoul 组的下三分之一段。该构造直径约 300m。由于碳酸盐岩露头良好,可进行三维空间研究。围绕该露头测量了 8 条间距相近、厚约 15~30m 的剖面,这些剖面会适当向上或向下延长以覆盖该露头出露的 85m 厚的地层,这些扩展剖面底部到达 Assoul 组下段。在"岛"露头附近悬崖上测量到涵盖相同层位的另外 10 个剖面(图 9-5)。利用其中几个剖面来展示"岛"露头中出露的非连续界面的侧向范围(图 9-5 和图 9-6)。

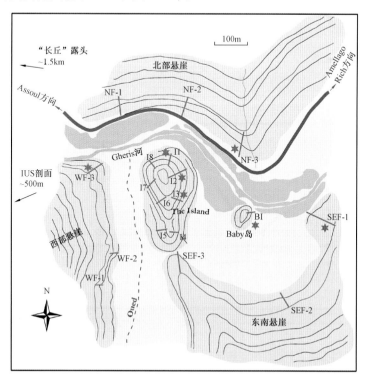

图 9-5 展示了主要露头和周边悬崖壁的"岛"露头简化图

图中已标示出地层剖面位置。红星指代图 9-6 中的剖面。研究区露头范围约为 700m×700m

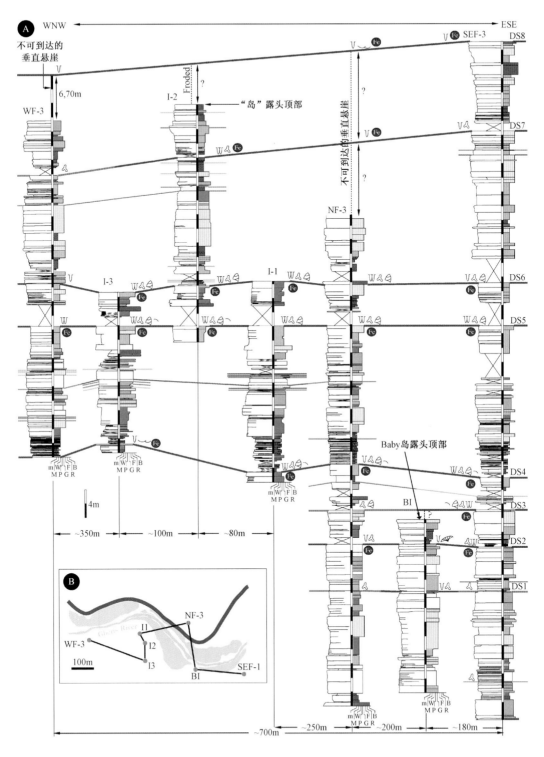

图 9-6 (A)"岛"露头的 7 个剖面中非连续界面横向范围和特征。(B)左下角插图指示剖面相对位置("岛"露头的剖面位置见图 9-5)。界面类型与特征,以及沉积相颜色编码参见图 9-8

(2)"长丘"露头。"长丘"露头位于"岛"露头西北1.5km(图9-7)处,处于Assoul组上三分之一层段,提供了侧向和垂向地层相变以及非连续界面的二维研究角度(图9-4B)。该露头横向距离约700m,地层厚度约20m。沙丘剖面底部位于"岛"露头顶部之上约90m。这两个露头之间Assoul组中部的层位总体偏泥质,非连续界面较罕见。在700m宽的悬崖面上测量了6个紧邻的剖面,每个厚度约20m。其中一个剖面(图9-4和图9-8中的S1)向上扩展24m,以便涵盖Assoul组的最顶部层位。

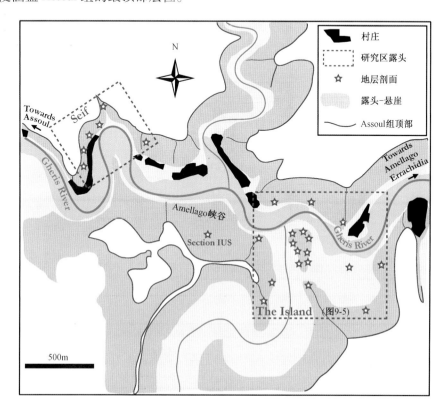

图9-7 Amellago 峡谷图
指示了研究区露头位置和范围,以及测量剖面的位置

由于"岛"露头剖面出露极好,尤其是沉积相变和非连续界面连续性三维视角良好,本研究主要采用"岛"露头资料,只有在必要时才利用"长丘"露头信息来补充部分数据。

9.4 方法

9.4.1 非连续界面

根据Bates和Jackson(1987)Clari等(1995)和Hillgärtner(1998),本研究中,非解释性术语"非连续界面"代表沉积间断、海相固底和硬底的成层、以及近地表暴露面或复合表面(Immenhauser等,2000;Satter等,2005)。为了刻画两个露头的侧向连续性和距离几米到上千米的变化,对非连续界面进行了详细刻画、地层记录和剖面间地层追踪。每个非连续界面,都对其形态(剥蚀剖面)、岩相特征和界面两侧的沉积相对比进行评价。在相关有区别的层次体系中对界面进行描述和等级排序:(1)具有初期凝缩特征的界面;(2)固底(S2);(3)发育完全的硬底(S3)。这种分级制度是基于侧向范围、生物扰动密度、钻孔或潜穴和母岩之间接触面特征(模

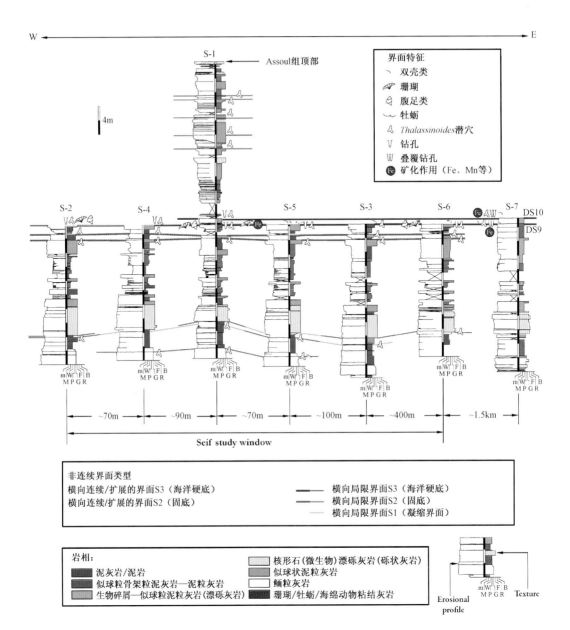

图 9-8 "长丘"露头的 7 个剖面中非连续界面横向范围和特征

位置参考图 9-4B 和图 9-7。剖面 S7 代表 IUS 剖面最顶部的 20m 地层,位于"长丘"露头东部 1km

糊还是清晰)、成岩作用和生物扰动影响的深度、矿化作用的发育和类型、穿过非连续界面的相变,以及调查界面的主要可视特性进行的。

露头的沉积相组合和岩相:通过观察露头、抛光岩片以及 200 多块薄片,对岩相和沉积相特征进行调查。根据 Immenhasuser 等(2004)应用过的颜色编码和碳酸盐岩命名法,用一种半量化方法定义非骨架和骨架成分的相对丰度(0 = 不存在;1 = 存在;2 = 经常;3 = 大量;4 = 主要)。用沉积构造和结构、生物群落、遗迹化石、水动力强度和成岩特征的相关术语来描述沉积环境。一共定义了 7 种相组合(表 9-1)。结构的描述参考了 Dunham(1962)的石灰岩分类法。

9.4.2 分析条件

碳、氧同位素分析和阴极射线发光分析：在 Ruhr-University Bochum 的研究室利用 Thermo Fisher Scientific Gasbench II 碳酸盐岩装置（Thermo Finnigan MAT GmbH，Bremen，Germany），结合 Thermo Fisher Scientific Delta S Isotope Ratio Mass Spectrometer 设备对 233 个碳酸盐岩粉末样品进行碳同位素和氧同位素分析。测量之前用磷酸与方解石在 70℃ 下反应 1 个小时。用来计算方解石 $\delta^{18}O_{carbonate}$ 值的分级因子 α 值为 1.008703。每个层序都包括 35 个样品，6 个国际参考物质（国际原子能机构 IAEA-CO-1，IAEA-CO-8）和 7 个内部标准（RUB-STD，方解石）。据已证实和已测量到的碳酸盐岩参考物（IAEA-CO-1，IAEA-CO-8）同位素得出的线性校正被应用到样品和内部标准中。对已证实碳酸盐岩标准（NBS 19 和 IAEA CO-1 和 CO-8）反复分析显示，$\delta^{18}O$ 的外部再生率 $\leqslant 0.1‰$，$\delta^{13}C$ 外部再生率 $\leqslant 0.06‰$。每 10 个样品中一个样品副本显示 $\delta^{13}C$ 散射率 $\leqslant 0.1‰$，$\delta^{18}O \leqslant 0.2‰$。所有同位素结果都用常规方法中 V-PDB（Vienna-Pee Dee Belemnite）标准的‰来表示。

利用"热阴极"CL 显微镜（HC1-LM 型）对 10 个薄片进行阴极射线发光检测。电子束加速电压是 14keV，设定射束电流的等级，以使其可以在样品表面获得约 $9\mu A\ mm^{-2}$ 电流密度。利用 EG 和 G 数字三格光谱仪（Princeton Applied Research，Acton，MA，USA）来获取阴极射线频谱，这种光谱仪通过 1.2m 长的石英灯引导将 Princeton Instrument Pixis@ 冷却的 CCD 镜头（Princeton Instruments，Tren-ton，NJ，USA）安装到 Cl 显微镜上。在 -120℃ ±0.1℃ 下利用 Peltier 效果对 CCD 镜头进行冷却以获得高信噪比。在 30s 的曝光时间内对光谱进行累加。由于分析区域可以集中于 $30\mu m$ 直径的圆点上，因此可得到较高的侧向分辨率。

9.5 碳酸盐斜坡地貌和沉积相

遵循前人的斜坡体系分类方案（Burchette 和 Wright，1992），从 Assoul 组沉积岩内部划分出三个主要的沉积环境：内斜坡、中斜坡、外斜坡。据 Burchette 和 Wright（1992）的斜坡模型，研究区内未出现包含近海相和潮坪沉积环境的最内部斜坡沉积环境，在区域上是否出露仍不清楚。根据 Pierre（2006）和 Pierre 等（2010），将沉积作用和沉积剖面归结为两个交互的碳酸盐斜坡模式（图 9-9）：(1) 鲕粒岩斜坡体系，伴随充当阻挡海洋扩张的障壁的内部潟湖和浅滩，浅滩沉积中有部分泥质互层（图 9-9A）；(2) 乏鲕粒石（Pierre 等，2010）的泥质斜坡沉积体系，特征是总体偏低的水动力强度，主要物质是泥灰岩、泥岩和粒泥灰岩，存在礁丘（图 9-9B）。

在 Amellago 峡谷的两个斜坡沉积体系中，最邻近的沉积相是远端内斜坡到近端中斜坡，主要标志是核形石（微生物）漂砾灰岩（F6）、生屑似球粒粒泥灰岩（F4）、生屑漂砾灰岩（F7）和泥灰岩（F2），水动力强度相对低到中等（表 9-1，图 9-9）。在泥质斜坡体系中，露头中最远端沉积相主要特征是相组合 F1 中的泥灰岩相（表 9-1，图 9-9B）。这些沉积成为远端中部斜坡和近端外部斜坡的主要特征，包括相组合 F12（表 9-1，图 9-9B）中生物礁，横向范围几米到几十米，高度几十厘米到高于 10m。泥质斜坡体系中远端中部斜坡的其他主要沉积特征包括生屑球状粒泥灰岩（F3），其中局部包含部分牡蛎和生物礁层的相组合 F12（表 9-1，图 9-9B）。鲕粒岩斜坡体系中最远端的部分主要是前浅滩中的生物碎屑蓝细菌砾状灰岩（F11），之后是似球粒（F8）和 F9 组合中的鲕粒岩浅滩（表 9-1，图 9-9A），F9 伴随有东北东

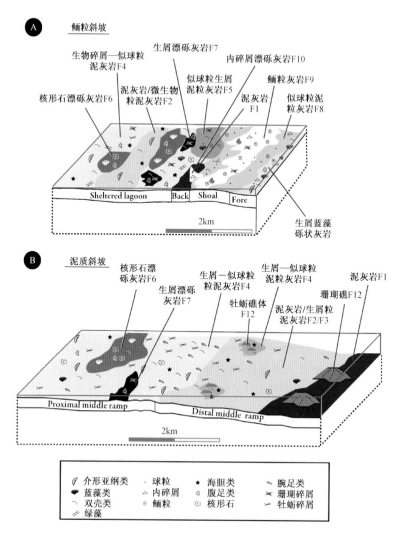

图 9-9 Assoul 组两个斜坡沉积体系的沉积剖面及其相关沉积相
(A)鲕粒斜坡;(B)泥质斜坡。不同沉积相类型的描述参考表 9-1

和东南东走向的滑塌岩沉积相,代表了正常浪基面之上的永久性高水动力强度扰动。后浅滩主要由相组合 F10 的鲕粒—内碎屑漂砾灰岩(砾屑)构成,随后是水动力强度向潟湖逐步退却的沉积相,比如似球粒生屑泥粒灰岩(F5)以及 F1 中的泥灰岩(表 9-1,图 9-9A)。

Assoul 组的外斜坡沉积体系,位于正常浪基面之下,特征是含黏土沉积相逐渐增多,伴随有灰岩和厚层泥质层序(相组合 F1 中的泥灰岩)的互层。石灰岩包括泥灰岩到粒泥灰岩以及更多的含黏土层序的互层。向古盆地方向(向东),灰岩层段逐渐减少,只有部分最厚的钙质标记层顶部才到达非连续界面。

单个剖面由厚度 2~3cm 至 2m 的碳酸盐岩地层组成,很少有标记层达到 4m 厚。在泥质斜坡中,含黏土沉积相占主导,单个层序显示了向上变洁净的旋回,即向上表现为泥质含量减少。相对于粒状斜坡,泥质斜坡的层理更加清晰,层厚与沉积相有关,具体表现为,厚层(约 1m)主要是碳酸盐岩沉积相,指代了最低水动力强度,比如泥灰岩相(F1)和(F2),F4 的生物碎屑球粒岩(粒泥灰岩),以及 F6 相组合的漂砾灰岩(表 9-1,图 9-9)。在颗粒斜坡总体偏泥质的夹层中局部可以观察到泥灰岩和泥岩（粒泥灰岩）互层。这些泥质夹层的累积厚度一般小于30cm。

9.6 非连续界面的刻划与解释

在 Assoul 组上下段两个露头以及垂向上与其有关联的剖面中，共识别出80个非连续界面（图9-3）。综合考虑横向连续和横向非连续剖面，露头观察到的非连续界面与其在地层中的分布在统计上是相关的。但合成的剖面（图9-3）只能提供一维的视角。在此，由于未考虑未进入剖面就已经尖灭的横向非连续界面，平均每米中非连续界面数量可能被低估。不同界面的主要特征描述如下。

9.6.1 界面形态

9.6.1.1 描述

剖面中 Assoul 组展示了三种界面形态：(1) 平坦，局部平滑（图9-10A）；(2) 轻度不规则（局部波浪起伏），坡降<5cm（图9-10B）；(3) 不规则形态，坡降可达几厘米（主要是在固底到硬底初期，图9-10C）。局部地区平坦界面横向过渡为更加起伏的界面形态。上覆于鲕粒滩和礁丘之上的非连续界面特征是其不平坦性。在研究区，经常观察到非连续界面形态在横向上显著变化（例如，在横向几百米距离内由平坦形态过渡为轻度不规则形态）。

9.6.1.2 解释

前人（Jaanuson，1961；Read 和 Grover，1977；Fursich，1979）认为非连续界面的形态可以反映同沉积成岩作用的发育特征。但必须注意的是，非连续界面形态通常是前成岩作用（侵蚀、胶结和生物扰动，Hillgärtner，1998）和后生成岩相关作用（包括侵蚀、生物扰动和差异压实；Fursich，1979）的产物。Hillgärtner（1998）认为平滑形态通常与高能冲刷环境相关，例如正常浪基面之上的深海环境和后生成岩期侵蚀的过程。Fursich（1979）提供了英国、法国、波兰和德国境内侏罗系平滑非连续界面的详细资料（1979），这些界面通常是薄层粗粒滞留沉积，当水流搬运卵石和粗砂冲刷水下硬底时使其变得更加粗糙，研究区内少见滞留沉积。极不规则界面可能代表了碳酸盐海底成岩作用程度的差别（补丁状成岩，图9-10C）或穴居生物对原始平滑界面的改造作用，换言之，海洋、大气或者埋藏区的后生成岩作用、差异压实作用以及露头的局部侵蚀和岩溶均可改造界面的形态。Kennedy 和 Garrison（1975）讨论了潜穴充填和沉积母质之间的成岩作用差异与潜穴作用之间的关系。轻度不规则非连续界面可能是同沉积期均匀沉积物被重要的前成岩作用侵蚀的结果（Fursich，1979；Hillgartner，1998）。潜在的侵蚀营力包括浪底轨迹、洋流或底栖生物扰动。正如非连续界面形态的横向显著变化所展示的，上述所有因素以高度复杂和区域各异的方式共同作用，还需注意的是，没有任何证据表明 Assoul 组非连续界面曾出露水面或发生地表岩溶。

9.6.2 非连续界面化石相的古生态特征

9.6.2.1 前生成岩期遗迹化石相（潜穴）：描述

前生成岩期化石相，也被称为早期遗迹缺失体系，常体现在软底和固底之中（Goldring 和 Kazmierczak，1974；Fursich，1979）。在非连续界面之下或者内部软底内很少能保存有海相遗迹化石。与研究区内浅海灰岩相的特征一致，早期遗迹缺失体系很可能属于 Psilonichnus 遗迹化石相（与松软基底有关）、Skolithos 遗迹化石相（与松软和坚硬基底有关）和 Cruziana 遗迹化石相（与松软基底有关）。

图 9-10 非连续界面形态

(A)平滑非连续界面("长丘"露头中的 S2 型),顶面视图,黄圈中的笔为参照比例,长 12cm;(B)不规则形态的非连续界面(DS6),顶面视图,黄圈中的锤子(长 28cm)为参照比例;(C)上图:极不规则形态的非连续界面(位于"岛"露头,S2 型),坡降达 15cm,剖面视图,注意凹槽中的再沉积碎屑;下图:上述露头照片的描绘图,注意非连续界面与非均质基底间的复杂形态

图9-11 同生成岩期遗迹化石相(潜穴)

(A)大型的 *Thalassinoides* 潜穴,被黄褐色含珊瑚碎屑的白云岩充填("长丘"露头的 S2 型非连续界面)。笔为参照物,长15cm。(B)上图:同一地层中的两个(或更多)清晰的潜穴;下图:以上露头照片的描绘图。锤子为参照物,长28cm。(C)潜穴壁(DS5)指示的成岩作用的不同阶段:(1)不规则潜穴壁,指示未固结成岩的基底,铅笔为参照物,长13cm;(2)清晰潜穴壁,指示了在部分固结成岩基底中的生物扰动。(D)被黄色后生无作用沉积所充填的具有清晰壁的大型"Y"状 *Thalassinoides* 潜穴("长丘"露头的 S2 界面)。锤子为参照物。(E)高密度潜穴网。箭头指示了 *Thalassinoides* 潜穴(参照 Seilache,2007 的图版73)(DS10),笔为参照物

　　研究区内,几乎在所有非连续界面下都观察到了与上覆和下伏沉积岩相关的前成岩作用期的强烈生物潜穴。潜穴偶尔从非连续界面向下部碳酸盐岩扩展30~35cm,从而构成一个密集的网络。生物扰动的程度通常从顶部向下减小。在这种背景下,潜穴壁的特征及其与周边碳酸盐母岩的接触界面就具有重要意义。潜穴跟碳酸盐相母岩之间通常是轻度模糊到清晰的潜穴壁和接触面(图9-11)。潜穴很少被上覆地层的沉积物所填充。不论在露头还是样品中,潜穴填充物通常都是黄褐色到黄色,这是常观察到的一个特征,通常跟潜穴选择性白云岩化有关(Gingras 等,2004;Rameil,2008)。骨架材料包括完整和破碎的双壳类贝壳或者珊瑚碎

屑(图 9 – 11A),碳酸盐岩屑是潜穴相的常见成分。潜穴偶尔显示局部的含铁矿物颜色的潜穴壁(图 9 – 12)。潜穴几个阶段的证据包括潜穴体系的交叉横切关系和化石相的不同类型。在 46~70 米处非连续界面下观察到特定层理中两个及以上连续的潜穴段(图 9 – 3 和9 – 11B)。相似地,在一个非连续界面中也能局部识别到潜穴的两个阶段(图 9 – 11C)。

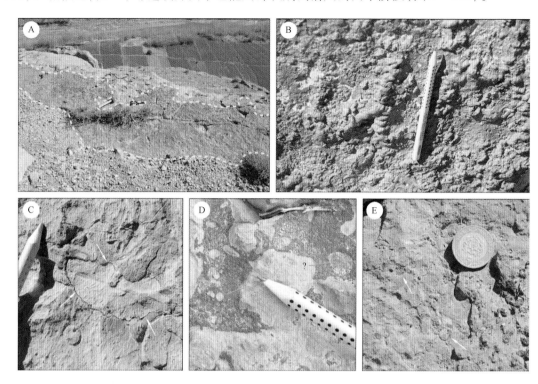

图 9 – 12 (A)具有普遍性铁矿化作用的非连续界面(DS5)。锤子为比例参照,长 28cm。
(B)非连续界面 DS5 之上几厘米厚的结核状铁壳。铅笔为比例参照物,长 13cm。
(C)非连续界面 DS5 中铁成矿作用影响潜穴壁(白色箭头)。铅笔为比例参照物。
(D)"岛"露头 DS5 非连续界面钻孔切割铁壳(红色箭头)。铅笔为参照物。
(E)铁染色优先影响填充钻孔顶部(白色箭头;DS5)。硬币为参照物,直径 2.5cm

Assoul 组固底和下伏的碳酸盐岩(主要是骨架粒泥灰岩或者骨架—球状泥粒灰岩)中从中等到广泛地发育代表固底的 *Glossifungites* 潜穴(Seilacher,1967)。*Glossifungites* 潜穴构造特征明显,但在研究区内则以 *Thalassinoides* 潜穴(与以沉积物为食的甲壳类生物相关的居住迹)为主。潜穴直径为 1~5cm 且向上减少(Seilacher 的术语,1964),(图 9 – 11A—E;Bromley,1975a;Goldring 和 Kazmierczak,1974;Palmer,1978;Brown 和 Farrow,1978)。研究区很多固底和硬底局部显示了强烈的 *Thalassinoides* 潜穴作用(图 9 – 11D)。

9.6.2.2 同生成岩作用和后生成岩作用动物和化石相:描述

研究区非连续界面(及其早期成岩阶段)固着生物包括双壳类(特别是牡蛎)、腕足类和珊瑚(图 9 – 13)。据 Taylor 和 Wilson(2002)研究,"表层动物群"指的是在硬底栖息的动物,而"层内动物群"指的是在固结成岩的地层中侵蚀和钻孔的动物(图 9 – 13A)。Bromley(1975a)曾经描述过更小的动物,比如在侏罗纪硬底动物群中典型的龙介虫,但在摩洛哥非连续界面中则没有发现。

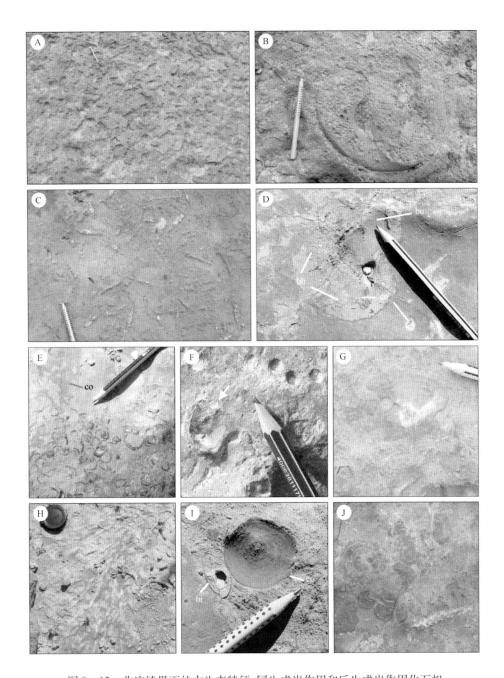

图 9-13 非连续界面的古生态特征:同生成岩作用和后生成岩作用化石相

(A)S3 型非连续界面,推测由层内生物群造成的普遍生物侵蚀导致不规则界面形态,铅笔为参照物,长 13cm。(B)发育大量钻孔的 DS5 界面中的双壳类化石,铅笔是参照物。(C)DS5 非连续界面顶部风蚀的腹足类化石地层。注意有一定的优势方位,铅笔为参照物。(D)潜穴(bu)、钻孔(bh)和板状珊瑚结壳(co)的非连续界面,铅笔为参照物。(E)被牡蛎和板状珊瑚结壳(co)的非连续界面 DS8,铅笔为参照物。(F)Gastrochaenolites 钻孔切割一个牡蛎(白色箭头),展示了非连续界面 DS8 上动物栖息的相对顺序,铅笔为参照物。(G)跟(F)相似,DS2 非连续界面中 Gastrochaenolites 钻孔切割一个板状珊瑚结壳,铅笔为参照物。(H)固结成岩之前,枝杈状珊瑚栖息的非连续界面 DS8。镜头盖为参照物,直径 5.5cm。(I)"岛"露头中腕足类(br)和类柱状珊瑚(co)造礁形成粗粒的 S3*型界面,铅笔为参照物。(J)非连续界面 DS5 横切面上的微生物构造,照片宽度为 8cm

牡蛎是固底动物群中最常见的表层动物(图9-13E、F),下伏沉积主要是岩化鲕粒砂岩浅滩(如图9-6剖面I3中硬底DS4)。牡蛎通常形成生物礁块或层,通常范围达到几十厘米,偶尔厚度可达数米(主要在Assoul组底部)。在大型生物礁层中,牡蛎跟腕足类、珊瑚和海绵动物共生,可以观察到古生物成带现象。在建隆底部和核部牡蛎和腕足类动物很多,块状珊瑚也时有出现。苔藓虫和微生物经常成为这些动物的外壳。海绵动物和珊瑚的丰度向上增加显著。珊瑚主要呈现丛状和枝杈状形态,含有小萼。以滤食动物为主以及小分叉珊瑚的出现表明这些建隆可能经历了压力环境,可能与水体中的高沉积物输入/富浊积体或高营养等级有关。部分腕足类动物还经常跟非连续界面的包壳动物群共生(图9-13I)。很明显并不是所有前成岩作用和硬底生物遗迹都跟其贝壳化石一起出现的。例如,丰富的 *Gastrochaenolites* 的双壳类痕迹化石都与钻孔共同保留而得以识别,但是在露头中却见不到它们贝壳的残余(图9-14)。另外,非钻孔双壳类动物(未确定)在非连续界面下的碳酸盐岩以及非连续界面中都有出现。非连续界面的另一种典型生物是板状和块状珊瑚(图9-13D、G),局地还发现有枝杈状珊瑚(图9-13H)。在"Baby Island"露头的DS2非连续界面中,直径5~25cm的块状珊瑚是主体(图9-13G)。

腹足类动物从常见(图9-13)到极为丰富(在DS5非连续界面54.1m处达每平方米500个腹足类动物;图9-13G)。在该处,腹足类动物长轴方向随机,但主要位于界面顶部。在其他沉积环境中,腹足类在非连续界面之下几十厘米厚的碳酸盐岩层顶端较常见。DS5硬底54.1m处(图9-3)出露了大约10个大型双壳类和腹足类动物,体型达20cm以上(图9-13B)。后者局限于54.1m处的非连续界面,尤其是在剖面I2、I3和I6之间的区域(图9-3和图9-5)。腹足类中文石常重结晶,被方解石交代。局部由文石溶解而形成的空洞被黄色—黄褐色细粒沉积物填充,可能与填充潜穴的物质类似。

完全岩化的基底(成岩作用完全)中的层内钻孔动物群以 *Trypanites* 化石相组合为特征(Frey 和 Seilacher,1980)。遗迹化石主要包括 *Gastrochaenolites* 钻孔(Leymerie,1842)和 *Trypanites* 钻孔(Magdefrau,1932),以及较罕见的 *Gnatichnus* 钻孔(Bromley,1975)。隶属于 *Trypanites* 化石相的岩层经常过渡为 *Gastrochaenolites* 化石相(横向上和地层上),还可能横切软底和硬底岩层(Bromley,1975a,1996)。

很多非连续界面显示石栖双壳类动物和海绵动物的向下钻孔和侵蚀的证据,局部可见石栖双壳类动物的外壳保存在生活区的钻孔中。不同界面的钻孔不同,同一界面的钻孔在横向上也有差异(图9-14A、C)。钻孔的直径2~20mm,平均5~10mm。露头中很难量化钻孔的大小,但是非连续界面岩层的垂向切片显示钻孔深度通常不超过1cm。而在局部有重要意义的非连续界面(比如图9-3、图9-6和图9-15中的DS4、DS5、DS6、DS7和DS8)中,钻孔经常出现,可以观察到最多达5代的钻孔叠置(图9-14E)。并非所有钻孔都同属一类,而且由于风化作用,它们也不能简单地归类为 *Gastrochaenolites*、*Trypanites* 或 *Gnatichnus* 钻孔迹。局部的最大分布频率估计可达每平方米10000个钻孔(图9-14D),即大约每$0.01m^2$ 100个钻孔。横向上几十米内,钻孔密度明显地增加或减小,或者钻孔完全消失。在有限的覆盖于高能鲕粒灰岩相之上的非连续界面中,钻孔比较罕见(图9-14A)。如果存在,钻孔填充物在岩性上跟潜穴的填充物很相似,但露头中的钻孔有时候是空的。

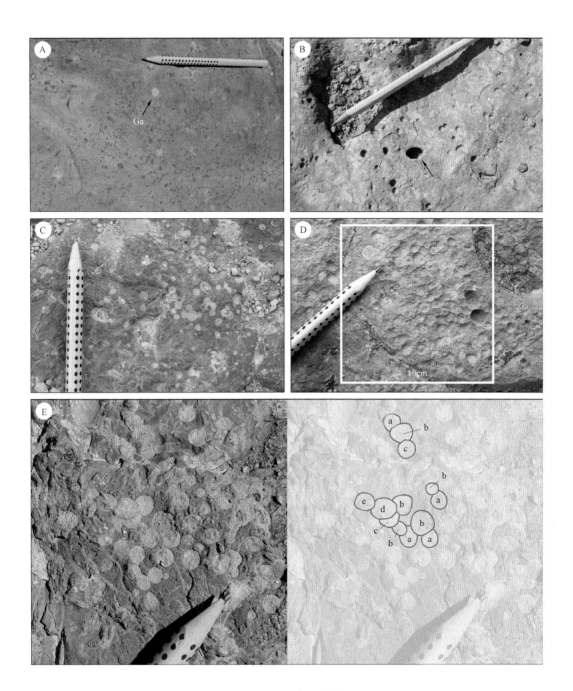

图 9-14 非连续界面的古生态特征:钻孔生物

(A)粗粒(浅滩砂体)非连续界面 DS4 上局部发育的 *Gastrochaenolites* 类(Ga)钻孔,表层视图(黑色箭头);(B)沙丘露头非连续界面 DS9 中的双壳类钻孔(黑色箭头);(C)具有 *Gastrochaenolite* 钻孔的非连续界面 DS4;(D)钻孔高度发育的非连续界面 DS5(每平方米 1000~10000 个钻孔);(E)左图:非连续界面特写,展示了几个钻孔的相对时期(a 最老,e 最年轻;右图)。铅笔为参照物,长 13cm

图 9-15 "岛"露头主要非连续界面(DS1 到 DS8)展布

主要非连续界面位于悬崖顶部,是局部的重要特征。(A)"岛"露头中识别出的 7 个非连续界面,DS8 遭受剥蚀;(B)东南悬崖壁上出现的 8 个非连续界面;DS1 和 DS2 也出现在"Baby Island"露头局部地区;(C)北部悬崖壁 8 条主要非连续界面(DS1 到 DS8)的全景分布,悬崖壁相对于"岛"露头的位置参见图 9-5

9.6.2.3 解释

解释非连续界面中的固着和活动后生动物的一个主要困难是区分前成岩作用、同生成岩作用和后生成岩作用期的动物群。海床侵蚀和刻划的几个显著阶段与沉积物减少时期的交替重叠也必须加以考虑。另外，海床上细粒沉积的淘选和下伏沉积物的剥蚀以及相关动物群可以产生跟双壳类栖息的非连续界面相似的贝壳层。如图 9 – 13C 中的腹足类动物在经历了上覆地层剥蚀后可能在界面上暴露出来。实际上，部分在非连续界面中占主导的后生动物也出现在界面之下几厘米处的最顶端。关于这些界面古生态学详细研究并非本文的范围，不在此过多研究，但对钻孔的出现（或不出现）的观察对于这些特征的解释有着重要意义。

类 *Thalassinoides* 潜穴是脱水、固底（但未胶结；Bromley，1975b；Ekdale 和 Bromley，1984；MacEachern 和 Burton，2000）或者软底沉积物的特征（Bromley 和 Frey，1974；Savrda 等，2001a）。Assoul 组中的遗迹化石暂时划归 *Thalassinoides* 化石相，主要表现为良好的清晰的无衬里化石壁，也是固底相关构造的特征。*Ophiomorpha*（Lundgren，1891）是软底中类 *Thalassinoides* 遗迹化石的专有名称，它有着小球构成的固结衬里化石壁（Bromley 和 Frey，1974），未在 Amellago 峡谷中观察到。研究区内软底海洋化石相的缺失可能是因为固底和硬底的形成机制具有损坏作用（潜穴和钻孔，生物剥蚀或压实）。在浅海环境中，表面不连续的遗迹化石可能很快消失，因为沉积物上层的强烈混合作用，形成了广泛分散的斑驳纹理（Bromley，1996；Jensen 等，2005）。

Gastrochaenolites 是由硬底钻孔生物形成的，因此是后生成岩作用的海床遗迹化石，但是很多作者（Kelly 和 Bromley，1984；Pemberton 等 2000；Ekdale 和 Bromley，2001；Carmona 等 2007；MacEachern 等 2007）都认为这些生物也可能在固底中占主导地位，因此也是同生成岩作用阶段的典型代表。珊瑚是某些非连续界面的典型栖息动物（图 9 – 13G），经常被 *Gastrochaenolites* 钻孔所钻穿。

在非连续界面形成生物礁块的牡蛎，被认为是后生成岩作用期动物的代表，比沉积再作用面更早。某些非连续界面（如图 9 – 6 的 DS5 界面）的牡蛎有着氧化铁外壳，然而其他的却没有任何次生铁化合物的外壳，寄居在氧化铁外壳上的第三族群覆盖了硬底。在多个界面中（如图 9 – 6 的 DS8），数个牡蛎化石被 *Gastrochaenolites* 钻孔所削截（图 9 – 13F）。这些削截关系解释了部分非连续界面长期复杂的栖居过程，并标明了长期的沉积缺失。其他特征，比如双壳类钻孔切割较老钻孔（图 9 – 14C、D、E）是持续无沉积和生物逐步定殖形成硬底界面的证据，这包括非连续界面 DS4、DS5、DS6、DS7 和 DS8（图 9 – 6）观察到的 *Gastrochaenolites*，它是界面连续侵蚀的标志（Fursich 等，1992）。腹足类生物碎屑中的钻孔是后无作用阶段的证据，但可能在贝壳中不稳定的文石发生成岩作用时消失。

研究区界面之下不到 1cm 深的 *Gastrochaenolites* 钻孔是成岩强化过程中显著的碳酸盐沉积物侵蚀和海床淘洗的证据。相比之下，Sattler 等（2005）发现了相似的钻孔深度，但 Benner 等（2004）认为是 3～4cm，Ekdale 和 Bromley（2001）认为达到 6cm，Watkins 描述的 *Gastrochaenolites* 钻孔达 7cm 深，而 El-Hedeny（2007）报告了深达 15cm 的钻孔；另外，Watkins（1990）、Benner 等（2004）和 Ekdale 等（2002）提供了遗迹化石及其孔径、孔腔和孔顶的精细描述（Kelly 和 Bromley，1984 的术语）。需要指出的是，距硬底界面更近的 *Gastrochaenolites* 钻孔总体相对较窄（1～3mm），也取决于其孔腔的长度（Wilson 等，1998）。研究区钻孔直径 5～10mm，可能与孔腔被下切侵蚀的部位有关，导致其直径比统计孔径更宽（图 9 – 1；Kelly 和 Bromley，1984）。空的钻孔最可能的解释为现今气候下填充物被侵蚀或移除。侏罗纪硬底中很常见的龙介虫（Bromley，1975a；Fursich，1979）在本区未有发现，海床显著的水下侵蚀和出露风化可能是其消失的主要原因。

考虑到潜穴相对于钻孔的相对时间，切割了潜穴的钻孔还没有观察到。相反，在几个主要界面中观察到了被高达5代叠合钻孔所钻穿的沉积物填充 *Gastrochaenolites* 钻孔(例如图9-14E的DS5),该特征表明了长期的无沉积作用阶段(Purser,1969;Shinn,1969)。根据上述作者的观点,叠合钻孔的形成要么需要连续的海底成岩作用,要么需要特别的海平面波动来形成淹没和暴露的反复旋回,从而形成钻孔和沉积物的充填。本区的主要非连续界面DS5(图9-3、图9-6和图9-15)中的钻孔被几种不同的碳酸盐相沉积填充(图9-14E),每种沉积相都表征着不同的水动力水平,这也表明了该界面经历了长期复杂的无沉积阶段。

9.6.3 非连续界面的浸染

9.6.3.1 描述

Assoul组非连续界面中与凝缩段相关的次生铁氧化物、锰氧化物或磷酸盐矿化作用是常见特征(图9-12),前人描述过相似的矿化作用(Bromley,1975b;Fürsich,1979;Carson 和 Crow-ley,1993;Gomez 和 Fernandez-Lopez,1994;Burkhalter,1995;Hillgärtner,1998)。部分横向展布有限的界面局部被次生铁氧化物和锰氧化物所浸染,区域上只有6个有重要意义的硬底显示了丰富且广泛分布的矿化作用特征(图9-6和图9-12A)。后者局部厚度可达3~4cm,有结核状的浸染结核(图9-12B)。在很多实例中,钻孔和潜穴的侧壁被次生矿物所浸染(图9-12C)。在其他例子中,钻孔清晰地钻穿了浸染壳(图9-12D的DS5)。局部观察到钻孔或潜穴的选择性矿化作用(图9-12E)。

9.6.3.2 解释

固结成岩的碳酸盐海床中的次生矿化通常与凝缩作用有关,以及与长期水下沉积间断形成的海洋硬底有关(Follmi 等,1991;Pomoni-Papaioan-nou,1994;Clari 等,1995;Watkins 等,1995)。需要指出的是,不能将同沉积矿化作用与晚期成岩作用再活化、铁锰氧化物的沉积以及地下流动障壁可溶物沉淀相混淆(Immenhauser 等,2000)。通常原生及次生浸染矿物并不是直接析出的,晚期埋藏阶段碳酸盐岩和页岩界面上可溶矿物系列的活化作用很常见(Immenhauser 等,2000)。通过野外露头观察可以研究浸染、钻孔和晚期矿脉切割矿化壳的相对时间,在研究区界面背景下观察钻孔和界面浸染的关系对其研究也有所帮助。海洋硬底环境中次生铁氧化物矿化作用导致的钻孔和潜穴壁浸染(图9-12C)指示了晚期同沉积过程或者埋藏相关矿化作用。钻穿了浸染外壳的钻孔成为生物钻孔晚于海底矿化的直接证据(图9-12D)。研究区未见硅结砾岩、钙质砾岩或红土褪色或染色等近地表暴露矿化作用的证据(Wright,1994;Immenhauser 等,2000;Sattler 等,2005)。

9.6.4 非连续界面两侧的沉积相对比

9.6.4.1 描述

本文利用系统化方法研究了该区39个出露的硬底和固底非连续界面两侧的水动力条件和相关沉积相变的差异。考虑到浅水环境中碳酸盐相的自然变化,把泥岩详细地划分为从黏土岩到粒泥灰岩几个等级及类似方法并没有被采用。不同碳酸盐相的意义是其体现了水体能量变化。实质上,39个非连续界面中有15个(15个固底)的特征是界面两侧没有(或无明显)发生相变(如界面之下的生屑漂砾灰岩(基质是粒泥灰岩)到界面之上的生屑粒泥灰岩)。两个非连续界面表现为中等相变(如界面之下的含球粒颗粒灰岩到界面之上的含球粒泥粒灰岩),还有22个界面表现为显著的相变化(如界面之下的鲕粒灰岩到界面之上的泥灰岩)。表9-2列出了Assoul组S2和S3型非连续界面的相变,并整理为统计图9-16。这些相变在储层封隔中具有重要意义,作为跟横向上广泛分布的硬底界面相关的成岩层位,结合上覆的低渗泥灰岩相,共同成为垂向运移的很好的遮挡层。

表 9－2 主要非连续界面岩相变化（S2 与 S3）

界面类型	地层位置*（m）	水动力强度变化	下部岩相/厚度（cm）	上部岩相/厚度（cm）
固底	76.30	无变化至较低	似核形石 Ws－Fs/45	似核形石 Ws－Fs/20
固底	82.80	无变化至较低	生物碎屑 Ps－Fs/90	生物碎屑 Ps－Fs/25
固底	105.20	无变化至较低	Ms/290	泥灰－Ms/20
固底	114.40	无变化至较低	Ms/160	泥灰/5
固底	143.30	无变化至较低	Ms/410	Ms/50
固底	158.80	无变化至较低	Ms－Ws/210	未知/20**
固底	176.60	无变化至较低	Ms－Ws/330	未知/5**
固底	195.50	无变化至较低	生物碎屑 Fs/35	生物碎屑 Ws/80
固底	196.30	无变化至较低	生物碎屑 Ws/80	生物碎屑 Ws/70
固底	197.00	无变化至较低	生物碎屑 Ws/70	生物碎屑 Ws/50
固底	210.30	无变化至较低	生物碎屑 Fs/130	生物碎屑 Fs/90
固底	211.20	无变化至较低	生物碎屑 Fs/90	生物碎屑 Fs/90
固底	212.10	无变化至较低	生物碎屑 Fs/90	似核形石 Fs/50
固底	212.60	无变化至较低	似核形石 Fs/50	生物碎屑 Fs/20
固底	212.80	无变化至较低	生物碎屑 Fs/20	生物碎屑 Fs/45
硬底	45.70	较低变化至较高	似核形石 Fs/40	鲕粒 Gs/90
固底	74.60	较低变化至较高	生物碎屑 Ws/25	球状 Ps－Gs/45
固底	84.90	较低变化至较高	Ms－Ws/120	球状 Ws－Ps/20
硬底	88.30	较低变化至较高	生物碎屑 Ws/25	牡蛎生物化石/15
固底	219.00	较低变化至较高	生物碎屑 Ws/25	鲕粒 Gs/25
硬底	14.60	较高变化至较低	生物碎屑 Ws－Fs/15	泥灰/10
硬底	22.00	较高变化至较低	球状 Ps－Gs/35	泥灰/40
硬底	29.40	较高变化至较低	生物碎屑 Ps/10	未知/80**
硬底	33.40	较高变化至较低	生物碎屑 Fs/60	泥灰/60
硬底	45.20	较高变化至较低	球状 Gs/15	球状 Ws－Ps/10
硬底	54.10	较高变化至较低	生物碎屑 Ps/300	未知/80**
硬底	60.50	较高变化至较低	生物碎屑 Fs/20	Ms－Ws/50
固底	63.20	较高变化至较低	生物碎屑 Ws/60	Ms/15
固底	73.80	较高变化至较低	生物碎屑 Ws－Ps/150	Ms/55
硬底	78.60	较高变化至较低	生物碎屑 Ws/50	未知/40**
固底	119.80	较高变化至较低	似核形石 Fs/60	未知/80**
固底	123.10	较高变化至较低	生物碎屑 Ps－Fs/20	Ms/40
固底	151.30	较高变化至较低	生物碎屑 Ps－Fs/20	泥灰/15
固底	178.80	较高变化至较低	似核形石 Ps－Gs/150	未知/15**
硬底	197.50	较高变化至较低	生物碎屑 Ws/50	Ms/20
硬底	198.40	较高变化至较低	生物碎屑 Ws/65	未知/30**
固底	208.40	较高变化至较低	生物碎屑 Fs/35	未知/15**
固底	215.50	较高变化至较低	生物碎屑 Fs/20	Ms/35
固底	216.00	较高变化至较低	生物碎屑 Fs/15	Ms/90

注：＊代表 Assoul 组综合剖面中非连续界面地层位置（图 9－3）；＊＊未知或"未被观察到"序列，可能为被风化的泥灰岩。岩相术语（例如 Ms 代表 Mudstone）参考表 9－1。

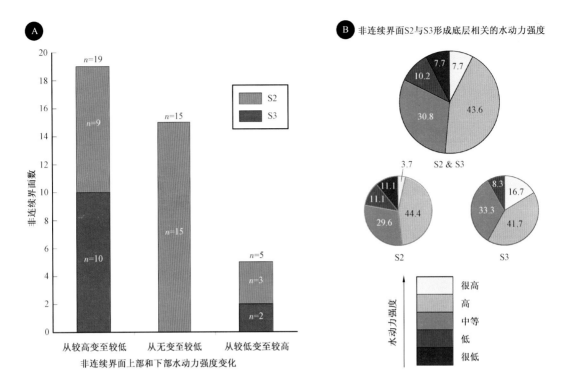

图9-16 （A）S2固底与S3硬底之间岩相变化对比；（B）S2固底与S3硬底以下沉积物相对丰度（单位：%）及其结构和相关水动力强度对比

注意大多数非连续界面形成于碳酸盐相，指示中—高度水动力强度

9.6.4.2 解释

在古环境解释中，碳酸盐沉积相可以与不同浪基面或者洋流能量（水动力强度）相联系起来。观察到三种主要类型：（1）通过沉积相对比，不连续代表无或者很有限的水动力强度变化；（2）以水动力强度从相对低转换为相对高为特征的非连续界面；（3）非连续界面代表相反趋势，即界面之下和之上沉积相的水动力强度从高变低。

显示水动力强度没有或只有很少变化的非连续界面典型实例包括114.4masb（masb：剖面底面之上的米数）显示为从泥岩变化为泥灰岩，或者，212.1masb处的固底显示从界面之下生物碎屑漂砾灰岩变为界面之上核形石漂砾灰岩（图9-3）。显示水动力强度从相对低变为相对高的非连续界面实例包括88.3masb处的硬底显示生物碎屑粒泥灰岩变为牡蛎粘结灰岩，以及219masb处的固底显示从生物碎屑到鲕粒灰岩的突变（图9-3，表9-2）。最后，显示水动力强度从相对高到相对低变化的非连续界面包括22masb处硬底，显示了从界面之下的球状泥粒灰岩/颗粒灰岩变为界面之上的海退的泥灰岩相。相似地，215.5masb处的固底特征是从生物碎屑漂砾灰岩变为上覆泥岩的非连续界面。

上述观察（图9-16A）与Fursich（1979）发布的在欧洲各地侏罗系硬底中的观察有着显著的不同。Fursich（1979）认为形成顶部高能鲕粒灰岩相是硬底的主要特征。相似地，Brett和Brookfield（1984）也观察到大量形成于浅滩环境粗粒基底的硬底。研究区Assoul组露头中有10个主要非连续界面（图9-3）形成于中—高能基底上，只有2个形成于球粒鲕粒灰岩之上（表9-2）。

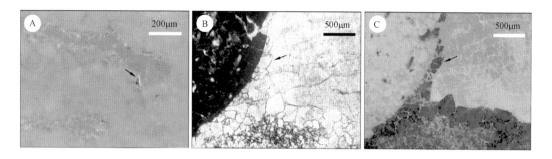

图9-17 非连续界面DS5之下的孔隙填充胶结的形成与埋藏成岩作用及其阴极发光(CL)显微镜下的特征
(A)具有发光边缘的三角形方解石胶结物(黑色箭头)。注意该胶结物也可能与早期海相成岩作用有关。
(B)透射光下鞍状白云石(黑色箭头),以及(C)阴极发光视域(黑色箭头)

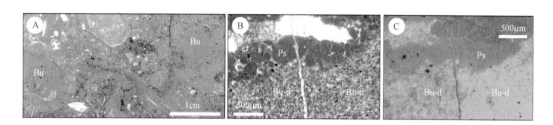

图9-18 (A)DS5以下地层的样品薄片,显见淡黄色—黄褐色潜穴充填物(Bu)。(B)孔隙充填物,
含球粒粒泥灰岩基质(Ps)与菱形方解石(Bu-d)相伴生,透射光下,两者皆被一条晚期方解石脉切割;
(C)阴极发光

9.6.5 胶结物岩相学及与非连续界面相关的碳酸盐的碳、氧同位素特征

9.6.5.1 碳酸盐胶结物:描述

发现了以下碳酸盐相:(1)片状偏三角面体胶结物;(2)填充于孔隙内的块状方解石胶结物;(3)鞍状白云石;(4)菱形方解石。非连续层中分布有偏三角面胶结物,少数情况下厚度较小,($<50\mu m$),通常为等厚沉积物(图9-17A)。在阴极发光条件下,边缘发光(图9-17A)。块状胶结物尺寸范围从几十微米至几毫米不等,通常认为是交代早期腹足类贝壳中文石的产物,并且后期填充于方解石脉。该胶结物相发微弱光至束状光。鞍状白云石较常见,其尺寸为$100\sim500\mu m$(图9-17B、C)。阴极发光片中主要呈深棕色(图9-17C)。菱形方解石主要赋存于生物钻孔与孔洞中填充的黄色—褐色沉积物中(图9-18A、B)。尚不清楚其是否代表了去白云石化作用或以方解石形式沉淀;阴极发光呈微弱—亮橘黄色(图9-18C),其尺寸从几微米至$100\mu m$。在研究薄片中尚未观察到月牙形以及重力胶结物。

9.6.5.2 碳酸盐胶结物:解释

浅埋至深埋成岩中均见孔隙填充胶结物(偏三角面体方解石以及鞍状白云石)。块状方解石胶结物可形成于多种成岩环境中,但是根据其发光类型,本区块状方解石胶结主要形成于埋藏期(Richter等,2003);因此,就提出了是否存在普遍性的海底形成的早期海相胶结物的问题。可能来源于早期海相沉积的沉淀物为轻微的偏三角面胶结物;目前最合理的解释为海相细粒孔隙充填,该胶结物甚至在阴极发光条件下也难以分辨。该非连续界面是海相还是埋藏成因不在本文讨论范围之内。

表 9-3 固底和硬底之下各种来源的碳酸盐碳与氧同位素

类型		泥晶灰岩 $\delta^{13}C$	潜穴/孔隙填充物 $\delta^{13}C$	埋藏胶结物 $\delta^{13}C$	泥晶灰岩 $\delta^{18}O$	潜穴/孔隙填充物 $\delta^{18}O$	埋藏胶结物 $\delta^{18}O$
未指定非连续界面	N 岩样	38	18	17	38	18	17
	N 样品	98	46	30	98	46	30
	最大(‰)	+3.7	+4.2	+4.3	-3.2	-3.0	-4.2
	最小(‰)	+0.8	-4.4	+2.7	-6.2	-8.4	-13.5
	平均(‰)	+3.1	+1.5	+3.2	-4.4	-6.3	-7.1
	标准偏差	0.5	2.0	0.3	0.6	1.5	2.2
固底	N 岩样	17	7	5	17	7	5
	N 样品	43	16	8	43	16	8
	最大(‰)	+3.6	+4.2	+4.3	-3.2	-3.0	-5.1
	最小(‰)	+1.7	-1.8	+2.7	-6.1	-8.0	-7.5
	平均(‰)	+3.2	+2.2	+3.2	-4.3	-5.1	-6.8
	标准偏差	0.4	1.7	0.5	0.5	1.4	0.8
硬底	N 岩样	21	11	12	21	11	12
	N 样品	55	30	22	55	30	22
	最大(‰)	+5.0	+3.4	+3.4	-2.9	-4.1	-4.2
	最小(‰)	+0.8	-4.4	+2.8	-6.2	-8.5	-13.5
	平均(‰)	+3.0	+1.0	+3.2	-4.6	-6.9	-7.2
	标准偏差	0.5	2.0	0.2	0.5	1.1	2.6

9.6.5.3 碳、氧同位素:描述

在碳酸盐岩地层中,稳定同位素为解释不整合界面的主要(虽然也与环境相关)工具(Allan 和 Matthews,1982;Joachimski,1994;Sattler 等,2005)。许多研究中发现(Immenhauser 等,2008,详细论述见参考文献),在相对海平面下降以及海底出露剥蚀的情况下,在碳酸盐岩地层中可以发现雨水中较轻的氧同位素以及土壤带较轻的碳同位素。

研究中采用了分别命名为地层和岩石的两种不同的地球化学方法。这里所采用的化学地层法主要基于从总计 9 个典型非连续界面中采集的基质泥晶灰岩样品,这些非连续界面涵盖了 Amellago 峡谷中所能观察到的所有非连续界面类型。化学地层段从地表以下几米的深度开始,一直到上覆于碳酸盐岩地层,约为 50cm。基质样品的碳同位素值为 +2.1‰ ~ +5.0‰,涵盖非连续界面及其底部($n=74$;平均值 = +3.2‰;标准偏差(σ) = 0.3)。然而,对非连续界面进行取样时,未发现清晰的同位素变化趋势。同一层段,基质样品氧同位素值范围为 -6.1‰ ~ -2.9‰($n=74$;平均值 = -4.5‰;σ = 0.5)。其中,20 个样品在非连续界面下方的同一层段采集(非连续界面以下 0 ~ 20cm 处)。在非连续界面下方采集的基质样品碳同位素值为 +2.1‰ ~ +3.4‰(平均值 = +3.0‰;σ = 0.3),同一层段 $\delta^{18}O$ 率为 -5.0‰ ~ -3.6‰(平均值 = -4.4‰;σ = 0.5)。

岩相方法中,从非连续界面以及其下(0 ~ 20cm)所有肉眼可见胶结物以及钻孔与生物钻孔中的充填物(细粒基质;图 9-18)采集子样品,用以评价隐藏成岩阶段的地球化学证据,而

这些证据由这些特点表明。进行地球化学分析的岩相子样品收集于10个分布于不同空间和地层的非连续界面(包括固底以及硬底),且对其进行了分析,分析结果在表9-3中列出。采集的进行地球化学分析的胶结相样品中有少量发弱光偏三角面(犬牙状)胶结物(图9-17A),束状—弱光块状胶结物以及深棕色光鞍状白云石(图9-17B和C)。

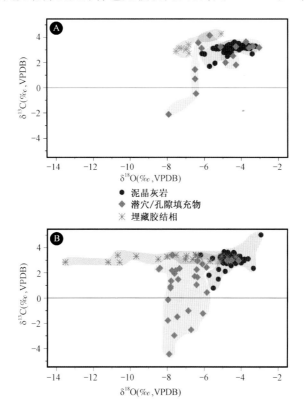

图9-19 (A)两个露头6类固底泥晶灰岩$\delta^{13}C$与$\delta^{18}O$值对比。图中显示不同胶结相和潜穴充填沉积物的数据。(B)S3硬底(DS4、DS5、DS6以及DS9)4种主要泥晶灰岩样品$\delta^{13}C$与$\delta^{18}O$值。图中显示不同胶结相和潜穴充填沉积物的数据

将这些子数据集中的地球化学数据进行比较,可以观察到以下现象:(1)基质泥晶灰岩碳同位素值分布范围相对狭小,从约+1‰变化至+4‰;(2)相反的,位于非连续界面内部及其下方的生物钻孔与钻孔充填沉积物$\delta^{13}C$和$\delta^{18}O$同位素值主要为$\delta^{18}O$负异常(低至-8.5‰)和$\delta^{13}C$负异常(低至-4.4‰;图9-19);固底差异较小(图9-19A),而硬底差异较显著(图9-19B)。碳酸盐胶结物主要为$\delta^{18}O$负异常(低至-13.5‰),而其$\delta^{13}C$值变化不大(图9-19)。

9.6.5.4 碳、氧同位素:解释

非连续界面下方和上方样品的泥晶灰岩$\delta^{13}C$和$\delta^{18}O$值都显示出较小差异,这与非连续界面未经历出露地表从而不具有与界面有关的较强的碳、氧同位素负异常趋势相一致(Allan与Matthews,1982;Humphrey等,1986;Beier,1987;Marshall,1992;Immenhauser等,2002)。该结论具有重要的意义,其与研究区地表没有与出露相关显著特征相一致。

不同碳酸盐岩胶结物的氧同位素值分布范围较大,为-14‰~-4‰,而$\delta^{13}C$变化较小。该特征很难用降水相关的贫^{18}O流体来解释,即便季节性降水有助于土壤和植被相关有机质

的形成,从而可以导致贫^{13}C的同位素(Lohmann,1988)。该同位素类型在浅部成岩(埋藏)胶结沉淀物中更为明显(Van Der Kooij等,2009),在岩石缓冲体系(δ^{13}C变化较小)中,由于埋藏流体温度升高而导致δ^{18}O值负异常,该特征与明显的缺少大气淡水成因胶结物相一致。

实验测得钻孔与生物钻孔填充物样品的δ^{13}C值介于$-4.5‰ \sim +3‰$,而δ^{18}O值($-9‰ \sim -6‰$)较低但稳定,难以解释。Rameil(2008)报道了Jura山脉侏罗纪早期成岩阶段缝洞白云岩。与Jura缝洞白云石相比(总δ^{13}C值介于$+2.0‰ \sim +2.6‰$;总δ^{18}O值介于$-4.3‰ \sim -1.9‰$),摩洛哥碳、氧同位素值皆偏低,主要与埋藏卤水中有机质分解以及含轻有机碳有关(Van Der Kooij等,2009),但也可能是由于更为复杂的共生史。值得注意的是,沉积、岩相以及地球化学证据都不能表明Assoul组非连续界面演化过程中有地表出露阶段。

9.7 Assoul组非连续界面分类

根据上述特征,Assoul组记录的所有非连续界面都具有水下固底或硬底以及海相界面缺失的特征。露头与薄片中没有观察到地表出露面证据,如土壤带CO_2影响引起明显的稳定同位素变化或地表岩溶、重力胶结物、钙质胶砾岩、根迹以及褪色等岩相证据(Lohmann,1988;Joachimski,1994;Clari等,1995;Immenhauser等,2000)。据此,可以得出,这些非连续界面代表单一或多个海相岩化界面,而非Immenhauser(2000)和Sattler等(2005)所描述的地表出露界面或复杂组合界面(既有水下岩化又有地表出露证据)。

以术语的精确性和可用性为目的,对Assoul组复杂的非连续界面进行归类,可分为3类,见表9-4。该分类根据所代表沉积间断、非连续界面横向范围、界面之下的岩化层段地层厚度以及生物穿孔和生物扰动强度的重要性增序排列。非连续界面重要性最低的群组主要为显示初期岩化或凝缩的界面(称为界面S1);比其重要的群组包括海相固底(称为S2);最重要的一类界面是区域性海相硬底(S3),其在时间上和横向上均有一定的延展性。在该界面分类中,又采用增加的星号(*)数量进行进一步细分。例如,S2*** 固底与S2* 固底相比,具有更高的空间重要性。各层面类型的特征在下文进行总结,详细情况如表9-4所示。

表9-4 研究区非连续界面特征

类型	生物扰动强度	钻孔以及潜穴与围岩的界面	生物扰动深度	界面横向连续性	矿化	研究界面主要特征
S1:凝缩层:若没有生物扰动,只具有轻微以及浅显的潜穴,生物碎屑从存在到频繁出现						
S1	生物钻孔或生物扰动不明显	非常微弱—微弱	浅薄(<2cm)	横向不连续(<10m)	无	局部且总体上不相关
S2:固底(部分岩化界面):频繁出现潜穴与生物碎屑						
S2*	轻度生物钻孔	微弱—适中	浅薄—浅(<10cm)	横向不连续(<100cm)	无	局部出现且总体上不相关
S2**	中度生物钻孔	适中—清晰	浅薄—中等深度(主要介于5~15cm)	可变(几十米至几千米)	无	构成层顶,或多或少有所关联
S2***	强烈生物钻孔	清晰	中等—深(介于5~30cm)	可变(几十米至几千米)	无	构成重要层位顶部,通常相关

续表

类型	生物扰动强度	钻孔以及潜穴与围岩的界面	生物扰动深度	界面横向连续性	矿化	研究界面主要特征
S3：硬底：发育孔隙或结壳（补充特征：FeO 矿化与大量生物碎屑）						
S3*	轻度—中度钻孔，无多代钻孔	非常清晰	钻孔浅薄	<1km	轻微—明显 FeO 浸染	构成形态上稳定的地层顶部，但局限于在（鲕粒）浅滩条件下
S3**	中度—强烈钻孔，存在多代钻孔（2）	非常清晰	钻孔浅薄	1km 或更长	轻微—明显 FeO 浸染	构成主要峭壁或形态稳定地层顶部
S3***	强烈钻孔，频繁出现多代钻孔（3-4）	非常清晰	钻孔浅薄	1km 或更长	轻微—明显 FeO 浸染（2~3cm 厚的小瘤）	构成主要峭壁或形态稳定地层顶部

9.7.1　凝缩面（S1）

凝缩面的特征是从下至上生物钻孔强度不断增大，其中强度最大的钻孔密度出现在层面以下最上部 2cm 处，形成一个小的非连续界面，成岩较弱。凝缩面主要出现在泥岩—粒泥灰岩相的顶部。90% 的凝缩面皆具有以上特征。腹足类与双壳类化石为主的生物碎屑存在于其中并在局部以较高频率出现。该类界面横向延伸不超过几十米，横向连续性差。就对储层影响而言，该界面类型出现频率较高，但由于其横向展布范围有限，其重要性不大。

9.7.2　固底（S2）

固底代表剥露的古海床，以低—中度岩化为特征（硬底的初期阶段），固底相当于部分岩化界面。其下方具有生物孔洞，界面上发育高频—丰富的生物碎屑，但岩化层下段也相当丰富，界面以下生物扰动深度范围从薄（约 5cm）至厚（30cm）。与凝缩面相比，固底横向范围较大，延伸距离可达几百米。该界面对碳酸盐岩储层中的流体流动有中度的影响。

9.7.3　海相硬底界面（S3）

海相硬底出现孔洞或结壳，次生铁矿化较频繁。常含有大量生物碎屑（双壳类、腹足类、珊瑚、腕足类以及牡蛎）。硬底多始于 Amellago 峡谷的 Assoul 组，主要覆盖稳定地层以及重要的地层层序。根据之前的论述，该特征代表持续较久的间断时期以及更为重要的缺失。部分 S3 界面形成晚于较小的非连续界面，这代表了岩化的几个阶段，因此地层向上具有凝缩程度上升的趋势（例如局部显示，图 9-20 非连续界面 DS5 之下）。硬底通常横向延伸超过几百米至几千米，影响浅滩鲕粒砂岩的硬底除外，其横向延伸范围小（少数 100m）。由于其具有空间延展范围大，受影响的岩层层段广，上覆的通常为页岩低渗透相等特征，认为 S3 界面对储层封隔的意义重大。

9.7.4　S1、S2 以及 S3 非连续界面相对地层分布

平均 220m 厚的地层层段中共发育了 44 个 S1 界面、26 个 S2 界面以及 10 个 S3 界面（图

9-3),界面/厚度比分别相当于每5m一个凝缩面(S1),每8m一个固底(S2),以及每18m约一个海相硬底(S3)。

但是更详细的评价显示界面在地层中呈不规则分布。例如,35~40m的地层层段发育了7个S1界面,201~204m则发育了9个S1界面(图9-3)。不过,若忽略相对于地层厚度的界面不规则分布,可以得到以下明显的分布规律。泥质斜坡体系发育13个S1界面(每10m含有1~6个S1界面),鲕粒斜坡体系发育了31个S1界面(每10m含有2~3个S1界面);泥质斜坡体系发育5个S2界面(每10m含有0~6个S2),而鲕粒斜坡体系发育21个S2界面(每10m含有1~6个S2);泥质斜坡体系不含S3界面,鲕粒斜坡体系发育10个S3界面(每10m含有0~7个S3界面)。每个地层层段(下部粒状层段、中部泥质层段以及上部粒状层段)中不同的非连续界面类型相对统计分布以柱状图形式列于图9-21。

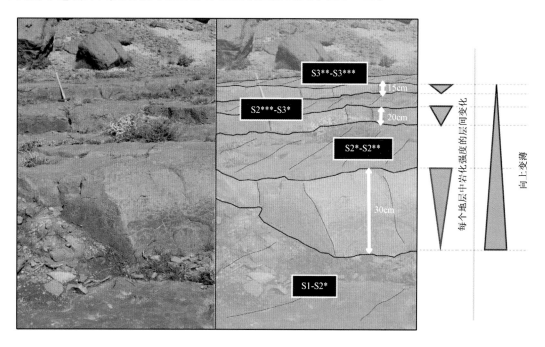

图9-20 三个顺序发育的非连续界面实例
每一个后续阶段长期无作用及海床岩化证据增多,最后一个阶段形成了主要非连续界面DS5

9.8 讨论

9.8.1 碳酸盐斜坡相、水深和非连续界面的关系

Pierre(2006)和Pierre等(2010)描述了两种不同的交替高频斜坡沉积体系:(1)主要是泥灰岩、泥岩、球状粒泥灰岩和牡蛎、珊瑚、腕足类粘结岩构成的低能泥质斜坡体系;(2)中高—高能量鲕粒斜坡沉积体系,特征是球状粒泥灰岩到泥粒灰岩、鲕粒—鲕粒生物碎屑泥粒灰岩到颗粒灰岩,以及牡蛎、珊瑚和腕足类粘结岩。Pierre等(2010)的术语中,这两个沉积模式代表米级高频沉积相模式。但要注意的是,Pierre等(2010)认为的高频沉积相模式比图9-3记录的总体偏泥质和总体偏颗粒质的斜坡沉积体系之间有着更大规模(几十米)的改变。为避免混淆,应该注意下文中出现的规模和术语上的不同。

观察了三套地层,分别是:(1)下部的,总体高能的,Assoul组底部的鲕粒层(图9-3的

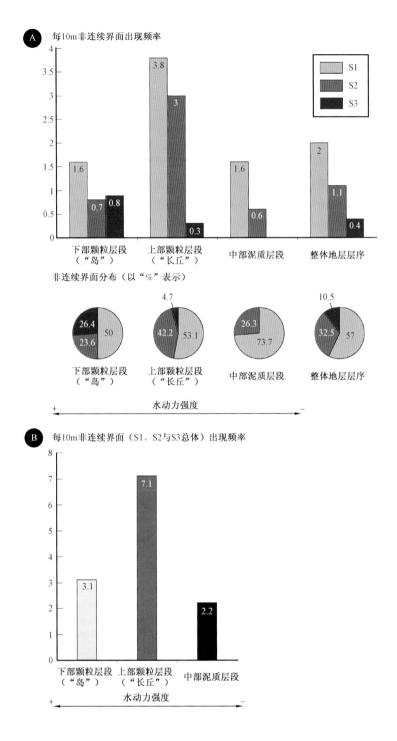

图9-21 (A)颗粒和泥质层段(图9-3)非连续界面S1、S2和S3频率和相对分布(单位:%)。注意真正的海相硬底(S3)下部颗粒层段中最为普遍,在泥质层段中缺失。(B)颗粒和泥质层段所有界面类型频率分布。注意界面的分布特征为在颗粒层段上部出现频率最高

0~90m);(2)中部泥质层位(图9-3的90~175m),水动力强度较低;(3)冲刷较强的鲕粒碳酸盐和开放潟湖相(图9-3的175~220m处)交替的上部层位,是研究区内Assoul组出露地层的顶部。下部颗粒相代表最高水动力强度,之后的上部颗粒质地层代表高—中高水动力强度,中部的泥质层位是典型低能相。人们对下部总体偏高水动力强度和上部颗粒质层位存在争议,包括冲刷较强的具交错层理的厚层和鲕粒沙丘沉积是否出现,后者在中部泥质层位完全缺失。

只考虑水深、波浪和洋流能量对颗粒质(浅水,较高的水动力强度)和泥质(深水,低水动力强度)相的解释可能过于简单化,还必须考虑其他非周期性因素,比如气候、陆地风化和地表径流、风力、风暴模式和导致的波浪气候、海水营养度和酸碱度等(Shinn,1969;Kempe 和 Kazmierczak,1993;Hill-gartner,1998;Immenhauser 等,2000)。同样,周期性因素比如鲕粒浅滩的移动、潮道、海水环流模式或者局部有机质数量(Strasser 等,1999)也有一定影响。

本文讨论的两个露头位于下部("岛"露头;图9-3、图9-4A、图9-5和图9-6)和上部颗粒质层位(沙丘露头,图9-3、图9-4B和图9-8)中。图9-3的概略剖面连接了两个颗粒质层位,涵盖了中部泥质层位,揭示出 Assoul 组中非连续界面的地层频率分布是双峰度的(图9-21)。总体上,代表长期无沉积阶段的S3硬底经常出现在上部和下部颗粒质层位中,而中部泥质层位则主要记录S1和S2型非连续界面的固底(图9-21)。这样就能得出沉积相、水深和非连续界面的出现及其类型间的关系。可以从下伏于Agoudim泥灰岩(图9-2)的外围斜坡剖面中寻得证据,该灰岩地层主要跟海侵有关。在这些剖面中,碳酸盐相分析显示水动力强度总体较低,与Assoul组的泥质层对应。因此,其中的非连续界面比 Assoul 组泥质层位中更少。这种观察结果是水深和非连续界面出现与否之间存在联系的证据。

图9-16和图9-21强调了硬底(S3界面)主要形成于下部颗粒质层段,为最高水动力环境(S3界面的80%处;图9-21A)。非连续界面S1在三个层位中都最为常见(图9-21A)。一个关键的问题是,S2和S3界面的比值显著不同,在下部颗粒质层位中是1:1,在上部颗粒质层位中是10:1(图9-21)。界面频率分布和斜坡沉积模式间的这种关系对古环境研究和实际应用都有重要意义。

9.8.2 非连续界面形成的驱动因素

9.8.2.1 浅海—陆缘海沉积体系中硬底的形成

作为水下岩化的硬底是海床出露的证据,在现代海洋和地质记录中都很常见。在海洋—浅海领域(水深<150m),大部分作者都把凝缩层和固底到硬底的形成归结为沉积缓慢或者沉积路过不留从而发生早期水下成岩作用的结果(Purser,1969;Kennedy 和 Garrison,1975;Bromley,1978;Brett 和 Brookfield,1984;Galloway,1989;Kidwell,1991;Gomez 和 Fernandez-Lopez,1994;Burkhalter,1995;Clariet 等,1995;Hillgärtner,1998;Imme-nhauser 等,2000)。

无沉积作用的成因有两种观点,一是沉积输入饥饿;或是碳酸盐岩海床上未固结沉积颗粒被洋流带走,导致未岩化沉积物和较老沉积物的剥蚀。无作用阶段可能发生在沉积饥饿的初始海侵和最大海侵期(Galloway,1989),或者在最大海退期(该时期还未出露但较低的浪基面导致风浪淘选和水下剥蚀(Osleger,1991))。这种观点过于简单化,因为全新世和晚更新世沉积显示其地质记录要复杂得多,有很多孔缝未充填、不规则充填和过度充填的实例(Aurell 等,1995)。

在滨海陆架和陆缘海沉积环境中,浪基面对松软海床沉积的影响巨大(Immenhauser,

2009)。同样,风暴是临滨到大陆架或者较深内陆架盆地中大量沉积物剥蚀和搬运的主要动力,风暴还与沿岸风成洋流和波浪产生的大型搬运流的形成有关,波浪还可在浅海环境中形成增强的下层逆流或离岸流。周边海水的地球化学参数,比如酸碱度变化(Kempe 和 Kazmierczak,1993)或海水营养等级变化,可能导致碳酸盐产量下降以及硬底的形成。

在滨岸到浅海—陆棚沉积环境中,潮汐活动也能支配洋流,尤其在潮道逐渐变窄和向上变浅的地方(Immenhauser,2009)。通常其中一条潮汐洋流可能速度高但持续时间短,短途沉积物搬运能力强。非对称潮汐洋流的证据为砂波中主要滑塌相被逆向潮流轻微改造(Rubin 和 McCulloch,1980;Van Der Veen 等,2006;Immenhauser,2009)。"岛"露头和有时封盖沙丘相(剖面 I3;图 9 – 6,DS4)的 S3* 型非连续界面(表 9 – 4)的特征相之一就是东北东和东南东向鲕粒沙丘沉积的主体滑塌相。有趣的是,这些沉积看起来与沙丘沉积末期以及另一个沉积体系开始的时期相对应。

9.8.2.2 深海和水动力强度

在解释 Amellago 峡谷 Assoul 组非连续界面的时候,应该注意三个问题:(1) Assoul 组底部和顶部中高到高能层位非连续界面的丰度;(2)总体上缺乏地表出露特征证据,比如岩溶现象;(3)Bajocian 期 Assoul 组巨大的地层厚度。

根据 Gradstein 等(2004),Bojacian 期 Assoul 组持续时间 3.9Ma。Assoul 组地层厚度约 220m(图 9 – 3),据此可以得出沉积速率约为每百万年 56.4m(= 56.4B(Bubnoff);Fischer,1969)。该沉积速率可能被低估,有两个原因:(1) Assoul 组只是 Bajocian 期的中晚期才沉积(图 9 – 2);(2)该平均值是根据很多包含沉积间断面的地层计算出来的。Schlager(1999,2005)记录到某些现代碳酸盐岩沉积环境的沉积速率可以达到 4000B(比如 Belize 的珊瑚礁体系),但现在的沉积速率不能与摩洛哥的侏罗纪斜坡沉积环境相类比。

与其他中生界热带碳酸盐沉积环境相比,Assoul 组沉积速率达到了已公布数据的上限(Schlager,2005)。瑞士和法国的上侏罗统和下白垩统浅海台地环境中,沉积速率为 70 ~ 590B,而且这些速率也不能跟 Assoul 组的速率相比,因为 Strasser 和 Samankassou(2003)给出的速率指的是薄层,即沉积时间短(<0.02Ma)而且缺乏间断面。或许 Assoul 组巨大地层厚度的最好解释是基底连续快速沉降背景下可容空间的不断增长,但在摩洛哥该推断已公布的证据很少。因此,采纳 Soreghan 和 Dickinson(1994)的术语,Assoul 组斜坡沉积体系被描述为"保持(keep up)"沉积模式。该术语指代的沉积环境特征是,基底快速沉降导致可容空间持续增长,同时被未达到基准面或海平面的沉积物不断充填,这种说法不仅解释了巨大的地层厚度,还能解释地表出露界面的缺失。

根据前人(Hardenbol 等,1998)的研究,侏罗纪时期海平面起伏中等,但是有人发表过大幅度海平面变化的证据(Hallam,1997;Hesselbo 和 Jenkyns,1998;Immenhauser,2005)。考虑到鲕粒灰岩相经常出现,代表水动力水平中等,即使中等幅度海平面变化也可以导致 Assoul 组内斜坡周期性出露,这在全世界中生代台地环境都能找到(Sadler,1981)。地表出露的缺失代表着海底持续沉降,沉积物持续沉积但不能补偿海平面相对上升带来的可容空间的增加。

9.8.2.3 古生物学和沉积学证据

古生物过程是同生沉积成岩作用最好的证据之一,包括钻孔和结壳以及生物化学剥蚀。浸染或结壳最常发生在硬底,偶尔在固底也有发现(图 9 – 12C;Bertling,1999)。需要注意的是,并非所有钻孔或者结壳生物都指示岩化基底(Goldring 和 Kazmierczak,1974),还需要其他证据,比如被钻孔切割的颗粒。上述观察说明必须符合多条标准才能确定硬底的存在,其中化石相证据意义最大。

研究区很多非连续界面(S2**、S2***和所有S3型)之下,*Thalassinoides*钻孔向上部地层密度增大,导致非连续界面呈不规则波状形态,代表沉积速率减小(Shinn,1969;Wilson,1975;Pienkowski,1985)。部分界面中*Thalassinoides*钻孔壁铁斑的可能成因是,在生物钻孔填充或出露于现今气候环境之前的硬底海底矿化作用(Sattler等,2005)。非连续界面的棕黄色浸染可能源自近期流动屏障附近的液体流动(之上是泥灰质沉积;Gomez和Fernandez-Lopez,1994)。

很多S2和S3型界面的不规则界面形态也可能是由于*Thalassinoides*钻孔间沉积物经历过早期岩化,因此比未岩化生物钻孔内部沉积物对压实作用有更强的抵抗力,从而导致差异压实而形成(图9-10C)。差异成岩作用导致了碳酸盐海底软粒沉积物的部分或全部搬移或消失(Fursich,1979;Fursich等,1992)。

*Thalassinoides*钻孔通常在沉积物的含氧孔隙水区带下形成(Coimbra等,2009)。特定地点的含氧量和含氧孔隙水下层过渡深度取决于多种因素,比如底水含氧等级、沉积速率和有机碳酸盐数量(Immenhauser等,2008)。但考虑到当地碳酸盐岩的特征是深海动物以及活跃水循环的沉积学证据,预计含氧底水和含氧孔隙水地层为几厘米到几十厘米厚。Amellago峡谷研究区中的很多非连续界面中,*Thalassinoides*钻孔在非连续界面清晰可见。该观察结果表明在硬底形成期间,含氧孔隙水区的厚层沉积被剥蚀和搬运走(Wetzel和Aigner,1986)。

9.8.3 非连续界面形成的初步模型

基于上述考虑,对于大部分Assoul组硬底而言,或许其最好的成因解释是相对海平面下降(Tucker和Wright,1990;Osleger,1991;Schlager,1993)。与相对海平面下降(尤其是最大海退期)同步的是,有效正常浪基面下降将海底和其松软沉积颗粒暴露在波浪和洋流的活动中。波浪对海底未岩化的砂粒级规模的碳酸盐沉积物的影响有以下三种类型:(1)近海波浪速率尚未达到使沉积物运动的门限值,故沉积物未受到干扰;(2)在最长波长波浪接触到的海底开始形成波痕标志(Allen,1979);该情况下近海底水道颗粒运动速度已超过门限值但仍未超过更高的门限值。(3)在逐渐增高的水动力强度下,在仍旧水平的层理中(上部水平层理)发生了颗粒活跃搬运而引发的局部流动。波浪和洋流在该水动力机制下冲洗碳酸盐海底。在持续正常浪基面作用下,或者在风暴和低风暴浪基面相关的更强干扰条件下,都可能产生上述效果(Immenhauser等,2009)。相似地,风成洋流也是浅海—陆棚环境中沉积物搬运的半持久性动力。不论是被波浪、洋流还是二者共同驱动,水流循环都从近海底沉积物的孔隙中汲取海水,与碳酸盐胶结沉淀(Grammer等,1999;Van Der Kooij等,2010)一起导致了固底向硬底转变。图9-22用一个简单模型展示了相对海平面下降和正常浪基面直接的相互影响、受浪基面影响的海底位置以及不同类型非连续界面的形成。应该注意,研究区内非连续界面的形成和类型并不是只受水动力强度和相对水深的控制。

在研究波斯湾水下成岩作用的一篇文章中,Shinn(1969)描述过沉积相与近海底沉积物成岩作用深度之间的关系。在砂粒级沉积物中,水下胶结是沉积—水界面之下最显著的作用。沿剖面向下,早期海洋胶结作用降低,在沉积基准面之下10~15cm消失。因为孔隙都被胶结物填充,连续的海相碳酸盐胶结作用导致渗透率下降。如果沉积速率比胶结速率低,则形成厚度达10~15cm的岩化地层。相反,在渗透率总体偏低的细粒泥质沉积物中,沉积物—水界面下胶结作用可影响的深度大大减小。

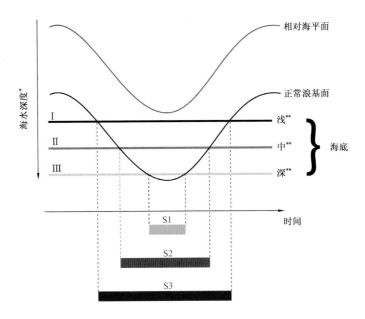

图 9-22　海平面、有效正常浪基面和海底等深线位置的推测概念图

总体浅水环境中,下降的海平面,碳酸盐海底(Ⅰ)经历了长期冲刷和真正的 S3 型海洋硬底。相反,碳酸岩海底处于相对深水环境(Ⅲ)只间歇性地经受降低浪基面的影响。这里形成了海洋固底(S2)。由于可容空间持续地被沉积物充填,这里未显示出沉降作用特征。注意,在自然沉积环境中,水动力强度、浪基面和沉积(或冲刷)作用的关系非常复杂,非沉积区域可能横向过渡为沉积区域

因此本文的初步模型表明 Assoul 组非连续界面的形成受四个因素主控。首先,相对海平面、有效正常和风暴浪基面以及波成洋流的变化,以及与此相关的碳酸盐岩海底的水动力强度;第二,海底碳酸盐相的类型(孔隙度和渗透率);第三,沉积速率;第四,斜坡横剖面的地理位置(内、中、外斜坡)。

因此,研究区的凝缩面/凝缩层(S1)就是相对海平面缓慢下降引起的水动力强度缓慢增高的结果,也使碳酸盐斜坡在剖面上延伸到更远更深的位置,并因泥质相发育阻碍水下胶结而且有总体偏低的渗透率,这些因素导致了局部(这些特征横向范围有限)冲刷或者沉积分异(Hillgartner,1998)。但不能排除的是,导致碳酸盐岩沉积后退的海侵可能致使凝缩层外部和深处斜坡环境碳酸盐生长的短暂停止(至少减少)(即沉积饥饿)。另一种解释是,该类型非连续界面也可能反映环境因素引起的碳酸盐岩生长变化,包括酸碱度(Kempe 和 Kazmierczak,1993)或周围海水营养物质改变(Hallock 和 Schlager,1986)。

固底 S2 界面可能源自更明显的相对海平面下降,以及与其相关的水动力强度升高或形成更近的(更浅的)浅海碳酸盐斜坡。该界面类型在碳酸盐环境中相对常见。与本文的解释不同,该界面类型通常被解释为海侵—海退旋回(准层序)中的海侵或海泛面(Pemberton 和 Frey,1985;Loutit 等,1989;Mitchum 和 VanWagoner,1991;Osleger,1991;Vail 等 1991;Goldhammer 等,1993;Brett,1995;Pemberton 和 Mac - Eachern,1995;Gingras 等,2001;Savrda 等,2001b)。尽管本文没有质疑将固底解释为海侵界面的正确性,作者还是希望强调在特定环境下,相对海平面下降(之前解释过)也可能形成固底;因此必须注意避免对碳酸盐沉积体系运用的模型过于简单。

在 Amellago 峡谷中,近海(最浅)环境中明显最常见的是海洋硬底 S3 界面,"岛"露头中出现了颗粒内斜坡相(图 9-3 和图 9-15)。作者观点是该环境体系与将这些特征解释为最大洪泛面的观点不一致。更可能的是,在最大海退期,由于基底快速沉降未发生出露,但是降低的浪基面和洋流导致了冲刷、水下剥蚀和成岩作用(Osleger,1991)。颗粒灰岩相使沉积地层上部可以形成明显的海水循环,从而导致了普遍性海底胶结作用(Shinn,1969)。图 9-23 用

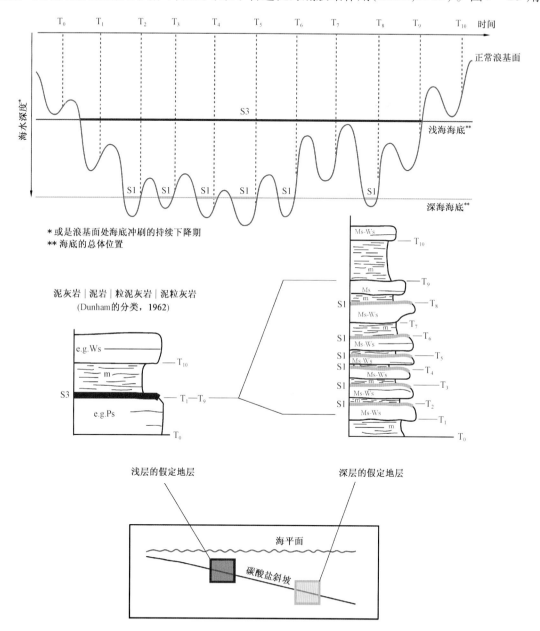

图 9-23 不同水深背景下相对海平面下降的两种类型的不同效果

在相对海平面下降导致可容空间增大情况下,浅海体系长期处于海底冲刷和无沉积作用之下,形成 S3 型海洋硬底。在较深水环境下,高幅度海平面的间歇性涨落导致短暂的海底冲刷,形成了众多次要非连续界面。上图表示的海底类型是基底沉降导致的可容空间恰好被沉积物所充填的情况

示意图说明在相对海平面下降过程,更近海(更浅)和更远海(更深)斜坡沉积体系中水深、水动力强度共同作用与非连续界面形成的关系。图9-23中描述的几套连续的S1和S2非连续界面的分布,与Assoul组非连续界面的分布一致(图9-3)。

9.9 结论

基于本文所采用的资料以及相关解释和讨论,得出下列结论:

(1)摩洛哥高阿特拉斯山Amellago峡谷出露的中上Bajocian阶Assoul组220m厚的地层中共发现了80个非连续界面。

(2)Assoul组非连续界面的主控因素有四个。首先,相对海平面、正常和风暴浪基面以及浪成洋流的变化,以及与此相关的碳酸盐海底的水动力强度;第二,海底碳酸盐相的类型(孔隙度和渗透率);第三,沉积速率;最后,斜坡横剖面的地理位置(内、中、外斜坡)。

(3)非连续界面被清晰地划分为水下固底、硬底和海洋无沉积作用界面,还可以分为三组:① 具有初始岩化特征的横向范围有限的非连续界面S1;② 横向上范围有限到很大的固底S2;③ 横向上范围很大到连续的硬底S3。地表出露特征明显缺乏可能与基底快速沉降导致的可容空间持续增大("保持"模式)相关。

(4)凝缩面(S1)源自相对海平面适度下降引起的水动力强度适度增大或者是由于总体沉积水深加深。固底(S2)代表大幅度相对海平面下降及其相关的水动力强度上升,或者碳酸盐斜坡的更近更浅环境。硬底(S3)代表最大海退阶段(但非地表出露),其特征为在研究区域内分布范围较广或位于研究区内最近最浅的碳酸盐斜坡沉积体系。

(5)在研究区识别出的80个非连续界面中,44个为凝缩面(S1),26个为固底(S2),还有10个为硬底(S3)。S3界面硬底是该地区的重要特征;形成于中—高能环境中,低能环境中未观测到S3面。大部分非连续界面在中等能量之下形成(每10m 7.1个界面),高能浅水环境每10m只有3.1个界面。在相对较深斜坡环境中,非连续界面更加少见,主要是S2和S1型界面(每10m 2.2个界面)。

(6)本文对于增进对中生代碳酸盐斜坡沉积体系非连续界面的理解具有重要意义,有助于建立更多量化储层模型。

参 考 文 献

Abdeljalil, L., Marcais, I., Choubert, G. and Fallot, P. (1959) Carte Géologique du Marocà 1:500,000, 4: Ouarzazate. Direction de la production industrielle des mines, 1950 – 1959, Rabat.

Allan, J. R. and Matthews, R. K. (1982) Isotopes signatures associated with early meteoric diagenesis. Sedimentology, 29, 797 – 817.

Allen, J. R. L. (1979) A model for the interpretation of wave ripple-marks using their wavelength, textural composition, and shape. J. Geol. Soc. London, 136, 673 – 682.

Aurell, M., McNeill, D. F., Guyomard, T. and Kindler, P. (1995) Pleistocene shallowing-upward sequences in New Providence, Bahamas: signature of high-frequency sea-level fluctuations in shallow carbonate platforms. J. Sed. Petrol., 65, 170 – 182.

Bassoullet, J. P., Poisson, A., Elmi, S., Cecca, F., Bellion, Y., Guiraud, R., Le Nindre, Y. M. and Manivit, J. (1993) Middle Toarcian palaeoenvironments. In: Atlas Tethys Palaeoenvironmental Maps (Eds J. Dercourt, L. E. Ricou and B. Vrielynck), pp. 63 – 80. Gauthier-Villars, Paris.

Bates, R. L. and Jackson, J. A. (1987) Glossary of Geology. American Geological Institute, Alexandria, VA, 788 pp.

Bathurst, R. G. C. (1975) Carbonate Sediments and Their Diagenesis. Elsevier, Amsterdam.

Beauchamp, J. (1988) Triassic sedimentation and rifting in the High Atlas (Morocco). Develop. Geotecton., 22, 477–497.

Beier, J. A. (1987) Petrographic and geochemical analysis of caliche profiles in a Bahamian Pleistocene dune. Sedimentology, 34, 991–998.

Benner, J. S., Ekdale, A. A. and De Gibert, J. M. (2004) Macroborings (Gastrochaenolites) in Lower Ordovician hardgrounds of Utah: sedimentologic, paleontologic, and evolutionary implications. Palaios, 19, 543–550.

Bertling, M. (1999) Taphonomy of trace fossils at omission surfaces (Middle Triassic, East Germany). Palaeogeogr. Palaeoclimatol. Palaeoecol., 149, 27–40.

Blomeier, D. P. G. and Reijmer, J. J. G. (1999) Drowning of a lower Jurassic carbonate platform: Jbel Bou Dahar, High Atlas, Morocco. Facies, 41, 81–110.

Brede, R. (1987) Strukturelle Entwicklung des zentralen Hohen Atlas nordwestlich von Errachidia (Marokko). Berl. Geowiss. Abh., A/85, 65.

Brett, C. E. (1995) Sequence stratigraphy, biostratigraphy, and taphonomy in shallow marine environments. Palaios, 10, 597–616.

Brett, C. E. and Brookfield, M. E. (1984) Morphology, faunas and genesis of Ordovician hardgrounds from southern Ontario, Canada. Palaeogeogr. Palaeoclimatol. Palaeoecol., 46, 233–290.

Bromley, R. G. (1975a) Trace fossils at omission surfaces. In: The Study of Trace Fossils (Ed. R. W. Frey), pp. 399–428. Springer, New York.

Bromley, R. G. (1975b) Comparative analysis of fossil and recent echinoid bioerosion. Palaeontology, 18, 725–739.

Bromley, R. G. (1978) Hardground diagenesis. In: The Encyclopaedia of Sedimentology (Eds R. W. Fairbridge and J. Burgeois), pp. 397–400. Dowen, Hutchinson & Ross, Stroudsburg, PA.

Bromley, R. G. (1996) Trace Fossils: Biology, Taphonomy and Applications, 2nd edn. Routledge, London, 384 pp.

Bromley, R. G. and Frey, R. W. (1974) Redescription of the trace fossil Gyrolithes and taxonomic evaluation of Thalassinoides, Ophiomorpha and Spongeliomorpha. Bull. Geol. Soc. Denmark, 23, 311–335.

Brown, B. J. and Farrow, G. E. (1978) Recent dolomitic concretions of crustacean burrow origin from Loch Sunart, west coast of Scotland. J. Sed. Res., 48, 825–833.

Burchette, T. P. and Wright, V. P. (1992) Carbonate ramp depositional systems. Sed. Geol., 79, 3–57.

Burkhalter, R. M. (1995) Ooidal ironstones and ferruginous microbialites: origin and relation to sequence stratigraphy (Aalenian and Bajocian, Swiss Jura mountains). Sedimentology, 42, 57–74.

Cander, H. (1995) Interplay of water-rock interaction efficiency, unconformities, and fluid flow in a carbonate aquifer: Floridan aquifer system. In: Unconformities and Porosity in Carbonate Strata (Eds D. A. Budd, A. H. Saller and P. M. Harris), Am. Assoc. Petrol. Geol., Tulsa, 63, 103–124.

Carmona, N. B., Mangano, M. G., Buatois, L. A. and Ponce, J. J. (2007) Bivalve trace fossils in an early Miocene discontinuity surface in Patagonia, Argentina: burrowing behavior and implications for ichnotaxonomy at the firmground-hardground divide. Palaeogeogr. Palaeoclimatol. Palaeoecol., 255, 329–341.

Carson, G. A. and Crowley, S. F. (1993) The glauconite phosphate association in hardgrounds: examples from the Cenomanian of Devon, southwest England. Cretaceous Res., 14, 69–89.

Catuneanu, O., Abreu, V., Bhattacharya, J. P., Blum, M. D., Dalrymple, R. W., Eriksson, P. G., Fielding, C. R., Fisher, W. L., Galloway, W. E., Gibling, M. R., Giles, K. A., Holbrook, J. M., Jordan, R., Kendall, C. G. S. C., Macurda, B., Martinsen, O. J., Miall, A. D., Neal, J. E., Nummedal, D., Pomar, L., Posamentier, H. W., Pratt, B. R., Sarg, J. F., Shanley, K. W., Steel, R. J., Strasser, A., Tucker, M. E. and Winker, C. (2009) Towards the standardization of sequence stratigraphy. Earth-Sci. Rev., 92, 1–33.

Clari, P. A., Della Pierre, F. and Martire, L. (1995) Discontinuities in carbonate successions: identification, interpretation and classification of some Italian examples. Sed. Geol., 100, 97–121.

Coimbra, R., Immenhauser, A. and Olóriz, F. (2009) Matrix micrit δ^{13}C and δ^{18}O reveals synsedimentary marine lithification in Upper Jurassic Ammonitico Rosso limestones. Sed. Geol., 219, 332–348.

Crevello, P. D. (1990) Stratigraphic Evolution of Lower Jurassic Carbonate Platforms: Record of Rift Tectonics and Eustasy, Central and Eastern High Atlas, Morocco. Unpublished PhD thesis, Colorado School of Mines, Golden,

456 pp.

Du Dresnay, R. (1971) Extension et développement des phénomènes récifaux jurassiques dans le domaine atlasique marocain, particulièrement au Lias moyen. Bull. Soc. Géol. France, 13/7, 46 – 56.

Du Dresnay, R. (1979) Sédiments Jurassiques du domaine des chaines Atlasiques du Maroc. Symposium "Sédimentation Jurassique W. Européen". Assoc. Sédimentol. Fr. Publ. Spéc. , 1, 345 – 365.

Dubar, G. (1962) Notes sur la paléogéographie du Lias marocain (Domaine atlasique). In: Livre Mém. P. Fallot (Ed. M. Durand-Delga), pp. 529 – 544. Soc. Géol. Fr. 1, Paris, Paris.

Dunham, R. J. (1962) Classification of carbonate rocks according to their depositional texture. In: Classification of Carbonate Rocks (Ed. W. E. Ham), AAPG Mem. , 1, 108 – 121. AAPG, Tulsa, OK.

Ehrenberg, K. (1944) Ergänzende Bemerkungen zu den seinerzeit aus dem Miozän von Burgschleintz beschriebenen Gangkeren und Bauten decapoder Krebse. Paläontol. Z. , Berlin, 23, 354 – 359.

Ekdale, A. A. and Bromley, R. G. (1984) Sedimentology and ichnology of the Cretaceous-Tertiary boundary in Denmark: implications for the causes of the terminal Cretaceous extinction. J. Sed. Petrol. , 54, 681 – 703.

Ekdale, A. A. and Bromley, R. G. (2001) Bioerosional innovation for living in carbonate hardgrounds in the Early Ordovician of Sweden. Lethaia, 34, 1 – 12.

Ekdale, A. A. , Benner, J. S. , Bromley, R. G. and De Gibert, J. M. (2002) Bioerosion of Lower Ordovician hardgrounds in southern Scandinavia and western North America. Acta Geol. Hisp. , 37, 9 – 13.

El-Hedeny, M. (2007) Encrustations and bioerosion on Middle Miocene bivalve shells and echinoid skeletons: paleoenvironmental implications. Rev. Paléobiol. , Genève, 26, 381 – 389.

Fischer, A. G. (1969) Geological time-distance rates: the Bubnoff unit. Geol. Soc. Am. Bull. , 80, 549 – 552.

Föllmi, K. B. , Garrison, R. E. and Grimm, K. A. (1991) Stratification in phosphatic sediments: illustrations from the Neogene of California. In: Cycles and Events in Stratigraphy (Eds G. Einsele, W. Ricken and A. Seilacher), pp. 492 – 507. Springer, New York.

Frey, R. W. and Seilacher, A. (1980) Uniformity in marine invertebrate ichnology. Lethaia, 13, 183 – 207.

Fürsich, F. T. (1979) Genesis, environments and ecology of Jurassic hardgrounds. Neues Jb. Geol. Paläontol. Abh. , 158, 1 – 63.

Fürsich, F. T. , Oschmann, W. , Singh, I. B. and Jaitly, A. K. (1992) Hardgrounds, reworked concretion levels and condensed horizons in the Jurassic of western India: their significance for basin analysis. J. Geol. Soc. London, 149, 313 – 331.

Galloway, W. E. (1989) Genetic stratigraphic sequences in basin analysis I: architecture and genesis of flooding-surface bounded depositional units. AAPG Bull. , 73, 125 – 142.

Gingras, M. K. , Pemberton, S. G. and Saunders, T. D. A. (2001) Bathymetry, sediment texture, and substrate cohesiveness: their impact on modern Glossifungites trace assemblages at Willapa Bay, Washington. Palaeogeogr. Palaeoclimatol. Palaeoecol. , 169, 1 – 21.

Gingras, M. K. , Pemberton, S. G. , Muelenbachs, K. and Machel, H. G. (2004) Conceptual models for burrow-related, selective dolomitization with textural and isotopic evidence from the Tyndall Stone, Canada. Geobiology, 2, 21 – 30.

Goldhammer, R. K. , Lehmann, P. J. and Dunn, P. A. (1993) The origin of high-frequency platform carbonate cycles and third-order sequences (Lower Ordovician El Paso Gp, West Texas): constraints from outcrop data and stratigraphic modeling. J. Sed. Petrol. , 63, 318 – 359.

Goldring, R. and Kazmierczak, J. (1974) Ecological succession in intraformational hardground formation. Palaeontology (Oxford), 17, 949 – 962.

Gomez, J. J. and Fernandez-Lopez, S. (1994) Condensation processes in shallow platforms. Sed. Geol. , 92, 147 – 159.

Gradstein, F. M. , Ogg, J. G. and Smith, A. G. (2004) A Geologic Time Scale. Cambridge University Press, Cambridge, 589 pp.

Grammer, G. M. , Crescini, C. M. , McNeill, D. F. and Taylor, L. H. (1999) Quantifying rates of syndepositional marine cementation in deeper platform environments-new insight into a fundamental process. J. Sed. Res. , 69, 202 – 207.

Hallam, A. (1997) Estimates of the amount and rate of sea-level change across the Rhaetian-Hettangian and Pliensbachian-Toracian boundaries (latest Triassic to early Jurassic). J. Geol. Soc. London, 154, 773 – 779.

Hallock, P. and Schlager, W. (1986) Nutrient excess and the demise of coral reefs and carbonate platforms. Palaios, 1, 389 – 398.

Handford, C. R. and Loucks, R. G. (1993) Carbonate depositional sequences and systems tracts-response of carbonate platforms to relative sea-level changes. AAPG Mem. ,57,3 – 42.

Hardenbol, J. , Thierry, J. , Farley, M. B. , Jacquin, T. , DeGraciansky, P. -C. and Vail, P. R. (1998) Mesozoic-Cenozoic sequence chronostratigraphic framework. In: Sequence Stratigraphy of European Basins (Eds P. -C. DeGraciansky, J. Hardenbol, T. Jacquin, P. R. Vail and M. B. Farley), SEPM Spec. Publ. ,60,3 – 13.

Hauptmann, M. (1990) Untersuchungen zur Mikrofazies, Stratigraphie und Paläogeographie jurassischer Karbonatgesteine im Atlas-System ZentrAl-Marokkos. Unpublished PhDthesis, Free University of Berlin, Berl. geow. Abh. , 119,90 pp.

Heim, A. (1924) Über submarine Denudation und chemische Sedimente. Geol. Rundsch. ,15,1 – 47.

Heim, A. (1934) Stratigraphische Kondensation. Eclogae Geol. Helv. ,27,372 – 383.

Hesselbo, S. P. and Jenkyns, H. C. (1998) British lower Jurassic sequence stratigraphy. In: Mesozoic and Cenozoic Sequence Stratigraphy of European Basins (Eds P. -C. de Graziansky, J. Hardenbol, T. Jacquin and P. R. Vail), Vol. 60, pp. 561 – 581. SEPM, Tulsa, OK.

Hillgärtner, H. (1998) Discontinuity surfaces on a shallowmarine carbonate platform (Berriasian, Valanginian, France and Switzerland). J. Sed. Res. ,68,1093 – 1108.

Humphrey, J. D. , Ranson, K. L. and Matthews, R. K. (1986) Early meteoric diagenetic control of Upper Smackover Production, Oaks Field, Louisiana. AAPG Bull. ,70,70 – 85.

Immenhauser, A. (2005) High-rate sea-level change during the Mesozoic: new approaches to an old problem. Sed. Geol. ,175,277 – 296.

Immenhauser, A. (2009) Estimating paleo-water depth from the physical rock record. Earth-Sci. Rev. ,96,107 – 139.

Immenhauser, A. , Schlager, W. , Burns, S. J. , Scott, R. W. , Geel, T. , Lehmann, J. , Gaast, S. v. d. and Bolder-Schrijver, L. J. A. (2000) Origin and correlation of disconformity surfaces and marked beds, Nahr Umr Formation, Northern Oman. SEPM Spec. Publ. ,69,209 – 225.

Immenhauser, A. , Kenter, J. A. M. , Ganssen, G. , Bahamonde, J. R. , Van Vliet, A. and Saher, M. H. (2002) Origin and significance of isotope shifts in Pennsylvanian carbonates (Asturias, NW Spain). J. Sed. Res. ,72,82 – 94.

Immenhauser, A. , Hillgärtner, H. , Sattler, U. , Bertotti, B. , Schoepfer, P. , Homewood, P. , Vahrenkamp, V. , Steuber, T. , Masse, J. -P. , Droste, H. H. J. , van Koppen, J. , van der Kooij, B. , van Bentum, E. C. , Verwer, K. , Hoogerduijn-Strating, E. , Swinkels, W. , Peters, P. , Immenhauser-Potthast, I. and Al Maskery, S. A. J. (2004) Barremian-lower Aptian Qishn Formation, Haushi-Huqf area, Oman: a new outcrop analogue for the Kharaib/Shu'aiba reservoirs. GeoArabia,9,153 – 194.

Immenhauser, A. , Holmden, C. and Patterson, W. P. (2008) Interpreting the carbon-isotope record of ancient shallow Epeiric seas: lessons from the recent. In: Dynamics of Epeiric Seas (Eds B. R. Pratt and C. Holmden), Geol. Assoc. Canada Spec. Publ. ,48,135 – 174.

Jaanuson, V. (1961) Discontinuity surfaces in limestones. Univ. Uppsala Geol. Inst. Bull. ,40,221 – 241.

Jensen, S. , Droser, M. L. and Gehling, J. G. (2005) Trace fossils preservation and the early evolution of animals. Palaeogeogr. Palaeoclimatol. Palaeoecol. ,220,19 – 29.

Joachimski, M. M. (1994) Subaerial exposure and deposition of shallowing upward sequences: evidence from stable isotopes of Purbeckian peritidal carbonates (basal Cretaceous), Swiss and French Jura Mountains. Sedimentology, 41,805 – 824.

Kauffman, E. G. , Elder, W. P. and Sageman, B. B. (1991) Highresolution correlation: a new tool in chronostratigraphy. In: Cycles and Events in Stratigraphy (Ed. G. Einsele), pp. 795 – 820. Springer, Berlin.

Kelly, S. R. A. and Bromley, R. G. (1984) Ichnological nomenclature of clavate borings. Palaeontology,27,793 – 807.

Kempe, S. and Kazmierczak, J. (1993) The role of alkalinity in the evolution of ocean chemistry, organization of living systems, and biocalcification processes. In: Past and Present Biomineralization Processes (Eds F. Doumenge, D. Allemand and A. Toulemont), Vol. 13, pp. 64 – 117. Musée Océanographique, Monaco.

Kennedy, W. J. and Garrison, R. E. (1975) Morphology and genesis of nodular chalks and hardgrounds in the Upper

Cretaceous of southern England. Sedimentology,22,311 – 386.

Kidwell,S. M. (1991) Condensed deposits in siliciclastic sequences: expected and observed features. In: Cycles and Events in Stratigraphy (Eds G. Einsele, W. Ricken and A. Seilacher), pp. 682 – 695. Springer, Berlin.

van der Kooij,B. ,Immenhauser,A. ,Csoma,A. ,Bahamonde,J. and Steuber,T. (2009) Spatial geochemistry of a Carboniferous platform-margin-to-basin transect: balancing environmental and diagenetic factors. Sed. Geol. , 219, 136 – 150.

van der Kooij,B. ,Immenhauser,A. ,Steuber,T. ,Bahamonde,J. R. and Merino Tomé,O. (2010) Precipitation mechanisms of volumetrically important early marine carbonate cement volumes in deep slope settings. Sedimentology,57, 1491 – 1525.

Leymerie,M. A. (1842) Suite de mémoire sur le terrain Crétacédu département de l'Aube. Soc. Gé ol. Fr. Mém. ,5,1 – 34.

Lohmann,K. C. (1988) Geochemical patterns of meteoric diagenetic systems and their application to studies of Paleokarst. In: Paleokarst (Eds N. P. James and P. W. Choquette), pp. 58 – 80. Springer, Berlin.

Loutit,T. S. , Hardenbol, J. , Vail, P. R. and Baum, G. R. (1989) Condensed sections: the key to age determination and. correlation of continental margin sequences. In: Sea-Level Changes: An Integrated Approach. (Eds C. K. Wilgus, B. S. Hastings, C. G. St. C. Kendall, H. W. Posamentier, C. A. Ross and J. C. Van Wagoner et al.), SEPM Spec. Publ. ,42,183 – 214.

Lundgren,S. A. B. (1891) Studier öfver fossilförande lösa block. Geol. Fören. Stockh. Förh. ,13,111 – 121.

MacEachern,J. A. and Burton,J. A. (2000) Firmground Zoophycos in the Lower Cretaceous Viking Formation, Alberta: a distal expression of the Glossifungites ichnofacies. Palaios,15,387 – 398.

MacEachern,J. A. ,Pemberton,S. G. ,Gingras, M. K. and Bann, K. L. (2007) The ichnofacies paradigm: a fifty-year retrospective. In: Trace Fossils: Concepts, Problems, Prospects (Ed. W. I. Miller), pp. 52 – 77. Elsevier, Amsterdam.

Mägdefrau,K. (1932) Über einige Bohrgänge aus dem Unteren Muschelkalk von Jena. Paläontol. Z. ,14,150 – 160.

Marshall,J. D. (1992) Climatic and oceanographic isotopic signals from the carbonate rock record and their preservation. Geol. Mag. ,129,143 – 160.

Mattis,A. (1977) Nonmarine Triassic sedimentation, Central High Atlas Mountains, Morocco. J. Sed. Petrol. , 47, 107 – 119.

Mitchum,R. M. and VanWagoner,J. C. (1991) High-frequency sequences and their stacking patterns: sequence-stratigraphic evidence of high-frequency eustatic cycles. Sed. Geol. ,70,131 – 160.

Mitchum,R. M. ,Vail,P. R. and Thompson,S. (1977) Seismic Stratigraphy and Global Changes of Sea Level, Part 2: The Depositional Sequence as a Basic Unit for Stratigraphic Analysis. Application of Seismic Reflection Configuration to Stratigraphic Interpretation. In: Seismic Stratigraphy-Applications to Hydrocarbon Exploration. AAPG Special Volumes,26,53 – 62.

Osleger, D. (1991) Subtidal carbonate cycles: implications for allocyclic vs autocyclic controls. Geology, 19, 917 – 920.

Palmer,T. J. (1978) Burrows at certain omission surfaces in the Middle Ordovician of the Upper Mississippi Valley. J. Paleontol. ,52,109 – 117.

Pemberton,S. G. and Frey,R. W. (1985) The Glossifungites ichnofacies: modern examples from the Georgia coast. In: Biogenic Structures: Their Use in Interpreting Depositional Environments (Ed. H. A. Curran), SEPM Spec. Publ. , 35,237 – 259.

Pemberton,S. G. and MacEachern,J. A. (1995) The sequence stratigraphic significance of trace fossils: examples from the Cretaceous Foreland Basin of Alberta, Canada. In: Sequence Stratigraphy of Foreland Basin Deposits (Eds J. C. Van Wagoner and G. T. Bertram), AAPG Mem. ,64,429 – 475.

Pemberton,S. G. ,Maceachern, J. A. , Gingras, M. K. and Jianping, Z. (2000) Significance of ichnofossils to genetic stratigraphy-examples from the Cretaceous of Alberta,Canada. Sci. China (Ser. D): Earth Sci. ,43,541 – 560.

Pemberton,S. G. ,McEachern,J. A. and Saunders,T. (2004) Stratigraphic applications of substrate-specific ichnofacies: delineating discontinuities in the rock record. In: The Application of Ichnology to Paleoenvironmental and

Stratigraphic Analysis, Lyell Meeting 2003 (Ed. D. McIlroy), Geol. Soc. London Spec. Publ., 228, 29–62.

Pienkowski, G. (1985) Early Liassic trace fossil assemblages from the Holy Cross Mountains, Poland: their distribution in continental and marginal marine environments. In: Biogenic Structures: Their Use in Interpreting Depositional Environments (Ed. H. A. Curran), SEPM Spec. Publ., 35, 37–51.

Pierre, A. (2006) Un exemple de reference pour les systèmes de rampe oolithiques: un affleurement continu de 37 km (falaises jurassiques d'Amellago, Haut-Atlas, Maroc). Unpublished PhD thesis, University of Burgundy, Dijon, 235 pp.

Pierre, A., Durlet, C., Razin, P. and Chellai, E. H. (2010) Spatial and temporal distribution of ooids along a Jurassic carbonate ramp: Amellago transect, High-Atlas, Morocco. Geol. Soc. London Spec. Publ., 329, 65–88.

Pomoni-Papaioannou, F. (1994) Paleoenvironmental reconstruction of a condensed hardground-type depositional sequence at the Cretaceous-Tertiary contact in the Parnassus-Ghiona zone, central Greece. Sed. Geol., 93, 7–24.

Purser, B. H. (1969) Syn-sedimentary marine lithification of Middle Jurassic limestones in the Paris basin. Sedimentology, 12, 205–230.

Rameil, N. (2008) Early diagenetic dolomitization and dedolomitization of Late Jurassic and earliest Cretaceous platform carbonates: a case study from the Jura Mountains (NW Switzerland, E France). Sed. Geol., 212, 70–85.

Read, J. F. and Grover Jr, G. A. (1977) Scalloped and planar erosion surfaces, Middle Ordovician limestones, Virginia; analogues of Holocene exposed karst or tidal rock platforms. J. Sed. Petrol., 47, 956–972.

Read, J. F. and Horbury, A. D. (1994) Eustatic and tectonic controls on porosity evolution beneath sequence-bounding unconformities and parasequence disconformities on carbonate platforms. In: Diagenesis and Basin Development (Eds A. D. Horbur and A. G. Robinson), AAPG Stud. Geol., 36, 155–197.

Richter, D. K., Götte, T., Götze, J. and Neuser, R. D. (2003) Progress in application of cathodoluminescence (CL) in sedimentary petrology. Mineral. Petrol., 79, 127–166.

Roch, E. (1950). Histoire stratigraphique du Maroc. Notes Mem. Serv. Geol. Maroc., 80, 440.

Rubin, D. M. and McCulloch, D. S. (1980) Single and superimposed bedforms: a synthesis of San Francisco Bay and flume observations. Sed. Geol., 26, 207–231.

Sadler, P. M. (1981) Sediment accumulation rates and the completeness of stratigraphic sections. J. Geol., 89, 569–584.

Sarg, J. F. (1988) Carbonate sequence stratigraphy. In: Sea-level Changes: An Integral Approach (Eds C. K. Wilgus, B. S. Hastings, C. G. St. C. Kendall, H. W. Posamentier, C. A. Ross and J. C. van Wagoner), SEPM Spec. Publ., 42, 155–181.

Sattler, U., Immenhauser, A., Hillgärtner, H. and Esteban, M. (2005) Characterization, lateral variability and lateral extent of discontinuity surfaces on a carbonate platform (Barremian to Lower Aptian, Oman). Sedimentology, 52, 339–361.

Savrda, C. E., Browning, J. V., Hesselbo, S. P. and Krawinkel, H. (2001a) Firmground ichnofabrics in deep-water sequence stratigraphy, Tertiary clinoform-toe deposits, New Jersey slope. Palaios, 16, 294–305.

Savrda, C. E., Krawinkel, H., McCarthy, F. M. G., McHugh, C. M. G., Olson, H. C. and Mountain, G. (2001b) Ichnofabrics of a Pleistocene slope succession, New Jersey margin: relations to climate and sea-level dynamics. Palaeogeogr. Palaeoclimatol. Palaeoecol., 171, 41–61.

Schlager, W. (1993) Accommodation and supply-a dual control on stratigraphic sequences. Sed. Geol., 86, 111–136.

Schlager, W. (1999) Scaling of sedimentation rates and drowning of reefs and carbonate platforms. Geology, 27, 183–186.

Schlager, W. (2005) Carbonate Sedimentology and Sequence Stratigraphy. SEPM, Tulsa, OK, 200 pp.

Seilacher, A. (1964) Biogenic sedimentary structures. In: Approaches to Paleoecology (Eds J. Imbrie and N. D. Newell), pp. 296–316. John Wiley & Sons, Inc., New York.

Seilacher, A. (1967) Bathymetry of trace fossils. Mar. Geol., 5, 413–428.

Seilacher, A. (2007) Trace Fossil Analysis. Springer-Verlag, Berlin, Heidelberg, New York, 226 pp.

Shinn, E. A. (1969) Submarine lithification of Holocene carbonate sediments in the Persian Gulf. Sedimentology, 12, 109–144.

Soreghan, G. S. and Dickinson, W. R. (1994) Generic types of stratigraphic cycles controlled by eustasy. Geology, 22, 759 – 761.

Southard, J. B., Lambie, J. M., Federico, D. C., Pile, H. T. and Weidman, C. R. (1990) Experiments on bed configurations in fine sands under bidirectional purely oscillatory flow, and the origin of hummocky cross-stratification. J. Sed. Petrol., 60, 1 – 17.

Strasser, A. and Samankassou, E. (2003) Carbonate sedimentation rates today and in the past: Holocene of Florida Bay, Bahamas and Bermuda vs. Upper Jurassic and Lower Cretaceous of the Jura Mountains (Switzerland and France). Geol. Croat., 56, 1 – 18.

Strasser, A., Pittet, B., Hillgärtner, H. and Pasquier, J. -B. (1999) Depositional sequences in shallow carbonate-dominated sedimentary systems: concepts for a high-resolution analysis. Sed. Geol., 128, 201 – 221.

Taylor, P. D. and Wilson, M. A. (2002) A new terminology for marine organisms inhabiting hard substrates. Palaios, 17, 522 – 525.

Tucker, M. E. and Wright, V. P. (1990) Carbonate Sedimentology. Blackwell Science, Oxford, 482 pp.

Vail, P. R., Audemard, F., Bowman, S. A., Eisner, P. N. and Perez-Cruz, C. (1991) The Stratigraphic Signatures of Tectonics, Eustasy and Sedimentology-An Overview. In: Cycles and Events in Stratigraphy (Eds G. Einsele, W. Ricken and A. Seilacher), pp. 617 – 659. Springer-Verlag, Berlin.

Van Houten, F. B. (1977) Triassic-Liassic deposits of Morocco and eastern North America: comparison. Am. Assoc. Petrol. Geol. Bull., 61, 79 – 99.

van der Veen, H. H., Hulscher, S. J. M. H. and Knaapen, M. A. F. K. (2006) Grain size dependency in the occurrence of sand waves. Ocean Dyn., 56, 228 – 234.

Wagner, P. D., Tasker, D. R. and Wahlman, G. P. (1995) Reservoir degradation and compartmentalization below subaerial exposure unconformities: limestone examples from West Texas, China and Oman. In: Unconformities and Porosity in Carbonate Strata (Eds. D. A. Budd, A. H. Saller and P. M. Harris), AAPG Special Volumes, 63, 177 – 194.

Warme, J. E. (1988) Jurassic carbonate facies of the Central and Eastern High Atlas rift Morocco. Lect. Notes Earth Sci., 15, 169 – 199.

Watkins, R. (1990) Paleoecology of a Pliocene rocky shoreline, Salton through region, California. Palaios, 5, 167 – 175.

Watkins, D. K., Premoli Silva, I. and Erba, E. (1995) Cretaceous and Paleogene manganese-encrusted hardgrounds from Central Pacific Guyots. In: Proceedings of the Ocean Drilling Program, Scientific Results (Eds J. A. Haggerty, I. Premoli Silva, F. Rack and M. K. McNutt), ODP, 144, 97 – 126.

Wetzel, A. and Aigner, T. (1986) Stratigraphic completeness: tiered trace fossils provide a measuring stick. Geology, 14, 234 – 237.

Wilmsen, M. and Neuweiler, F. (2008) Biosedimentology of the Early Jurassic post-extinction carbonate depositional system, central High Atlas rift basin, Morocco. Sedimentology, 55, 773 – 807.

Wilson, I. G. (1975) Carbonate Facies in Geologic History. Springer, Berlin, 471 pp.

Wilson, M. A., Curran, H. A. and White, B. (1998) Paleontological evidence of a brief global sea-level event during the last interglacial. Lethaia, 31, 241 – 250.

Wright, V. P. (1994) Paleosols in shallow marine carbonate sequences. Earth-Sci. Rev., 35, 367 – 395.

第 10 章 海相碳酸盐岩晚古生代冰期的地层特征

JESSE T,KOCH,TRACY D. FRANK 著

陈瑞银 译,吴因业 校

摘　要　二叠系记录了显生宙最极端气候变迁中的一个,并且期间的气候与环境动力学过程以及它们对浅海碳酸盐系统的影响都很少受到约束。美国新墨西哥州中南部奥罗格兰德盆地上宾夕法尼亚统至下二叠统的地层中,新的地层与碳同位素记录为洞察古热带碳酸盐系统对冈瓦纳冰原生长消亡的响应提供了新素材。奥罗格兰德盆地记录了冰川作用的早二叠世顶点:以 $\delta^{13}C$ 接近于 $-6‰$ 的高变化量为特征的系列陆上暴露面。这些底层与同位素格局解释认为反映冰川扩张期间海洋从盆中消退。总体上,在晚古生代冰期缓慢衰退期间沉积的更年轻地层(下萨克马尔阶—孔谷阶)是以相对更高和更均匀的 $\delta^{13}C$ 值($+2‰ \sim +5‰$)为特征,也没有显示长期陆上暴露的迹象。然而这些地层保留了 $\delta^{13}C$ 值更负的两个短期偏离。这两个偏离在奥罗格兰德盆地内发生在记录向更浅更差环境变迁的整个过渡期。奥罗格兰德盆地记录向更浅更差环境的变迁过程,以上两个偏离贯穿始终。这些变迁在时间上与萨克马尔晚期到阿丁斯克期—孔谷早期贯穿东澳洲的两期冰川扩张对应。这些关系表明,东冈瓦纳冰川扩张的脉动可足以导致奥罗格兰德盆地相对海平面的下降和环境变化的可能。本研究结果表明,奥罗格兰德盆地的海洋化学、沉积环境以及海平面深受冈瓦纳冰川作用期的影响。这些推论帮助(我们)增进理解古热带碳酸盐系统如何响应在晚古生代冰期顶峰与消退阶段冰冻圈变化的影响。

关键词　碳酸盐岩　晚古生代　宾夕法尼亚统　二叠系　稳定同位素

10.1　概况

至今,晚古生代冰期(LPIA)被看作一个历经 60～70Ma 长时间的冰川时期,其冰原渐长渐消,几乎永久覆盖着冈瓦纳古陆(Veevers 和 Powell,1987;Crowell,1999)。很大程度上基于欧美旋回层的时空分布,认为该冰期在晚宾夕法尼亚世达到顶峰在早二叠世结束(Veevers 和 Powell,1987;Gonzalez-Bonorio 和 Eyles,1995;Wright 和 Vanstone,2011;Heckel,2008,及其文内参考文献)。因此有关 LPIA 的同位素研究主要集中于石炭系,特别是集中于冰室条件出现时期的地层(Bruckschen 等,1999;Mii 等,1999,2001;Veizer 等,1999;Saltzman,2003)。

然而最近对冈瓦纳古陆近冰沉积物的调查研究形成了一种新观点,认为 LPIA 是动态、多相的,在早二叠世某一冰川作用顶峰达到极致(Fielding 等,2008a,b)。虽然稳定同位素记录为 LPIA 的气候变化认识提供了很多依据(Bruckschen 等,1999;Mii 等,1999,2001;Veizer 等,1999;Saltzman,2003;Grossman 等,2008),但记录早二叠世环境条件的依据却极少(Frank 等,2008)。之后的稳定同位素研究发现,碳氧同位素记录的漂移与冈瓦纳古老地层学记录的冰川期具有很好的对应性(Montanez 等,2007;Frank 等,2008;Grossman 等,2008;Birgenheier 等,2010)。虽然这些记录趋于解释全球性的冈瓦纳冰川效应,但仍存在一点明显的争议。即在已知的同位素证据中,有关二叠纪的记录缺乏代表性,并且可用的记录是由少量数据组成,它们又来自多个地层差异深远的地区(Korte 等,2005;Montanez 等,2007;Frank 等,2008)。这些

拼凑而成的记录显然难以把全球性的信号从任何变化中区分出来,这些变化可能是由构造背景、古海洋环境和后沉积历史引起的。

浅水环境的碳酸盐体系是海洋环境、气候及海平面变化的敏感记录体。在这点上,上古生界碳酸盐岩层序中的地层和稳定同位素格局应该记录冈瓦纳冰川作用的全球范围影响,因此也提供了一种检验近来发表的关于冈瓦纳地层研究正确性的手段(Fielding 等,2008a,b)。拿美国新墨西哥中南部奥罗格兰德盆地海相碳酸盐岩层系的同位素与地层学研究,作为该假设的一个初步检验实例(图 10-1)。这整个古生代都位于赤道 10°带内的盆地(Ross 和 Ross,1990;Scotese 和 Langford,1995;Golonka 和 Ford,2000),发育有美国西南部上宾夕法尼亚统到下二叠统最连续的海相碳酸盐岩剖面中的几条(Jordan,1975;Raatz,2002;Wahlman 和 King,2002)。该盆地提供了一个理想的场所,可以检验晚古生代冰川作用在浅水碳酸盐体系中同位素和地层变化方面的远场印记(far-field imprint)。目前论文报告了该调查研究的结果。该层系通过一些地表露头和选自在美国新墨西哥州索科罗的新墨西哥地质矿产资源局收藏的钻井岩心来检验。依据沉积环境,对剖面进行了描述、同位素分析的调查和剖面解释。绘制了地层和同位素记录及 Isbell 等(2003)和 Fielding 等(2008a,b)获知的冰期/非冰期时刻的相关性图件。

10.2 地质背景

10.2.1 奥罗格兰德盆地

位于新墨西哥中南部的奥罗格兰德盆地形成于中密西西比纪,是一个浅而呈长轴展布的盆地(图 10-1;Seager 等,1976;Candelaria,1988;Raatz,2002)。晚宾夕法尼亚世—早二叠世期间,盆地位于赤道 10°带内(Ross J. R. P. 和 Ross C. A.,1990;Scotese 和 Langford,1995;Golonka 和 Ford,2000)。区域研究表明,长期以来,奥罗格兰德盆地东边与二叠纪盆地(如西北大陆架/特拉华盆地)相连,西南边与佩德雷戈萨盆地相连(Jordan,1975;Candelaria,1988;Raatz,2002)。宾夕法尼亚纪的区域构造活动造就了邻近盆地的隆起带(佩德纳尔隆起、迪亚拜罗隆起、佛罗里达隆起),这些隆起与奥齐塔—马拉松造山带、联合古陆最终拼合相关(图 10-1;Candelaria,1988;Raatz,2002)。在晚宾夕法尼亚世—早二叠世期间,隆起区域得到最大程度的缓和,导致数量惊人的碎屑沉积物沿着盆地边缘沉积下来。到狼营期末,由于陆源碎屑和海洋碳酸盐类的填充以及沉降速率的降低,盆地沉积容纳空间受到了限制(Jordan,1975;Candelaria,1988;Raatz,2002)。在伦纳德期,浅海和蒸发相沉积在全盆盛行,在东部盆地与特拉华盆地和西北大陆架相连接(Candelaria,1988;Raatz,2002)。

10.2.2 地层

奥罗格兰德盆地中可以见到多个宾夕法尼亚系—下二叠统连续剖面(Jordan,1075;Wahlman 和 King,2002)。沉积展布从南部的广海碳酸盐地层到北部的硅质碎屑—碳酸盐混合型地层(Jordan,1975;Candelaria,1988;Rankey 等,1999)。下二叠统地层呈不整合接触覆盖于宾夕法尼亚地层之上(Schoderbek 和 Chafetz,1988;Wahlman 和 King,2002);该不整合目前穿越了美国西南部的大部分地区(Ross,1986)。图 10-2 展示了年代地层格架。

在此讨论的宾夕法尼亚系—二叠系的界限是以定于哈萨克的二叠系 GSSP 标准为基础的(Davydov 等,1998)。牙形石与䗴的研究已经对哈萨克宾夕法尼亚系—二叠系界限和整个北美地层剖面进行了比对(Ross,1963;Ritter,1995;Wahlman 和 King,2002)。该界限迁移重新把

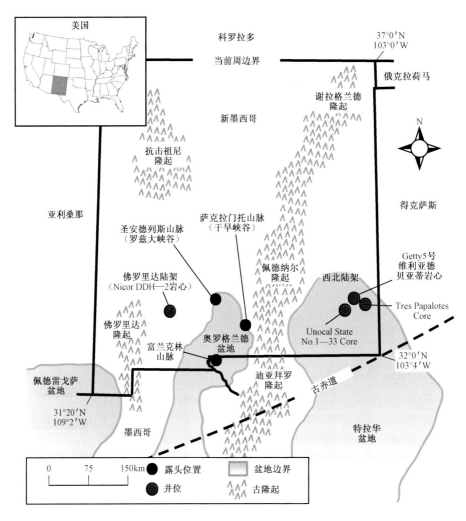

图10-1 奥罗格兰德盆地区域图(据Candelaria,1988,修改)

该研究中检测露头位置用黑点标记,岩心位置用红点标记。每个剖面的准确位置在图10-3中以标题给出。区域图范围在美国插图中用蓝色突出

下狼营统划为宾夕法尼亚统顶部地层。如博萨穆组和博奥段这些原认为是二叠系的地层单元,现在划分为宾夕法尼亚系顶部(Wahlman和King,2002)。

10.2.2.1 宾夕法尼亚顶部地层

宾夕法尼亚顶部地层划分为Virginlian Beeman组、上Virgilian Holder组(相当于Panther Seep)和顶部Virginlian Bursum(Laborcita)组(图10-2)。在Sacramento山脉,Holder和Beaman地层出露相当好,那里由广海石灰岩、陆相碎屑岩层和大型叶状藻礁的交互旋回层构成(Wilson,1967;Rankey等,1999)。与Panther Seep组稍深水相当,Holder组为浅水、上倾的(Schoderbek和Chafetz,1988;Rankey等,1999)。利用鎝的生物地层学资料揭示了Holder和Beeman地层的一个弗吉尔阶时代(Cline,1959;Schoderbek和Chafetz,1988)。

Bursum地层的广海石灰岩和陆相碎屑岩层在盆地中零星分布(Jordan,1975;Candelaria,1988;Raatz,2002)。通过鎝的研究获取了晚宾夕法尼亚世Bursum地层的限定年代(Steiner和Williams,1968;Wahlman和King,2002)。

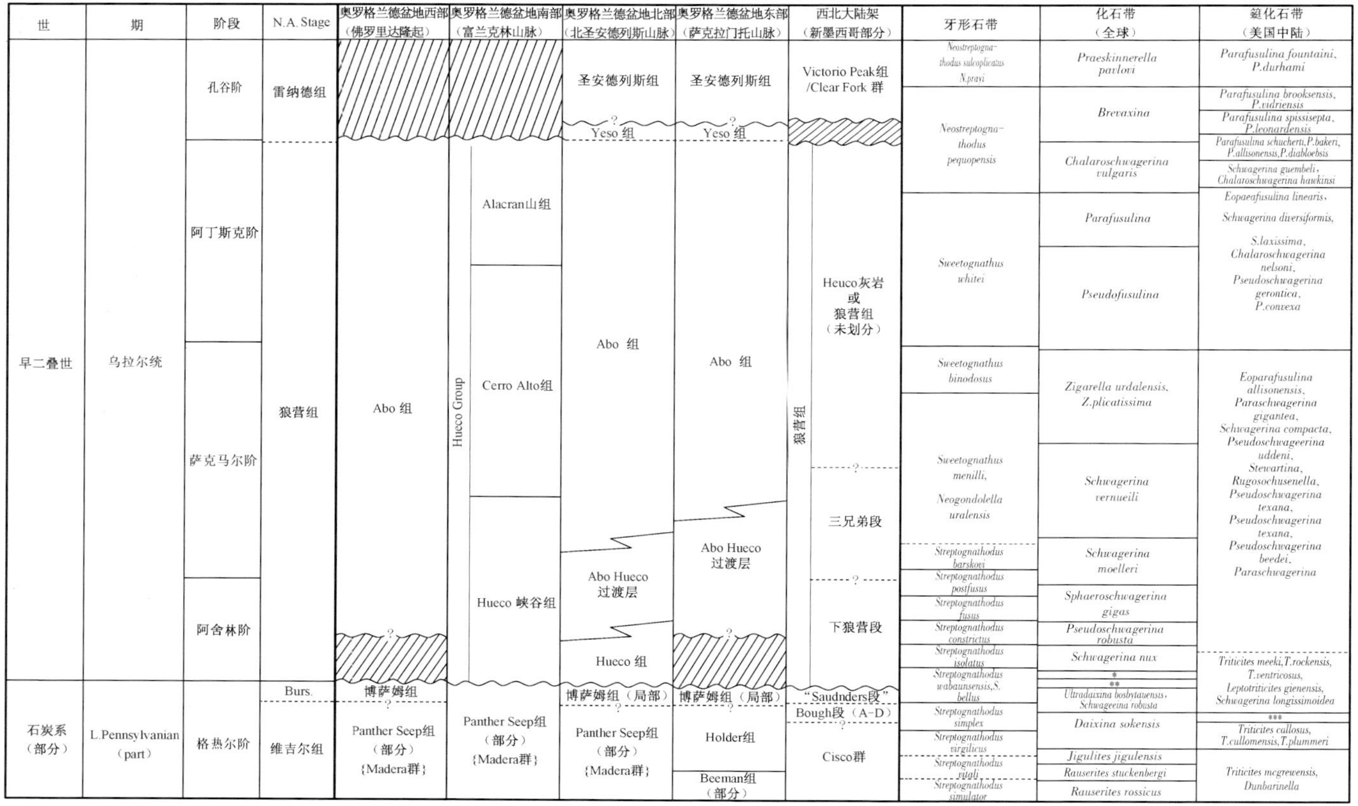

图10-2 基于前人提供的研究区年代地层学格架（Jordan, 1975; Cys和Mazzullo, 1985; Malek–Aslani, 1985; Candelaria, 1988; Rankey等, 1999; Raatz, 2002; Mack等, 2003）

时间刻度依据Gradstein等（2004）。牙形石与全球化石带据Gradstein等（2004及其内参考文献）；美国中陆化石带据Ross和Ross（1988, 1995）。注解 *: Sphaeroschwagerina fusiformis; **: Sphaeroschwagerina vulgaris; ***: Triticites subventricosus, T. Whetstonensis, Schubertella schetensis

特拉华盆地的一部分(如西北陆架)延伸进入新墨西哥州的东南部,萨克拉门托山脉的东部(图10-1,图10-2)。非正式的Bough地层单元(A到D)和Saunders段已描述为下二叠统Hueco石灰岩的部分(Cys和Mazzullo,1985;Malek-Aslani,1985)。然而,宾夕法尼亚系—二叠系界限的重新定义(Dabydov等,1998)导致了这些地层单元对宾夕法尼亚晚期地层的再分配。Bough段主要是由包括叶状藻礁在内的广海碳酸盐和广海风暴为主的大陆架沉积(Cys和Mazzullo,1985;Malek-Aslani,1985)。

10.2.2.2 下二叠统地层

奥罗格兰德盆地南部下二叠统碳酸盐岩地层主要由有严格限定的Hueco群海相地层单元组成(图10-2;Jordan,1975;Seager等,1976;Mack和James,1986;Candelaria,1988;Mack等,1988;Wilson和Jordan,1988;Raatz,2002;Wahlman和King,2002;Mack等,2003;Mack,2007)。盆地北部对等地层由以碎屑岩为主的Abo组构成。在盆地中部地层单元衔接之处(Robledo山脉),即Hueco地层和Abo地层间存在一过渡带。在该过渡带内,Hueco群的海相碳酸盐岩表现为与一种更为严格的浅水环境一致的特征。包括河口沉积在内,Abo组的碎屑岩单位含有边缘海沉积相(Jordan,1975;Wahlman和King,2002;Mack等,2003)。筳数据表明,Hueco群地层沉积于早二叠纪(Willianms,1963,1966;Simo等,2000;Wahlman和King,2002;Krainer等,2005)。

在盆地中部到北部地区,下二叠统的陆源到边缘海相Abo组地层上部为碎屑岩/局限的碳酸盐岩Yeso组地层(Kottlowski,1975;Candelaria,1988;Lindsay和Reed,1992;Mack等,2003)。在雷纳德世,该地区向San Andres组地层浅水碳酸盐沉积主导的海相环境转变(图10-2;Kottlowski,1975;Milner,1976)。San Andres组由不同程度白云岩化的、含本地化石(海百合类、伸长腕足类动物等)的灰质泥岩—颗粒灰岩组成,常含石膏团块(Milner,1976;Lindsay和Reed,1992)。这些碳酸盐岩单元被认为是在一个受限的广海和风暴为主的环境中沉积下来的(Kottlowski,1975;Raatz,2002)。

10.3 样品与分析方法

研究样品为取自奥罗格兰德盆地的岩心和露头(见图10-1,图10-3)。不同沉积相的286块样品被切成薄片,并用包括阴极发光在内的标准岩相方法进行了分析。利用胶结物(如泥晶灰岩)、腕足动物类和其他物质(如海百合类、冰洲石等)进行了碳氧同位素的分析(见表10-1,图10-4A,图10-5)。由于缺乏足够的、保存完好的腕足类贝壳实物,样品多数取自胶结物。所有同位素分析是在堪萨斯大学Keck古环境与环境稳定同位素实验室(KPESIL)完成的,所用仪器为Kiel Carbonete Device III + Finnigan MAT253同位素比值质谱仪(ThermoFinnigan,德国)。数据资料的年代是基于已发表的奥罗格兰德盆地的生物地层格架(Williams,1966;Jordan,1975;Simo等,2000;Wahlman和King,2002;Krainer等,2005),并根据Gradstein等(2004)对时间刻度进行了校正。

表 10-1 本研究所用实测剖面和岩心原始稳定同位素数据

样品	剖面	岩性	距剖面底部(m)	$\delta^{13}C(‰, VPDB)$	$\delta^{18}O(‰, VPDB)$
Nicor DDH-2 岩心					
35	博萨穆组	泥晶灰岩	2.56	-2.3	-6.9
36	博萨穆组	泥晶灰岩	7.4	-5.6	-7.4
37	博萨穆组	泥晶灰岩	15.2	-2.4	-7.0
38A	博萨穆组	泥晶灰岩	16.6	-0.9	-6.8
38A	博萨穆组	腕足动物	16.6	-0.5	-7.0
39A	博萨穆组	泥晶灰岩	17.9	-3.8	-7.8
39B	博萨穆组	腕足动物	17.9	-3.2	-7.1
40A	博萨穆组	泥晶灰岩	18.96	-5.1	-7.6
40B	博萨穆组	腕足动物	18.96	-5.3	-6.4
41	博萨穆组	泥晶灰岩	19.35	-5.6	-7.5
42A	博萨穆组	泥晶灰岩	24.9	-4.7	-7.9
42B	博萨穆组	腕足动物	24.9	-3.9	-8.4
富兰克林山脉					
FM-18	Hueco 峡谷组	泥晶灰岩	18	3.1	-6.8
FM-38	Hueco 峡谷组	泥晶灰岩	38	1.1	-5.3
FM-38a	Hueco 峡谷组	晶石充填物	38	0.7	-5.5
FM-54.38	Hueco 峡谷组	泥晶灰岩	54.38	3.2	-6.5
FM-62.3	Hueco 峡谷组	泥晶灰岩	62.3	4.2	-5.0
FM-62.3a	Hueco 峡谷组	裂缝充填物	62.3	1.8	-7.8
FM-62.3b	Hueco 峡谷组	软体动物	62.3	4.0	-5.0
FM-79	Hueco 峡谷组	泥晶灰岩	79	0.9	-4.7
FM-87	Hueco 峡谷组	泥晶灰岩	87	-1.6	-5.5
FM-90	Hueco 峡谷组	泥晶灰岩	90	-0.6	-0.8
FM-93	Hueco 峡谷组	泥晶灰岩	93	0.8	-4.9
FM-98	Hueco 峡谷组	泥晶灰岩	98	2.2	-6.8
FM-103	Hueco 峡谷组	泥晶灰岩	103	4.3	-5.9
FM-115	Hueco 峡谷组	泥晶灰岩	115	2.3	-4.2
FM-122	Hueco 峡谷组	泥晶灰岩	122	4.5	-3.9
FM-135.8	Hueco 峡谷组	泥晶灰岩	135.8	2.8	-4.8
FM-141.8	Hueco 峡谷组	泥晶灰岩	141.8	-0.7	-5.2
FM-154	Hueco 峡谷组	泥晶灰岩	154	1.7	-5.9
FM-180	Hueco 峡谷组	泥晶灰岩	180	2.8	-6.1
FM-190	Hueco 峡谷组	泥晶灰岩	190	1.9	-4.4
FM-190a	Hueco 峡谷组	泥晶灰岩	190	1.9	-3.8
FM-201	Hueco 峡谷组	泥晶灰岩	201	4.4	-6.7
FM-210	Hueco 峡谷组	泥晶灰岩	210	4.6	-4.3

续表

样品	剖面	岩性	距剖面底部(m)	$\delta^{13}C(‰,VPDB)$	$\delta^{18}O(‰,VPDB)$
FM-210a	Hueco 峡谷组	海百合类	210	4.7	-4.2
FM-222	Hueco 峡谷组	泥晶灰岩	222	4.5	-7.5
FM-225	Hueco 峡谷组	泥晶灰岩	225	4.1	-5.1
FM-237	Hueco 峡谷组	泥晶灰岩	237	4.2	-3.8
FM-237a	Hueco 峡谷组	海百合类	237	4.7	-3.6
FM-237b	Hueco 峡谷组	晶石充填物	237	4.1	-3.0
FM-241.5	Hueco 峡谷组	泥晶灰岩	241.5	3.9	-4.9
FM-244.5	Hueco 峡谷组	泥晶灰岩	244.5	3.8	-2.8
FM-244.5a	Hueco 峡谷组	腕足动物	244.5	3.7	-2.6
FM-244.5b	Hueco 峡谷组	软体动物	244.5	3.3	-3.1
FM-249	Hueco 峡谷组	泥晶灰岩	249	4.2	-4.0
FM-272	Cerro Alto 组	泥晶灰岩	272	3.8	-3.2
FM-286	Cerro Alto 组	泥晶灰岩	286	3.6	-3.9
FM-299	Cerro Alto 组	泥晶灰岩	299	4.4	-3.8
FM-317	Cerro Alto 组	泥晶灰岩	317	4.1	-4.6
FM-322	Cerro Alto 组	泥晶灰岩	322	4.6	-3.0
FM-334	Cerro Alto 组	泥晶灰岩	334	4.4	-4.1
FM-348	Cerro Alto 组	泥晶灰岩	348	4.4	-4.5
FM-359.5	Cerro Alto 组	泥晶灰岩	359.5	4.7	-5.2
FM-372	Cerro Alto 组	泥晶灰岩	372	4.8	-4.4
FM-424	Cerro Alto 组	泥晶灰岩	424	3.1	-4.1
FM-452	Cerro Alto 组	泥晶灰岩	452	2.5	-3.8
FM-483	Cerro Alto 组	泥晶灰岩	483	2.6	-5.1
FM-483a	Cerro Alto 组	海百合	483	2.4	-4.8
FM-518	Cerro Alto 组	泥晶灰岩	518	2.8	-2.3
FM-534	Cerro Alto 组	泥晶灰岩	534	3.4	-3.2
FM-595	Cerro Alto 组	泥晶灰岩	595	3.0	-4.0
FM-603	阿拉克兰山组	泥晶灰岩	603	3.3	-3.9
FM-603a	阿拉克兰山组	腕足动物	603	2.6	-2.5
FM-628	阿拉克兰山组	泥晶灰岩	628	1.4	-6.3
FM-653	阿拉克兰山组	泥晶灰岩	653	2.1	-3.9
FM-653a	阿拉克兰山组	腕足动物	653	1.8	-3.9
FM-667	阿拉克兰山组	泥晶灰岩	667	2.9	-4.5
FM-698	阿拉克兰山组	泥晶灰岩	698	2.9	-2.7
FM-702	阿拉克兰山组	泥晶灰岩	702	3.8	-5.6
FM-731.5	阿拉克兰山组	泥晶灰岩	731.5	3.6	-5.3
FM-731.5a	阿拉克兰山组	腕足动物	731.5	3.8	-3.9

续表

样品	剖面	岩性	距剖面底部(m)	$\delta^{13}C(‰, VPDB)$	$\delta^{18}O(‰, VPDB)$
FM-745	阿拉克兰山组	泥晶灰岩	745	5.4	-3.1
FM-780	阿拉克兰山组	泥晶灰岩	780	4.0	-5.6
下圣安德列斯山脉					
SA-1.0	Panther Seep 组	泥晶灰岩	1	-0.5	-7.7
SA-1.6	Panther Seep 组	泥晶灰岩	1.6	-2.5	-7.2
SA-1.8	Panther Seep 组	泥晶灰岩	1.8	-2.5	-7.6
SA-1.8a	Panther Seep 组	腕足动物	1.8	-2.4	-7.7
SA-1.8b	Panther Seep 组	腕足动物	1.8	-2.4	-7.7
SA-1.8c	Panther Seep 组	晶石充填物	1.8	-2.1	-8.2
SA-2.9	Panther Seep 组	泥晶灰岩	2.9	-2.5	-7.9
SA-2.9a	Panther Seep 组	泥晶灰岩	2.9	-2.5	-7.9
SA-3.1	Panther Seep 组	泥晶灰岩	3.1	-3.3	-8.5
SA-3.1a	Panther Seep 组	腕足动物	3.1	-1.6	-8.1
SA-3.1b	Panther Seep 组	腕足动物	3.1	-1.7	-9.2
SA-3.1c	Panther Seep 组	裂缝充填	3.1	-2.8	-10.4
SA-4.0	Panther Seep 组	泥晶灰岩	4	-2.2	-6.0
SA-4.5	Panther Seep 组	泥晶灰岩	4.5	-3.2	-7.5
SA-6.0	Panther Seep 组	泥晶灰岩	6	1.3	-8.9
SA-7.0	Panther Seep 组	泥晶灰岩	7	0.2	-8.8
SA-9.0	Hueco 组	泥晶灰岩	9	-0.8	-9.5
SA-9.0a	Hueco 组	晶石充填物	9	-0.2	-8.7
SA-12	Hueco 组	泥晶灰岩	12	-0.4	-9.4
SA-12a	Hueco 组	腕足动物	12	-0.1	-7.4
SA-14	Hueco 组	泥晶灰岩	14	-0.9	-8.9
SA-21.5	Hueco 组	泥晶灰岩	21.5	0.4	-7.4
SA-24	Hueco 组	泥晶灰岩	24	-3.4	-8.7
SA-24#2	Hueco 组	泥晶灰岩	24	-3.1	-8.4
SA-21#2a	Hueco 组	裂缝充填物	24	-2.5	-13.0
SA-28	Hueco 组	泥晶灰岩	28	2.5	-5.4
SA-30.5	Hueco 组	泥晶灰岩	30.5	-0.8	-8.4
SA-30.5a	Hueco 组	腕足动物	30.5	-0.6	-6.4
SA-30.5b	Hueco 组	海百合	30.5	-0.3	-5.1
SA-32#2	Hueco 组	泥晶灰岩	32	-2.1	-7.3
SA-32	Hueco 组	泥晶灰岩	32	-1.1	-7.8
SA-33	Hueco 组	泥晶灰岩	33	-2.2	-9.3
SA-37.5	Hueco 组	泥晶灰岩	37.5	1.6	-8.3
SA-39	Hueco 组	泥晶灰岩	39	0.6	-5.1

续表

样品	剖面	岩性	距剖面底部(m)	$\delta^{13}C(‰,VPDB)$	$\delta^{18}O(‰,VPDB)$
SA-50	Hueco/Abo 过渡带	泥晶灰岩	50	-1.3	-6.5
SA-64	Hueco/Abo 过渡带	泥晶灰岩	64	-0.2	-7.6
SA-64a	Hueco/Abo 过渡带	腕足动物	64	-0.8	-8.4
SA-78	Hueco/Abo 过渡带	泥晶灰岩	78	-5.0	-1.5
SA-79	Hueco/Abo 过渡带	泥晶灰岩	79	3.1	-5.7
上圣安德列斯山脉					
SA2-1.01	圣安德列斯组	泥晶灰岩	1.01	3.5	-4.9
SA2-4.01	圣安德列斯组	泥晶灰岩	4.01	4.2	-1.9
SA2-8.2	圣安德列斯组	泥晶灰岩	0.2	-5.8	-7.6
SA2-11.01	圣安德列斯组	泥晶灰岩	11.01	1.7	-8.9
SA2-15.1	圣安德列斯组	泥晶灰岩	15.1	2.1	-8.6
SA2.18	圣安德列斯组	泥晶灰岩	18	2.6	-8.0
SA2-24	圣安德列斯组	泥晶灰岩	24	2.4	-7.0
SA2-24a	圣安德列斯组	海百合	24	3.5	-4.7
SA2-24b	圣安德列斯组	三叶虫	24	1.8	-5.7
SA2-27	圣安德列斯组	泥晶灰岩	27	2.5	-5.4
SA2-27a	圣安德列斯组	海百合	27	3.6	-4.0
SA2-39	圣安德列斯组	泥晶灰岩	39	3.2	-6.9
SA2-44	圣安德列斯组	泥晶灰岩	44	2.9	-4.7
SA2-46	圣安德列斯组	泥晶灰岩	46	3.3	-4.9
SA2-49	圣安德列斯组	泥晶灰岩	49	2.7	-6.4
SA2-58	圣安德列斯组	泥晶灰岩	58	2.6	-7.0
SA2-58a	圣安德列斯组	腕足动物	58	2.6	-7.2
SA2-58b	圣安德列斯组	海百合	58	3.3	-4.9
SA2-58c	圣安德列斯组	腕足动物	58	3.2	-5.4
SA2-66	圣安德列斯组	泥晶灰岩	66	3.5	-4.9
SA2-69	圣安德列斯组	泥晶灰岩	69	3.8	-5.1
SA2-75	圣安德列斯组	泥晶灰岩	75	2.3	-8.8
SA2-75a	圣安德列斯组	腕足动物	75	2.5	-8.2
SA2-75b	圣安德列斯组	腕足动物	75	2.5	-7.6
SA2-83	圣安德列斯组	泥晶灰岩	83	3.4	-4.9
SA2-83a	圣安德列斯组	腕足动物	83	3.1	-5.2
SA2-83b	圣安德列斯组	腕足动物	83	3.6	-4.3
SA2-83c	圣安德列斯组	腕足动物	83	3.5	-4.9
SA2-87	圣安德列斯组	泥晶灰岩	87	3.5	-7.8
SA2-87a	圣安德列斯组	腕足动物	87	4.0	-6.7

续表

样品	剖面	岩性	距剖面底部(m)	$\delta^{13}C(‰,VPDB)$	$\delta^{18}O(‰,VPDB)$
下萨克拉门托山脉					
SN-0	Beeman 组	泥晶灰岩	0	-1.3	-4.0
SN-3.75	Beeman 组	泥晶灰岩	3.75	-0.8	-2.6
SN-4.0	Beeman 组	泥晶灰岩	4	1.0	2.6
SN-8.3	Beeman 组	泥晶灰岩	8.3	-2.2	-3.0
SN-9.8	Beeman 组	泥晶灰岩	9.6	-0.8	-8.3
SN-11.0	Beeman 组	泥晶灰岩	11	-1.3	-2.6
SN-24	Beeman 组	泥晶灰岩	24	-0.2	-3.3
SN-46.47	Beeman 组	泥晶灰岩	46.47	-4.5	-4.2
SN-65	Holder 组	泥晶灰岩	65	1.2	-6.0
SN-66	Holder 组	泥晶灰岩	66	2.8	-6.4
SN-67	Holder 组	泥晶灰岩	67	1.2	-3.8
SN-73	Holder 组	泥晶灰岩	73	0.3	-3.8
SN-73a	Holder 组	腕足动物	73	1.2	-2.9
SN-73b	Holder 组	晶石充填物	73	1.6	-3.7
SN-73.5	Holder 组	泥晶灰岩	73.5	-0.6	-8.1
SN-75	Holder 组	泥晶灰岩	75	-0.7	-3.9
SN-77	Holder 组	泥晶灰岩	77	-4.1	-4.4
SN-77a	Holder 组	腕足动物	77	-1.2	-6.7
SN-83	Holder 组	泥晶灰岩	83	-5.4	-4.1
SN-84	Holder 组	泥晶灰岩	84	1.5	-3.8
SN-84a	Holder 组	海百合	84	2.0	-1.9
SN-86	Holder 组	泥晶灰岩	86	-5.9	-5.7
SN-89	Holder 组	泥晶灰岩	89	-1.7	-5.1
SN-93	Holder 组	泥晶灰岩	93	1.1	-4.6
SN-103	Holder 组	泥晶灰岩	103	0.5	-5.9
SN-109	Holder 组	泥晶灰岩	109	2.5	-5.6
SN-113	Holder 组	泥晶灰岩	113	1.7	-5.0
SN-113a	Holder 组	腕足动物	113	3.1	-5.7
SN-114	Holder 组	泥晶灰岩	114	2.0	-4.8
SN-115	Holder 组	泥晶灰岩	115	1.5	-7.3
SN-118	Holder 组	泥晶灰岩	118	1.7	-6.5
SN-125	Holder 组	泥晶灰岩	125	-3.6	-6.6
SN-125a	Holder 组	腕足动物	140	2.4	-6.6
SN-140	Holder 组	泥晶灰岩	140	-3.9	-5.7
SN-140a	Holder 组	晶石充填物	144	-2.9	-7.2
SN-144	Holder 组	泥晶灰岩	144	-5.6	-5.7

续表

样品	剖面	岩性	距剖面底部(m)	$\delta^{13}C(‰,VPDB)$	$\delta^{18}O(‰,VPDB)$
SN-144#2	Holder 组	泥晶灰岩	144	-1.9	-6.8
SN-144#2a	Holder 组	腕足动物	144	-1.0	-6.9
SN-144#2b	Holder 组	腕足动物	144	-1.0	-6.8
SN-144#2c	Holder 组	裂缝充填物	144	-1.9	-7.6
SN-163	Holder 组	泥晶灰岩	163	-0.8	-4.9
SN-170	Holder 组	泥晶灰岩	170	-4.5	-5.1
SN-170a	Holder 组	腕足动物	170	-4.7	-6.1
SN-170b	Holder 组	腕足动物	170	-4.6	-5.7
SN-170c	Holder 组	软体动物	170	-4.6	-6.6
SN-173	Holder 组	泥晶灰岩	173	-1.4	-3.4
SN-177	Holder 组	泥晶灰岩	177	-4.6	-5.0
SN-198	Holder 组	泥晶灰岩	198	-5.8	-5.6
SN-207	Holder 组	泥晶灰岩	207	0.0	-5.8
SN-208	Holder 组	泥晶灰岩	208	-0.8	-5.3
SN-218	Holder 组	泥晶灰岩	218	-4.8	-5.1
SN-235	Abo 组	泥晶灰岩	235	-4.8	-5.5
SN-262	Abo 组	泥晶灰岩	262	-4.8	-3.6
SN-262a	Abo 组	腕足动物	262	-5.2	-4.4
SN-282	Abo 组	泥晶灰岩	282	-5.0	-4.0
上萨克拉门托山脉					
SN-B-2	圣安德列斯组	泥晶灰岩	15	1.5	-4.6
SN-B-9	圣安德列斯组	泥晶灰岩	22	3.0	-4.5
SN-C-0	圣安德列斯组	泥晶灰岩	465	3.9	-2.2
SN-C-3	圣安德列斯组	泥晶灰岩	495	4.3	-2.9
SN-D-10	圣安德列斯组	泥晶灰岩	61	3.5	-5.7
SN-D-15	圣安德列斯组	泥晶灰岩	69	3.3	-4.6
SN-D-22	圣安德列斯组	泥晶灰岩	74	4.1	-3.1
SN-D-3	圣安德列斯组	泥晶灰岩	84	2.5	-4.6
SN-E-0	圣安德列斯组	泥晶灰岩	134	1.3	-4.9
SN-E-5	圣安德列斯组	泥晶灰岩	139	3.1	-2.6
SN-F-0	圣安德列斯组	泥晶灰岩	153.5	4.1	-2.7
SN-F-8	圣安德列斯组	泥晶灰岩	161	3.3	-4.2
SN-G-1	圣安德列斯组		185.5	3.9	-3.3
其他岩心资料					
1	Getty No.5 Williard Beatty 岩心(Bough D)	泥晶灰岩	0.4	-0.2	-4.6
2	Getty No.5 Williard Beatty 岩心(Bough D)	泥晶灰岩	2.0	0.6	-4.7
3	Getty No.5 Williard Beatty 岩心(Bough D)	泥晶灰岩	4.4	-0.3	-5.3

续表

样品	剖面	岩性	距剖面底部(m)	$\delta^{13}C(‰, VPDB)$	$\delta^{18}O(‰, VPDB)$
4	Getty No.5 Williard Beatty 岩心(Bough D)	泥晶灰岩	5.0	0.3	-5.8
5	Getty No.5 Williard Beatty 岩心(Bough D)	泥晶灰岩	6.6	0.7	-4.0
6A	Getty No.5 Williard Beatty 岩心(Bough D)	泥晶灰岩	8.8	1.0	-5.4
6B	Getty No.5 Williard Beatty 岩心(Bough D)	筳化石	8.8	-0.5	-5.4
7A	Getty No.5 Williard Beatty 岩心(Bough D)	泥晶灰岩	9.1	-0.5	-5.1
7B	Getty No.5 Williard Beatty 岩心(Bough D)	腕足动物	9.1	0.5	-4.8
8A	Getty No.5 Williard Beatty 岩心(Bough D)	泥晶灰岩	9.7	0.1	-5.0
8B	Getty No.5 Williard Beatty 岩心(Bough D)	腕足动物	9.7	0.5	-5.3
9	Getty No.5 Williard Beatty 岩心(Bough D)	泥晶灰岩	11.8	-0.2	-4.8
10	Getty No.5 Williard Beatty 岩心(Bough C)	泥晶灰岩	13.5	0.0	-5.1
11	Getty No.5 Williard Beatty 岩心(Bough C)	泥晶灰岩	14.2	1.9	-4.3
12A	Getty No.5 Williard Beatty 岩心(Bough C)	泥晶灰岩	18.3	0.2	-4.1
12B	Getty No.5 Williard Beatty 岩心(Bough C)	腕足动物	18.3	0.1	-5.1
13	Getty No.5 Williard Beatty 岩心(Bough C)	泥晶灰岩	19.5	0.8	-4.5
14	Getty No.5 Williard Beatty 岩心(Bough C)	泥晶灰岩	21.3	1.2	-4.4
15	Getty No.5 Williard Beatty 岩心(Bough C)	泥晶灰岩	25.2	-0.6	-5.5
16	Getty No.5 Williard Beatty 岩心(Bough C)	泥晶灰岩	29.0	-0.6	-5.2
43	UNOCAL State No.1-33 Core(Bough B)	泥晶灰岩	30.0	3.8	-4.8
44	UNOCAL State No.1-33 Core(Bough B)	泥晶灰岩	30.3	4.6	-3.0
45	UNOCAL State No.1-33 Core(Bough B)	泥晶灰岩	31.2	3.7	-4.0
46	UNOCAL State No.1-33 Core(Bough B)	泥晶灰岩	32.1	2.8	-4.3
47	UNOCAL State No.1-33 Core(Bough B)	泥晶灰岩	32.4	3.9	-3.8
48	UNOCAL State No.1-33 Core(Bough B)	泥晶灰岩	35.5	2.8	-4.6
17A	Getty No.5 Williard Beatty Core(Bough B)	泥晶灰岩	40.6	0.7	-4.5
17B	Getty No.5 Williard Beatty Core(Bough B)	腕足动物	40.6	0.9	-4.3
49	UNOCAL State No.1-33 Core(Bough B)	泥晶灰岩	41.0	2.2	-4.1
50	UNOCAL State No.1-33 Core(Bough B)	泥晶灰岩	41.3	3.9	-4.4
18	Getty No.5 Williard Beatty Core(Bough B)	泥晶灰岩	42.3	0.8	-4.9
51	UNOCAL State No.1-33 Core(Bough B)	泥晶灰岩	43.7	2.2	-3.7
52	UNOCAL State No.1-33 Core(Bough B)	泥晶灰岩	44.6	2.0	-4.4
53	UNOCAL State No.1-33 Core(Bough B)	泥晶灰岩	45.3	1.0	-3.8
54	UNOCAL State No.1-33 Core(Bough B)	泥晶灰岩	46.2	1.1	-3.9
19	Getty No.5 Williard Beatty Core(Bough B)	泥晶灰岩	47.0	0.2	-4.9
20	Getty No.5 Williard Beatty Core(Bough B)	泥晶灰岩	48.6	-1.3	-5.0
55	UNOCAL State No.1-33 Core(Bough B)	泥晶灰岩	51.1	2.3	-3.8
56	UNOCAL State No.1-33 Core(Bough B)	泥晶灰岩	51.4	4.2	-3.5
57	UNOCAL State No.1-33 Core(Bough B)	泥晶灰岩	52.0	3.9	-3.7

续表

样品	剖面	岩性	距剖面底部(m)	$\delta^{13}C(‰, VPDB)$	$\delta^{18}O(‰, VPDB)$
58	UNOCAL State No. 1-33 Core(Bough B)	泥晶灰岩	52.3	4.3	-3.3
59	UNOCAL State No. 1-33 Core(Bough B)	泥晶灰岩	54.4	4.3	-3.0
60	UNOCAL State No. 1-33 Core(Bough B)	泥晶灰岩	54.7	4.0	-3.6
61	UNOCAL State No. 1-33 Core(Bough B)	泥晶灰岩	55.6	3.7	-4.1
62	UNOCAL State No. 1-33 Core(Bough A)	泥晶灰岩	56.2	3.5	-4.3
63	UNOCAL State No. 1-33 Core(Bough A)	泥晶灰岩	56.5	3.4	-4.1
64	UNOCAL State No. 1-33 Core(Bough A)	泥晶灰岩	56.8	2.7	-4.7
65	UNOCAL State No. 1-33 Core(Bough A)	泥晶灰岩	57.8	2.6	-4.0
66	UNOCAL State No. 1-33 Core(Bough A)	泥晶灰岩	58.4	2.8	-3.9
67	UNOCAL State No. 1-33 Core(Bough A)	泥晶灰岩	59.3	3.0	-4.9
68	UNOCAL State No. 1-33 Core(Bough A)	泥晶灰岩	59.9	1.5	-4.6
69	UNOCAL State No. 1-33 Core(Bough A)	泥晶灰岩	60.8	0.8	-4.0
69B	UNOCAL State No. 1-33 Core(Bough A)	腕足动物	60.8	1.1	-5.3
70A	UNOCAL State No. 1-33 Core(Bough A)	泥晶灰岩	62.0	0.4	-4.2
70B	UNOCAL State No. 1-33 Core(Bough A)	晶石充填物	62.0	-0.1	-4.8
71A	UNOCAL State No. 1-33 Core(Bough A)	泥晶灰岩	62.3	-0.6	-4.9
71B	UNOCAL State No. 1-33 Core(Bough A)	海百合	62.3	-0.4	-4.8
71C	UNOCAL State No. 1-33 Core(Bough A)	泥晶灰岩	62.3	-0.3	-4.1
24	Tres Papalotes Care(Bough A)	泥晶灰岩	62.6	-2.5	-6.0
72	UNOCAL State No. 1-33 Core(Bough A)	泥晶灰岩	62.6	-0.3	-5.1
25	Tres Papalotes Care(Bough A)	泥晶灰岩	62.6	-3.2	-5.9
21	Cetty No. 5 Williard Beatty Core(Bough B)	泥晶灰岩	65.0	-1.6	-4.6
26	Tres Papalotes Care(Bough A)	泥晶灰岩	65.7	-1.0	4.9
22	Cetty No. 5 Williard Beatty Core(Bough B)	泥晶灰岩	68.2	-4.8	-4.4
27A	Tres Papalotes Care(Bough A)	泥晶灰岩	66.8	-0.5	-4.1
27B	Tres Papalotes 岩心	腕足动物	66.8	-0.4	-4.4
23	Getty No. 5 Williard Beatty 岩心	泥晶灰岩	66.9	-2.6	-5.4
28	Tres Papalotes 岩心	泥晶灰岩	68.3	-0.9	-4.4
29	Tres Papalotes 岩心	泥晶灰岩	70.1	-1.7	-4.6
30	Tres Papalotes 岩心	泥晶灰岩	70.8	-0.2	-4.7
31	Tres Papalotes 岩心	泥晶灰岩	73.1	0.3	-4.4
32	Tres Papalotes 岩心	泥晶灰岩	73.9	-0.9	-4.3
33	Tres Papalotes 岩心	泥晶灰岩	77.7	3.4	-3.7
34A	Tres Papalotes 岩心	泥晶灰岩	79.0	3.5	-3.9
34B	Tres Papalotes 岩心	腕足动物	79.0	3.4	-3.8

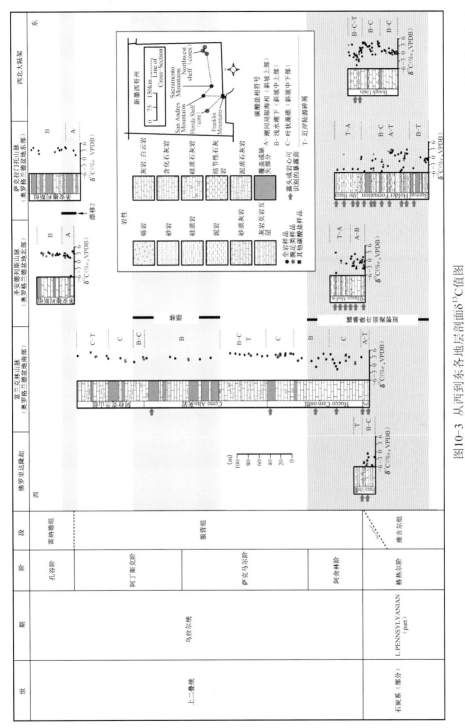

图10-3 从西到东各地层剖面δ¹³C值图

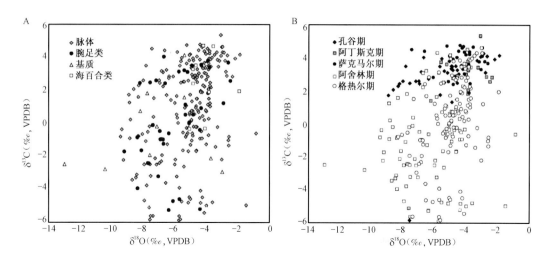

图 10-4 碳氧同位素交会图
（A）按照组分绘制的碳氧关系；（B）按照时间（阶段）绘制的碳氧关系

用于同位素分析的微量样品（20~80μg）是在抛光的平板上用微观固定的钻具和 20~500μm 直径的牙钻齿片钻取的。微量样品在 200℃ 的真空条件下烘烤 1 小时，以去除任何易挥发的有机组分。钙质微量样品和标样用磷酸在 75℃ 的真空条件下让其反应。磷酸是根据斯坦福大学稳定同位素实验室在线手册准备的。结果以相对于 VPDB 的碳氧同位素按照标准 Δ 计数法完成的报告。国家标准局（NBS）NBS-18 碳酸盐岩[国家标准与技术院（NIST）标准物质 8543]和 NBS-19 灰岩（NIST 标准物质 8544）作为质量控制标准；碳同位素的分析精度高于 0.01‰，氧同位素的分析精度高于 0.02‰。

10.4 沉积相与沉积趋势

沉积相配置关系表明，奥罗格兰德盆地从上部宾夕法尼亚系到下二叠统的碳酸盐岩是在一个均斜缓坡上沉积的（Koch，2010）。如多数其他晚古生代热带碳酸盐系统（如 James，1983；Wahlman，2002），这些沉积缺乏镶边台地和礁格架建造机制。取而代之的是，常见有相对浅水叶状藻类和富 Tubiphytes 的藻丘（Wahlman，2002；Forsythe，2003）；这与奥罗格兰德盆地的沉积物类型一致（Jordan，1975；Candelaria，1988）。

针对该文章的目的，把碳酸盐岩相分为了三个相组合：（A）泥岩和粒泥灰岩，沉积于局限海到潮间带环境；（B）粒泥灰岩—颗粒灰岩，沉积于浅水潮下带环境；以及（C）粘结岩/漂砾灰岩，为叶状藻——以 Tubiphytes 为主的生物礁。Koch（2010）给出了一个较为详细的沉积相与层序地层学分析结果。有关奥罗格兰德盆地特殊沉积相组合和沉积趋势的其他详情，参看图 10-3 和表 10-2。

10.4.1 相组合 A：泥岩与粒泥灰岩（局限海—潮间带环境）

被解释为局限海—潮间带环境的沉积，钙质泥岩和粒泥灰岩在 Beeman、Holder 和 Panther Seep 组合上宾夕法尼亚层系内普遍发育。该沉积相组合在下 Hueco Canyon 地层的二叠系最底部层系也普遍发育（图 10-1、图 10-3）。在富兰克林山脉剖面（如 Cerro Alto—Alacran 山脉地层）的上部未见有该沉积相发育，表明在中萨克乌尔到阿丁斯克时期，盆地南部的潮下带环境保留了多数的碳酸盐岩沉积（图 10-1、图 10-3）。相组合 A 也在下圣安德列斯地层中

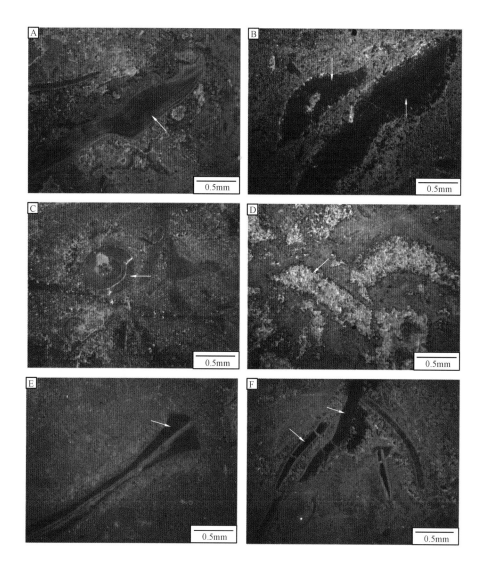

图 10-5 研究区阴极发光图像

(A)下萨克拉门托山脉剖面(73m,Holder 组)图像;箭头直指发光的腕足类壳体,表明变化;(B)下萨克拉门托山脉剖面(262m,Hueco/Abo 组)图像;箭头所指不发光的腕足类壳体,表明微变化或无变化;(C)Nicor DDH-2 岩心(166m,佛罗里达陆架—博萨穆组)图像;箭头指的是部分变化了的标识腕足类的棘;(D)Getty No.5 Williard Beatty 岩心(19.5m,西北陆架,Bough Unit)图像;箭头所指的是一个变化了的叶状藻片段;(E)富兰克林山脉剖面(653m,阿拉克兰组)图像;箭头所指为以几乎不发光的腕足类壳体,表明微或无变化;(F)富兰克林山脉剖(237m,Cerro Alto 组)图像;箭头所指为不发光的腕足类壳体,表明微变化或无变化

发育,表明在早孔谷期发生了一次向极浅水环境的回转。

相组合 A 包括泥晶—粒泥灰岩单元(厚度小于 1m)和微生物纹层(厚度小于 10cm)。格状组构与上部斜坡背景下的极浅水、局限环境一致,且为常见(Wilson,1975;Flügel,2004;见表 10-2)。碎屑石英、陆源泥质的存在以及腹足类、介形类的低多样性/高富集性表明,该相组合也记录了在受压、可能局限的、靠近碎屑物源环境条件下的沉积(Flügel,2004;Koch,2010)。

10.4.2 相组合 B:粒泥灰岩—颗粒灰岩(浅水潮下环境)

虽然被认为是在浅水潮下环境中沉积的碳酸盐岩单元在整个研究区内常见,但它在富兰克林山系 Cerro Alto 组和部分 Alacran 山组中却是较厚,而且相当丰富(图 10-1,图 10-3)。这些组合表明,在中萨克马尔到阿丁斯克时期,奥罗格兰德盆地南部主要表现为碎屑输入量相对低的广海潮下环境。

浅水潮下沉积物(1~30m 厚)由生物扰动岩、光能生物粒泥灰岩—颗粒灰岩组成(表 10-2)。化石目录富有异国多种多样的广海动物群,包括:叶状藻、腕足类、珊瑚虫类、有孔虫类、䗴类、腹足类、介形虫类、棘皮类、苔藓虫类以及三叶虫类。泥质含量的变化揭示了一个从相对低能(粒泥灰岩)到高能(颗粒灰岩)的宽泛能量条件。包粒和鲕粒也相当多,显示为高能碳酸盐浅滩环境(Flügel,2004)。

10.4.3 相组合 C:粘结灰岩/漂砾灰岩(叶状藻—*Tubiphytes* 生物礁)

相组合 C 在富兰克林山系的下二叠统和西北大陆架的 Bough 地层单元中是相当普遍(图 10-1,图 10-3)。在富兰克林山系中,下 Hueco Canyon 和 Alacran 山地层以厚层生物礁沉积为主。与下伏 Alacran 山组相比,Cerro Alto 组的地层以相组合 B 的潮下单元为主,揭示了微浅的沉积条件(图 10-3)。

叶状藻—富 *Tubiphytes* 漂砾灰岩/粘结灰岩单元(1~30m 厚)解释为生物礁沉积(表 10-2)。目前的研究人员推测,这些以叶状藻和 *Tubiphytes* 为主的沉积可能是起抗浪、缓冲作用的类礁构造,在晚宾夕法尼亚世—早二叠世期间较普遍(Janmes,1983;Wahlman,2002;Forsythe,2003)。叶状藻的出现暗示灰岩生物礁或许是在一个中-低斜坡环境的透光带内沉积下来的(Wahlman,2002;Forsythe,2003;Flügel,2004)。

表 10-2 奥罗格兰德盆地上古生界碳酸盐岩地层的相组合

相组合	特征	解释环境
相组合 A: 泥岩与粒泥灰岩	块状—薄层状沉积(厚度<1m),含微生物纹层结构和纹层泥晶灰岩带(厚度<10cm);常见格状组构;除了腹足动物、介形动物和有孔虫类外,少见化石物质;多为石英砂和陆相泥	(局限的)浅海潮下到低潮间环境;受影响的碎屑
相组合 B: 粒泥灰岩—颗粒灰岩	层状—块状沉积(厚度1~30m),含多样的广海动物群,包括海百合、有孔虫、䗴、腕足类、介形虫类和苔藓虫类;颗粒常有磨圆;可含有内碎屑和偶尔灰岩颗粒—卵石角砾岩;可见本地藻核灰岩和鲕粒	高能中—上斜坡位置;广海盐度;透光带内;可能是有影响的风暴
相组合 C: 粘结灰岩/漂砾灰岩	块状—厚层单元(厚度1~30m),以叶状藻和蓝绿藻为主;常见其他广海门类,如海百合类、三叶虫、腕足类和苔藓虫类	透光带内生物礁或碳酸盐岩丘环境;中—低斜坡位置;广海盐度;上斜坡见补丁礁

10.4.4 结果

包括骨架物质、胶结物和细粒基质在内的所有分析组分,其碳氧同位素组成都存在明显的变化(图 10-4A;表 10-1)。总体上来说,$\delta^{13}C$ 范围为 $-5.9‰ \sim 5.4‰$,跨度约 $11‰$。$\delta^{18}O$ 值总体上显示了较低的变化:除了两个胶结物样品外,都在 $-10‰ \sim -0.8‰$ 间波动。

地层背景下的同位素资料检测,揭示了 $\delta^{13}C$ 值变化幅度和程度在垂向上的趋势(图10-3)。比起层序中(均值为 $+3.3$;标准差为 $0.9‰$;阈值为 $4.0‰$)较新地层(萨克马尔—孔谷阶)的数值($n=86$),格热尔—阿舍林阶地层的数值($n=197$)更低(平均为 $-0.3‰$),变化更大(标准差为 $2.6‰$;值域为 $10.5‰$)。虽然不那么明显,$\delta^{18}O$ 的值显示了类似的地层趋势,相对于萨克马尔—孔谷阶地层,格热尔—阿舍林阶地层的数值总体上更低变化更大(图10-4B)。

10.5 成岩作用与地层趋势

10.5.1 对成岩作用的思考

由于海相碳酸盐岩的活性属性,在暴露于成分变化的间质流体期间,它们对同位素的调整反应敏感,并且在埋深和温度上经历后沉积作用的改变(Morse 和 Mackenzie,1990)。相对于海洋环境的水体,大多数的成岩流体的 $\delta^{18}O$ 和 $\delta^{13}C$ 大大降低。因此,海相碳酸盐岩与成岩流体间的化学交换有海相碳酸盐岩组分向低 $\delta^{18}O$ 和 $\delta^{13}C$ 变化的趋势。$\delta^{18}O$ 值对于成岩作用的变化尤其敏感。由于一定量的水和碳酸盐岩石中的氧分子量大体上相等,且仅少量碳酸盐可以溶入一定量的水内,成岩沉淀物的氧同位素组分更可能反应的是流体的氧同位素组成。另一方面,原始 $\delta^{13}C$ 值的保存受制于相对于水的碳酸盐岩石中碳的高度聚积。由于该悬殊,成岩流体的 $\delta^{13}C$ 值可能很快减小到原岩中的 $\delta^{13}C$ 值,使得成岩相的 $\delta^{13}C$ 值与原始值更相近(Lohmann,1988;Banner 和 Hanson,1990)。当溶解碳的含量以有机或热成因的 CO_2(具有比海相碳酸盐岩更贫化的 ^{13}C)为主导时,可能发生例外。海相碳酸盐岩与该流体间的化学交换可以导致碳酸盐岩中 ^{13}C 的贫化。

来自当前研究的沉积学检测和同位素数据表明,$\delta^{18}O$ 值已重新设定(图10-4至图10-6)。这些数值比那些记录在保存完好的二叠系腕足类(Korte 等,2005)和古热带环境的生物礁腔体海相胶结物(Given 和 Lohmann,1985)的数据要低几个千分数,古热带环境下的数值范围为大约 $-2.5‰ \sim -3‰$(图10-4A)。因此,本研究认为 $\delta^{18}O$ 值不是可靠的古环境指标。

海相 $\delta^{13}C$ 的保存情况是随地层变化的。在较老的地层中(格热尔—阿舍林阶),$\delta^{13}C$ 值变化较大,变化范围从大约 $±5‰$(与二叠系碳酸盐岩相中保存完好的记录数据值相似,Given 和 Lohmann,1985;Veizer 等,1999;Korte 等,2005;Grssman 等,2008)低至 $-6‰$。该交替部分的地层学和沉积学研究表明存在多个地表暴露界面(图10-3和图10-6C),这在奥罗格兰德盆地宾夕法尼亚系—二叠系的边界尤其普遍,世界的其他地方也是如此(Algeo 等,1992;Algeo,1996;Immenhauser 等,2002;Elrick 和 Scott,2010;Koch,2010)。在的格热尔—阿舍林阶地层中,C-O 同位素交会图显示,$\delta^{13}C$ 值相对于 $\delta^{18}O$ 值变化很大(图10-4B),这与地表暴露期间因贫化 ^{13}C(注入的土壤 CO_2)径流地表水成岩作用调整的模式相符(Lohmann,1988;Patterson 和 Walter,1994;Algeo,1996;Theiling 等,2007)。在剖面的这一段,仅 $\delta^{13}C$ 的高值可以作为可靠的古环境指标,低值用以了解各地表暴露事件的转换。

序列中较新的部分(萨克马尔—孔谷阶)缺乏大范围地表暴露的地层学和沉积学的证据(图10-3)。碳氧同位素交会图显示地层剖面(图10-4B)中的该段具有 $\delta^{18}O$ 急速降低并伴有 $\delta^{13}C$ 轻微降低的趋势。在相对于主体岩石 $\delta^{13}C$ 值已经降低了的流体前,相对恒定的 $\delta^{13}C$ 值和变化的 $\delta^{18}O$ 值,与经历了埋藏成岩作用的碳酸盐岩相符(Brand 和 Veizer,1981;Choquette

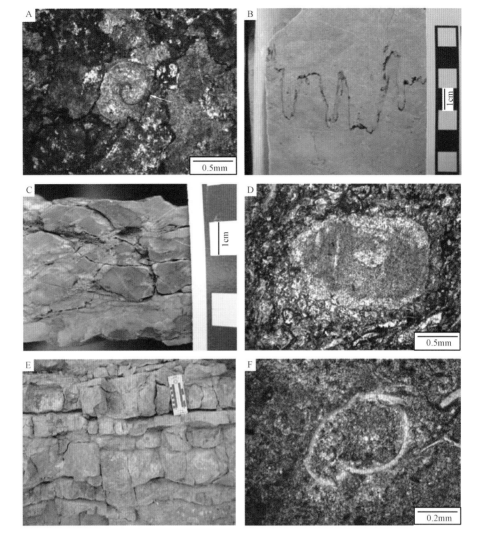

图 10-6 岩相与露头图像展示的成岩变化

(A)下圣安德列斯山脉剖面(4.0m,Panther Seep 组顶部)岩相图像,箭头所指为腹足类壳体周围缝合接触所示的埋藏成岩作用;(B)Unocal State No.1-33 岩心(13.9m,Bough B 段)图像,显示了腹足类所表明的埋藏成岩作用;(C)Getty No.5 Williard Beatty 岩心(64.8m,Bough B 段)图像,显示角砾化暴露地表所示的大气变化;(D)富兰克林山脉剖面(483m,Cerro Alto 组)岩相图像,显示部分白云岩化和压实的海百合颗粒(图像中心部分),缝合接触,显示埋藏成岩作用;(E)富兰克林山脉剖面(285m,Cerro Alto 组)图像,显示波状层理,表明埋藏成岩过程中的压溶作用;刻度总程度约为 16.5cm;(F)上萨克拉门托山脉剖面(48m,圣安德列斯组)岩相图像,显示破裂压实的介形类颗粒所示的埋藏成岩作用

和 James,1990;Brand,2004)。而且,阴极射线发光分析获得的明显的阴极基质物质(图 10-5E,图 10-5F)显示了 Mn^{2+} 含量的转变,也与还原的埋藏环境一致。沉积学证据(例如缝合作用、物理压实等)进一步显示了埋藏成岩环境的转变(图 10-6D 至图 10-6F)。对比已发表的关于保存完好的海相碳酸盐岩同位素数据的刊物,表明整个序列部分的 $\delta^{13}C$ 值紧密反映了原始组分,并且可以作为可靠的古环境指标(Given 和 Lohmann,1985;Veizer 等,1999;Korte 等,2005;Batt 等,2007;Grossman 等,2008)。

10.5.2 $\delta^{13}C$ 值的地层趋势

虽然 $\delta^{13}C$ 值上存在大量的全面变化,地层上下部分的数据检测表明其中的许多变化是系统性的垂直方向上发生的。图 10-3 可见单个剖面中同位素数据显示的地层趋势;图 10-7 可以找到一条复合曲线。

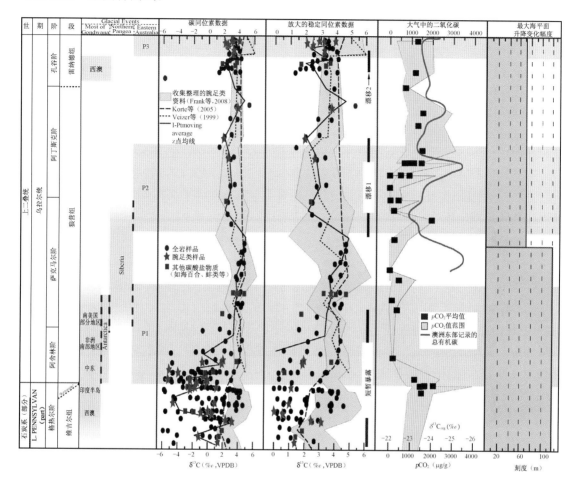

图 10-7 源于图 10-3 所示剖面的奥罗格兰德盆地 $\delta^{13}C$ 组成记录

时间刻度基于 Gradstein 等(2004)提供的数据;冈瓦纳冰期分布据 Fielding 等(2008c,及其内文献);大气 pCO_2 值据 EKart 等(1999)和 Montanez 等(2007);有机碳含量数据来自于 Brigenherer(2010);海面数据来自 Rygel 等(2008)

10.5.2.1 格热尔—阿舍林阶地层学趋势

分析了跨盆地的格热尔—阿舍林阶地层的地层剖面(图 10-3)。在佛罗里达陆架 Nicor DDH-2 岩心中,存在一个并不明朗的地层趋势组合剖面,其浅水潮下和叶状藻生物礁相(相组合 B 和 C)的 $\delta^{13}C$ 值分布范围为 $-5.6‰ \sim -0.5‰$(表 10-1,图 10-3)。$\delta^{13}C$ 值上的清晰趋势组合剖面的缺失或许是地层间隔研究不全面的结果(图 10-3)。佛罗里达陆架的岩相手标本也展示了重要埋藏成岩作用影响的证据(图 10-5C,图 10-5D,图 10-6B)。

相当多的同位素散布在富兰克林山系剖面的底部(Asselian),其特征为 Hueco Canyon 组下部 $\delta^{13}C$ 值为 $-1.6‰ \sim +4.7‰$(图 10-3)。这些数据中同位素的散布主要在浅水潮下与叶状藻生物礁相(相组合 B、C)。同样,圣安德烈斯山系剖面下段(Bursumian—Asselian)的

$δ^{13}C$ 值也有较大变化（-5.0‰~+3.1‰）。圣安德烈斯山系剖面下段的沉积相比富兰克林山系剖面下段的更浅、更局限（相组合 A、B），但另外剖面显然不存在清晰的地层趋势（图10-3）。萨克拉门托山系剖面下部包括 Beeman、Holder 和休科/Abo 组（格热尔—阿舍林阶）。这些地层包含了不同的 $δ^{13}C$ 值（-5.9‰~+3.1‰），也未见存在明显地层趋势的证据（图10-3）。不论沉积环境如何，这些变化的同位素数据存在于所有已知的碳酸盐岩相中（相组合 A 至相组合 C）。岩相分析也解释了埋藏成岩作用的可能影响（图10-5A）。

西北陆架的岩心中，$δ^{13}C$ 值也是多变的，在浅水潮下和叶状藻生物礁相中的范围为 -4.8‰~+4.6‰（相组合 B、C）。在宾夕法尼亚统近底部，存在 $δ^{13}C$ 值增大的一种可能的地层趋势，其后在二叠纪早期向 $δ^{13}C$ 值降低的趋势变迁（图10-3）。不清楚的是，这到底是一个实际的地层趋势，还是反映了宾夕法尼亚—二叠系边界附近的地表暴露和成岩作用的转化。

格热尔—阿舍林阶地层展示了大量的碎屑—碳酸盐岩混合地层沉积旋回性。每 5~10m 的旋回中包含一个通常由暴露面覆盖的向上变浅的层序。例如，在 Holder 组（萨克拉门托山系，见图10-3）中，记录了17个这样的旋回，它们与 Rankey 等（1999）同地区报道的22个层序比较。这些旋回很可能反映了与北美其他地区类似的相对海平面高频变化（例如 Algeo 等，1992；Algeo，1996；Heckel，2008）。

10.5.2.2 萨克马尔—孔谷阶地层趋势

主要通过富兰克林、圣安德列斯和萨克拉门托山剖面分析了萨克马尔—孔谷阶层系地层剖面（图10-3）。富兰克林山脉的萨克马尔—阿丁斯克地层（上 Hueco Canyon，Cerro Alto 和 Alacran 山地层）显示 $δ^{13}C$ 值从 +1.4‰到+5.4‰小范围变化，但均值为 +1.4‰~+5.4‰。富兰克林山系剖面上段也展现了 $δ^{13}C$ 值降低的小趋势（漂移1；见图10-3），这与一期向浅水相（相组合 B）的迁移吻合。该同位素漂移与偏移过程中的一期与浅水潮下向生物礁相沉积（相组合 B 和 C）的相变相吻合，该相变位置是在以浅水潮下沉积（相组合 B）为主的漂移之下。漂移末端与回归与略深的相组合 B 和 C 单元的一期相变相吻合（图10-3）。

上圣安德列斯山系剖面孔谷阶地层的 $δ^{13}C$ 值变化更小，大约平均值在 +3‰~4‰（图10-3）。该趋势与萨克拉门托山系圣安德列斯组的同位素数据完全一致（表10-1，图10-3）。圣安德列斯组下部存在一个 $δ^{13}C$ 值低至 -5.8‰的低度递减趋势（漂移2）。漂移2发生在相组合 A 潮间到局限单元内。漂移2之上，沉积相恢复到相组合 B 的浅水潮下沉积（图10-3）。

萨克马尔—孔谷阶地层的旋回性不同于下伏的格热尔—阿舍林阶单元。大量的（>10m 厚）叶状藻生物礁沉积（相组合 C）更加普遍。虽然向上变浅的成套地层的确存在（5~20m 厚），但地表暴露的证据不多（图10-3，图10-4）。每个旋回由潮下沉积构成，通常是相组合 C，它们由浅水相（如相组合 C）或者近岸碎屑沉积相所覆盖；萨克马尔—孔谷阶地层中少见有丘状暴露面（cyclecapping）（图10-3）。萨克马尔期有可能是区域构造变化、全球海平面升降模式或两者共同引起了这一沉积旋回性的变化。

10.5.2.3 综合数据

用目前已发表的年龄约束每个地层单元后（Cline，1959；Williamns，1963，1966；Schoderbek 和 Chafetz，1988；Rankey 等，1999；Simo 等，2000；Wahlman 和 King，2002；Krainer 等，2005），所有数据合成一个综合剖面（图10-7）时，上述趋势更加明显。相对于该层序的其他部分来说，格热尔—阿舍林阶地层离散性更大（scatter）和 $δ^{13}C$ 值分布更宽（宽几个千分数）（图10-7）。这些地层的 $δ^{13}C$ 值一般来说都较低，值域为 -5.9‰~+4.6‰。即便是各相明显不同，格热尔—阿舍林阶地层 $\delta^{13}C$ 值较低且多变的整体趋势在研究区内各处也都如此。例如，下 Hueco Canyon 地层的浅水潮下带到生物礁沉积相带（相组合 B 和 C），其 $\delta^{13}C$ 值的均值在 +1‰ 到 +3‰ 之间；同期的 Hueco 组（下圣安德列斯山系）内主体局限沉积单元（相组合 A）的 $\delta^{13}C$ 值平均在 -2‰ 到 0‰ 之间（图 10-3）。

从下萨克马尔阶贯彻至孔谷阶地层的 $\delta^{13}C$ 值比其下伏地层变化更小，值域为 +1.3‰~5.4‰（有一例外，上圣安德列斯山系剖面 -5.8‰ 的外露层）。加之，萨克马尔—孔谷期的沉积相趋于记录比老地层更深的环境。尤其是，萨克马尔—孔谷阶地层比格热尔—阿舍林阶单元更缺乏潮间相（相组合 A，图 10-3）。上述两个向低 $\delta^{13}C$ 值的少量漂移，其漂移量都为 1.5‰（见图 10-3）。图 10-7 中列出了只展示 $\delta^{13}C$ 值在 0~+6‰ 间的一幅图，突出了这些较小的漂移（标记了"放大的同位素数据"）。漂移 1 和漂移 2 期间的 $\delta^{13}C$ 均值为 +2.5‰；非漂移期的 $\delta^{13}C$ 均值大约为 +4.0‰（图 10-3，图 10-7）。

结合上述成岩作用的影响，地层趋势显示，在整个海平面一段时期内，格热尔—阿舍林阶地层低而多变的 $\delta^{13}C$ 值受到地表暴露的明显影响（图 10-3 至图 10-7）。相反，整个萨克马尔阶到孔谷阶地层（包括漂移 1 和 2）不可能在沉积中（或之后）重要时间段内遭受暴露。该解释是以相对稳定的 $\delta^{13}C$ 值和缺少能识别的暴露面为依据的（图 10-3 至图 10-7）。虽然漂移 1 和 2 伴随有略微低值的 $\delta^{13}C$ 值，但要充分证实快速多样的成岩，总体数值域（大约 4‰）不够大。况且，格热尔—阿舍林阶期间低 $\delta^{13}C$ 值仍比报道的漂移 1 和 2 的 $\delta^{13}C$ 值（见图 10-3 至图 10-7）要低（几个千分点）。

Saller 等（1999）展示了邻近二叠纪盆地具有一个与之类似的同位素趋势：格热尔—阿舍林阶期间低 $\delta^{13}C$ 值，萨克马尔—孔谷阶期间高 $\delta^{13}C$ 值。其研究中，格热尔地层同位素值域为 -5.5‰~-3.5‰，下二叠统（狼营组）碳酸盐岩地层的 $\delta^{13}C$ 值域为 -2.5‰~+3.5‰。

10.6 讨论

10.6.1 目前对晚古生代冰期的认识

晚古生代冰期（LPIA）是显生宙最长的冰室期，并且至今为止，工作人员已把这个冰室世界视为一个唯一的延长的冰期，持续了 60~80Ma（例如 Veevers 和 Powell，1987；Crowley 和 Baum，1991，1992；Frakes 等，1992）。石炭系欧美旋回层格架长期被认为是解释 LPIA 期海平面波动变化的最可靠的记录（Heckel，1986，1994，2008；Wright 和 Vanstone，2001）。该解释的前提是，欧美陆表海记录形成于冰末端位置，与流动冰的均衡负载量无关。例如，Wright 和 Vanstone（2001）用古热带旋回层记录证明 LPIA 开始是发生于密西西比期。欧美旋回层沉积的解释展示了晚宾夕法尼亚期冰川作用的一个高峰，该高峰基于更大的可见的海平面波动（如 Veevers 和 Powell，1987；Heckel，1994，2008；Gonzalez-Bonorino 和 Eyles，1995；Crowell，1999）。

Isbell 等（2003）回顾了整个冈瓦纳古陆的地层学资料，证明 LPIA 可以划分为三个独立的冰川期，其中两期在石炭纪（冰期 I 和 II），另一期在晚石炭世—萨克马尔世（冰期 III）。这些冰川期被非冰川环境的时间隔断。最近，工作人员对冈瓦纳古陆近冰沉积研究形成了一个 LPIA 时期冰期精细的地层格架（Fielding 等，208b，c，以及其内文献）。这些地层学研究表明，冰川最大冰量发生于早二叠世，而非长期所认为的晚宾夕法尼亚世顶峰期（Veevers 和 Powell，1987；Gonzalez-Bonorino 和 Eyles，1995；Heckel，2008）。间接记录用陆相和海相碳酸盐岩稳定同位素数据表明,格热尔时期为相对温和的全球气候条件,之后在早二叠世转换为一段明显变冷的气候条件;此次气候变化与冈瓦纳古陆冰量的急剧增大同期(Mnontanez 等,2007)。Heckel(2008)根据近期发表的全球间接记录和修订的冈瓦纳地层格架,总结认为石炭纪旋回层最可能记录了低冰量和所有的高海平面的次数,这些低冰量事件造成了欧美克拉通主体被海水所淹没。最近对海平面升降趋势的整理资料也支持了 Heckel(2008)所持的这一观点。该资料表明,海平面升降最大振幅(高达约 120m)发生于晚宾夕法尼亚世—中萨克马尔期,它可能是由冰河—海平面升降变异引起的(Rhgel 等,2008;见图 10-7)。中萨克马尔期之后,冰河—海平面升降的变化限于 70m 以内,可能反映了冈瓦纳冰川冰的整体消减。

冈瓦纳大陆的下二叠统冰川沉积为人所周知,它包括南美、非洲南部、中东、印度、南澳、南极以及西澳(Veevers 和 Tewari,1995;Fielding 等,2008a;Holz 等,2008;Isbell 等,2008a,b;Martin 等,2008;Mory 等,2008;Stollhoffen 等,2008)。早二叠世的冰川沉积也因西伯利亚而知名,但这些沉积物严格的年代地层位置有点模糊(见 Raymond 和 Meta,2004,以及其内文献)。

人们已知冈瓦纳几个区域的上宾夕法尼亚统为冰川沉积,它包括南美、非洲南部、中东、印度和西澳(Veevers 和 Tewari,1995;Holz 等,2008;Isbell 等,2008a,b;Martin,2008;Mory 等,2008;Stollhoffen 等,2008)。已知晚宾夕法尼亚世的南极洲和澳洲东部为非冰川沉积。整个冈瓦纳一些地区缺少地层学的控制,使得人们难以准确确定冰期的时间。

澳洲西部晚古生代冰川记录是冈瓦纳目前最明确的地层格架;这首先是厚层完整的层序列加上由放射性测年和生物年代资料很好限定的一个结果(Fielding 等,2008a)。识别了八个清晰的冰川时期,每个持续时间约 1~8Ma。每个冰川时期包含多个发展和消退的旋回(Fielding 等,2008a,b;Birgenheier 等,2009)。分割每个冰川时期的是间隔一定冰川时期的非冰川时期。从晚密西西比世到中宾夕法尼亚世的时间内发生了四个明确的石炭纪冰川时期(C1—C4)(Fielding 等,2008a)。二叠纪记录由四个非叠加的冰川时期(P1—P4)组成,这四个冰川时期发生在早二叠世至中二叠世,其中 P1 和 P2 的冰川时期规模最大(Fielding 等,2008a)。由于澳洲东部冰川记录的高地层学分辨率,有可能被作为一个格架,用于比较新墨西哥州奥罗格兰德盆地的 $\delta^{13}C$ 趋势。

10.6.2 全球早二叠世间接地化趋势的简单回顾

Veizer 等(1999)、Korte 等(2005)和 Grossman 等(2008)发表了上古生界稳定同位素专辑,被认为是代表了全球的趋势,但它仅依据对于二叠系相对不多的数据点。Frank 等(2008)收集了包括 Veizer 等(1999)和 Korte 等(2005)在内的目前发表的数据集。以上作者指出,全球气候环境在二叠纪伊始变得相当冷,特别是由于 $\delta^{18}O$ 值升高了大约 2‰,揭示了全球冰量的一期重要扩张。除此之外,Frank 等(2008)指出在宾夕法尼亚纪—二叠纪转换期间全球 $\delta^{13}C$ 值约 1‰~2‰ 的一次增大,与一期全球 pCO_2 水平减弱相符合。早二叠世的 pCO_2 水平减弱可能部分由海洋生产力的提升和冰川条件所伴随的 ^{12}C 隔离引起的。

从成土的碳酸盐物质推断得知低 pCO_2 值(大气二氧化碳),也表明了在二叠纪伊始期间向更冷的全球环境变迁(Ekart 等,1999;图 10-7)。Montanez 等(2007)用海相碳酸盐岩中的腕足动物资料、成土的碳酸盐和化石有机物质,证明了晚宾夕法尼亚世—中二叠世期间 $\delta^{13}C$ 和 $\delta^{18}O$ 的短期变化。这些学者认为,pCO_2 的变化、海平面升降、全球温度以及冰量与证实冈瓦纳冰川时期的地层学记录相关(Isbell 等,2003;Fielding 等,2008a)。最近,Birgenherer 等(2010)的有机相构建了澳洲东部的总 $\delta^{13}C_{org}$ 记录;他们的结果表明,$\delta^{13}C_{org}$ 的浮动(有机质埋藏比例的变化)与目前研究(如 Montanez 等,2007)中记载的 pCO_2 全球趋势相关。Birgenherer 等(2010)的总 $\delta^{13}C_{org}$ 记录变化与同地层的高分辨地层记录(Fielding 等,2008a)相关,这表明澳洲东部的冰川时期与二叠纪的 pCO_2 全球变化相对应(Montanez 等,2007)。

在同位素记录中,$\delta^{13}C$ 和 $\delta^{18}O$ 值的总体降低发生于近早二叠世末,又延续到晚二叠世(Korte 等,2005;Frank 等,2008)。该趋势被解释为记录了主冈瓦纳冰川期结束以来渐高的 pCO_2。

10.6.3 是保存在奥罗格兰德盆地的一个气候信号?

奥罗格兰德盆地在晚宾夕法尼亚世至二叠纪早期的大规模地表暴露显示相对基准面的长期回落,这是以增加沉积旋回的高频海平面变化为特征的。可能的积极因素包括构造抬升或海平面升降。由于泛古陆的最后拼合,晚宾夕法尼亚世至二叠纪早期是欧美西部地区的构造活动期(Ross,1986;Scotese 和 Langford,1995;Golonka 和 Ford,2000;Blakey,2008)。然而对奥罗格兰德盆地的目前研究表明,大规模的构造活动已经在宾夕法尼亚纪结束前整体上减缓了下来(Jordan,1975;Candelaria,1988)。Virgilian 期间得到最大缓解的主隆起区在狼营期末几乎都遭受了剥蚀(如 Pedernal Landmass 等)。到狼营期末,盆地中部和北部地区为陆相碎屑所填平,而碳酸盐岩沉积则局限于像富兰克林山脉这样的南部区域(Jordan,1975;Candelaria,1988)。相对海平面的升高发生在孔谷期,整个地区恢复到浅海环境(Kottlowski,1975;Milner,1976;Lindsay 和 Reed,1992;Raatz,2002)。进入早萨克马尔期,构造抬升不可能与碳酸盐岩地层地表的暴露完全一致响应。

根据海平面升降和同位素趋势特征(Frank 等,2008;Rygel 等,2008)以及冈瓦纳地层学记录(Fielding 等,2008a,b,c),可以推断认为,冰川海平面升降变化是控制晚宾夕法尼亚世至早二叠世奥罗格兰德盆地地表暴露的最可能候选因素。特别是,奥罗格兰德盆地的格热尔—阿舍林阶地层变化的 $\delta^{13}C$ 值,可能代表了主要受全球过程(如冰川海平面升降)控制的一个局部或区域(地表暴露)信号。在时间上,萨克马尔期间奥罗格兰德盆地主要地表暴露的末期对应冈瓦纳部分地区主体冰川时代的末期(Fielding 等,2008b),包括澳洲东部 P1 冰川时期的末期在内(Fielding 等,2008a)。中萨克马尔期结束的主要冰川—海平面升降波动,这一推论结果也与证实二叠纪海平面变化趋势的全球整理统计资料相吻合(Rygel 等,2008)。

主冈瓦纳冰川结束很久之后,小规模的冰心驻留在澳洲东部和非洲,直到晚二叠世(Fielding 等,2008a);小规模的冰心在中二叠世也驻留在西伯利亚(见 Raymond 和 Metz,2004 和其中参考文献)。这些冈瓦纳的小规模事件可用于解释奥罗格兰德盆地中一些其他的 $\delta^{13}C$ 趋势。例如,澳洲东部包括漂移1在内的时间间隔大体上与 P2 冰川时期吻合(图10-7)。漂移2期间 $\delta^{13}C$ 的降低大体上与西澳的冰川时期吻合,也与澳洲东部 P3 冰川时期的开始时间吻合(图10-7中漂移2)。这些结论表明,小规模的冰川作用不仅仅是区域性的,也可能影响到了边远地区,如奥罗格兰德盆地。此外,奥罗格兰德盆地的这些结论与全球同位素研究结果一致(如 Isozaki 等,2007a,2007b;Montanez 等,2007),同位素研究结果表明,澳洲东部 P2 和 P3 冰川时期具有全球性意义。

P2 和 P3 冰川时期两个负漂移的 $\delta^{13}C$ 值未表明奥罗格兰德盆地的地表暴露,这与 P1 冰川时期的沉积地层形成了对照(图10-7)。这些小规模事件发生期间,$\delta^{13}C$ 值大约 1.5‰ 的回落的确表明盆地内的环境条件与冈瓦纳的冰川波动存在间接的对应关系。Patterson 和 Walter(1994)证实,碳酸盐台地内部的 $\delta^{13}C$ 值因暴露或盐度的增高而使 ^{13}C 大大降低。该作者认为，甚至是在略微局限的条件下，$\delta^{13}C$ 值可能比在广海条件下低 4‰。漂移 1 和 2 期间 $\delta^{13}C$ 降低 1.5‰正好在 Patterson 和 Walter(1994)所描述的局限水体 $\delta^{13}C$ 可能值域内。此外，这些漂移期间的沉积相显示了从叶状藻生物礁较深水斜坡到碳酸盐斜坡之上的风暴为主沉积的变浅过程，其中的碳酸盐斜坡更容易遭受主要地表暴露事件(Jordan,1975；Kottlowski,1975；Milner,1976；Lindsay 和 Reed,1992；Koch,2010)。不存在漂移 1 期间极端超盐度水体沉积环境的证据。该时间间隔期间 $\delta^{13}C$ 值 1.5‰的降低可能代表比广海环境略微局限的环境条件，而非超盐度水体环境。相比之下，圣安德列斯组下段地层显示它沉积于比 Cerro Alto 组更浅的环境(如相组合 A 的潮间—局限环境)。漂移 2 期间 $\delta^{13}C$ 值的降低，可能是由于该时间间隔期盆内的超盐度水体环境。

另一种选择是，淡水大量流入盆地的加剧也可以说明漂移 1 和 2 的原因。相对于海相海水环境来说淡水的 $\delta^{13}C$ 大大减小，由此可使得靠近淡水水源沉淀的海相碳酸盐岩 $\delta^{13}C$ 变低(Patterson 和 Walter,1994；Panchuk 等,2005)。如果漂移 1 和 2 期间的区域沉淀量增加，增加的淡水流入奥罗格兰德盆地，有可能会轻微降低海相碳酸盐岩沉积物的 $\delta^{13}C$ 值。最近的晚古生代研究表明，冰川作用增强时期的热带地区更加潮湿(Cecil 等,2003；Tabor 和 Montanez,2004；Montanez 等,2007；Poulsen 等,2007)。与冰川时期 P2 和 P3 有关的强热带沉淀量，可能可以说明奥罗格兰德盆地内与漂移 1 和 2 有关的低 $\delta^{13}C$ 值的原因。

局限环境或者是流入盆内淡水的增加可以解释漂移 1 和 2 的 $\delta^{13}C$ 趋势特征。P2 和 P3 冰川时期的冰川作用增强了热带沉积量，导致了盆地相对海平面的降低，但这不是重要时期碳酸盐台地地表暴露的原因。对小规模漂移的 $\delta^{13}C$ 趋势特征解释表明，澳洲东部 P2 和 P3 冰川时期比发生在二叠纪早期的冰川时期规模要小。要不然，应该可以找出重要的地表暴露证据，该地表暴露是研究区内上宾夕法尼亚世到早二叠世地层的事件(图 10 - 3)。该证据与澳洲东部地层和同位素研究不符；澳洲东部的研究结果表明，P2 冰川时期与之前的 P1 冰川时期一样广泛(Fielding 等,2008a；Birgenheier 等,2010)。如果晚萨克马尔—孔谷期奥罗格兰德盆地沉陷加剧，就会限制碳酸盐岩沉积所受地表暴露(冰川—海平面变化减弱)的程度，由此会使得 P2 冰川时期显得比 P1(规模)更小。然而，奥罗格兰德盆地的地层研究表明，晚萨克马尔—孔谷期的沉降量实际上是降低的；宾夕法尼亚地层总厚度比相同时间内沉积的上覆二叠系地层厚 300~500m，表明二叠纪区域沉积量的降低(Jordan,1975；Candelaria,1988)。P2 冰川时期奥罗格兰德盆地与冈瓦纳之间的这一明显差异，仍然是众多悬而未决的问题之一，需要进一步研究。

10.7 结论

(1)贯穿奥罗格兰德盆地的整个格热尔—萨克马尔底部地层，展示了多期地表暴露的证据。该沉积间断的同位素趋势特征表明，快速成岩作用期的 $\delta^{13}C$ 值被重新调整。暴露的时刻与冈瓦纳冰川扩张期和晚古生代冰期高峰一致。例如，奥罗格兰德盆地地表暴露是以二叠纪早期冰川量增加导致的冰川—海平面降低为特征的(如澳洲东部 P1 冰川时期)。

(2)比早萨克马尔期更新的地层，其 $\delta^{13}C$ 值域全球趋势特征一致。奥罗格兰德盆地无地层学证据能证实该时间段内存在广泛的地表暴露，由此 $\delta^{13}C$ 值代表了全球同位素的信号。

(3)在上萨克马尔阶—下阿丁斯克阶(漂移 1)和孔谷阶地层中，发生了两期 $\delta^{13}C$ 值的负漂移。两漂移与向浅水和轻微局限环境的相变有关，与广泛的地表暴露无关。由于两期漂移和(水体)变浅的证据与纵穿冈瓦纳部分地区的冰川冰再次扩展对应(如澳洲东部 P2 和 P3 冰川时期),认为它们记录了低幅度的冰川—海平面的回落。

(4)奥罗格兰德盆地的地层模式与$\delta^{13}C$值证据显示,冰川—海平面波动的幅度和频率在格热尔期至萨克马尔早期是相当大的,它导致了广泛的地表暴露。相反地,大规模的地表暴露不可能发生在比中Sakmarian期更新的地层中;这表明,冈瓦纳的年轻冰川期(如澳洲东部的P2和P3)难以形成足够的冰川—海平面变化,以使得奥罗格兰德盆地的碳酸盐岩地层在中萨克马尔期—孔谷期期间的长时间内暴露地表。

(5)本研究获得的资料为早二叠世提供了许多必需的同位素数据。然而,下二叠统碳酸盐环境的其他同位素研究成果,还需要确定奥罗格兰德盆地的$\delta^{13}C$趋势特征到底是主要源于局部到区域的构造影响(例如沉降趋势),或者还是真正代表了一个全球性信号。

参 考 文 献

Algeo, T. J. (1996) Meteoric water/rock ratios and the significance of sequence and parasequence boundaries in the Gobbler Formation (Middle Pennsylvanian) of south-central New Mexico. In: Palaeozoic Sequence Stratigraphy: Views from the North American Craton(Eds B. J. Witzke, G. A. Ludvigson and J. Day), Geol. Soc. Am. Spec. Pap., 306, 359 – 371.

Algeo, T. J., Wilkinson, B. H. and Lohmann, K. C. (1992) Meteoric-burial diagenesis of Pennsylvanian carbonate: water/rock interactions and basin geothermics. J. Sed. Petrol., 62, 652 – 670.

Banner, J. L. and Hanson, G. N. (1990) Calculation of simultaneous isotopic and trace element variations during water-rock interaction with applications to carbonate diagenesis. Geochim. Cosmochim. Acta, 54, 3123 – 3137.

Batt, L. S., Montañez, I. P., Isaacson, P., Pope, M. C., Butts, S. H. and Abplanalp, J. (2007) Multi-carbonate component reconstruction of mid-Carboniferous (Chesterian) seawater $\delta^{13}C$. Palaeogeogr. Palaeoclimatol. Palaeoecol., 256, 298 – 318.

Birgenheier, L. P., Fielding, C. R., Rygel, M. C., Frank, T. D. and Roberts, J. (2009) Evidence for dynamic climate change on sub – 10^6-year scales from the late Palaeozoic glacial record, Tamworth Belt, New South Wales, Australia. J. Sed. Res., 79, 56 – 82.

Birgenheier, L. P., Frank, T. D., Fielding, C. R. and Rygel, M. C. (2010) Coupled carbon isotopic and sedimentological records from the Permian system of eastern Australia reveal the response of atmospheric carbon dioxide to glacial growth and decay during the late palaeozoic ice age. Palaeogeogr. Palaeoclimatol. Palaeoecol., 286, 178 – 193.

Blakey, R. C. (2008) Gondwana paleogeography from assembly to breakup: a 500 m. y. odyssey. In: Resolving the Late Palaeozoic Ice Age in Time and Space (Eds C. R. Fielding, T. D. Frank and J. L. Isbell), Geol. Soc. Am. Spec. Pap., 441, 1 – 28.

Brand, U. (2004) Carbon, oxygen and strontium isotopes in Palaeozoic carbonate components: an evaluation of original seawater-chemistry proxies. Chem. Geol., 204, 23 – 44.

Brand, U. and Veizer, J. (1981) Chemical diagenesis of a multicomponent carbonate system-2: stable isotopes. J. Sed. Petrol., 51, 987 – 997.

Bruckschen, P., Oesmann, S. and Veizer, J. (1999) Isotope stratigraphy of the European Carboniferous: proxy signals for ocean chemistry, climate and tectonics. Chem. Geol., 161, 127 – 163.

Candelaria, M. P. (1988) Synopsis of the Late palaeozoic depositional history of the Orogrande Basin, New Mexico and Texas. In: Basin to Shelf Facies Transition of the Wolfcampian Stratigraphy of the Orogrande Basin (Eds S. R. Robichaud and C. M. Gallick), 1988 Permian Basin Section-SEPM Annual Field Seminar, Publication No. 88 – 28, pp. 1 – 5.

Cecil, C. B., Dulong, F. T., West, R. R., Stamm, R., Wardlaw, B. A. and Edgar, N. T. (2003) Climate controls on the stratigraphy of a Middle Pennsylvanian Cyclothem in North America. In: Climate Controls on Stratigraphy (Eds

C. B. Cecil and T. N. Edgar) ,SEPM Spec. Publ. ,77,151 – 180.

Choquette, P. W. and James, N. P. (1990) Limestones: the burial diagenetic environment. In: Diagenesis (Eds I. A. Mcilreath and D. W. Morrow), Geosci. Can. Reprint Series,4,75 – 112.

Cline, L. M. (1959) Preliminary studies of the cyclical sedimentation and paleontology of the upper Virgilian strata of the La Luz area, Sacramento Mountains, New Mexico. Field Conference Guidebook. Permian Basin SEPM-Roswell Geol. Soc. ,172 – 185.

Crowell, J. C. (1999) Pre-Mesozoic ice ages; their bearing on understanding the climate system. Geol. Soc. Am. Mem. ,192,106.

Crowley, T. J. and Baum, S. K. (1991) Estimating carboniferous sea-level fluctuations from Gondwana ice extent. Geology,19,975 – 977.

Crowley, T. J. and Baum, S. K. (1992) Modeling late Palaeozoic glaciation. Geology,20,507 – 510.

Cys, J. M. and Mazzullo, S. J. (1985) Depositional and diagenetic history of a Lower Permian(Wolfcamp) Phylloid-Algal Reservoir, Hueco Formation, Morton Field, Southeastern New Mexico. In: Carbonate Petroleum Reservoirs(Eds P. O. Roehl and P. W. Choquette), Springer-Verlag, New York,279 – 288.

Davydov, V. I., Glenister, B. F., Spinosa, C., Ritter, S. M., Chernykh, V. V., Wardlaw, B. R. and Snyder, W. S. (1998) Proposal of Aidaralash as Global Stratotype Section and Point (GSSP) for base of the Permian System. Episodes,21,11 – 18.

Ekart, D. D. , Cerling, T. E. , Montañez, I. P. and Tabor, N. J. (1999) A 400 million year carbon isotope record of pedogenic carbonate: implications for paleoatmospheric carbon dioxide. Am. J. Sci. ,299,805 – 827.

Elrick, M. and Scott, L. A. (2010) Carbon and oxygen isotope evidence for high-frequency($10^4 - 10^5$ yr) and My-scale glacio-eustasy in Middle Pennsylvanian cyclic carbonates (Gray Mesa Formation), central New Mexico. Palaeogeogr. Palaeoclimatol. Palaeoecol. ,285,307 – 320.

Fielding, C. R., Frank, T. D., Birgenheier, L. P., Rygel, M. C., Jones, A. T. and Roberts, J. (2008a) Alternating glacial and non-glacial intervals characterize the late Palaeozoic Ice Age: stratigraphic evidence from eastern Australia. J. Geol. Soc. London,165,129 – 140.

Fielding, C. R., Frank, T. D. and Isbell, J. L. (2008b) The late Palaeozoic ice age: a review of current understanding and synthesis of global climate patterns. In: Resolving the Late Palaeozoic Ice Age in Time and Space (Eds C. R. Fielding, T. D. Frank and J. L. Isbell), Geol. Soc. Am. Spec. Pap. ,441,343 – 354.

Fielding, C. R., Frank, T. D. and Isbell, J. L., Eds. (2008c) Resolving the late Palaeozoic Ice Age in time and space. Geol. Soc. Am. Spec. Pap. ,441,354.

Flügel, E. (2004) Microfacies of Carbonates: Analysis, Interpretation and Application. Springer Publishing, New York, 976 pp.

Forsythe, G. T. W. (2003) A new synthesis of Permo-Carboniferous phylloid algal reef ecology. In: Permo-Carboniferous Carbonate Platforms and Reefs (Eds W. M. Ahr, P. M. Harris, W. A. Morgan and I. D. Somerville), SEPM Spec. Publ. ,78 and AAPG Mem. ,83,171 – 188.

Frakes, L. A. , Francis, J. E. and Syktus, J. I. (1992) Climate Modes of the Phanerozoic. Cambridge University Press, Cambridge,274 pp.

Frank, T. D. , Birgenheier, L. P. , Montañez, I. P. , Fielding, C. R. and Rygel, M. C. (2008) Controls on late Palaeozoic climate revealed by comparison of near-field stratigraphic and farfield stable isotopic records. In: Resolving the Late Palaeozoic Ice Age in Time and Space (Eds C. R. Fielding, T. D. Frank and J. L. Isbell), Geol. Soc. Am. Spec. Pap. ,441,331 – 342.

Given, K. R. and Lohmann, K. C. (1985) Derivation of the original isotopic composition of Permian marine cements. J. Sed. Petrol. ,55,430 – 439.

Golonka, J. and Ford, D. (2000) Pangean (late Carboniferous-Middle Jurassic) paleoenvironment and lithofacies. Palaeogeogr. Palaeoclimatol. Palaeoecol. ,161,1 – 34.

Gonzalez-Bonorino, G. and Eyles, N. (1995) Inverse relation between ice extent and the late Palaeozoic glacial record of Gondwana. Geology,23,1015 – 1018.

Gradstein, F. M. , Ogg, J. G. and Smith, A. G. , Eds. (2004) A Geologic Time Scale 2004. Cambridge University Press, Cambridge, pp. 222 – 270.

Grossman, E. L. , Yancey, T. E. , Jones, T. E. , Bruckschen, P. , Chuvashov, B. , Mazzullo, S. J. and Mii, H. (2008) Glaciation, aridification, and carbon sequestration in the Permo-Carboniferous: the isotopic record from low-latitudes. Palaeogeogr. Palaeoclimatol. Palaeoecol. ,268,222 – 233.

Heckel, P. H. (1986) Sea-level curve for Pennsylvanian eustatic marine transgressive-regressive depositional cycles along Midcontinent outcrop belt, North America. Geology,14,330 – 334.

Heckel, P. H. (1994) Evaluation of evidence for glacio-eustatic control over marine Pennsylvanian cyclothems in North America and consideration of possible tectonic effects. In: Tectonic and Eustatic Controls on Sedimentary Cycles (Eds J. M. Dennison and F. R. Ettensohn), SEPM, Concepts Sedimentol. Paleontol. ,4,65 – 87.

Heckel, P. H. (2008) Pennsylvanian cyclothems in Midcontinent North America as far-field effects of waxing and waning of Gondwana ice sheets. In: Resolving the Late Palaeozoic Ice Age in Time and Space (Eds C. R. Fielding, T. D. Frank and J. L. Isbell), Geol. Soc. Am. Spec. Pap. ,441,275 – 289.

Holz, M. , Souza, P. A. and Iannuzzi, R. (2008) Sequence stratigraphy and biostratigraphy of the late Carboniferous to early Permian glacial succession (Itararé subgroup) at the eastern-southeastern margin of the Paraná Basin, Brazil. In: Resolving the Late Palaeozoic Ice Age in Time and Space (Eds C. R. Fielding, T. D. Frank and J. L. Isbell), Geol. Soc. Am. Spec. Pap. ,441,115 – 130.

Immenhauser, A. , Kenter, J. A. M. , Ganssen, G. , Bahamonde, J. R. , Vliet, A. V. and Saher, M. H. (2002) Origin and significance of isotopic shifts in Pennsylvanian carbonates (Asturias, NW Spain). J. Sed. Res. ,72,82 – 94.

Isbell, J. L. , Miller, M. F. , Wolfe, K. L. and Lenaker, P. A. (2003) Timing of late Palaeozoic glaciation in Gondwana: was glaciation responsible for the development of Northern Hemisphere cyclothems? In: Extreme Depositional Environments: Mega End Members in Geologic Time (Eds M. A. Chan and A. A. Archer), Geol. Soc. Am. Spec. Pap. , 370,5 – 24.

Isbell, J. L. , Cole, D. I. and Catuneanu, O. (2008a) Carboniferous-Permian glaciation in the main Karoo Basin, South Africa. In: Resolving the Late Palaeozoic Ice Age in Time and Space (Eds C. R. Fielding, T. D. Frank and J. L. Isbell), Geol. Soc. Am. Spec. Pap. ,441,71 – 82.

Isbell, J. L. , Koch, XX, Zelenda, J. , Szablewski, G. M. and Lenaker, P. A. (2008b) Permian glacigenic deposits in the Transantarctic Mountains, Antarctica. In: Resolving the Late Palaeozoic Ice Age in Time and Space (Eds C. R. Fielding, T. D. Frank and J. L. Isbell), Geol. Soc. Am. Spec. Pap. ,441,59 – 70.

Isozaki, Y. , Kawahata, H. and Minoshima, K. (2007a) The Capitanian (Permian) Kamura cooling event: the beginning of the Paleozoic-Mesozoic transition. Palaeoworld,16,16 – 30.

Isozaki, Y. , Kawahata, H. and Ayano, O. (2007b) A unique carbon record across the Guadalupian-Lopingian (Middle-Upper Permian) boundary in mid-oceanic paleo-atoll carbonates: The productivity "Kamura event" and its collapse in Panthalassa. Glob. Planet. Change,55,21 – 38.

James, N. P. (1983) Reefs. In: Carbonate Depositional Environments (Eds P. A. Scholle, D. G. Bebout and C. H. Moore), AAPG Mem. ,33,345 – 462.

Jordan, C. F. (1975) Lower Permian (Wolfcampian) Sedimentation in the Orogrande Basin, New Mexico. In: Guidebook of the Las Cruces Country (Eds W. R. Seager, R. E. Clemons and J. F. Callender), New Mexico Geological Society Guidebook, 26 Field Conference, pp. 109 – 117.

Koch, J. T. (2010) A Sequence Stratigraphic and Isotopic Study of Uppermost Pennsylvanian-Lower Permian Carbonate Strata, Orogrande Basin, New Mexico. PhD thesis, University of Nebraska, Lincoln, 236 pp.

Korte, C., Jasper, T., Kozur, H. W. and Veizer, J. (2005) $\delta^{18}O$ and $\delta^{13}C$ of Permian brachiopods: a record of seawater evolution and continental glaciation. Palaeogeogr. Palaeoclimatol. Palaeoecol., 224, 333 – 351.

Kottlowski, F. E. (1975) Stratigraphy of the San Andres Mountains in South-Central New Mexico. In: Guidebook of the Las Cruces Country (Eds W. R. Seager, R. E. Clemons and J. F. Callender), New Mexico Geological Society Guidebook, 26th Field Conference, pp. 95 – 104.

Krainer, K., Lucas, S. G. and Spielmann, J. A. (2005) Hueco Group (Lower Permian) stratigraphy in the Dona Ana Mountains, southern New Mexico. In: The Permian of Central New Mexico (Eds S. G. Lucas, K. E. Zeigler and J. A. Spielmann), New Mexico Mus. Nat. Hist. Sci. Bull., 31, 60 – 73.

Lindsay, R. F. and Reed, C. L., Eds. (1992) Sequence Stratigraphy Applied to Permian Basin Reservoirs: Outcrop Analogs in the Caballo and Sacramento Mountains of New Mexico. 1992 Field Seminar Guidebook, West Texas Geological Society, Publication No. 92 – 92, 131 pp.

Lohmann, K. C. (1988) Geochemical patterns of meteoric diagenetic systems and their application to studies of Paleokarst. In: Paleokarst (Eds N. P. James and P. W. Choquette), pp. 58 – 79. Springer-Verlag Publishing, New York.

Mack, G. H. (2007) Sequence stratigraphy of the Lower Permian Abo member in the Robledo and Dona Ana Mountains near Las Cruces, New Mexico. New Mex. Geol., 29, 3 – 12.

Mack, G. H. and James, W. C. (1986) Cyclic sedimentation in the mixed siliciclastic-carbonate Abo-Hueco transitional zone (Lower Permian), southwestern New Mexico. J. Sed. Petrol., 56, 635 – 647.

Mack, G. H., James, W. C. and Seager, W. R. (1988) Wolfcampian (Early Permian) stratigraphy and depositional environments in the Dona Ana and Robledo Mountains, South-Central New Mexico. In: Basin to Shelf Facies Transition of the Wolfcampian Stratigraphy of the Orogrande Basin (Eds S. R. Robichaud and C. M. Gallick), 1988 Permian Basin Section-SEPM Annual Field Seminar, Publication No. 88 – 28, pp. 97 – 106.

Mack, G. H., Leeder, M., Perez-Arlucea, M. and Bailey, B. (2003) Sedimentology, paleontology, and sequence stratigraphy of Early Permian Estuarine Deposits, South-Central New Mexico, USA. Palaios, 18, 403 – 420.

Malek-Aslani, M. (1985) Permian Patch-Reef Reservoir, North Anderson Ranch Field, Southeastern New Mexico. In: Carbonate Petroleum Reservoirs (Eds P. O. Roehl and P. W. Choquette), pp. 267 – 276. Springer-Verlag, New York.

Martin, J. R., Redfern, J. and Aitken, J. F. (2008) A regional review of the late Palaeozoic glaciation in Oman. In: Resolving the Late Palaeozoic Ice Age in Time and Space (Eds C. R. Fielding, T. D. Frank and J. L. Isbell), Geol. Soc. Am. Spec. Pap., 441, 175 – 186.

Mii, H-S, Grossman, E. L. and Yancey, T. E. (1999) Carboniferous isotope stratigraphies of North America; implications for Carboniferous paleoceanography and Mississippian glaciation. Geol. Soc. Am. Bull., 111, 960 – 973.

Mii, H-S, Grossman, E. L., Yancey, T. E., Chuvashov, B. and Egorov, A. (2001) Isotopic records of brachiopod shells from the Russian Platform; evidence for the onset of Mid-Carboniferous glaciation. Chem. Geol., 175, 133 – 147.

Milner, S. (1976) Carbonate Petrology and Syndepositional Facies of the Lower San Andres Formation (Middle Permian), Lincoln County, New Mexico. J. Sed. Petrol., 46, 463 – 482.

Montañez, I. P., Tabor, N. J., Niemeier, D., DiMichele, W. A., Frank, T. D., Fielding, C. R., Isbell, J. L., Birgenheier, L. P. and Rygel, M. C. (2007) CO_2 – forced climate and vegetation instability during Late Palaeozoic Deglaciation. Science, 315, 87 – 91.

Morse, J. W. and Mackenzie, F. T. (1990) Geochemistry of Sedimentary Carbonates. Elsevier, Amsterdam, 707 pp.

Mory, A. J., Redfern, J. and Martin, J. R. (2008) A review of Permian-Carboniferous glacial deposits in Western Australia. In: Resolving the Late Palaeozoic Ice Age in Time and Space (Eds C. R. Fielding, T. D. Frank and J. L. Isbell), Geol. Soc. Am. Spec. Pap., 441, 29 – 40.

Panchuk, K. M., Holmden, C. and Kump, L. R. (2005) Sensitivity of the epeiric sea carbon isotope record to locAl-scale carbon cycle processes: Tales from the Mohawkian Sea. Palaeogeogr. Palaeoclimatol. Palaeoecol., 228, 320 – 337.

Patterson, W. P. and Walter, L. M. (1994) Depletion of ^{13}C in seawater $\sum CO_2$ on modern carbonate platforms: significance for the carbon isotopic record of carbonates. Geology, 22, 885 – 888.

Poulsen, C. J., Pollard, D., Montañez, I. P. and Rowley, D. (2007) Late Paleozoic tropical climate response to Gondwanan deglaciation. Geology, 35, 771 – 774.

Raatz, W. D. (2002) A stratigraphic history of the Tularosa Basin Area, South-Central New Mexico. In: Geology of White Sands (Eds V. W. Lueth, K. A. Giles, S. G. Lucas, B. S. Kues, R. Myers and D. S. Ulmer-Scholle), New Mexico Geological Society Guidebook, 53rd Field Conference, pp. 141 – 157.

Rankey, E. C., Bachtel, S. L. and Kaufman, J. (1999) Controls on stratigraphic architecture of icehouse mixed carbonate-siliciclastic systems: a case study from the Holder Formation (Pennsylvanian, Virgillian), Sacramento Mountains, New Mexico. In: Advances in Carbonate Sequence Stratigraphy: Application to Reservoirs, Outcrops and Models (Eds P. M. Harris, A. H. Saller and J. A. Simo), SEPM Spec. Publ., 63, 127 – 150.

Raymond, A. and Metz, C. (2004) Ice and its consequences: glaciation in the Late Ordovician, Late Devonian, Pennsylvanian-Permian, and Cenozoic Compared. J. Geol., 112, 655 – 670.

Ritter, S. M. (1995) Upper Missourian-Lower Wolfcampian (Upper Kasimovian-Lower Asselian) Conodont Biostratigraphy of the Midcontinent, USA. J. Paleontol., 69, 1139 – 1154.

Ross, C. A. (1963) Standard Wolfcampian series (Permian), Glass Mountains, Texas. Geol. Soc. Am. Mem., 88, 205.

Ross, C. A. (1986) Palaeozoic evolution of southern margin of Permian Basin. Geol. Soc. Am. Bull., 97, 536 – 554.

Ross, C. A. and Ross, J. R. P. (1988) Late Paleozoic transgressiveregressive deposition. In: Sea Level Changes: An Integrated Approach (Eds C. K. Wilgus, B. J. Hastings, H. Posamentier, J. C. van Wagoner, C. A. Ross and C. G. Kendall), SEPM Spec. Publ., 42, 227 – 247.

Ross, J. R. P. and Ross, C. A. (1990) Late Palaeozoic bryozoan biogeography. In: Palaeozoic Palaeogeography and Biogeography (Eds W. S. McKerrow and C. R. Scotese), Geol. Soc. London Mem., 12, 353 – 361.

Ross, C. A. and Ross, J. R. P. (1995) Foraminiferal zonation of late Paleozoic depositional sequences. In: Forams 94-Selected Papers from the Fifth International Symposium of Foraminifera (Eds M. R. Langer, J. H. Lipps, J. C. Ingle and W. V. Sliter), Mar. Micropaleontol., 26, 469 – 478.

Rygel, M. C., Fielding, C. R., Frank, T. D. and Birgenheier, L. P. (2008) The magnitude of late Palaeozoic glacioeustatic fluctuations: a synthesis. J. Sed. Res., 78, 500 – 511.

Saller, A. H., Dickson, J. A. D., Rasbury, E. T. and Ebato, T. (1999) Effects of long-term accommodation change on shortterm cycles, Upper Palaeozoic Platform Limestones, West Texas. In: Advances in Carbonate Sequence Stratigraphy: Application to Reservoirs, Outcrops and Models (Eds P. M. Harris, A. H. Saller and J. A. Simo), SEPM Spec. Publ., 63, 227 – 246.

Saltzman, M. R. (2003) Late Palaeozoic ice age: oceanic gateway or pCO_2? Geology, 31, 151 – 154.

Schoderbek, D. A. and Chafetz, H. S. (1988) Sedimentological and stratigraphical relationships of the panther seep formation, (Virgilian, Pennsylvanian) Southern San Andres Mountains, New Mexico. In: Basin to Shelf Facies Transition of the Wolfcampian Stratigraphy of the Orogrande Basin (Eds S. R. Robichaud and C. M. Gallick), 1988 Permian Basin Section-SEPM Annual Field Seminar, Publication No. 88 – 28, pp. 89 – 96.

Scotese, C. R. and Langford, R. P. (1995) Pangea and the Paleogeography of the Permian. In: The Permian of Northern Pangea, Volume 1 (Eds P. A. Scholle, T. M. Peryt and D. S. Ulmer-Scholle), pp. 3 – 19. Springer-Verlag, New York.

Seager, W. R., Kottlowski, F. E. and Hawley, J. W. (1976) Geology of Dona Ana Mountains, New Mexico. Cir. New Mex. Bur. Min. Miner. Resour., 147, 36.

Simo, J. A., Wahlman, G. P., Beall, J. B. and Stoklosa, J. L. (2000) Lower Permian (Wolfcampian) in the Hueco Mountains; stratigraphic and age relations. In: The Permian Basin: Proving Ground for Tomorrow's Technologies (Eds W. D. DeMis, M. K. Nelis and R. C. Trentham), West Texas Geological Society Publication No., 00 – 109, pp. 41 – 50.

Steiner, M. B. and Williams, T. E. (1968) Fusulinidae of the Laborcita Formation (Lower Permian), Sacramento Mountains, New Mexico. J. Paleontol. ,42,51 – 60.

Stollhoffen, H. , Werner, M. , Stanistreet, I. G. and Armstrong, R. A. (2008) Single-zircon U-Pb dating of Carboniferous-Permian tuffs, Namibia, and the intercontinental deglaciation cycle framework. In: Resolving the Late Palaeozoic Ice Age in Time and Space(Eds C. R. Fielding, T. D. Frank and J. L. Isbell) , Geol. Soc. Am. Spec. Pap. ,441,83 – 96.

Tabor, N. J. and Montañez, I. P. (2004) Morphology and distribution of fossil soils in the Permo-Pennsylvanian Wichita and Bowie Groups, north-central Texas, USA: implications for western equatorial Pangean palaeoclimate during icehouse-greenhouse transition. Sedimentology,51,851 – 884.

Theiling, B. J. , Railsback, B. , Holland, S. M. and Crowe, D. (2007) Heterogenetiy in geochemical expression of subaerial exposure in limestones, and its implications for sampling to detect exposure surfaces. J. Sed. Res. , 77, 159 – 169.

Veevers, J. J. and Powell, C. M. C. A. (1987) Late Palaeozoic glacial episodes in Gondwanaland reflected in transgressive-regressive depositional sequences in Euramerica. Geol. Soc. Am. Bull. ,98,475 – 487.

Veevers, J. J. and Tewari, R. C. (1995) Gondwana Master Basin of Peninsular India. Geol. Soc. Am. Mem. ,187,72.

Veizer, J. , Ala, D. , Azmy, K. , Bruckschen, P. , Buhl, D. , Bruhn, F. , Carden, G. , Diener, A. , Ebneth, S. , Godderis, Y. , Jasper, T. , Korte, C. , Pawellek, F. , Podlaha, O. and Strauss, H. (1999) $^{87}Sr/^{86}Sr$, $d^{13}C$ and $d^{18}O$ evolution of Phanerozoic seawater. Chem. Geol. ,161,59 – 88.

Wahlman, G. P. (2002) Upper Carboniferous-Lower Permian(Bashkirian-Kungurian) Mounds and Reefs. In: Phanerozoic Reef Patterns(Eds W. Kiessling, E. Flügel and J. Golonka) , SEPM Spec. Publ. ,72,271 – 338.

Wahlman, G. P. and King, W. E. (2002) Latest Pennsylvanian and earliest Permian fusulinids biostratigraphy, Robledo Mountains and adjacent ranges, south-central New Mexico. Cir. New Mex. Bur. Min. Mineral Resour. ,208,26.

Williams, T. E. (1963) Fusulinidae of the Hueco Group(Lower Permian) , Hueco Mountains, Texas. Peabody Museum of Natural History, Yale University, Bull ,18,123.

Williams, T. E. (1966) Permian Fusulinidae of the Franklin Mountains, New Mexico-Texas. J. Paleontol. , 40, 1142 – 1156.

Wilson, J. L. (1967) Cyclic and reciprocal sedimentation in Virgilian strata of southern New Mexico. Geol. Soc. Am. Bull. ,78,805 – 818.

Wilson, J. L. (1975) Carbonate Facies in Geologic History. Springer-Verlag, Berlin,471p.

Wilson, J. L. and Jordan, C. F. (1988) Measured Sections in the Pennsylvanian and Permian of the Sacramento Mountains(Fresnal Canyon) and the Hueco Mountains. In: Basin to shelf facies transition of the Wolfcampian stratigraphy of the Orogrande Basin(Eds S. R. Robichaud and C. M. Gallick) ,1988 Permian Basin Section-SEPM Annual Field Seminar, Publication No. 88 – 28, pp. 75 – 77.

Wright, V. P. and Vanstone, S. D. (2001) Onset of Late Palaeozoic glacio-eustasy and the evolving climates of low latitude areas: a synthesis of current understanding. J. Geol. Soc. London,158,579 – 582.

第11章 台地沉积物的稳定碳同位素组分与全球碳循环重建

AMANDA M. OEHLERT, KATHRYN A. LAMB-WOZNIAK, QUINN B. DEVLIN, GRETA J. MACKENZIE, JOHN J. G. REIJMER, PETER K. SWART 著

赵 霞 译,周永胜 校

摘 要 在远洋深海碳酸盐沉积物中,有机与无机组分的 $\delta^{13}C$ 值的协同变化的相关程度(即相关性)已被用来解释有机碳的生产、埋藏和分解速率。这种相关关系随着时间变化却保持相对不变,从而能够评估有机碳的生产和保存。然而,由于绝大部分年代早于200Ma的远洋深海沉积已经被俯冲消减掉了,因此常常就用形成于近陆浅海和台地的碳酸盐沉积替代远洋深海的碳酸盐,进行古代全球碳循环的分析。使用浅海碳酸盐物质面临众所周知的陷阱,包括:成岩作用、沉积环境的半封闭、无机 $\delta^{13}C$($\delta^{13}C_{无机}$)值迥异的不同沉积物的输入,这些陷阱可以抹消任何全球性的信号。评价 $\delta^{13}C_{无机}$ 记录是否准确反映全球 $\delta^{13}C$ 变化的一种方法是检查共生有机质的 $\delta^{13}C$($\delta^{13}C_{有机}$)记录的变化。如果 $\delta^{13}C_{有机}$ 记录与共生的 $\delta^{13}C_{无机}$ 记录一起变化,那么就会认为,这一定是与全球碳循环相关的信号。本研究对大洋钻探计划(ODP)第166航次从大巴哈马浅滩西缘采集的样品进行了研究,通过分析保存于顶部150m台缘沉积物中有机碳的同位素组成,对上述假定进行了调查。将本次研究测得的 $\delta^{13}C_{有机}$ 值与先前已发表的、在相同样品上测得的 $\delta^{13}C_{无机}$ 记录进行对比,从而得以对这两个记录之间的相关性进行研究。这些研究分析表明,$\delta^{13}C_{无机}$ 与 $\delta^{13}C_{有机}$ 之间的相关系数由近源处(1005 站位,$r^2=0.1$)到远源处(1006 站位,$r^2=0.63$)增大。在近源处(1005 站位),台地输入碳酸盐和有机物的重要性体现在无机与有机 $\delta^{13}C$ 记录之间缺乏共变关系,因为在台地上二者就是不相关的。相反,在大洋盆地(1006 站位),$\delta^{13}C$ 值的共变性可以用两点混合模型来解释,该模型证明,在促成有机与无机 $\delta^{13}C$ 值之间的正相关的过程中,远洋深海的和台地输入的碳酸盐—有机碳都起到了重要作用,因此在1006 站位 $\delta^{13}C_{无机}$ 与 $\delta^{13}C_{有机}$ 记录之间的相关性与全球碳循环无关。这些数据所提出的质疑是,用 $\delta^{13}C_{有机}$ 值来支持碳酸盐中 $\delta^{13}C_{无机}$ 值记录全球碳循环的能力是否合适,因为在这些碳酸盐环境中,可能有多个碳酸盐—有机碳的来源参与了总的 $\delta^{13}C$ 信号的形成。

关键词 碳循环　碳同位素　碳酸盐台地　有机质　pCO_2

11.1 概况

随着地质时期的变迁,碳酸盐沉积的 $\delta^{13}C$ 发生着改变,这些变化已经被解释成是代表了有机碳生产速率相对于有机碳埋藏和保存速率的变化(Hayes 等,1999)。一般来说,碳酸盐中的 $\delta^{13}C$ 值升高反映有机物生产兼或有机物埋藏的增加,而 $\delta^{13}C$ 值的降低则表明有机物的氧化增强,因而有机碳埋藏的比例降低(Veizer 和 Hoefs,1976;Schidlowski,1979;Shackleton,1985)。在重建过去大约200Ma间的全球碳循环研究中,远洋碳酸盐沉积和化石,特别是有孔虫,是获得 $\delta^{13}C$ 值而优先选择的碳酸盐类型;这些类型的沉积和化石被认为准确反映海洋当中溶解无机碳(DIC)$\delta^{13}C$ 值的变化(Zachos 等,2001)。例如:在过去 50Ma 间,有孔虫和远洋沉积物的 $\delta^{13}C$ 值变得越来越负了(Shackleton 和 Hall,1980;Broecker 和 Woodruff,1992),这种偏负值的变化趋势被认为是 10^{20}g 数量级的碳从有机碳储库向无机碳储库转移的结果(Shackleton,1985)。

由于缺乏年代大于200Ma的远洋深海记录,人们不得不用宏体化石,像腕足类和箭石类化石,或者全岩沉积样品来获得$\delta^{13}C$记录。这些沉积物和生物往往是沉积在较为受限的环境,像近缘浅海、缓坡或者碳酸盐台地。虽然在某些实例中这些记录可以与远洋$\delta^{13}C$记录的变化趋势相对应(Vahrenkamp,1996),然而从这些环境的生物骨骼和全岩样品测得的$\delta^{13}C$记录可能会出现重大的解释问题,因为在这些环境单元可能不曾有过真正的开放大洋条件,也可能发生了重大的、或矿物影响而生成具有不同$\delta^{13}C$值的碳酸盐,又或者沉积物和化石经历了成岩改造。一种常见的、验证大于200Ma碳酸盐$\delta^{13}C_{无机}$记录是否具有全球性的方法就是对比几个盆地同时代的$\delta^{13}C$记录。如果这些$\delta^{13}C$变化趋势是可比的,那么就认为,该记录代表一种准确的全球性信号,尤其在这些盆地相距甚远的情况下。在最近的一项研究中,从大巴哈马浅滩(GBB)环台地沉积物中测得的全新世到中中新世$\delta^{13}C$值与从非洲海岸外的远洋深海沉积物中测得的$\delta^{13}C$值进行了对比(Swart和Eberli,2005),其中,有五口取心井分布在一条由近源到远源的剖面上,该研究对其岩心做了碳酸盐沉积物$\delta^{13}C$值的测定分析。取心深达渐新统的顶,并且沿剖面进行了等时层序的对比;钻井跨越了上斜坡到大洋盆地环境,岩心样品由远洋和台地物质混合而成(即环台地沉积,Schlager和Ginsburg,1978)。尽管在各层序碳酸盐沉积中测得的$\delta^{13}C$值在各站位间可以对应,但是这些$\delta^{13}C$值与其他深海记录所呈现的全球碳循环却没有显示任何关联性(Shackleton,1985;Zachos等,2001;Swart和Eberli,2005)。事实上,这些环台地沉积的$\delta^{13}C$值显示出相反的变化趋势,即在过去25Ma间,$\delta^{13}C$值变得越高了。因此有研究推断,来自台地顶面、具有相对较重$\delta^{13}C$值(+4.5‰~+6‰,Swart等,2009)的碳酸盐与具有相对较轻$\delta^{13}C$值(0~+1‰,Milliman,1974)的远洋碳酸盐相混合了。

由于发现在大巴哈马浅滩(GBB)的各个站位的$\delta^{13}C$值是对应的(Swart和Eberli,2005),因此研究结论认为,在各站位所观察到的地层变化是由海平面变化控制的。在高海平面期,有大量台地表面沉积物和有机物供给斜坡,而在低海平面期,更多的沉积是来自于远洋深海物质。在全球其他临近碳酸盐台地和边缘的沉积物当中,所测得的$\delta^{13}C$记录也都有这一现象,包括:印度洋(马尔代夫)、大堡礁(昆士兰深海高原)、南澳大利亚湾(Swart,2008)。这些调查结果显示全球范围存在同时代的无机$\delta^{13}C$值的变化。然而,这些变化与保存于远洋输入型深海碳酸盐沉积当中的全球碳循环的长期变化是没有关系的。因此,在全球不同地方存在类似的$\delta^{13}C$记录不能用来确认$\delta^{13}C_{无机}$值就代表了全球碳循环的变化(Swart,2008)。

台地沉积物输入环台地环境的重要性并非局限于经历了高幅、高频海平面变化的更新世层序。在老的沉积,比如沃康提盆地的晚侏罗世沉积,早已清楚地证明,盆地中碳酸盐灰泥的聚集主要取决于与瑞士汝拉山区浅海环境碳酸盐生产循环变化相关的沉积物输出(Colombie和Strasser,2005)。在以下地区的沉积中也观察到类似的关系:翁布利亚—马尔凯盆地和卢西塔尼亚盆地的托尔阶(Pittet和Mattioli,2002;Mattioli和Pittet,2004),西南德国盆地的牛津阶—基末利阶(Pittet等,2000;Pittet和Mattioli,2002),阿曼的乌姆尔河盆地(阿尔比阶—阿普弟阶;Immenhauser等,1999),多洛米蒂山中—上三叠统的半远洋沉积岩(Preto等,2009)。这些结果表明,$\delta^{13}C_{无机}$值依赖海平面变化的模型不仅适用于现代浅海环境,也适用于古代浅海环境。

确定台地输入沉积物测得的无机$\delta^{13}C$记录($\delta^{13}C_{无机}$)是否准确记录全球碳循环还是有问题的。通常,共生有机质的$\delta^{13}C$($\delta^{13}C_{有机}$)被用来印证$\delta^{13}C_{无机}$信号的变化,典型的情形是,如果在一个沉积层序当中有机和无机组分的$\delta^{13}C$一起变化的话,那么就认为$\delta^{13}C_{无机}$的变化是可靠的,并且指示环境中$\delta^{13}C_{无机}$的真实变化(Margaritz等,1986;Gale,1993;Underwood等,1997;

Jarvis 等,2006)。此外,也有研究指出,有机与无机组分 $\delta^{13}C$ 之间的差异指示沉积环境的 pCO_2(Popp 等,1989)。尽管人们普遍认同有机质的成因对 $\delta^{13}C_{有机}$ 值起着绝对控制作用(因此,特征有机化合物的 $\delta^{13}C$ 值要比总有机质的 $\delta^{13}C_{有机}$ 更受青睐,Hayes 等,1990),但是仍有为数众多的研究给出的只是总有机质的 $\delta^{13}C_{有机}$ 值。这些研究用总有机质的 $\delta^{13}C_{有机}$ 值来支持它们对碳酸盐 $\delta^{13}C_{无机}$ 值长期变化以及全球 pCO_2 变化的解释。它们这么做是基于一个内在假定,即总有机碳和总无机碳的 $\delta^{13}C$ 起初是相关的,并且随着时间推移这种相关性仍保持一致。

本文将已发表的 $\delta^{13}C_{无机}$ 记录(Swart 和 Eberli,2005)与在相同样品新测的总有机质的 $\delta^{13}C_{有机}$ 记录进行比较,对这些假定进行验证。另外,还将已发表的 $\delta^{13}C_{无机}$ 分布与大巴哈马浅滩(GBB)顶面沉积物中沉积有机物的 $\delta^{13}C_{有机}$ 分析进行比较;还与降解形成沉积有机质的各种单一有机成员(藻、海草等)的 $\delta^{13}C_{有机}$ 分析进行比较;利用本研究所做的分析测量和对比,本文对目前在全球碳循环研究中的两个假定进行了验证。其中,第一个假定认为浅海碳酸盐台地输入沉积物中 $\delta^{13}C_{有机}$ 与 $\delta^{13}C_{无机}$ 之间的共变关系表明 $\delta^{13}C_{无机}$ 的变化与全球碳循环变化有关;第二个假设认为 $\delta^{13}C_{有机}$ 与 $\delta^{13}C_{无机}$ 之间的差异与有机物和沉积物形成时大气圈中的 CO_2 分压有关。

11.2 样品

本项研究所分析的样品是在大洋钻探计划(ODP)第 166 航次以及贝洛斯号科研船的航行调查过程中采集的。

11.2.1 第 166 航次

大洋钻探计划(ODP)第 166 航次在大巴哈马浅滩西缘外进行七站钻井取心(Eberli 等,1997)。其中有五个取心站位钻在西地震测线(Western Seismic Line)的延伸线上(Eberli 和 Ginsburg,1987;图 11-1)。从 1003、1005、1006 和 1007 站位采集到的沉积物用于本项研究。表 11-1 列出上述站位的坐标位置、水深、以及所钻遇最老沉积物的时代。在第 166 航次的剖面上识别出了有十七个地震层序,并且利用垂直地震剖面与沉积物做了对照,还利用生物地层学和磁性地层学的方法计算了层序界面的年代数据(Eberli 等,1997;Eberli,2000)。已有研究发现,在站位之间时代相当的层序当中,无机组分的 $\delta^{13}C_{无机}$ 是对应的,但与全球 $\delta^{13}C$ 的变化并不对应(Swart 和 Eberli,2005)。

表 11-1 取心站位表(按由近源到远源的顺序排列)
表中由近台地边缘的 1005 到远离台地边缘的 1007 依次列出各站位及其位置坐标、水深、钻到的最老沉积

站位	位置坐标	水深(m)	钻到的最老沉积
1005	24°33.772′N,79°14.141′W	350.7	晚中新世
1003	24°32.746′N,79°15.642′W	481.4	早中新世
1007	24°30.264′N,79°19.34′W	650.3	晚渐新世
1006	24°23.989′N,79°27.541′W	657.9	中中新世

各站位的平均取样间隔从 1~1.6m 不等。在 1003 站位,取样间隔从 0.4~8.1m 不等,平均取样间隔为 1.6m;1005 站位的取样间隔由 0.1~24.5m,平均为 1.6m;1007 站位的取样间隔介于 0.3~19.5m,平均为 1.2m;出现较大取样间隔的原因是钻井过程中岩心物质的漏失。

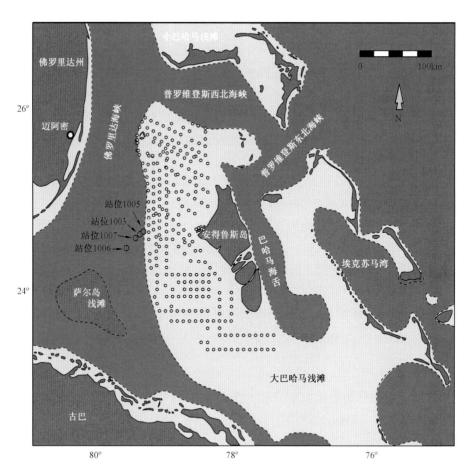

图 11-1 研究区及其取心站位、取样点的位置图

橙色圈为大洋钻探计划（ODP）第 166 航次取心站位：1003、1005、1006、1007；
黄色圈为大巴哈马浅滩顶部的取样点位

1006 站位的取样间隔介于 0.8~5.8m，平均为 1.6m；1006 站位记录到的更新世海平面变化最为连续，并且最少受浊积岩沉积的影响。

11.2.2 浅水碳酸盐沉积

2001 和 2004 年之间，贝洛斯号科研船在进行的四个航次中从大巴哈马浅滩（GBB）表面采集了 291 个样品（图 11-1），随后对样品进行了总 $\delta^{13}C_{无机}$ 值的分析（Swart 等，2009）。样品类型从安得鲁斯岛背风面的富灰泥颗粒泥晶灰岩到台地边缘的砾屑灰岩（Reijmer 等，2009）。虽然样品的分布与前人的工作雷同（Purdy，1963a，1963b；Traverse 和 Ginsburg，1966），但是此次的取样更细密，从而能够对沉积物进行细致的相解析和有机地球化学研究。其中识别出的两类端元相，即颗粒灰岩和富灰泥颗粒泥晶灰岩，分布在外台地和内台地，之间泾渭分明：粗粒沉积沿台地边缘分布，环绕在其中的是安得鲁斯岛西面背风地的富灰泥区（图 11-2）。该分布特征反映了洋流和屏蔽区对控制台地上面沉积物沉积的相对重要性。沉积物的矿物组成明显以文石为主，具有少量高镁方解石和低镁方解石。沉积物全组分和细组分的矿物学差异说明富文石的灰泥优先从该台地向周围的大洋盆地搬运（Reijmer 等，2009）。所有样品的 $\delta^{13}C_{无机}$ 值为较高的正值，范围是 +4.0‰ ~ +6.0‰（m = +4.5‰），并且与相的类型无关（Swart 等，2009）。

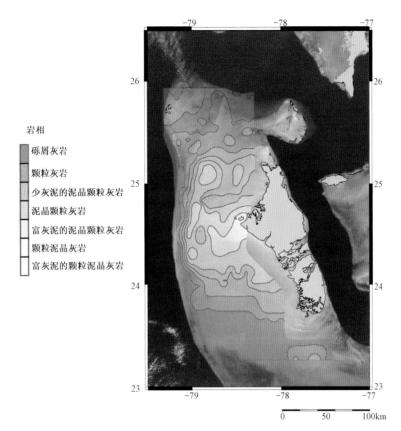

图 11-2 大巴哈马浅滩(GBB)顶部的相分布图(Reijmer 等,2009)
相的划分和描述是根据改进后的邓哈姆分类,详见 Reijmer 等(2009)

11.3 方法

11.3.1 沉积物的有机 $\delta^{13}C$ 组成

从每个位置采集的沉积物中称取大约 400g、用 10% 的盐酸(HCl)进行隔夜酸化处理,再用真空泵通过玻璃微纤维滤纸过滤。随后将滤纸置于干燥器中烘干至少 48h,或直至达到恒重为止。收集滤纸上的物质分装进锡囊中以分析 $\delta^{13}C_{有机}$ 的组成。

11.3.2 藻和海草的有机 $\delta^{13}C$ 组成

样品用去离子水(DI)冲洗去除盐分并干燥。干燥样品再用 5% 的盐酸(HCl)浸泡除去任何附生动物的无机碳酸盐,然后再用去离子水(DI)冲洗,置于烘箱中 40℃ 烘干直至达到恒定干重;称取干燥样品分装进锡囊中。

11.3.3 同位素分析

采用杜马燃烧法对装着样品的锡囊进行燃烧(使用 ANCA—自动氮碳分析仪,厂商是在英国克鲁的欧洲科技公司;或者是使用 ECS 4010 型 CHNSO 元素分析仪,厂商是在美国加利福尼亚州瓦伦西亚的 Costech 分析技术有限公司)。生成的 CO_2 在连续流同位素比值质谱仪(CFIRMS 20-20 仪,厂商是在英国克鲁的欧洲科技公司;或者是 Delta V Advantage 仪,厂商是在德国不来梅的赛默飞世尔科技公司)上进行分析。分析数据已经对常见的等重离子干扰进行了校正,数据结果相对于维也纳皮迪箭石(V-PDB)标准采用常规符号制式来表示。对已知重量和成分的内部标样进行重复分析,所得重现性误差为 ±0.1‰。样品中的有机碳浓度则是通过采用已知碳含量的标准样品和标准线估测的,其中,标准线标定了元素分析(ANCA 分析仪或 Costech 分析仪)检测出的峰面积与标准样品碳含量的对应关系。标准样品重量的选择是依照样品中预估有机碳浓度的范围而选定的。从对标准样品的重复分析结果评定,该方法的重现性误差为 ±1.5‰。

11.4 结果

11.4.1 大洋钻探计划(ODP)第 166 航次 1003 到 1007 站位

在第 166 航次所采集的环台地沉积物中,有机物的 $\delta^{13}C$(即 $\delta^{13}C_{有机}$)值平均为 −15.4‰(最高值 −14.4‰、最低值 −16.8‰,图 11 −3)。其中,最低值见于离台地边缘最远的 1006 站位,随着靠近台地边缘,$\delta^{13}C_{有机}$ 值逐步变重(表 11 −2)。$\delta^{13}C_{有机}$ 与 $\delta^{13}C_{无机}$ 之间的关系由在 1005 站位的不相关($r^2=0.09$, $P>0.1$)到在 1006 站位的显著统计正相关($r^2=0.64$, $P<0.01$;表 11 −2,图 11 −4)。1006 站位的更新统(仅此站位采集到了相当连续的更新统记录)显示出一个总体趋势就是较重的 $\delta^{13}C_{有机}$ 值与重的 $\delta^{13}C_{无机}$ 值相对应。根据有孔虫—红似抱球虫(*Globigerinoides ruber*)的 $\delta^{18}O$ 负值和文石的高含量所做的界定(Kroon 等,2000)表明,这些正值(相关)出现在间冰期(图 11 −5)。

表 11 −2 大洋钻探计划(ODP)第 166 航次采集的台地外围沉积物的平均同位素组成

站位	$\delta^{13}C_{无机}$	$\delta^{13}C_{有机}$	r^2	总有机碳(%)	酸不溶物(%)
1005	4.4 ± 0.5‰	−14.4 ± 0.5‰	0.09($P>0.1$)	0.5	1.4
1003	3.7 ± 0.8‰	−14.7 ± 0.9‰	0.04($P>0.1$)	0.4	1.5
1007	4.2 ± 1.0‰	−15.6 ± 0.4‰	0.41($P<0.02$)	0.3	1.6
1006	2.8 ± 0.9‰	−16.8 ± 0.2‰	0.64($P<0.01$)	0.2	9.1

注:表中由离台地边缘最近(1005)到最远(1006)依次列出各站位;无机的数据取自 Swart 和 Eberli(2005);r^2 栏代表由 $\delta^{13}C_{无机}$ 值与 $\delta^{13}C_{有机}$ 值之间相关性计算出的线性相关系数;总有机碳(%)是保存在沉积物中的有机碳百分比,由整个实测有机碳的平均重量除以被溶解沉积物的初始重量求得;酸不溶物(%)则是过滤后残留物的重量除以初始重量。

11.4.2 大巴哈马浅滩的表面

11.4.2.1 台地沉积物

台地顶部沉积有机物的 $\delta^{13}C$ 值为 −14.3‰ ~ −10.9‰,平均为 −12.2‰(表 11 −3)。相对较高的 $\delta^{13}C$ 值见于安得鲁斯岛的背风面。越接近台地边缘和位于安得鲁斯岛以北和以南的部位,$\delta^{13}C_{有机}$ 值变得越低(−23.0‰ ~ −14.0‰;图 11 −6A)。其空间展布与已经发表的粒度分布相对应(Reijmer 等,2009;图 11 −6B)。总沉积有机碳平均为 0.10%,区间为 0.002% ~ 0.97%(表 11 −3)。沉积物总有机碳呈现出类似于沉积有机物 $\delta^{13}C$ 的分布特征,其中,测得最高有机碳的样品也出现在安得鲁斯岛的背风面(图 11 −6C)。非参数统计检验(即克鲁斯凯—沃利斯检验)表明,Reijmer 等(2009)所描述的五种沉积相与总有机碳含量之间($P<0.01$)、与 $\delta^{13}C_{有机}$ 值之间($P<0.01$)都具有统计差异性。但是,在浅滩顶部的任何沉积相当中,$\delta^{13}C_{有机}$ 与 $\delta^{13}C_{无机}$ 组分之间却都不具有统计相关性($r^2=0.02$, $P>0.1$);唯有在富含灰泥的颗粒泥晶灰岩相当中是例外($r^2=0.70$, $P<0.01$;图 11 −7)。

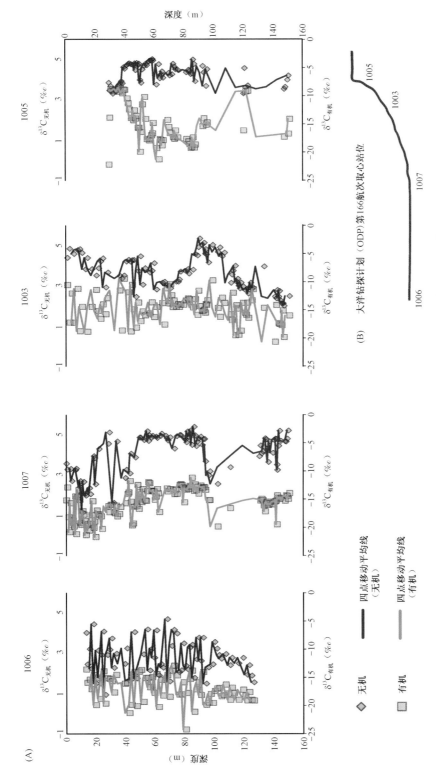

图11-3 (A)大洋钻探计划（ODP）第166航次海底以下0~150m取心样品测得的$\delta^{13}C_{无机}$值与$\delta^{13}C_{有机}$值之间的对应关系图其中，符号代表实测数据点，折线代表该航次断面上各站位取心的四点位移动平均线；(B)取心站位的海底深度，与台地边缘的距离示意图。任各取心站位当中，1006站位是最靠向大洋盆地的，远离台地边缘约30km

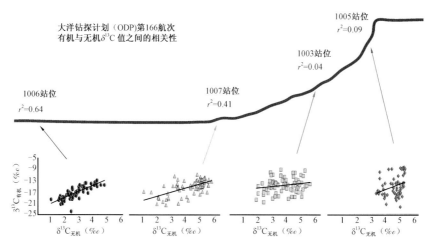

图 11-4 大洋钻探计划(ODP)第166航次各站位(1003到1007)有机与无机 $\delta^{13}C$ 值之间的相关关系图

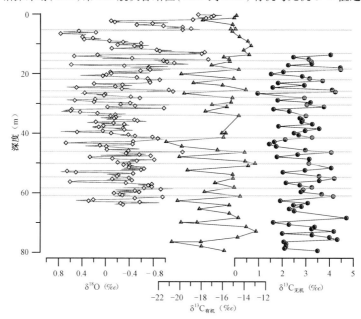

图 11-5 1006站位实测 $\delta^{13}C_{有机}$(蓝色三角)和 $\delta^{13}C_{无机}$ 值(红色圆圈)与深度(海底以下)关系图

$\delta^{18}O$ 记录(黄色菱形)由 Kroon 等(2000)从有孔虫测得,以及在岩心观测到的几个沉积间断;灰色横杠用于标记负的 $\delta^{18}O$ 值所指示的高水位期

表 11-3 浅滩顶部沉积物的相类型、分析样品数、无机和有机组分的 $\delta^{13}C$ 值、样品中有机碳所占重量的百分比、样品中酸不溶物所占重量的百分比

相	样品	$\delta^{13}C_{无机}$	$\delta^{13}C_{有机}$	总有机碳(%)	酸不溶物(%)
富灰泥的颗粒泥晶灰岩	21	4.5 ± 0.5‰	−10.9 ± 1.4‰	0.41 ± 0.40	1.59 ± 0.6
颗粒泥晶灰岩	34	4.7 ± 0.2‰	−10.9 ± 1.8‰	0.27 ± 0.26	1.27 ± 0.5
富灰泥的泥晶颗粒灰岩	11	4.8 ± 0.2‰	−11.3 ± 1.9‰	0.14 ± 0.08	1.04 ± 0.3
泥晶颗粒灰岩	47	4.8 ± 0.2‰	−12.1 ± 1.8‰	0.09 ± 0.07	0.63 ± 0.3
少灰的泥晶颗粒灰岩	41	5.0 ± 0.3‰	−13.2 ± 2.1‰	0.06 ± 0.06	0.63 ± 0.5
颗粒灰岩	109	4.9 ± 0.3‰	−14.3 ± 2.6‰	0.04 ± 0.03	0.47 ± 0.4
砾屑灰岩	2	3.7 ± 0.1‰	−12.9 ± 3.0‰	0.23 ± 0.27	0.79 ± 0.5

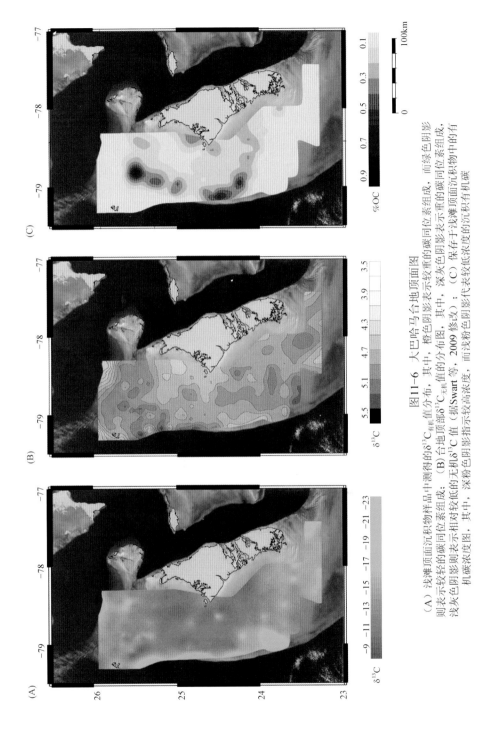

图11-6 大巴哈马台地地质图

（A）浅滩顶面沉积物样品中测得的$\delta^{13}C_{有机}$值分布，其中，橙色阴影表示较重的碳同位素组成，而绿色阴影则表示较轻的碳同位素组成；（B）台地顶部$\delta^{13}C_{无机}$值的分布图，其中，深灰色阴影表示重的碳同位素组成，浅灰色阴影表示相对较低的无机$\delta^{13}C$值（据Swart等，2009修改）；（C）保存于浅滩顶面沉积物中的有机碳浓度图，其中，深粉色阴影指示较高浓度，而浅粉色阴影代表低浓度的沉积有机碳

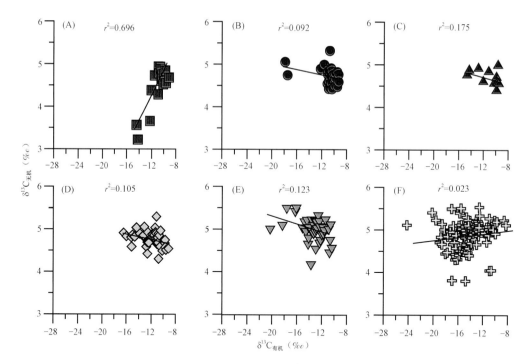

图 11-7 台地顶部沉积物样品中测得的 $\delta^{13}C_{无机}$ 值与 $\delta^{13}C_{有机}$ 值的相关图

(A)富灰泥的颗粒泥晶灰岩;(B)颗粒泥晶灰岩;(C)富灰泥的泥晶颗粒灰岩;(D)泥晶颗粒灰岩;(E)少灰泥的泥晶颗粒灰岩;(F)颗粒灰岩;$\delta^{13}C_{无机}$ 值与 $\delta^{13}C_{有机}$ 值之间只有在富灰泥的颗粒泥晶灰岩中($P<0.01$)是统计相关的

11.4.2.2 绿藻

对下述绿藻种属的样品做了有机物的 $\delta^{13}C$ 值分析：仙掌藻(*Halimeda* sp.)、须刷藻(*Penicillus* sp.)、伞藻(*Acetabularia* sp.)、钙扇藻(*Udotea* sp.)、松球藻(*Rhipocephalus* sp.)、绒枝藻(*Batophora* sp.)和蕨藻(*Caulerpa* sp.)。所得的数据见表 11-4 和图 11-8。全部绿藻样品的平均 $\delta^{13}C_{有机}$ 值为 -13.5‰，变化范围从 -16.0‰ 到 -8.0‰。

表 11-4 从大巴哈马浅滩表面采集到的沉积物和有机物生产者的平均 $\delta^{13}C_{有机}$ 组成

类型	$\delta^{13}C_{有机}$	标准差,样数
沉积物平均	-12.9‰	(±2.6,277)
海草		
泰来藻	-5.7‰	(±1.0,90)
二药藻	-7.0‰	(±1.7,30)
绿藻		
仙掌藻	-11.4‰	(±3.2,24)
须刷藻	-10.5‰	(±2.3,11)
伞藻	-11.2‰	(±1.5,4)
钙扇藻	-11.9‰	(±2.5,8)
松球藻	-10.3‰	(±1.2,4)
蕨藻	-23.6‰	(±3.3,3)

注:沉积物的平均 $\delta^{13}C$ 值是由浅滩顶部采集的各种粒级合计而来的。

11.4.2.3 海草

对龟裂泰来藻（*Thalassia testudinum*）和二药藻（*Halodule* sp.）的海草样品进行了有机物的 $δ^{13}C_{有机}$ 值分析。海草的 $δ^{13}C_{有机}$ 值范围为 -8.7‰ ~ -4.2‰，平均为 -6.4‰（表 11-4）。

11.5 讨论

11.5.1 现代台地的变化

11.5.1.1 大巴哈马浅滩上的有机 $δ^{13}C$ 值

由生长在大巴哈马浅滩（GBB）上的海草和大型海藻所产生的有机物中的 $δ^{13}C_{有机}$ 不仅比远洋环境产生的有机物中的明显要重，而且比生长在其他热带浅海环境的相似种类中的 $δ^{13}C_{有机}$ 值要重（图11-9），例如：泰来藻（*Thalassia* sp.）在大巴哈马浅滩（GBB）的 $δ^{13}C_{有机}$ 值（平均值 = -5.7‰；本研究测得）与以下区域同比，平均要高出 2‰ ~ 5‰：佛罗里达湾和佛罗里达

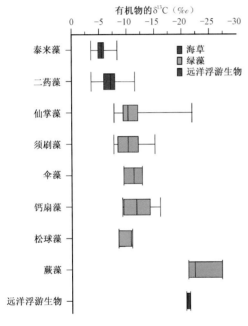

图 11-8 本研究从大巴哈马浅滩采集绿藻和海草样品测得的 $δ^{13}C$ 值箱须图

远洋浮游生物的 $δ^{13}C$ 值是由 Laws 等（1995）测得的；数据组的极差（即最大和最小值）用误差线条标定，彩色箱则代表上下四分位数，其中的黑竖线指数据组的中值

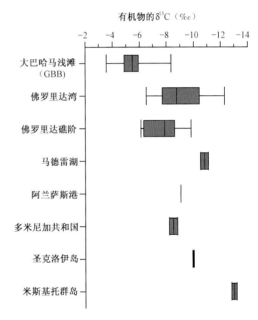

图 11-9 其他加勒比地区泰来藻（*Thalassia* sp.）的 $δ^{13}C$ 值箱须图

误差线条表示数据组的极差（即最大和最小值），彩色箱则代表上下四分位数，其中的黑竖线为数据组的中值。大巴哈马浅滩（GBB）的泰来藻 $δ^{13}C$ 值由本次研究测得；佛罗里达湾和佛罗里达礁阶的泰来藻 $δ^{13}C$ 值由 Anderson 和 Fourqurean（2003）测得；美国得克萨斯马德雷湖的泰来藻 $δ^{13}C$ 值由 Jones 等（2003）测得；美国得克萨斯阿兰萨斯港的 $δ^{13}C_{有机}$ 值由 Benedict 和 Scott（1976）报道；多米尼加共和国的 $δ^{13}C_{有机}$ 值是由 Tewfik 等（2005）分析的；美属维尔京群岛圣克罗伊岛和尼加拉瓜米斯基托群岛的 $δ^{13}C_{有机}$ 值由 Fry 等（1982）报道。如果在前人研究报告中只有一个数据，则在此绘成一个线条，其余情况下，都用标准差和极差绘制"触须"

礁阶（-10.4‰~-7.2‰，Anderson 和 Fourqurean,2003）、得克萨斯马德雷湖（-10.8±0.3‰，Jones 等,2003）、得克萨斯阿兰萨斯港（-9.0‰，Benedict 和 Scott,1976）、多米尼加共和国（-8.5±0.3‰，Tewfik 等,2005）。本项研究实测泰来藻（*Thalassia* sp.）的 $\delta^{13}C_{有机}$ 值落于另一项研究在圣克洛伊岛所测的数值范围，即-11‰~-4‰（据 Fry 等,1982）。这些藻类和海草偏高的 $\delta^{13}C_{有机}$ 值也反映在大巴哈马浅滩（GBB）顶部采集沉积物中测到的较重无机 $\delta^{13}C$ 值（见 Swart 等,2009）。这些工作者推断，溶解无机碳（DIC）较高的 $\delta^{13}C$ 值是由于大巴哈马浅滩（GBB）内浅水区相对受阻的水循环类型造成的，即台地内浅水区各种初级生产者的大量光合作用活动使局部溶解无机碳（DIC）碳库产生分馏，从而导致浅滩顶部沉积物中有机和无机 $\delta^{13}C$ 值都偏高。

沉积的 $\delta^{13}C_{有机}$ 值反映栖息在大巴哈马浅滩（GBB）浅水区的海洋植物和生物的降解有机物的输入。在大巴哈马浅滩（GBB）区，沉积有机物 $\delta^{13}C_{有机}$ 的最高值与安德鲁斯岛背风面的最小粒度（<125μm）相对应；在安德鲁斯岛北面和南面的边缘和开阔海区，沉积有机物的 $\delta^{13}C_{有机}$ 值较低。这种分布样式类似与在大巴哈马浅滩（GBB）顶面的粒度分布特征（Reijmer 等,2009），并且显示较细的沉积物具有较高的有机碳浓度，其同位素也比从平均粒度较粗相提取的沉积有机物要高（表 11-3）。水域越开阔，细粒沉积物及伴生有机物越可能被强的流动体制所冲走，因此有机质物浓度也越低。由于台地顶部的沉积 $\delta^{13}C_{有机}$ 值是由各种来源的有机物所贡献，并且 $\delta^{13}C_{无机}$ 相对均一，因此，在 $\delta^{13}C_{有机}$ 与 $\delta^{13}C_{无机}$ 记录之间并没有相关性；其中有例外的就是以灰泥为主的相（图 11-7），在此类沉积相当中，$\delta^{13}C_{有机}$ 与 $\delta^{13}C_{无机}$ 值之间呈正相关（$r^2=0.7$，$P<0.01$）。

在沉积物生产活跃期，由于风暴活动、潮汐涨落和信风吹袭，沉积物和有机物组分被搬运离开浅滩（Wilber 等,1990；Reijmer 等,2009）。这些从台地顶部输出的沉积物具有较高的 $\delta^{13}C_{有机}$ 和 $\delta^{13}C_{无机}$ 值（分别为-12.2±1.3‰和+4.6±0.4‰）。大巴哈马浅滩（GBB）斜坡沉积物中的 $\delta^{13}C_{有机}$ 值会是台地输入有机碳（大约-12.0‰；本项研究）和远洋深海输入有机碳（大约-21.0‰；Laws 等,1995）相互混合的反映。有机与无机组分之间的相关程度就会相对于台地边缘的距离而变化。例如：如果岩心中的绝大部分沉积物是来自台地顶面的话，那么岩心中的 $\delta^{13}C_{有机}$ 记录和 $\delta^{13}C_{无机}$ 记录应该不会协同变化，因为在台地顶面沉积物中本身就没有记录到这种相关关系。然而，随着与台地边缘的距离增加、台地输入沉积物所占比例减少，应该会看到 $\delta^{13}C_{无机}$ 与 $\delta^{13}C_{有机}$ 之间出现正相关性，这是由于台地输入同位素重的碳与远洋输入同位素轻的碳相互混合的结果。

11.5.1.2 向下游站位，取心中的有机 $\delta^{13}C$ 变化

各站位取心的平均 $\delta^{13}C_{有机}$ 值随着与台地边缘的距离增加负值越来越大（表 11-2）。在最远站位的取心，即 1006 站位，显示最低的平均 $\delta^{13}C_{有机}$ 值（-16.8‰），而离台地最近站位的取心，即 1005 站位，实测到-14.4‰ 的 $\delta^{13}C_{有机}$ 值。这些数据表明，在最靠大洋盆地的 1006 站位，来自远洋深海的有机物所占比例较高，相反，在最临近台地的 1005 站位，来自远洋深海的有机物所占比例较低（图 11-4）。

$\delta^{13}C_{有机}$ 与 $\delta^{13}C_{无机}$ 记录之间的相关性也随着与台地边缘的距离增加而明显改变。在近源站位的取心当中，即在 1003 和 1005 站位，$\delta^{13}C_{有机}$ 与 $\delta^{13}C_{无机}$ 值之间要么是极弱相关，要么是非统计相关；相反，在斜坡底部（即 1007 站位）和在大洋盆地（即 1006 站位），$\delta^{13}C_{有机}$ 与 $\delta^{13}C_{无机}$ 记录之间呈现强的正相关性（图 11-4）。正是海平面变化的影响决定着这些随空间位置变化所呈现的差异；这在 1006 站位表现得尤其突出，因为该站位有相当连续的更新统记录，其中可见

高水位期以相对较重的 $\delta^{13}C_{有机}$ 和 $\delta^{13}C_{无机}$ 值为特征、而低水位期则同位素偏低(图 11-5)。这样的观测结果也与别的研究结果相一致,即在高水位期边缘环境接受大量来自台地表面的沉积物(Schlager 等,1994;Roth 和 Reijmer,2004,2005)。

11.5.1.3 源自大巴哈马浅滩(GBB)体系内的系统沉积作用

在 30km 长、从台地到大洋盆地同时代取心的剖面范围内,观测到了 $\delta^{13}C_{无机}$ 与总 $\delta^{13}C_{有机}$ 记录之间的相关性变化不一。这一结果与认为二者之间始终应该是相关的、二者的同步变化反映全球碳循环变化的假定相矛盾。这些结果表明,海平面控制了台地输入碳酸盐的贡献多少,而输入的碳酸盐控制着 $\delta^{13}C_{无机}$ 与 $\delta^{13}C_{无机}$ 记录之间的关系。用模拟的方法可以进一步佐证这两个独立的碳酸盐和有机碳来源(远洋深海和台地)的重要性。如果大巴哈马浅滩(GBB)的表面被认为是同位素重的有机质物来源,而远洋深海是同位素轻的有机物来源,那么,环台地沉积物中的总 $\delta^{13}C_{有机}$ 值就可以用这两个来源的简单混合模式加以解释。用该模型就会预测到:(1)离台地近的沉积物比离台地远的沉积物同位素要重;(2)离台地较近的沉积物有机质碳浓度应较高;(3)在离台地较远的岩心当中,$\delta^{13}C_{有机}$ 与 $\delta^{13}C_{无机}$ 之间的相关性会增大。沿着大巴哈马浅滩(GBB)外围四个站位实际观测到了上述分布规律:从最靠近台地的 1005 站位到在大洋盆地的 1006 站位,平均 $\delta^{13}C_{有机}$ 值从 -14.4‰ 变到了 -16.8‰,平均总有机碳值从 0.5% 减少到了 0.2%,$\delta^{13}C_{有机}$ 与 $\delta^{13}C_{无机}$ 之间的相关性从 $r^2=0.1$ 增大到了 $r^2=0.6$。所观测到的这些变化趋势用简单混合模型便可充分模拟(图 11-10)。在此模型中设定的取值范围如下:大洋沉积物的 $\delta^{13}C_{无机}$ 值介于 0‰~1‰ 之间,$\delta^{13}C_{有机}$ 值介于 -22‰~-20‰ 之间;台地沉积物的 $\delta^{13}C_{无机}$ 值介于 +3‰~+6‰ 之间,$\delta^{13}C_{有机}$ 值介于 -16‰~-7‰ 之间。通过调节这两个物源输入的比例以拟合实际观测到的相关性(图 11-10)。就 1006 站位而言,沉积物组成当中来自台地的物质所占比例变化范围在 10% 至 80% 之间;在斜坡底部(即在 1007 站位),该比例的变化范围在 40% 至 90% 之间,到最靠近台地边缘的 1005 站位,该比例的变化范围则为 75% 至 99%。

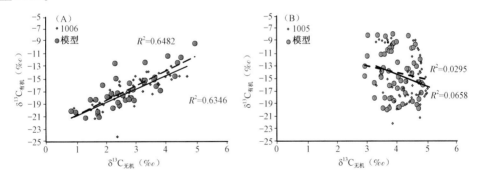

图 11-10 1006 站位(A)和 1005 站位(B)沉积物的 $\delta^{13}C_{无机}$ 与 $\delta^{13}C_{有机}$ 值模拟

每种情形都假定是由以下二源物质组分的混合,即:远洋深海物质组分($\delta^{13}C_{无机}$ 值的范围在 0~+1‰,$\delta^{13}C_{有机}$ 值的范围在 -22.0‰~-20.0‰)与台地物质组分($\delta^{13}C_{无机}$ 值的范围在 +3.0‰~+6.0‰,$\delta^{13}C_{有机}$ 值的范围在 -16.0‰~-7.0‰)。模型随机从远洋深海和台地物质组分中选取 $\delta^{13}C_{无机}$ 和 $\delta^{13}C_{有机}$ 值。(A)是针对远端的 1006 站位,模拟结果(蓝色大圆点)与实测数据(蓝色菱形点)绘制在同一坐标轴上,在该站位实际观测的相关性($r^2=0.64$)可以通过设定台地有机物和沉积物的比例占 20%~90% 得到模拟(模型的相关度 = 0.66,虚线);(B)是针对近端的 1005 站位,模拟结果(绿色大圆点)与实测数据(绿色菱形点)绘制在同一坐标轴上,观测到的相关度($r^2=0.07$,实线)通过设定台地有机物和沉积物的比例占 95%~99% 得到模拟(模型的相关度 = 0.04,虚线),二者都说明在 1005 站位有机与无机 $\delta^{13}C$ 记录之间不具有统计相关关系

11.5.1.4 其他可能的变化来源

$\delta^{13}C_{有机}$与$\delta^{13}C_{无机}$记录之间相关性出现变化,除了上述解释的原因外,至少还有另外两种可能性也许会引起$\delta^{13}C_{有机}$记录的变化。其一是,在1006和1007站位沉积的沉积物或许是来自别的台地,而在该台地浅滩顶部的沉积物和有机物中,有机与无机组分是协同变化的;此类浅水沉积物的可能来源包括:萨尔岛浅滩或者古巴周围的碳酸盐岩边缘(图11-1);尽管,类似于在大巴哈马浅滩(GBB)的自然环境作用可能也在萨尔岛浅滩或古巴碳酸盐岩边缘发生,但是关于这些地方$\delta^{13}C_{有机}$和$\delta^{13}C_{无机}$的分布尚未有任何公开发表的数据。其二是,在1003和1005站位,$\delta^{13}C_{有机}$与$\delta^{13}C_{无机}$值之间原本也有类似在1006和1007站位的相关性,但是由于更靠近巴哈马台地,可能被从剖面深部运移来的沥青干扰和压制掉了。据报道,曾在边缘站位钻井的过程中检测到经过运移的沥青(Eberli等,1997)。这些沥青有可能会增加沉积物当中的有机物量,并且改变其$\delta^{13}C$值。但是,在本项研究分析过的任何取心处,最上部的150m都未见有这样的迹象出现。尽管目前还无法判断这些来源的重要性,但是,现有的证据符合由远洋深海沉积物与台地输入沉积物之间形成的简单两组分混合。

11.5.1.5 在古代浅海碳酸盐台地体系中的应用

随着各种地质事件的发生和地质时代的变迁,有机和无机$\delta^{13}C$值发生了同步漂移。这种同步漂移已被用来评估全球碳循环的许多方面,提供了有关生产率与保存、环境条件和大气二氧化碳浓度变化的信息。这一原理已经应用于二叠纪—三叠纪的界限过渡段,进而来重建古代的pCO_2和了解导致该交界之际大规模灭绝事件的条件(Margaritz等,1992;Krull等,2004)。同样,有研究还使用$\delta^{13}C_{无机}$与$\delta^{13}C_{有机}$的同步变化来解释全球碳循环以及评价埃迪卡拉纪和元古代冰期溶解无机碳(DIC)碳库的缓冲作用(Kaufman等,1997;Swanson-Hysell等,2010)。影响$\delta^{13}C_{无机}$与$\delta^{13}C_{有机}$记录之间脱耦的因素还被用来评估地球海洋的氧化过程以及后续有机物氧化对碳循环的效应(Karhu和Holland,1996;Fike等,2006)。在重建地质时期全球碳循环的波动研究中,$\delta^{13}C_{无机}$与$\delta^{13}C_{有机}$记录的相关性已经成为广为使用的工具。

前人的工作表明,在远洋深海层序中测得的有机碳$\delta^{13}C$值与无机碳$\delta^{13}C$值之间的相关性反映全球范围有机碳的埋藏变化(Popp等,1989;Hayes,1993;Hayes等,1999;Zachos等,2001)。由于没有早于200Ma的深海层序可供研究,因此,许多研究者就用沉积在边缘环境的碳酸盐来代替,这些环境包括:碳酸盐台地、近缘浅海和缓坡(Hotinski等,2004;Saltzman,2002;Saltzman等,2004a,2004b;Underwood等,1997)。这种替代所依据的主要假定是:(1)溶解的无机碳(DIC)碳库与由浅水碳酸盐台地或缓坡衍生生物沉淀的无机碳酸盐处于同位素平衡状态;(2)并且这些有机物的$\delta^{13}C$特性与在深海的特性是相似的。然而,本项研究却得出一些不同的结论。在大巴哈马浅滩(GBB)输入的环台地沉积物中,$\delta^{13}C_{无机}$与$\delta^{13}C_{有机}$之间的协变性在各个部位并非协调一致。Ludvigson等(1996)提出$\delta^{13}C_{无机}$与$\delta^{13}C_{有机}$之间在一个位置可能发生脱耦的几种方式,包括:有机生产率的提高,高的保存潜力,环流、温度、盐度和pCO_2的变化,成岩作用改造,海平面变化。本项研究的结果支持了Ludvigson等提出的部分方式,但也得出了更具针对性的模型来描述海平面升降变化如何能够启动或终止在浅海台地的碳酸盐工厂。在近源处,如在1005和1003站位,之所以观测到$\delta^{13}C_{无机}$与$\delta^{13}C_{有机}$记录之间脱耦是由于岩心主要是由台地物质组成的,而在该现代台地的顶部本身就没有记录到$\delta^{13}C_{无机}$与$\delta^{13}C_{有机}$之间的协同变化;在远源处,如在1006和1007站位,之所以观测到$\delta^{13}C_{无机}$与$\delta^{13}C_{有机}$值之间明显的协变性是由于远洋深海物质与台地产生的物质之间的两点混合体系所致。因此,在对古代

浅海碳酸盐台地 $\delta^{13}C_{无机}$ 与 $\delta^{13}C_{有机}$ 记录之间的耦合或脱耦做出解释时,有关所采集沉积物的沉积环境以及局部海平面历史是必须要了解的背景信息。

当我们把该项研究成果与时代较老的沉积和解释古代全球碳循环相联系的时候,涉及到的重要问题是,新近系是否真的可以用来"将今论古"。目前尚不清楚:在早期有机物类型的同位素特性是怎样的? 是否早期也像现在一样,在深海与浅水之间有机物的 $\delta^{13}C$ 存在差异呢? 无机成分的 $\delta^{13}C$ 也存在差异吗? 虽然这些问题至今还是未知数,但是,在解释早期 $\delta^{13}C_{有机}$ 与 $\delta^{13}C_{无机}$ 记录之间的关联模型时,不能排除这样的可能性,即:有各种不同的生物在给这些古代浅海碳酸盐沉积提供有机的碳和无机的碳,正像在巴哈马所呈现的那样。

11.6 总结和结论

本研究的结果对于用浅海碳酸盐环境的 $\delta^{13}C_{有机}$ 值作为确认无机碳酸盐 $\delta^{13}C$ 值变化的方法提出了质疑。本文提供的数据表明,在离台地边缘比较近的范围内,$\delta^{13}C_{有机}$ 与 $\delta^{13}C_{无机}$ 值之间存在截然不同的关联模式。临近台地边缘,环台缘沉积物以台地输入物为主,$\delta^{13}C_{有机}$ 与 $\delta^{13}C_{无机}$ 记录之间并无相关性,这是台地输入沉积物中 $\delta^{13}C_{有机}$ 与 $\delta^{13}C_{无机}$ 值之间本身就没有相关性的结果;远离台地边缘,$\delta^{13}C_{有机}$ 与 $\delta^{13}C_{无机}$ 记录之间呈现出强的协变性;事实上,这种假相关是台地沉积物与远洋沉积物之间的混合线,其中,台地沉积物的 $\delta^{13}C_{有机}$ 和 $\delta^{13}C_{无机}$ 都偏高,远洋沉积物的 $\delta^{13}C_{有机}$ 和 $\delta^{13}C_{无机}$ 都偏低。因此,在针对浅海背景,如台地、缓坡、近缘浅海等环境,分析关于有机碳埋藏、pCO_2 等全球碳循环波动的解释时,应当清楚地知道:总的记录是来源于浅海和远洋环境中各种各样的生产者。然而,目前尚不清楚,这些成果是否也适用于古代沉积,因为构成古代沉积物的有机物和无机物可能是由不同于第三纪沉积(本文所研究)的生物或作用产生的。尽管如此,重要的是要认识到,在之前的浅海环境可能也存在着同样复杂的同位素关系;此外,有关古代环境的沉积条件和海平面历史等方面的相关背景信息,对于准确解释古代 $\delta^{13}C_{无机}$ 与 $\delta^{13}C_{有机}$ 记录之间的关系也是至关重要的。

参 考 文 献

Anderson, W. T. and Fourqurean, J. W. (2003) Intra-and interannual variability in seagrass carbon and nitrogen stable isotopes from South Florida, a preliminary study. Org. Geochem., 34, 185 – 194.

Benedict, C. R. and Scott, J. R. (1976) Photosynthetic carbon metabolism of a marine grass. Plant Physiol., 57, 876 – 880.

Broecker, W. S. and Woodruff, F. (1992) Discrepancies in the oceanic carbon isotope record for the last 15 million years. Geochim. Cosmochim. Acta, 56, 3259 – 3264.

Colombie, C. and Strasser, A. (2005) Facies, cycles, and controls on the evolution of a keep – up carbonate platform (Kimmeridgian, Swiss Jura). Sedimentology, 52, 1207 – 1227.

Eberli, G. P. (2000) The record of Neogene sea-level changes in the prograding carbonates along the Bahamas transect-Leg 166 synthesis. In: Proceedings of the Ocean Drilling Program, Scientific Results (Eds P. K. Swart, G. P. Eberli, M. J. Malone and J. F. Sarg), 166, pp. 167 – 177. Ocean Drilling Project, College Station TX.

Eberli, G. P. and Ginsburg, R. N. (1987) Segmentation and coalescence of Cenozoic carbonate platforms, northwestern Great Bahama Bank. Geology, 15, 75 – 79.

Eberli, G. P., Swart, P. K. and Malone, M. J. (Eds) (1997) Proceedings of Ocean Drilling Program Initial Reports, 166. Ocean Drilling Program, College Station, TX.

Fike, D. A., Grotzinger, J. P., Pratt, L. M. and Summons, R. E. (2006) Oxidation of the Ediacaran ocean. Nature, 444, 744 – 747.

Fry, B., Lutes, R., Northam, M. and Parker, P. L. (1982) A ^{13}C/^{12}C comparison of food webs in Caribbean seagrass meadows and coral reefs. Aquat. Bot., 14, 389–398.

Gale, A. (1993) Chemostratigraphy vs biostratigraphy: data from around the Cenomanian-Turonian boundary. J. Geol. Soc. London, 150, 29–33.

Hayes, J. M. (1993) Factors controlling C-13 contents of sedimentary organic-compounds-principles and evidence. Mar. Geol., 113, 111–125.

Hayes, J. M., Freeman, K. H., Popp, B. N. and Hoham, C. H. (1990) Compound-specific isotopic analyses-a novel tool for reconstruction of ancient biogeochemical processes. Org. Geochem., 16, 1115–1128.

Hayes, J. M., Strauss, H. and Kaufman, A. J. (1999) The abundance of ^{13}C in marine organic matter and isotopic fractionation in the global biogeochemical cycle of carbon during the past 800 Ma. Chem. Geol., 161, 103–125.

Hotinski, R. M., Kump, L. R. and Arthur, M. A. (2004) The effectiveness of the Paleoproterozoic biological pump: a δ^{13}C gradient from platform carbonates of the Pethei Group(Great Slave Lake Supergroup, NWT). Geol. Soc. Am. Bull., 116, 539–554.

Immenhauser, A., Schlager, W., Burns, S. J., Scott, R. W., Geel, T., Lehmann, J., Van der Gaast, S. and Bolder-Schrijver, L. J. A. (1999) Late Aptian to late Albian sea-level fluctuations constrained by geochemical and biological evidence (Nahr Umr formation, Oman). J. Sed. Res., 69, 434–446.

Jarvis, I., Gale, A. S., Jenkyns, H. C. and Pearce, M. A. (2006) Secular variation in Late Cretaceous carbon isotopes: a new δ^{13}C carbonate reference curve for the Cenomanian-Campanian (99.6-70.6 Ma). Geol. Mag., 143, 561–608.

Jones, W. B., Cifuentes, L. A. and Kaldy, J. E. (2003) Stable carbon isotope evidence for coupling between sedimentary bacteria and seagrasses in a sub-tropical lagoon. Mar. Ecol. Prog. Ser., 255, 15–25.

Karhu, J. A. and Holland, H. D. (1996) Carbon isotopes and the rise of atmospheric oxygen. Geology, 24, 867–870.

Kaufman, A., Knoll, A. H. and Narbonne, G. M. (1997) Isotopes, ice ages, and terminal Proterozoic earth history. Geology, 94, 6600–6605.

Kroon, D., Reijmer, J. J. G. and Rendle, R. (2000) Mid-to late quaternary variations in the oxygen isotopic signature of Globigernoides ruber at Site 1006 in the western subtropical Atlantic. In: Proceedings of the Ocean Drilling Program Scientific Results (Eds P. K. Swart, G. P. Eberli and M. J. Malone), 166, pp. 13–22. Ocean Drilling Program, College Station, TX.

Krull, E. S., Lehrmann, D. J., Druke, D., Kessel, B., Yu, Y. and Li, R. (2004) Stable carbon isotope stratigraphy across the Permian-Triassic boundary in shallow marine carbonate platforms, Nanpanjiang Basin, south China. Palaeogeogr. Palaeoclimatol. Palaeoecol., 204, 297–315.

Laws, E. A., Popp, B. N., Bidigare, R. R., Kennicutt, M. C. and Macko, S. A. (1995) Dependence of phytoplankton carbon isotopic composition on growth–rate and [CO_2](Aq)-theoretical considerations and experimental results. Geochim. Cosmochim. Acta, 59, 1131–1138.

Ludvigson, G. A., Jacobsen, S. R., Witzke, B. J. and Gonzalez, L. A. (1996) Carbonate component chemostratigraphy and depositional history of the Ordovician Decorah Formation, Upper Mississippi Valley. In: Paleozoic Sequence Stratigraphy: Views from the North American Craton (Eds B. J. Witzke, G. A. Ludvigson and J. Day), pp. 67–86. GeologicalSociety of America Special Paper 306, , Boulder, CO.

Margaritz, M., Holser, W. T. and Kirschvink, J. L. (1986) Carbon–isotope events across the Precambrian/Cambrian boundry on the Siberian Platform. Nature, 320, 258–259.

Margaritz, M., Krishnamurthy, R. V. and Holser, W. T. (1992) Parallel trends in organic and inorganiccarbon isotopesacross the Permian/Triassic boundary. Am. J. Sci., 292, 727–739.

Mattioli, E. and Pittet, B. (2004) Spatial and temporal distribution of calcareous nannofossils along a proximal–distal transect in the Lower Jurassic of the Umbria–Marche Basin (central Italy). Palaeogeogr. Palaeoclimatol. Palaeoecol., 205, 295–316.

Milliman, J. D. (1974) Marine Carbonates. Springer, Berlin, 375 pp.

Pittet, B. and Mattioli, E. (2002) The carbonate signal and calcareous nannofossil distribution in an Upper Jurassic section (Balingen-Tieringen, Late Oxfordian, southern Germany). Palaeogeogr. Palaeoclimatol. Palaeoecol., 179, 71–96.

Pittet, B., Strasser, A. and Mattioli, E. (2000) Depositional sequences in deep-shelf environments: a response to sea-level changes and shallow-platform carbonate productivity (Oxfordian, Germany and Spain). J. Sed. Res., 70, 392–407.

Popp, B. N., Takigiku, R., Hayes, J. M., Louda, J. W. and Baker, E. W. (1989) The post-Paleozoic chronology and mechanism of C-13 depletion in primary marine organic-matter. Am. J. Sci., 289, 436–454.

Preto, N., Spotl, C. and Guaiumi, C. (2009) Evaluation of bulk carbonate $\delta^{13}C$ data from Triassic hemipelagites and the initial composition of carbonate mud. Sedimentology, 56, 1329–1345.

Purdy, E. (1963a) Recent calcium carbonate facies of the Great Bahama Bank. 1. Petrography and reaction groups. J. Geol., 71, 334–335.

Purdy, E. (1963b) Recent calcium carbonate facies of the Great Bahama Bank. 2. Sedimentary facies. J. Geol., 71, 472–497.

Reijmer, J. J. G., Swart, P. K., Bauch, T., Otto, R., Roth, S. and Zechel, S. (2009) A reevaluation of facies on Great Bahama Bank I: new facies maps of Western Great Bahama Bank. In: Perspectives in Carbonate Geology: A Tribute to the Career of Robert Nathan Ginsburg, IAS Special Publication (Eds P. K. Swart, G. P. Eberli and J. A. McKenzie), 41, pp. 29–46. Wiley-Blackwell, Oxford.

Roth, S. and Reijmer, J. J. G. (2004) Holocene Atlantic climate variations deduced from carbonate periplatform sediments (leeward margin, Great Bahama Bank). Paleoceanography, 19, 231–244.

Roth, S. and Reijmer, J. J. G. (2005) Holocene millennial to centennial carbonate cyclicity recorded in slope sediments of the Great Bahama Bank and its climatic implications. Sedimentology, 52, 161–181.

Saltzman, M. R. (2002) Carbon isotope ($\delta^{13}C$) stratigraphy across the Silurian-Devonian transition in North America: evidence for a perturbation of the global carbon cycle. Palaeogeogr. Palaeoclimatol. Palaeoecol., 187, 83–100.

Saltzman, M. R., Cowan, C. A., Runkel, A. C., Runnegar, B., Stewart, M. C. and Palmer, A. R. (2004a) The Late Cambrian SPICE ($\delta^{13}C$) event and the Sauk II–Sauk III regression: new evidence from Laurentian basins in Utah, Iowa, and New-foundland. J. Sed. Res., 74, 366–377.

Saltzman, M. R., Groessens, E. and Zhuravlev, A. V. (2004b) Carbon cycle models based on extreme changes in $\delta^{13}C$: an example from the lower Mississippian. Palaeogeogr. Palaeoclimatol. Palaeoecol., 213, 359–377.

Schidlowski, M. (1979) Carbon isotope geochemistry of the 3.7×10^9 yr–old Isua sediments, West Greenland: implications for the Archean carbon and oxygen cycles. Geochim. Cosmochim. Acta, 43, 189–199.

Schlager, W. and Ginsburg, R. N. (1978) Influence of platformderived sediment on facies, diagenesis, and deformation in slope and basinal deposits, Tongue of Ocean, Bahamas. Am. Assoc. Pet. Geol. Bull., 62, 560–560.

Schlager, W. R., Reijmer, J. J. G. and Droxler, Andre. W. (1994) Highstand shedding of carbonate platforms. J. Sed. Res., 64, 270–281.

Shackleton, N. J. (1985) Atmospheric carbon dioxide, orbital forcing, and climate. In: The Carbon Cycle and atmospheric CO_2: Natural Variations Archaen to Present: GeophysicalMonograph (Eds E. T. Sundquist and W. S. Broecker), 32, pp. 412–417. AGU, Washington D. C..

Shackleton, N. J. and Hall, M. (1980) Oxygen and carbon isotope data from Leg 74 sediments. In: Initial Reports Deep Sea Drilling Project (Eds J. T. Moore and P. Rabinowitz), 74, pp. 613–619. U. S. Government Printing Office, WashingtonD. C..

Swanson-Hysell, N. L., Rose, C. V., Calmet, C. C., Halverson, G. P., Hurtgen, M. and Maloof, A. C. (2010) Cryogenian glaciation and the onset of carbon-isotope decoupling. Science, 328, 608–611.

Swart, P. K. (2008) Global synchronous changes in the carbon isotopic composition of carbonate sediments unrelated to changes in the global carbon cycle. Proc. Natl. Acad. Sci. ,105, 13741 – 13745.

Swart, P. K. and Eberli, G. P. (2005) The nature of the δ^{13}C of periplatform sediments: implications for stratigraphy and the global carbon cycle. Sed. Geol. , 175, 115 – 130.

Swart, P. K. , Reijmer, J. J. and Otto, R. (2009) A reevaluation of facies on Great Bahama Bank II: variations in the δ^{13}C, d^{18}O and mineralogy of surface sediments. In: Perspectives in Carbonate Geology: A Tribute to the Career of Robert Nathan Ginsburg, IAS Special Publication (Eds P. K. Swart, G. P. Eberli and J. A. McKenzie), 41, pp. 47 – 60. Wiley-Blackwell, Oxford.

Tewfik, A. , Rasmussen, J. B. and McCann, K. S. (2005) Anthropogenic enrichment alters a marine benthic food web. Ecology, 86, 2726 – 2736.

Traverse, A. and Ginsburg, R. N. (1966) Palynology of the surface sediments of Great Bahama Bank, as related to water movement and sedimentation. Mar. Geol. , 4, 417 – 459.

Underwood, C. J. , Crowley, S. F. , Marshall, J. D. and Brenchley, P. J. (1997) High-resolution carbon isotope stratigraphy of the basal Silurian Stratotype (Dob's Linn, Scotland) and its global correlation. J. Geol. Soc. London, 154, 709 – 718.

Vahrenkamp, V. C. (1996) Carbon isotope stratigraphy of the Upper Kharaib and Shuaiba Formations: implications for the Early Cretaceous evolution of the Arabian Gulf region. AAPG Bull. , 80, 647 – 662.

Veizer, J. and Hoefs, J. (1976) The nature of ^{18}O/^{16}O and ^{13}C/^{12}C secular trends in sedimentary carbonate rocks. Geochim. Cosmochim. Acta, 40, 1387 – 1395.

Wilber, R. J. , Milliman, J. D. and Halley, R. B. (1990) Accumulation of bank-top sediment on the western slope of Great Bahama Bank: Rapid progradation of a carbonate megabank. Geology, 18, 970 – 974.

Zachos, J. , Pagani, M. , Sloan, L. , Thomas, E. and Billups, K. (2001) Trends, rhythms, and aberrations in global climate 65 Ma to present. Science, 292, 686 – 693.

第 12 章 碳酸盐岩—碎屑岩混积环境的晚始新世珊瑚建造

MICHELE MORSILLI, FRANCESCA R. BOSELLINI, LUIS POMAR,
PAMELA HALLOCK, MARC AURELL, CESARE A. PAPAZZONI 著

张天舒 译,郭彬程 校

摘 要 下普里阿邦阶(Priabonian)珊瑚生物丘和生物层,出现在中南部比利牛斯山地区 Aínsa-Jaca piggyback 盆地,发育在前三角洲泥灰岩/黏土中。详细的岩相和边界面相片镶嵌地图揭示了珊瑚建造的结构。珊瑚岩石体或者孤立出现,或者联合成更大的建造。珊瑚岩石体 1~8m 厚,几百米宽;在珊瑚集群里黏土含量很高。成堆的生物丘形成了低起伏的建造,一般厚度为 20~30m,局部达到 50m。这些生物丘随着三角洲复合体的前积,向西年代逐渐变新。最底部富集骨骼的岩层由具粒泥灰岩到泥粒灰岩基质的苔藓虫浮石构成,含有大量浮游有孔虫,缺失与光照相关的有机生物体。底部的珊瑚生物层,以及很多生物丘的基底,由板状珊瑚"漂浮"在细粒基质中构成,富集红藻。带穹顶的或大型块状的珊瑚,局部混合了分枝珊瑚和丛状珊瑚,构成建造的核部。这些珊瑚被基质包围,并且缺乏有机构架。基质由粒泥灰岩和泥粒灰岩构成,由局部出现的浮石、大量的红藻、珊瑚碎片、苔藓虫、浮游和底栖有孔虫以及局部出现的海绵组成。珊瑚砾屑碳酸盐岩和骨骼泥粒灰岩,基质为粒泥灰岩至泥粒灰岩,在邻近建造边缘以楔状体形式出现。岩石结构、骨骼组分、建造解剖以及相结构的综合分析,揭示了这些珊瑚建造发育在前三角洲环境,在那里三角洲朵叶体的迁移或者降雨周期的变化偶尔会形成适合造礁珊瑚生长的水体透明度。建造的水深位置受依赖光的群落和岩相分布的控制。这里应用过程—结果分析支持了至少从晚始新世直到晚中新世造礁珊瑚在中光度环境繁盛的假设。对比分析北地中海地区上始新统的珊瑚建造表明,这两个地区的珊瑚建造在相、组分和结构上有相似之处,并且表明它们也可以在相对低光度(中光度)和低水动力环境中生长。

关键词 珊瑚建造 三角洲 始新世 中光度珊瑚 混合沉积体系

12.1 概况

在地质记录中,以石珊瑚(Scleractinian)为主体的建造发育在陆源序列中的现象很普遍。然而,人们普遍倾向于认为珊瑚和硅质碎屑彼此相互排斥(Sanders 和 Baron-Szabo,2005)。现代"热带碳酸盐工厂"通常与主体生物控制的碳酸盐生产相关,这种碳酸盐岩生产出现在热带和亚热带温暖、光照好、贫养的、近表层的水域(Hallock 和 Schlager,1986;Schlager,2000,2003;Hallock,2005)。这些环境普遍由光合作用自养生物(比如,钙性绿藻)以及光合作用共生有机生物体[比如珊瑚和较大型的底栖有孔虫(LBF)]占主体,它们的沉积组合被称为光养生物(James,1997)。一个重要的问题是:全新世环境在其他地质时期里具有多大代表性。而在陆源主导的边缘栖息地以及在中光度深度的海水中,现代珊瑚和珊瑚礁的分布的广度、生态学的重要性和演化的意义(Mass 等,2007;Chan 等,2009;Lesser 等,2009;Kahng 等,2010,以及相关参考文献)则是另外一个关键的争论。

Potts 和 Jacobs(2000)提出石珊瑚—虫黄藻共生体发育在混浊的、半自养到富养环境,并指出大量的混浊的、近岸的栖息地在地质时期中持续存在。因此,值得注意的是,由于陆源注

入相对普遍,在地质记录中,目前珊瑚和珊瑚礁繁盛在偶尔或者长期混浊的水域(Woolfe 和 Larcombe,1999;Potts 和 Jacobs,2000;Perry,2005;Sanders 和 Baron-Szabo,2005;Perry 和 Smithers,2010)。混合碳酸盐—硅质碎屑沉积环境与光照好的、贫养的环境,在许多重要的生态参数方面有所不同,包括浑浊度、太阳能、盐度、沉积作用和营养有效性;在这些混合沉积体系中短暂的环境变化,使珊瑚暴露在生理压力极限环境。Potts 和 Jacobs(2000)假设适应近岸多变环境的珊瑚必须具有较大的生理忍耐力,并且指出这些早期适应能力被解释为珊瑚得以在海洋栖息地兴盛的优势。Sanders 和 Baron-Szabo(2005)和 Lokier 等(2009)近来的研究讨论了在混积体系珊瑚礁的化石记录,但注意到在具体的关键间歇期中,关于它们详细的结构、组分、沉积类型和主要控制因素的数据很有限(Braga 等,1990;Wilson,2005;Lokier 等,2009)。

始新世一般被认为是发生在 K/T 危机之后礁体生态系统缓慢复苏和重组的古新世与礁体繁盛的渐新世—中新世之间的珊瑚礁演化转换时期(Perrin,2002)。这个时期与古新世—始新世热量最大值(PETM)(Zachos 等,2001)以及始新世—渐新世转换(EOT)相关联(Prothero 等,2003)。在始新世,重要的古地理和气候变化,比如海洋热力梯度的增强,(Hallock 等,1991;Hallock 和 Pomar,2008),可能影响造礁珊瑚的演化和珊瑚礁的体积。文献记载的渐新世生物礁主要分布在古地中海中央地区(Perrin,2002),尤其是环绕地中海地区(Santisteban 和 Taberner,1988;Darga,1990,1992;Eichenseer 和 Luterbacher,1992;Alvarez 等,1994,1995;Taberner 和 Bosence,1995;Schuster,1996;Bassi,1998;Bosellini,1998;Nebelsick 等,2005;Lokier 等,2009)。

Pochon 等(2006)假设现存的被珊瑚包裹的共生藻源自早渐新世,有一些源自 50Ma 前,而且多样化作用与渐新世全球变冷时期相吻合。根据这些作者的研究,第一次重要的分化发生在 PETM 之后的渐新世全球变冷时期,而大多数现存的共生藻从中新世开始分化。根据 Pochon 等(2006)对于生物共生体演化的解释,Pomar 和 Hallock(2007)假设:地中海地区晚中新世珊瑚栖息地变浅,以及珊瑚建造体积的变化,是全球变冷和藻类同期分化导致水深温度梯度增强,以及在 Messinian 盐度危机之前,与地中海逐渐被隔离过程相关的碳酸盐饱和度可能增加的结果。Pomar 和 Hallock(2007)推测,在晚中新世之前半自养—贫养区为造礁珊瑚提供了温暖的理想栖息地。

有文献详细描述了在北西班牙中—上始新统生物礁出现在不同的沉积环境:陆架内部—台地内部、陆架或者台地边缘、碳酸盐岩斜坡、三角洲以及扇三角洲沉积体系。一般认为,这些始新世生物礁发育在晴天浪基面以上,在地貌上,生物礁主要以补丁礁为特征,很少以后礁到前礁相的边缘礁形式出现。现代似礁体在珊瑚组分和生长形式上的古水深分带也被应用于建立沉积模式(Santisteban 和 Taberner,1988;Alvarez 等,1994,1995;Millán 等,1994;Taberner 和 Bosence,1995;Romero 等,2002;Lokier 等,2009)。

本文基于南中部比利牛斯山地区发育的生物礁建造的研究,提出上始新统珊瑚建造的沉积模式;通过对生物礁建造的相结构、骨骼组分和岩石结构的分析来解释古环境,包括透明度和透光性、水动力能量和营养有效性等因素。

12.2　地质背景

研究区位于 Aínsa-Jaca 盆地,在南中部比利牛斯山地区,狭窄的东西向背驮型(piggyback)盆地(图 12-1A)。这个盆地由古近系向南进积的南比利牛斯山褶皱冲断带形成(Millán 等,1994;Muñoz 等,1994;Castelltort 等,2003;Huyghe 等,2009)。前陆挠曲形成 Aínsa-Jaca 盆地主体,并形成前缘冲断背驮型盆地。向东产生的同期隆起为沉积物卸载和三角洲沉积体系提供了物源:Campodarbe 组(Heard 和 Pickering,2008)或者 Sobrarbe 三角洲复合体(Dreyer 等,1999;图 12-1B)。三角洲复合体在 Aínsa-Jaca 盆地南部向西进积(图 12-1B),逐渐覆盖在 Hecho Group 中下始新统浊积体系上(Mutti 等,1985;Remacha 等,2003,2005)。

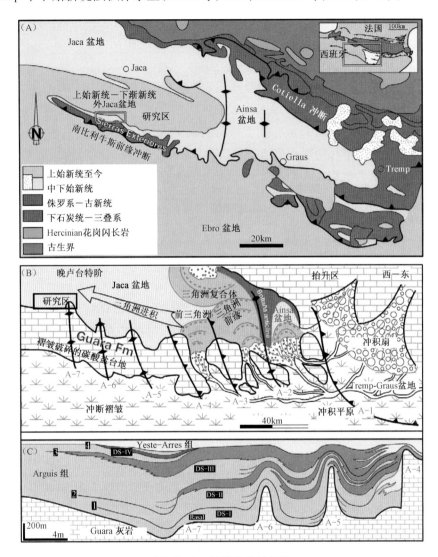

图 12-1 地质和地层情况

(A)Aínsa-Jaca 盆地在南部中央比利牛斯盆地的位置;据 Dreyer 等(1999)做简化。(B)晚卢台特阶 Sorrel 三角洲复合体古地理重建(深蓝色:Boltaña 背斜;较浅的蓝色:三角洲前缘和 Sorrel 三角洲复合体前三角洲)。一系列倾斜的褶皱和冲断斜坡(A-1 至 A-7)在卢台特阶时期开始形成,逐渐向西蔓延直到巴尔顿期到早普里阿邦期;根据 Millán 等(1994)和 Dreyer 等(1999)。(C)Yeste-Rasal-Pico del Aguila 地区 Arguis 组四个沉积层序(DS)地层结构图以及层序边界("1"到"4");据 Millán 等(1994)做简化;珊瑚建造夹层为红色。A-4 至 A-7:倾斜的褶皱和冲断和(B)一致

总体上,Aínsa-Jaca 盆地被 4km 厚的深海沉积充填,横跨早—中始新世伊普尔期(Ypresian)—鲁帝特期(Lutetian),记录了大约 10Ma 的深海沉积过程(Heard 和 Pickering,2008),以及中—晚始新世超过 2km 厚的三角洲—冲积扇沉积物(晚鲁帝特期—普里阿邦期)(Millán 等,1994,2000;Dreyer 等,1999;Sztrákos 和 Castelltort,2001;Callot 等,2009)。

与盆地充填同期的是一系列冲断斜坡和轴向为南北向的斜卧褶皱[齿状山外貌（Sierras Exteriores），也称为外部齿状山（External Sierras）]，开始形成于鲁帝特期，逐渐向西进积，直到巴尔顿期（Bartonian）至早普里阿邦期（Millán 等，1994，2000；Poblet 和 Hardy，1995；Poblet 等，1998；Dreyer 等，1999；Castelltort 等，2003），具有多变的缩短速率（Huyghe 等，2009）。这样，齿状山外貌向南侵位，形成一系列从东到西发展的几千米宽的南北向褶皱（图 12-1B，12-1C），这些背斜连续抬升和区域性相对海平面变化，控制着沉积剖面的分布、沉积层序厚度和沉积层序中的次旋回（Millán 等，1994）。

在研究区，始新世地层主要由四个组构成（详细划分根据 Millán 等，1994；图 12-2）。Guara 组，鲁帝特时期，是一个 800m 厚的浅水碳酸盐斜坡序列，富集货币虫 nummulitids 和 alveolinids。上覆 Arguis 组，为近 1km 厚黏土和泥灰质序列，夹厚层碳酸盐岩，解释为在鲁帝特期最晚和普里阿邦期最早时期累积形成的前三角洲沉积。上覆的 Yeste-Arrés 组，横向等同于 Belsué-Atarés 组，为近 200m 厚的砂岩和泥灰岩序列，解释为三角洲前缘沉积体系，部分属于普里阿邦阶地质时期。Campodarbe 组，部分属于普里阿邦时期到渐新世，为第四个地层单元，最大厚度超过 1km，主要由冲积相砾岩、砂岩和黏土构成（Millán 等，1994，2000；Castelltort 等，2003）。

沿着 Ebro 盆地东侧对花粉谱的研究（西班牙东北部）揭示了这样一个变化过程：从巴尔顿期温暖的环境（红树林沼泽）到普里阿邦时期的旱季气候（各种草本植物增长），再到早渐新世干旱和稍冷的气候（Cavagnetto 和 Anadón，1996）。Postigo Mijarra 等（2009）也指出了在始新世后半部分古热带种类的缺失，包括典型的红树林种类，与在 EOT 达到高峰的明显气候变化相关。Huyghe 等（2009）指出，在普里阿邦期-鲁培尔期（37~30Ma），源自比利牛斯山的沉积流体的增加与在始新世到渐新世界面处由气候引起的侵蚀作用增加有关。

本次研究的珊瑚建造，在早普里阿邦时期，发育在 Arguis 组上部泥灰质岩层中（图 12-3）。在这个地层组中，Millán 等（1994）基于最大洪泛面、角度不整合和上超几何构型划分了四个层序（见图 12-1C），并认为层序受控于区域性的构造周期。每个层序有明显的向西进积的趋势，由下部的泥灰质夹层（外斜坡）和上部的浅水环境的，硅质碎屑和碳酸盐岩混合夹层（中部—内斜坡）构成。构造下沉导致了洪泛和随之蓝色泥灰岩厚层序列沉积，而构造减弱时期，碳酸盐岩沉积（Millán 等，1994）。根据这些作者的研究，在研究区（层序Ⅱ—Ⅳ）识别出的碳酸盐岩相的骨骼组分主要为苔藓虫，其次为双壳类、LBF[货币虫（*Nummulites*）和盒果藤属（*Operculina*）]以及层序Ⅱ的海胆；层序Ⅲ主要为扇贝，其次为海胆、苔藓虫、底栖有孔虫和单体珊瑚；层序Ⅳ为生物碎屑岩集层与泥灰岩互层混合而成的珊瑚生物丘（图 12-1C）。

Arguis 组生物地层数据（Canudo 等，1988，1991；Millán 等，1994；Sztrákos 和 Castelltort，2001）表明：层序Ⅰ发育在晚鲁帝特期到早巴尔顿时期（*Guembelitrioides higginsi* 和下部的 *Turborotalia cerroazulensis pomeroli* 浮游有孔虫生物群；界面在 NP-16 微型浮游生物群和 NP-17 微型浮游生物群之间，层序上界面）；层序Ⅱ为中部到上部巴尔顿时期（*T. pomeroli* 生物群 *p. p.*），层序Ⅲ发育在早普里阿邦期下部（*Globigerinatheka semiinvoluta* 生物群 *p. p.*）；层序Ⅳ发育在早普里阿邦期上部。在 Arguis 组顶部，Canudo 等（1988）以及 Papazzoni 和 Sirotti（1995）记录了归属普里阿邦期于货币虫 *fabianii* 的货币虫（SBZ 19 = 早普里阿邦期，Serra-Kiel 等，1998）。

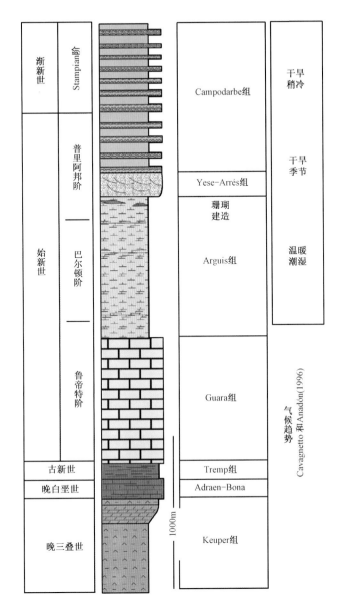

图 12-2 Rasal 地区 Aínsa-Jaca 盆地岩性地层单元(据 Millán 等,1994)
以及主要气候趋势(据 Cavagnetto 和 Anadón,1996)

12.3 方法

根据在 Rasal 和 Yeste 村相对连续的暴露面(图 12-3A,图 12-3B)可以详细分析出现在 Arguis 组最上部的一系列珊瑚建造的岩相、地层几何样式和相结构(层序Ⅳ,Millán 等,1994)。

本次研究的野外工作由以下几部分构成:(1)在正色摄影上(几何学校正的航空照片),对珊瑚建造进行编图和 GIS 分析(ESRI 的 ArcGIS 软件);(2)绘制 Arguis 组 13 个过珊瑚建造的地层剖面(图 12-3C);(3)编制两个珊瑚建造详细的相图(生物相、岩相和边界面),编图过程是在装有 ArcPad 软件(ESRI)的移动设备上连接 GPS,将所获得的数据点显示在倾斜的相片

— 351 —

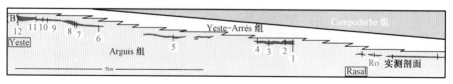
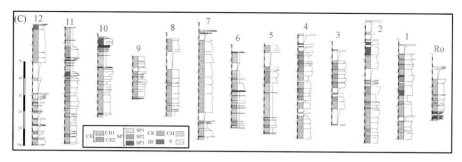

图12-3 (A)在Yeste和Rasal村之间正色摄影上的简化地质图(红:珊瑚建造;
黄线:Yeste-Arrés组显著的砂岩序列);(B)从(A)重建起来的横剖面;
(C)Yeste-Rasal地区实测地层剖面(测量编号)(相代码见表12-1)

镶嵌图和矢量图上,在此基础上编制相图。在调查和增加数据点过程中,GPS的预计误差小于5m,直接被限制在移动设备上正色摄影的可视化区域范围内。运用相同的方法编制了Yeste地区珊瑚建造和相邻相详细的地质图,以便用来在各种出露的岩相之间重建珊瑚礁建造的原始几何样式。所用数据是从 *Gobierno de Aragón* 的 SITAR 官方网点获得的公开发表的元数据和正色摄影照片。

微相分析,包括对岩石结构和骨骼成分的200个薄片分析。珊瑚相研究,包括主要的珊瑚分类单元、生长样式和珊瑚大小的二维测量(在生长地点珊瑚的最大直径和最大高度)。珊瑚分类单元研究,主要在种属层次上,在野外或者通过薄片和抛光片分析。

12.4 岩相

识别出6种主要岩相和各种亚相,包括硅质碎屑和碳酸盐岩(表12-1)。根据岩石结构、成分、层理和几何关系进行岩相识别。因为这些珊瑚建造并不能造出坚硬的框架,所以Insalaco(1998)提出的"生长结构"命名运用于原地和原生的钙化有机物的骨骼分类。Insalaco(1998)解释说,基本的生长结构定义为后生动物(多细胞动物)骨骼的形状和相对位置,结合分类来充分描述生长结构,从而避免从形态上解释生物作用而产生的问题,比如,Embry和Klovan骨架岩、粘结岩和障积岩——分别为骨架建造、粘结/结壳或者障积作用。Insalaco(1998)按照主要的生长形态来定义基本的生长结构:盘状岩(盘状—板状骨架),层状岩(片状和层状骨架),圆顶岩(圆顶状的和不规则块状骨架),枝状岩(分枝状、杆状和相对受限而侧面生长的管状单个形态)以及混合岩(不受一种生长形态主导,而由多种生长形态组成)。然而,Dunham(1962)定义的粘结岩被保留下来,用来定义 Embry 和 Klovan(1971)描述的原地灰岩,这种原地灰岩不能被识别出有机粘结的类型。在无法区分生长结构种类的情况下,用粘结岩这一成因术语来定义原地灰岩。

表12-1 主要岩相和亚相分布

相		亚 相	描 述
CB	珊瑚粘结岩(圆顶岩、盘状岩、柱状岩和混合岩)	CB1	原地和原生珊瑚,黏土—颗粒质灰泥灰岩基质(细粒),大量红藻
		CB2	原地和原生珊瑚,灰泥质颗粒灰岩基质(粗粒),骨骼碎屑和大量红藻
SP	含骨骼灰泥质颗粒灰岩(生物碎屑钙化)	SP1	好到中等分选,细粒泥质为主的含骨骼灰泥质颗粒灰岩
		SP2	差分选,粗粒(局部漂砾岩)泥质为主的含骨骼灰泥质颗粒灰岩
		SP3	差分选,灰泥质颗粒灰岩,含 LBF,局部含陆源颗粒
CR	珊瑚砾屑灰岩		珊瑚和红藻碎屑,LBF,苔藓虫,双壳类
BF	苔藓虫漂砾岩		苔藓虫和龙介虫漂砾岩,含颗粒质灰泥灰岩到灰泥质颗粒灰岩基质
CM	黏土到泥灰岩		硅质碎屑砂岩

12.4.1 珊瑚粘结岩(CB)

珊瑚粘结岩相由大部分原地生长的石珊瑚群落构成(图12-4)。在珊瑚粘结岩中没有识别出由紧密堆积的珊瑚群落构成的密集的连锁骨架。而珊瑚在骨架基质或/和在黏土—泥灰质中分布很少。在珊瑚骨架间的基质厚度范围从几个毫米至几个厘米,一般由漂砾岩/砾屑碳酸盐岩伴生细粒、分选较差的灰泥质颗粒灰岩(亚相CB1,图12-4E)到颗粒质灰泥灰岩组成(Lucia,1995)(亚相CB2;图12-4F)。这种岩相出现在结节状生物层(图12-5A)(15cm至几米厚)或者小型生物丘中(图12-5B),可以合并形成达到40m厚的建造。这种岩相,虽然多变,但一般依黏土—泥灰岩的含量而呈现结节状。珊瑚生物层,生物丘和更大的建造通常含有蓝色的灰泥质黏土。

原生的珊瑚群落,或者稍稍移位,但是仍然在原地,包括不同形状和大小的下列属的生物:*Actinacis*, *Colpophyllia*, *Caulastrea*, *Astrocoenia*, *Goniopora*, *Cyathoseris*, *Astreopora*, *Siderastrea*, *Alveopora*, *Agathiphyllia*, *Leptoria* 和 *Plocophyllia*。生长形态可以是盘状(主要是 agariciids,比如 *Cyathoseris*,也有一些 *Actinacis*, *Goniopora* 和 *Astreopora*;图12-4B)、块状(*Actinacis*, *Goniopora*, *Astreopora*, *Siderastrea*, *Colpophyllia*, *Leptoria*, *Agathiphyllia*;图12-4C)、粗枝状(*Actinacis*)或者柱状(phaceloid - like)(主要 *Caulastrea*;图12-4D)。

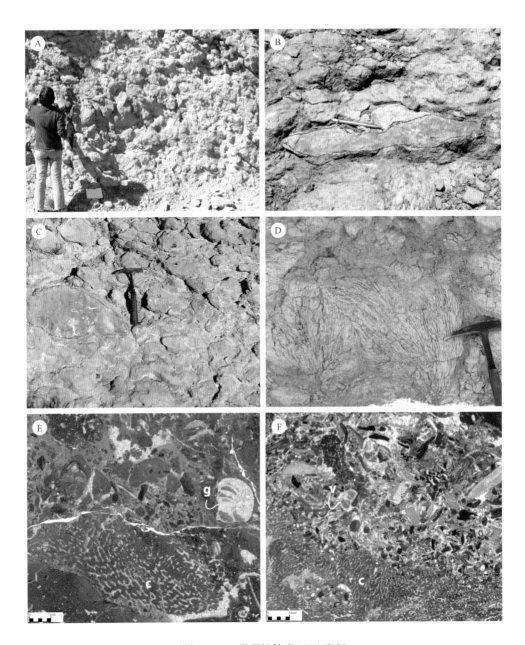

图 12-4 珊瑚粘结岩(CB)岩相

(A)以结节状形态为特征,这与珊瑚之间的黏土和细粒基质有关(测量编号 11)。比例尺人高 1.7m。(B)原生 *Actinacis* 盘状珊瑚(Yeste 地区;比例尺钢笔长为 14cm)。(C)珊瑚圆顶岩(Yeste 地区;比例尺地质锤长 31cm)。(D)原生 Phaceloid 珊瑚群落(*Caulastrea*)(测量编号 2;图 12-3)。(E)珊瑚—砾屑灰岩之间的基质含有灰泥质颗粒灰岩基质(CB-1)("c":poritid 珊瑚碎片;"g":*Gyroidinella* 蚤。(F)珊瑚—砾屑灰岩之间的基质含有灰泥质颗粒灰岩基质(CB-2)("c":*Actinacis* 珊瑚碎片;"v":*Victoriella* 种属)

根据 Insalaco(1998)的定义,可以识别出盘状岩、圆顶岩、柱状岩以及局部发育混合岩 4 种生长形式。珊瑚盘状岩是最显著的生长形式,Rosen 等(2002)通过宽度/高度比为 4/1 或以上的横切面图,定义为未粘结的、分散的、盘状的群落。这些群落呈层状或者板状[宽度(cm)/高度(cm)比为 10~20/2 和 100/2],一般出现在 0.5m 至几米厚,几百米宽的生物层中,并且优

图 12-5 （A）珊瑚生物层包含在黏土中（CM 岩相；测量编号 6，图 12-3）；
（B）Rasal 地区出露的单个珊瑚丘（测量编号 2，图 12-3）

虚线标出的是层理面，比例尺人高 1.9m；这个珊瑚丘是更大建造的一部分（见图 12-3 和 12-9A；比例尺人高 1.8m）

先出现在剖面的下部（图 12-4B）。珊瑚圆顶岩由圆顶状的和不规则块状珊瑚骨架构成，在珊瑚建造中大量出现。在二维露头点，圆顶群落的宽度范围为 20~70cm，大多为 40~50cm，宽度/高度比大约 2/1。珊瑚群落堆积并不紧密；它们零星分布在岩石中，被各种沉积物包围，缺少有机骨架的原地共生特征（图 12-4C）。珊瑚柱状岩出现在薄地层中，柱状珊瑚（主要为 *Caulastrea*），平均直径为 60cm，高度达到 45cm，或者出现在 1~2m 厚的单殖动物珊瑚地层，以原地分枝状的 *Actinacis* 为特征。*Actinacis* 的分枝被多基因层状珊瑚藻和有孔虫生物侵蚀；它们多数出现在中部和上部建造（图 12-4D）。除了上述描述的珊瑚生长结构以外，珊瑚混合岩也在局部发育，通过实测剖面识别出来。生物侵蚀作用很普遍；虽然海绵和蠕虫钻孔也有出现，尤其是在柱状群落和分枝珊瑚的碎片上，但最重要的生物钻孔是在块状珊瑚上的 *Lithophaga*。

在基质中，红藻明显以层状和分枝形式出现，并且出现大量珊瑚和海胆碎片。宏观和微观分析都揭示了分枝珊瑚 *Stylophora*，*Acropora*，*Pocillopora*，*Astreopora*，*Alveopora*，*Astrocoenia* 和 *Bacarella* 的碎片，这些碎片在原生地点很少保存。也识别出大量的水螅碎片。苔藓虫、双壳类（主要是 ostreid）、龙介虫、介形类和局部出现的刺毛海绵普遍发育。出现浮游类和较小的底栖类有孔虫，以及局部出现的 LBF 比如 *Nummulites* spp.，*Heterostegina* sp.，*Asterocyclina* sp.，*Operculina* ex gr. *Alpina*。多基因的层状红藻和结壳有孔虫（*Miniacina* sp.，*Haddonia* sp.，*Carpenteria* sp.，*Victoriella* sp.，*Fabiania* sp.，*Gyroidinella magna*，*Acervulina linearis* 和 *Gypsina* sp.）在珊瑚上普遍发育。

这些珊瑚生物层和生物丘可以认为是 Riding（2002）定义的"近丛礁"；骨架相邻，但不相连，形成了基质支撑结构。相对高的基质/骨架比率和低的外骨架早期胶结物含量是这种生物建造的特征。在这些生物礁中，沉积圈闭是骨架生长重要的必然结果，而且，丛礁有机体被认为对疏松沉积物有忍耐力。真正意义上的骨架缺失明显限制了地形起伏，从而限制了丛礁可以达到的空间展度，以及在礁内部的发育。

12.4.2 骨架颗粒质灰泥灰岩（SP）

骨架颗粒质灰泥灰岩出现在厚度变化的地层中（20cm~1.5m），局部主要在侧翼以及建造的上部与其他岩相混合。与正常地层的分界是突变的，呈平面到波状形态。在内部，这些地层无粒序或者局部递变粒序（向上变细）。甚至层理都很少见。基于粒度、分选和成分可区分出三种亚相：（1）SP1，由好到中等分选的细粒泥质为主的骨架颗粒质灰泥灰岩组成，富含红藻、双壳类、海胆和无法识别的生物碎屑(图12-6A);(2)SP2 由粗粒的、分选差的骨架颗粒质灰泥灰岩组成,含有珊瑚、红藻(大量分枝状的)的碎片,以及少量的苔藓虫、海胆、粟孔虫和其他有孔虫(图12-6B);(3)SP3 由分选差的颗粒质灰泥灰岩组成,含有 LBF(nummulitid)、双壳类、海胆、苔藓虫和少量的腹足类;其他较小的底栖类(Rotaliidae,Gypsina sp.,Acervulina linearis,Planorbulina sp. 和 Pellatispira sp.)以及浮游有孔虫也有出现(图12-6C)。

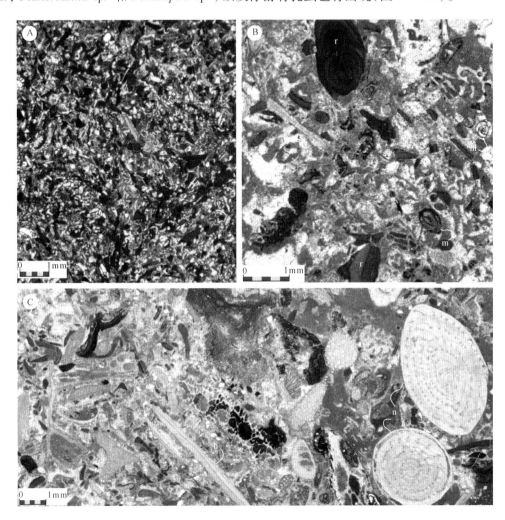

图12-6 含骨骼的灰泥质颗粒灰岩(SP)微相
(A)中等到好分选的细粒含骨骼灰泥质颗粒灰岩。(B)分选差,粗粒含骨骼灰泥质颗粒灰岩("r":红藻碎片;"m":粟孔虫)。(C)分选差,粗粒骨骼灰泥质颗粒灰岩含 LBF("b":双壳类;"n":Nummulites ex gr. fabianii)

12.4.3 珊瑚砾屑碳酸盐岩(CR)

珊瑚砾屑碳酸盐岩,含有颗粒质灰泥灰岩基质,出现在薄到厚地层中(20~50cm 厚,少数1m 厚),一般在侧翼和建造的顶部,通常与 SP 岩相组合,局部出现在一些小丘的底部,插入黏土—泥灰岩(岩相 CM)中。通常的成分为珊瑚碎片(多数为分枝和柱状的珊瑚;图12-7A,图12-7B),层状红藻和红藻碎片,以及苔藓虫、有孔虫(Gypsina sp.,Acervulina linearis,Gyroidinella sp.)、腹足类、海胆、双壳类、货币虫(nummulitids)、粟孔虫结壳,局部发育大量红藻石。

12.4.4 苔藓虫漂砾岩(BF)

一些碳酸盐岩富集的夹层由含灰泥质颗粒灰岩—颗粒质灰泥灰岩基质的苔藓虫漂砾岩(图12-7C)组成,并与蓝色黏土—泥灰岩(岩相 CM)构成互层和夹层。苔藓虫出现在整个生物群落和碎片中(图12-7D),并且龙介虫和双壳类也很普遍。在有孔虫中,浮游类大量发育,局部出现较小的底栖类。LBF 以小货币虫(*Nummulites* sp.)、*Spiroclypeus* sp. 和 *Pellatispira* sp. 为代表。在这种岩相中,珊瑚碎片出现很少,缺失红藻。

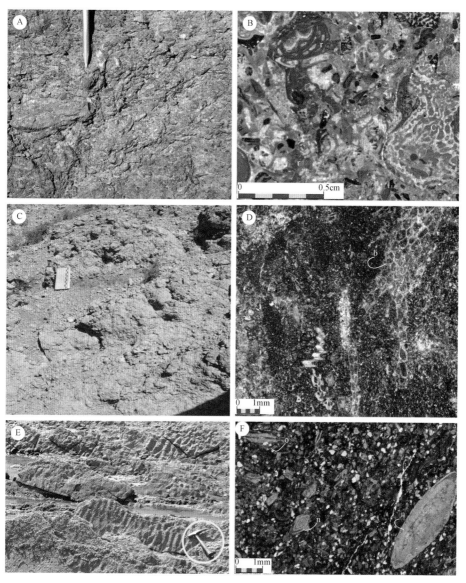

图12-7 (A)珊瑚砾屑灰岩(测量编号1;图12-3)。图中显示的铅笔作为比例尺长7cm。(B)砾屑灰岩—基质由没有分选的灰泥质颗粒灰岩组成,含有大量珊瑚和红藻碎片(CR 岩相)("r":红藻;"c":珊瑚)。(C)苔藓虫漂砾岩(BF 岩相)发育在蓝色黏土/泥灰岩中,Rasal 地区(测量编号1;图12-3;见图12-9A);笔记本长20cm。(D)苔藓虫漂砾岩微相(BF 岩相)("b":苔藓虫)。(E)Yeste 地区砂岩层(S 岩相)发育浪成波纹(测量编号10;图12-3)(比例尺地质锤长31cm)。(F)灰岩碎屑砂岩(S 岩相),含有石英颗粒和 LBF("n":货币虫属 *Nummulites* spp.;"o":盒果藤属 *Operculina* sp.)

12.4.5 蓝色黏土到泥灰岩相（CM）

蓝色泥灰岩到黏土相，缺失大型底栖动物，构成 Arguis 组；它们以成层的或无构造（局部生物扰动形式出现），或者局部包含几厘米厚的砂岩夹层，局部发育交错层理（流水或者波状构造）。上一节讲到的泥灰质黏土中含大量骨骼碎屑的泥灰岩夹层。

12.4.6 砂岩相（S）

成层性很好的砂岩，地层厚度变化大（5cm～1m），局部出现蓝色泥灰质黏土夹层（岩相CM），并且砂岩层也出现在基底或者上超到珊瑚建造侧翼。砂岩层主要由含有大量石英颗粒的灰岩碎屑（Bates 和 Jackson，1987）组成。含有各种成分的骨骼颗粒也出现在建造周围的砂岩层中；一些地层包含 LBF（*Nummulites* 和 *Amphistegina*）、苔藓虫和珊瑚碎片（图12-7F），而其他地层包含大量小的无法分辨的生物碎屑。颗粒粒度从细粒到粗粒，并且局部有极粗粒。粒序层理和平行层理普遍发育，在一些薄层的顶部发育对称波痕（图12-7E）。局部可见不对称波痕和丘状交错层理。孔洞虽然少，但也有发现。在珊瑚建造顶部，砂岩层向 Arguis 组顶部逐渐增加，并且在上覆 Yeste-Arrés 组中占主导。

12.5 相结构和沉积模式

碳酸盐岩主导的岩相（CB，SP，CR 和 BF）为周期性组合，形成建造，通常被包裹在硅质碎屑蓝色 CM 相中。岩相沉积的相对水深的估计对沉积相解释和模式的建立很关键，但是有难度；这里，通过依赖光能的骨骼成分结构分析和类型推测出水动力能量，从而得到相对水深。在波浪主导的体系中，水动力能量和光能穿透力都随深度以指数方式减少，造成生态水深梯度，导致生物带在浅水比深水狭窄（Kain 和 Norton，1990）。

12.5.1 水动力作用

由于控制浪基面深度的过程很复杂，所以估算水动力—水深梯度是一门不精确的学科。波浪受控于风力体系和方向，以及盆地的风浪区范围。风和潮汐引起的水流受控于复杂的盆地地理、古海洋学机制和气候之间的相互作用。而且，观测现代海洋沉积对于古海洋来说并不是可用的先验方法（Immenhauser，2009）。然而，对于形成珊瑚建造必要环境的水深估计可以从间接的证据中获得。

三角洲环境的珊瑚丘出现在前三角洲黏土的夹层中。在珊瑚丘上，剖面出现第一套砂岩层之前，存在 10～40m 厚的黏土（见图12-3）。在距离珊瑚建造相似的地层位置，缺失砂岩层，这种缺失消除了在朵体之间，珊瑚建造侧向上相当于活跃的三角洲朵叶体的可能性。砂—黏土之间的过渡位置虽然变化很大，但一般出现在三角洲水深大约 10～20m 范围内（表12-2）。因此，初次估算包括了一些珊瑚丘上面黏土层的厚度加上前三角洲黏土和三角洲前缘砂之间过渡的深度。可能的相对海平面的高频率变化的影响不能根据露头情况进行估计。

表12-2 现代河流三角洲体系三角洲前缘砂体和前三角洲泥之间的过渡深度(据多个作者)

位置	过渡深度	参考文献
印度洋—太平洋地区		
Mahakam三角洲河流	70m(前三角洲底)	Storms等,2005
扬子三角洲河流	18~25m	Hori等,2002
Mekong三角洲河流	18~20m	Ta等,2002
Mekong三角洲河流	15m	Xue等,2010
黄河三角洲河流	12~18m	Chun-ting等,1995
Fly三角洲河流	50m	Harris等,2004
红河三角洲河流	18~25m	Tanabe等,2006
印度洋		
Ganges-Brahmaputra	70m(前三角洲底)	Michels等,1998
Indus三角洲河流	20m	Giosan等,2006
大西洋		
Orinoco三角洲河流	10~40m	Van Andel,1967
亚马逊三角洲河流	60m(前三角洲底)	Sternberg等,1996
Paraíba do Sur三角洲河流	8~10m	Murillo等,2009
尼日尔三角洲河流	18到20m	Allen,1964
地中海		
Po三角洲河流	15m	Fox等,2004
Po三角洲河流	18~25m	Correggiari等,2005
Ombrone三角洲河流	20m	Tortora,1999
Rhône三角洲河流	20~25m	Sabatier等,2009
Ebro三角洲河流	30~40m	Somoza等,1998

在珊瑚丘周围的黏土沉积中没有发现风暴沉积。在珊瑚建造周围砂岩层并没有出现典型的向上变厚变粗的现象。向上变厚变粗的砂岩层开始出现在珊瑚丘以上几十米处,而且与三角洲前缘进积作用有关。仅有几个孤立的薄层砂体,或者发育递变粒序或者发育丘状交错层理,局部出现在一些建造周围。这些薄砂层反映了由洪水引起风暴或者高密度流而形成的密度流沉积(Mutti等,2003)。货币虫(*Nummulites*)很少出现,但是局部出现在珊瑚丘的周围和上面,和砂岩层共生(外源沉积物),或者出现在递变层(浊流)。这些异地的货币虫(*Nummulites*)指示了珊瑚建造所处的深度比适合*Nummulites*生长的深度要深。

12.5.2 光的穿透性

光的穿透性受到影响水体透明度的多种因素的调节(比如,陆源物质卸载、营养注入、河流溶解的有机物和腐殖质物质等等)。透光区的下限(光照区域)也与纬度的变化相关(图12-8A),在热带—亚热带的深度较深,而在较高纬度地区,随着表面光照强度的减少和光线入射角的增加而变浅(Liebau,1984;Lüning,1990)。绿藻和红藻分布深度是古沉积深度最好的指示之一,这是因为它们对光照具有强依赖性(Flügel,2004)。虽然三个宏观藻群(绿藻、褐藻和红藻)含有叶绿素,在蓝色和红色光谱范围出现吸收峰值,但是绿藻、褐藻和红藻的典型差异取决于最大值时的附加颜色和在光谱绿色部分不同的吸收作用(Lüning,1990)。由于物种对颜色的适应性不同,红藻的生长深度通常比绿藻和海草深。然而,正如其他颜色可以补充叶绿素吸收光谱一样,三种藻群都可以出现在深度和光线强度范围很大的区域,红藻表现尤其显著(Lüning,1990;Lee,1999)。没有经过搬运的绒枝绿藻指示了非常浅的深度(几十厘米到几米),但是codiacean 绿藻生长在更广的深度范围。比如,*Halimeda* 生物丘,可以出现在高水体透明度60m 的深度(Hallock,2001,以及相关参考文献)。这样,现代海草的生长范围,考虑到某些颜色的适应性,非绒枝绿藻可以用来界定透光层的范围。

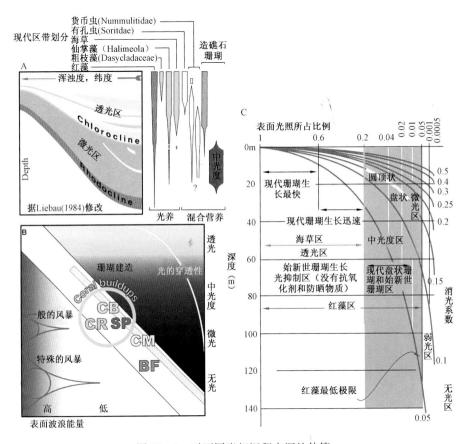

图 12-8 对不同岩相沉积水深的估算

(A)根据原地光能自养和混合营养生物的发育情况获得光区水深和它们的深度下限。Chlorocline 是原地海草和非绒枝绿藻中发育水深最深的生物。Rhodocline 是原地红藻中发育水深最深的生物。由于表面光照随纬度增加而降低,透光区的深度下限随着浑浊度和纬度而变化,在热带/亚热带变深,向亚寒带地区变浅(据 Liebau,1984,有修改)。当骨骼成分允许更细微的水深/光穿透性比值差异时,中光度区可以从透光区和微光区中区分出来,见图 12-12。(B)本文中描述的岩相水深位置与浪基面和光的穿透性有关;在波浪主导体系中,两者都随深度增加而呈指数递减;相代码如表 12-1 所示(绿色圆圈:珊瑚建造)。(C)不同消光系数的表面光随深度变化的比例,以及现代珊瑚的生长速率。不同消光系数的光穿透性的曲线,根据 Kanwisher 和 Wainwright(1967)、Huston(1985)、Hallock 和 Schlager(1986)及 Kahng 等(2010)。始新世珊瑚占据了中光度水深区域,这个区域被盘状珊瑚占据。黄色的线表示本文根据现代三角洲体系光的穿透性来估算始新世珊瑚的水深。

值得注意的是,水平比例尺不是线性的,以便更好地展示最低光能范围的曲线多年生海藻的最低正常深度界限(通常为层状红藻)出现在相当于接收0.05%~0.1%表面光照的深度;红藻生长深度比其他所有藻类都深,这是因为它们可以更好地利用在水中穿透最远的蓝光(Lee,1999)。另外,深水藻类的生存策略包括缓慢生长(使呼吸作用最小化),以及避免被吃掉,这些生存策略是让叶状植物长期生存的重要前提条件。在所有地区,较低藻类深度界限和海底洞穴中,层状珊瑚红藻是仅有的显著的藻类生长形式(Lüning,1990)。因此,作为出现较深的原地红藻(图12-8),足以界定微光区的深度下限。

因此,对透光区(光线较好,开阔海域,通常为高波浪能量)、中光度区(满足珊瑚生长的光线,通常在波浪作用面以下)、微光区(满足珊瑚红藻的光线)以及无光区(无光合作用可用的光线)进行区分,成为在分析碳酸盐台地和估算成因过程中的有效方法。

对于混合营养生物,比如珊瑚和LBF,关于它们的水深位置具有一定程度的不确定性,这源自现代可对比生物的缺失(如在一些LBF)、藻类共生体的演化(LaJeunesse,2005;Pochon等,2006),以及珊瑚中对于共生体进化枝变化的可塑性(Baker等,2004;Fautin和Buddemeier,2004;Rowan,2004;Chen等,2005;Stanley,2006)。然而,总体上,现代珊瑚生长(Yentsch等,2002)的三种最常见形态,即,分枝、圆顶和盘状珊瑚(Hallock和Schlager,1986,以及相关参考文献)对光照的要求是众所周知的。而且,对于在低光照条件下,部分依赖于光合作用的生物(比如,为捕获最大光线而形成的薄盘状)在形态上的适应性,与捕食浮游生物的异养生物(比如,直立的分枝或者网状的形态,如八放珊瑚或者苔藓虫)完全不同,以至于这种从中光度—混合营养到无光区—异养型的过渡非常明显。

12.5.3 岩相累积的水深估算

苔藓虫漂砾岩,含有颗粒质灰泥灰岩—灰泥质颗粒灰岩基质,大多出现在生物层中,厚度变化大(通常0.5~2m,局部达到3~4m),并且与生物丘不相连,含有黏土,地层发育在珊瑚建造之下(图12-7C)。这种岩相包含大量滤食型有机体(异养生物),而缺失与光作用相关的组分(自养和混合营养生物)。浮游有孔虫普遍出现,小的底栖有孔虫很少。苔藓虫漂砾岩出现在生物丘中,侧向过渡到黏土。这些特征揭示了风暴浪基面以下的无光区沉积(图12-8B)。

珊瑚粘结岩岩相,发育盘状岩生长结构,大多以生物层(10~20cm到2m厚)形式出现,与黏土互层,但是它们在一些建造的底部也普遍出现。盘状珊瑚(比如,*Cyathoseris*)被灰泥(颗粒质灰泥灰岩)和粉砂—黏土基质包围,基质中含有大量层状红藻和浮游有孔虫,并且苔藓虫碎片和结壳有孔虫 *Miniacina* spp. 也普遍出现。细粒颗粒质灰泥灰岩基质的出现表明该岩相在浪基面以下沉积,但是层状红藻和盘状珊瑚一起出现,表明盘状岩生长结构出现在红藻生长深度以上的中光度区—微光区(图12-8B)。

圆顶生长结构大多出现在几米厚的生物丘中(见图12-5B),不仅可以堆积形成40m厚的复合不对称的建造(图12-9),而且可以出现向侧向延伸的50cm到几米厚的生物层(图12-5A);生物丘和生物层与黏土互层。由于受到露头连续性影响,对这些生物层总体延展的直接观察比较局限,但是可以观察到几米至几百米。在这些建造中,珊瑚—砾屑碳酸盐岩和骨架—泥灰质颗粒灰岩岩相一般在其边缘以楔状形态与建造相接触(图12-9C、D)。薄的珊瑚柱状岩层也在局部出现,但是优先发育在一些生物丘的顶部。在珊瑚粘结岩中,基质通常由细粒、分选差的颗粒质灰泥灰岩(CB1)至泥质主导的灰泥质颗粒灰岩(CB2)构成,含有不同比例的珊瑚碎片,指示了主要为低能沉积环境,但是偶尔发育事件沉积。较小的底栖有孔虫普遍出现,浮游有孔虫也有出现。不依赖光能的滤食型生物也很普遍。在一些露头,*Acanthochaetetes*,一种海绵(demosponge)也有出现,表明是光线减少的隐秘环境或者礁前环境(Reitner和Engeser,1987)。在与光能有关的有机体中,红藻大量出现,层状和分枝形式的绿藻缺失。LBF,如 *Heterostegina*,*Operculina* 以及一些铁饼状的货币虫(*Nummulites*)出现在基质中。这些成分特征指示了珊瑚圆顶岩和柱状岩生长结构的沉积深度比盘状岩浅,虽然也发育在中光度区。

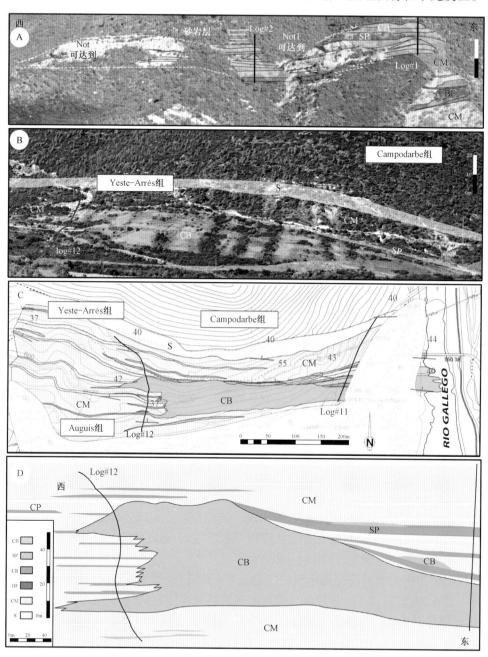

图12-9 建造结构(相代码如表12-1所示)

(A)在 Rasal 附近大型、复合建造的结构(位置见图12-3)。透视畸变已经被部校正,仅有可达到的地区编入图内(比例尺是近似的,并不是常量)。值得注意的是,横向不均匀性阻碍了横向剖面的对比。(B)Yeste 附近一个珊瑚建造的全貌(位置见图12-3;比例尺是近似的,并不是常量)。(C)(B)图中出露的珊瑚建造详细的相图;岩相边界由 GPS 测量成图。(D)从以上地质图件中重建的珊瑚建造的 2D 几何结构(比例尺2:1)。值得注意的是,侧翼的不对称性;含骨骼的灰泥质颗粒灰岩在东边一侧更发育,在西边侧翼,珊瑚砾屑灰岩与砂岩层共同出现。这个建造发育在黏土中(CM 岩相),砂岩层局部超覆在侧翼上

12.5.4 沉积模式

这些碳酸盐岩相在 Arguis 组蓝色黏土—泥灰岩中的空间分布特征,和它们的结构特征和骨架成分,表明了一种沉积模式(图 12-10 和图 12-11)。Arguis 组与前三角洲黏土一样,被三角洲前缘砂体覆盖(Yeste-Arrés 组)。在这个三角洲复合体中,碳酸盐岩工厂在中光度区至无光区,浪基面以下黏土主导的沉积环境中生产积累沉积物,这种沉积环境发育在活跃的三角洲朵体或与其有一定距离的位置与分流河道之间。而在浅水透光区生产碳酸盐沉积物的绿藻或者与海草相关的有孔虫在珊瑚建造中并没有发现。

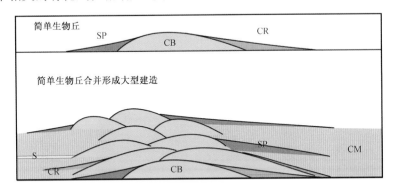

图 12-10 草图展示的是珊瑚建造的起源

在中光度区,珊瑚同红藻一起,形成了小型的生物丘(见图 11-5 和图 11-9A)。偶尔发生的高水动力脉冲使建造上骨骼碎片脱落下来并产生砾屑灰岩和侧翼的灰泥质颗粒岩楔状体。大型珊瑚建造见图 12-3 和图 12-9,由单个生物丘的合并和垂向堆积形成。CB:珊瑚粘结岩;CR:珊瑚砾屑灰岩;SP:骨骼灰泥质颗粒灰岩;CM:黏土和泥灰岩;S:砂岩

在中光度区,chlorocline(Liebau,1984 定名)发育深度以下,珊瑚生存,并与红藻共生(见图 12-8)。在上部,珊瑚形成小型珊瑚丘,联合形成建造(图 12-10)。在这种环境中,透光的高水动力条件为重新形成骨骼沉积物以及珊瑚砾屑碳酸盐岩和骨架灰泥质颗粒灰岩提供了能量(见图12-9)。光照随深度减少,盘状珊瑚发育,与层状红藻和苔藓虫共生。这个碳酸盐工厂并不如上部工厂高效,形成了生物层而不是生物丘。这个区域缺失重新形成的岩相指示了生产/沉积作用发生在安静的环境。更深的是无光区,碳酸盐生产只能依赖于异养生物。定位和露头质量无法描述中光度、微光区和无光区碳酸盐工厂的连续性/不连续性(图 12-11)。

相变化与深度和水体透明度的变化有关,水体透明度与三角洲朵体的迁移或气候周期有联系。气候周期中,湿润的环境导致了更高的分水岭和海岸植被覆盖,从而减少了沉积物的注入。据 Yentsch 等(2002)估算,藻珊瑚的生长最少需要 1% 的表面光。比如,如果平均水体透明度为典型的三角洲水体,用于光合作用活跃辐射的消光系数(ECL)为 0.3~0.4,那么无光组合的深度在 15m 以下(图 12-8C)。如果平均水体透明度增加,那么 ECL 减少到 0.25,一些盘状珊瑚就会生长在近 20m 的深处。如果平均水体透明度增加至近海环境(ECL 为 0.18),盘状珊瑚就可以生长在约 25m 深处。

黏土/泥灰岩岩相总是出现在珊瑚建造的周围和内部,作为整个三角洲环境的沉积背景(图 12-9)。在 Arguis 组内部,砂岩层局部以薄夹层形式发育在黏土—泥灰岩中,局部发育交错层理(对称波痕和少量丘状交错层理)。在这些砂岩层中发育平行层理,正常递变层理,以及局部发育波状层理,说明发育风暴引起的密度流沉积或者由洪水引起的事件沉积(Mutti 等,

2003)。这些事件,无论怎样都不能使活跃的珊瑚丘窒息而死。然而,在建造上面几十米,砂岩层(Yeste-Arrés 组至 Belsué-Atarés 组)增多、增厚,这是三角洲进积和风暴活动的结果,表明珊瑚丘在风暴浪基面以下,但是偶尔被沉积事件搅动。

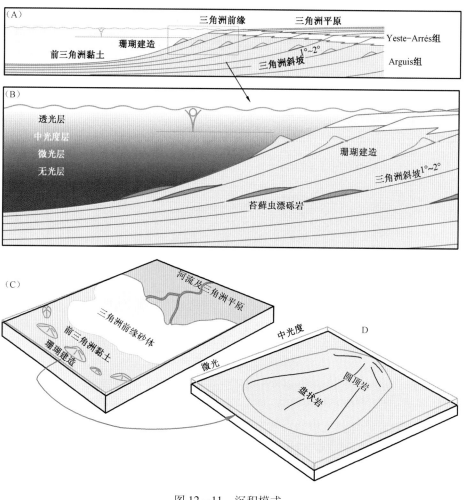

图 12-11 沉积模式

前三角洲环境中,发育在波浪作用面以下的珊瑚建造(见图 12-3)。珊瑚圆顶岩生长结构发育在中光度区上部。在这个环境中,偶尔发生的强水动力袭击建造,为形成珊瑚砾屑灰岩和骨骼灰泥质颗粒灰岩楔状体提供了能量。珊瑚盘状岩发育在中光度区下部,苔藓虫漂砾岩发育在无光区

12.6 讨论

除了肯定均变论方法的作用,很多作者都反对均变论在碳酸盐岩沉积学中的滥用。同样,很多人认为沉积环境、大气、海水化学成分以及它们与碳酸盐有机体之间的关系在显生宙是独特的(Sandberg,1983;Morse 等,1997;Stanley 和 Hardie,1998;Schlager,2005;Wright 和 Burgess,2005;Berner,2006;Pomar 和 Hallock,2008;Kiessling,2009)。

本文的资料展示了谨慎和批判地运用均变论的必要性。此前对始新世珊瑚建造的解释,除了 Bassi(1998),基于"现代"浅水的,光线很好(透光的/微光的)的珊瑚礁范例,其他都运用了"障壁礁—潟湖"以及"补丁礁"模式。为避免倾向解释为均变论,要注意以下几个方面的分析:(1)对珊瑚建造的解剖分析;(2)建造内部相结构编图;(3)建造和建造内部所包含的相之

间的几何关系分析;(4)对骨骼成分及其生态需求的考虑,尤其是自养生物的光能和异养生物的食物需求;(5)岩石结构,尤其是基质结构,对水动力条件的指示作用。

这些分析可以证明,透光的/微光的珊瑚礁模式不能应用于始新世珊瑚建造。相反,Yeste-Rasal 发育在黏土为主的前三角洲环境,在低光能环境,浪基面以下。这些珊瑚不能形成坚固的、抗浪的骨架结构。但是,它们在底泥上居住,形成了"近丛礁",这种底泥或者源自生物丘上的碳酸盐生产,或者源自三角洲低密度流带的黏土沉淀(Riding,2002)。

12.6.1 北部地中海地区其他的始新世例子

12.6.1.1 Igualada-Vic 地区西班牙东北部

巴尔顿阶(Bartonian)至普里阿邦阶(Priabonian)珊瑚建造大量出现在 Igualada-Vic 地区,位于南比利牛斯前陆盆地东部边缘(Busquets 等,1985;Santisteban 和 Taberner,1988;Alvarez 等,1994;Hendry 等,1999;Romero 等,2002)。除了石珊瑚以外,红藻和 LBF 在建造中显著出现,它们发育在浅海硅质碎屑进积体上,早于广泛分布的陆相沉积。

Igualada-Vic 建造与 Yeste-Rasal 建造有一系列相似之处:(1)它们都与三角洲体系共生;(2)它们都具有结节状特征;(3)没有识别出由近距离堆积的珊瑚群落形成的连锁骨架;(4)珊瑚个体或者嵌入黏土或者嵌入成岩较好的粉砂—泥灰质颗粒灰岩基质中;(5)除了珊瑚,铰接的层状红藻、苔藓虫、双壳类、海胆、LBF 和浮游有孔虫是普遍的骨骼成分;(6)缺失绿藻,除了在 Calders 建造,Hendry 等(1999)描述存在 dasycladacean 藻;(7)岩相相似性值得注意,尤其是与 Alvarez 等(1994;表12-3)的描述相比较时。考虑到珊瑚群落的成分,两个地区(Igualada-Vic 和 Yeste-Rasal)沉积底部都发育由 agariciid 珊瑚 *Cyathoseris* 主导的盘状珊瑚组合,而珊瑚粘结岩也有类似的特征,由大量的 *Actinacis* 和 phaceloid 珊瑚组成,比如 *Caulastrea* 和 cf. *Cereiphyllia*。以前对 Igualada-Vic 珊瑚藻建造的解释,除了几何形态不能清晰可见之外,都是基于现代珊瑚礁模式,比如似加勒比海礁(Busquets 等,1985;Santisteban 和 Taberner,1988;Alvarez 等,1994;Hendry 等,1999;Romero 等,2002)。

表 12-3 Igualada-Vic 地区珊瑚建造的相(据 Alvarez 等,1994)与 Yeste-Rasal 地区(本次研究)的对比

Igualada-Vic 地区(Alvarez 等,1994)	Yeste-Rasal 地区(本次研究)
(a)灰泥质的颗粒质灰泥灰岩中的原地层状珊瑚	珊瑚盘状岩(CB 的亚相)
(b)在颗粒质灰泥灰岩基质中的结壳的、分枝的红藻和珊瑚	珊瑚粘结岩(CB)
(c)骨骼灰泥质颗粒灰岩—砾屑灰岩	珊瑚砾屑灰岩
(d)结节状的颗粒质灰泥灰岩,含有原地珊瑚	珊瑚粘结岩(CB)
(e)混合的碳酸盐—硅质碎屑灰泥质颗粒灰岩,含有红藻结壳	骨骼灰泥质颗粒灰岩

然而,本文提出南比利牛斯前陆盆地 Igualada-Vic 地区的珊瑚藻建造发育与 Aínsa-Jaca 背驮式盆地 Yeste-Rasal 地区的建造生长环境相似。两个地区建造的相似之处为:(1)生物组合相似;(2)建造内部岩石结构指示了低水动力环境;(3)独特的是,红藻出现在原地自养群落中(出现在 Calders 建造绿藻碎片指示异地沉积),表明在两个地区建造都生长在 chlorocline 属和 rhodocline 属之间的中光度区(Liebau,1984);(4)在两个地区,建造都和三角洲体系共生。

12.6.1.2　意大利北部 Nago 灰岩

巴尔顿阶（Bartonian）至普里阿邦阶（Priabonian）意大利北部的 Nago 灰岩，由泥灰质灰岩和红藻/珊瑚灰岩组成一个向上变浅的序列（Luciani 等,1988;Luciani,1989）。在这套灰岩中，Bosellini(1998)识别出两大类主要的岩相:（1）微晶灰岩结构为主,生物体破碎程度较低;（2）丘状块状珊瑚层,侧翼发育骨架灰泥质颗粒灰岩—颗粒岩。

在微晶灰岩结构为主的岩相中,Bosellini(1998)在向上变浅的层序底部区分出泥灰岩至结节状含泥灰的颗粒质灰泥灰岩,随后向上是颗粒质灰泥灰岩/灰泥质颗粒灰岩岩层。结节状含泥灰的颗粒质灰泥灰岩包含红藻石和层状红藻,小型的有孔虫、LBF,以及少量的苔藓虫、海胆、小型双壳类、牡蛎、龙介虫和一些介形类和浮游有孔虫。局部出现原生盘状珊瑚。颗粒质灰泥灰岩/灰泥质颗粒灰岩岩层,局部呈丘状,富含红藻石和层状红藻、LBF 和盘状珊瑚、少量的苔藓虫、龙介虫、扇贝和海胆碎片。

在沉积序列顶部,块状岩层含有合并的珊瑚丘,被 Riding(2002)定义为"丛礁结构"。珊瑚为沉积物基质支撑;近距离的骨骼并没有相连,而是被沉积物基质包围,没有孔洞。基质由灰泥质颗粒灰岩组成,局部出现珊瑚漂砾岩,富含大量珊瑚碎片、红藻（以 mastophoroid 为主）、LBF、苔藓虫、海胆和红藻石。所处位置最深的珊瑚相特征为珊瑚多样性低,珊瑚覆盖率低,发育盘状珊瑚形态类型（以 *Cyathoseris* 属珊瑚为主）。珊瑚多样性最高的情况出现在生物丘的核心,在生物丘顶部珊瑚多样性降低,出现结壳形态类型以及珊瑚角砾岩。侧翼相由粗粒的、生物扰动的灰泥质颗粒灰岩/颗粒质灰泥灰岩组成,局部发育砾屑灰岩,含有 LBF、红藻石和红藻碎屑,以及少量的软体动物、海胆、苔藓虫和珊瑚碎片（Bosellini,1998）。

Bassi(1998)主要研究红藻和 LBF 的成分,在 Nago 灰岩中识别出五个岩相和生物相类型:(1) 红藻石铺石状构造;(2) 红藻石丘状颗粒质灰泥灰岩/灰泥质颗粒灰岩,含有层状珊瑚;(3) 层状和分枝状红藻砾屑灰岩;(4) 珊瑚—藻类粘结岩;(5) 藻类/*Discocyclina* 灰泥质颗粒灰岩。这些岩相并非与 Bosellini(1998)描述的有所不同:"珊瑚—藻类粘结岩"岩相在侧翼被"藻类结壳—分枝状砾屑灰岩"包围,而"红藻石铺石状构造"、"红藻石丘状颗粒质灰泥灰岩/灰泥质颗粒灰岩"和"藻类结壳—分枝状砾屑灰岩"相当于 Bosellini(1998)描述的微晶结构为主的岩相。Bosellini(1998)解释了 Nago 灰岩作为礁体系的产物,以补丁礁为主要特征,具有古水深的相分带性,这种分带性中,礁前到浅水陆架边缘礁相普遍与珊瑚成分的变化和形态类型相关。然而,Bassi(1998)预见了可以用碳酸盐斜坡进积过程来解释珊瑚藻相的分布;在斜坡上,Mastophoroideae 主导浅水环境,而 Melobesioideae 和 Sporolithaceae 主导深水相,珊瑚—藻建造发育在斜坡中部环境。

12.6.1.3　德国 Eisenrichterstein 碳酸盐斜坡

普里阿邦阶（Priabonian）含有珊瑚的 Eisenrichterstein 灰岩复合体发育在 Bavaria 南部，Reichenhall 附近,那里也发育珊瑚建造,缺少珊瑚骨架结构（Darga,1990）。在这个下普里阿邦阶碳酸盐斜坡的例子中,滨岸砾岩和含砾砂岩发育在沉积序列底部,被含有大量较小底栖有孔虫和软体动物的细粒泥质砂屑灰岩覆盖。在这些砂屑灰岩中,陆源植物残骸与 LBF、单个珊瑚和几个群落一同出现。保存下来的潜穴构造和微小的铸模及居住在海草中的双壳类和有孔虫一同出现,证明了当时海草大面积覆盖。上覆"碎屑状的礁灰岩",底部由珊瑚—有孔虫砾屑灰岩构成,这些砾屑灰岩含有大量红藻（*Peyssonneliaceae* 和 *Corallinaceae*）,基质以微晶为主,包含小的"礁灰岩"碎片。礁灰岩由非常密集堆积的球状和分枝状珊瑚组成,很少有骨架建造,生长在低水动力环境,穿插有高能量事件沉积。含有孔虫灰泥质颗粒灰岩/颗粒岩、*Discocyclina* 颗粒质灰泥灰岩和泥质为主的红藻石夹层,含珊瑚碎片的颗粒质灰泥灰岩也和"礁灰岩"建造一同出现。珊瑚建造和相关的相向盆地方向相互交叉,发育含珊瑚碎屑、小型底栖和浮游有孔虫以及一些扁平的 *Discocyclina* 和 *Heterostegina* 的泥灰岩。

Darga(1990)设想在一个低角度碳酸盐斜坡沉积环境发育的珊瑚建造中,几乎所有的相类型(除了近岸)都形成于低能环境间歇穿插短期事件沉积。

12.6.1.4 比较分析

对 Yeste-Rasal 地区和西班牙 Igualada-Vic 地区的珊瑚建造,以及意大利的 Nago 灰岩和德国的 Eisenrichterstein 灰岩比较分析表明,它们在岩相、成分和相的空间关系上有很多相似之处。珊瑚丘不连续或者合并,侧翼发育灰泥质颗粒岩和砾屑岩—漂砾岩,含有大量的红藻,基质富含泥质至以泥质为主。在水深解释中,光能自养生物中占主导的红藻和少量绿藻说明珊瑚藻建造发育在中光度区到微光区,这与在基质中出现大量的泥指示低水动力环境相一致。在 Eisenrichterstein 灰岩中,光能自养生物并没有被详细描述。但是,岩石结构也表明建造生长在低能环境,含有 *Discocyclina*,说明属于中光度环境,而透光的海草主导的地区分布在较浅的近岸地区。

然而,它们也存在差异:Nago 灰岩和 Eisenrichterstein 灰岩为碳酸盐岩主导的台地,有少量的硅质碎屑注入(在较深的岩相出现泥灰质灰岩),而 Yeste-Rasal 和 Igualada-Vic 建造都发育在前三角洲黏土中。硅质碎屑对比碳酸盐环境其他的差异是纬度、水动力和珊瑚密度差异。然而,Yeste-Rasal 建造发育在活跃的进积三角洲,Nago 建造出现在 Lessini 陆架较狭窄的边缘(Bosellini,1989)。这个陆架较狭窄的边缘可以解释为较高的水动力条件,引起在建造上部较高的珊瑚密度。Yeste-Rasal 相比 Nago 具有更强烈的生物侵蚀作用和结壳作用,含有大量苔藓虫,而 LBF 含量更少,这些也可以认为是较高营养有效性的结果。

除了硅质碎屑注入和随之营养运输在这些实例中的差异以外,几乎在 Nago 和 Eisenrichterstein 分析的"纯"碳酸盐台地以及始新世珊瑚建造都明显发育在低能中光度环境。透光区的碳酸盐岩沉积物(比如,海草附生有机体)仅仅出现在 Eisenrichterstein 灰岩中的"碎屑礁灰岩"和滨岸砾岩、砂岩之间的海侵序列(Darga,1990)。在 Nago 灰岩中,虽然罕见绿藻(Bassi,1998),但是没有发现浅水透光岩相。在前三角洲黏土中出现的建造,缺失绿藻和海草组分,除了几个在 Calders 建造发现的 Dasycladacean 藻碎片(Hendry 等,1999);这说明,与现代沉积一样,透光区的碳酸盐工厂并不能发育在有大量陆源供给的浅水区域。

12.6.2 陆源碎屑为主的沉积体系中的珊瑚建造

在 Mount(1984)的重要论文中分析了除硅质碎屑对分泌碳酸盐的有机体的抑制作用以外,也分析了解释"混合"沉积组分光谱的理论体系。作者强调混合沉积物异化成分由珊瑚藻和大部分为有孔虫的沉积组合构成。Mount(1984)解释了有孔虫组分的优势地位是浊流事件、基质的不稳定性以及滤食生物组合的障碍作用,加之陆源物质注入的结果。Woolfe 和 Larcombe(1999)认为沉积物在物理上通过三个主要方式影响珊瑚的生长和存活:(1)沉积物堆积作用通过在激发沉积物注入机制上消耗过多能量从而减少珊瑚发育;(2)水体中高含量的悬浮沉积物可以减少光照,这样光合作用就受到限制(透光区减少)或者停止;(3)珊瑚的软体组织被沉积物颗粒磨蚀或者影响而被损坏。

当通观浊流沉积化石的时候,Sanders 和 Baron-Szabo(2005)发现石珊瑚(Scleractinian)占主体的浊流沉积很普遍,这说明在某一临界值,一些生物群落与陆源浊流沉积共生,以增加异养生物来适应环境。分析来自 Yeste-Rasal 和 Igualada-Vic 始新世的实例证实了 Sanders 和 Baron-Szabo(2005)的结论,他们认为出现在陆源沉积体系中的珊瑚建造很常见:珊瑚岩相是孤立的,并且没有形成礁复合体、丛至段的结构(Riding,2002)占主体,罕见或不发育骨架孔充填海相胶结物,没有碳酸盐斜坡,因而珊瑚群落在沉积物注入过程中恢复成块状和盘状生长形式。

在相对混浊的水域中,现代珊瑚补丁礁的珊瑚生长被限制发育在相对较浅的位置(Titlyanov 和 Latypov,1991;Wilson 和 Lokier,2002;Hallock,2005;Wilson,2005,2008)。始新世 Aínsa-Jaca 盆地的浑浊度与高密度携带黏土的流体相关,而这一过程在礁生长阶段中很短暂。在 Arguis 组进积的前三角洲黏土中,珊瑚建造出现在单个的丛里(见图 12-3),说明或者是在三角洲朵体转换阶段,或者在增加陆源输送与在碳酸盐岩生产过程中三角洲进积之间的交替周期,很有可能的情况是,潮湿的陆源环境植被覆盖很好,红树林在岸线上占主体,它们捕获了大部分陆源沉积物。

12.6.3　现代中光度珊瑚可以和始新世珊瑚相对比吗?

依赖光能的虫黄藻珊瑚(z-corals)、无共生藻石珊瑚(azooxanthellate scleractinian)、大型藻类和海绵主导现代"中光度珊瑚生态系统"。这样的珊瑚群落最深可达透光区的一半深度,发育在非常清澈的从 40m 向下至 150m 热带和亚热带海水中(见 Kahng 等,2010,以及相关参考文献)。科学潜水和水下机器人技术点燃了人们对这些深水礁的兴趣。本文描述了这个中光度区的环境(图 12-12)。这个地区的虫黄藻珊瑚展现了独特的光照适应性策略来加强在低光能环境中的光合作用(Mass 等,2007;Chan 等,2009;Lesser 等,2009):很多,并不是全部,虫黄藻珊瑚的存在对于这些深度来说是特有的;它们发育盘状形态去捕获尽可能多的光能;它们形成圆锥形的突起作为"光能圈闭";一些珊瑚产生特殊的反光颜色或者高叶绿素含量的虫黄藻。

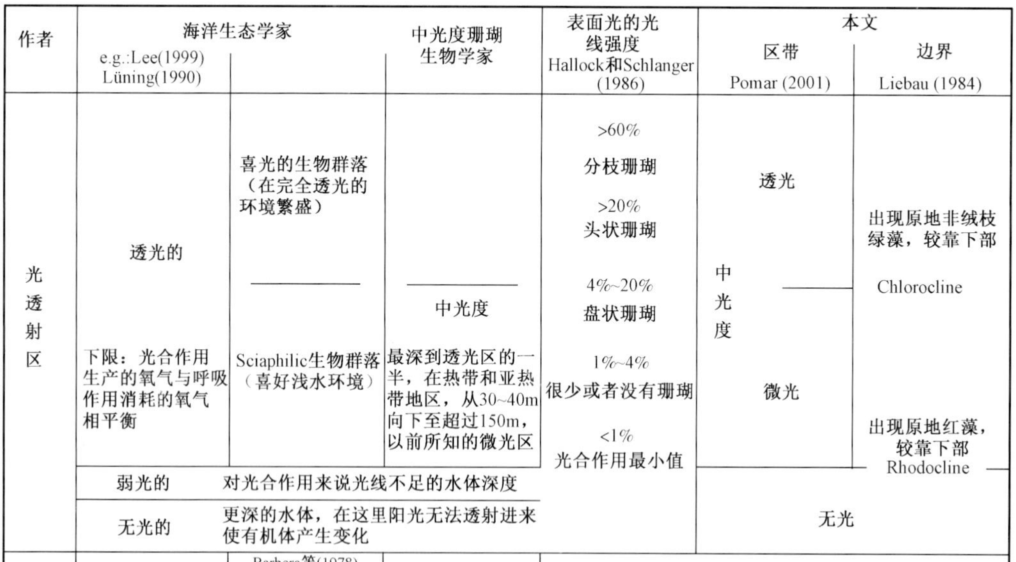

作者	海洋生态学家 e.g.:Lee(1999) Lüning(1990)	中光度珊瑚 生物学家	表面光的光线强度 Hallock和Schlanger (1986)	本文	
				区带 Pomar (2001)	边界 Liebau (1984)
光透射区	透光的	喜光的生物群落 (在完全透光的环境繁盛)	>60% 分枝珊瑚 >20% 头状珊瑚	透光	
		中光度		中光度	出现原地非绒枝绿藻,较靠下部 Chlorocline
	下限:光合作用生产的氧气与呼吸作用消耗的氧气相平衡	Sciaphilic生物群落 (喜好浅水环境)	4%~20% 盘状珊瑚		
		最深到透光区的一半,在热带和亚热带地区,从30~40m向下至超过150m,以前所知的微光区	1%~4% 很少或者没有珊瑚	微光	
			<1% 光合作用最小值		出现原地红藻,较靠下部 Rhodocline
	弱光的	对光合作用来说光线不足的水体深度			
	无光的	更深的水体,在这里阳光无法透射进来使有机体产生变化		无光	
使用在	Schlager (1992)	Barbera等(1978) Simone和Carannante (1985) Carannante等(1996)	http://www.mesophotic.org/	Flüger (2004)	

图 12-12　从光透射角度定义的不同水深区的对比

光线强度沿深度梯度减小。从"透光"到"微光"的过渡代表了要求充足太阳能的碳酸盐生产者(比如,丘状的虫黄藻珊瑚和大量的钙质绿藻)被珊瑚藻主导的低光生物代替的深度。当骨骼成分解释了更细微的水深/光穿透性差异的时候(比如,最初的盘状形态珊瑚),一个中间的"中光度"区也被定义出来。这个中光度区的深度界限随水体透明度和中光度生物的适应性而变化另外,珊瑚可以随着深度增加光能减少,从开始的自养型转化为更依赖异养型(Muscatine 等,1989;Mass 等,2007;Alamaru 等,2009;Chan 等,2009;Lesser 等,2009)。一些作者也指出一些珊瑚利用异养方式作为对高浑浊度环境的响应(Anthony 和 Fabricius,2000;Larcombe 等,2001)。

在 Yeste - Rasal 地区珊瑚发育带,甚至于在 Igualada - Vic 地区、Nago、以及 Eisenrichterstein 地区,研究表明这些始新世珊瑚是依赖光能的。沉积物的成分、结构、地层环境,进一步表明了这些珊瑚在中光度区局限的水动力条件里繁盛的情况。如现代中光度区珊瑚,对异养方式的依赖程度可以貌似有理的被推断出来。在低光能环境,这些始新世珊瑚可以形成建造。海底的抬升增加了食悬浮物质的生物的食物供给,同时提供了产生紊流的条件(图 12 - 13A),以及增加了水流携带超微型浮游生物和浮游植物的作用效率(Bilger 和 Atkinson,1992;Hearn 等,2001;Ribes 等,2003;Pomar 和 Hallock,2008)。这种机制已经在现代亚得里亚海得以记录,食悬浮物质的蛇尾类动物和扇贝利用局部地貌高点(比如多毛环节蠕虫和大型半内底栖双壳类)来获得海底抬升所增加的捕食效率(McKinney 等,2007)。此外,高营养梯度和反光颜色的最大富集,代表了叶绿素富集的最大值(比如在浮游植物中),叶绿素和叶绿素降解产生的脱镁叶绿素,通常在密度跃层的底部同处一地(图 12 - 13B),一般出现在透光区的较低部位(Steele 和 Yentsch,1960;Anderson,1969;Hallock 等,1991),在那里,通过上升流或者内波,富养(更深)水体进入营养耗尽的透光区(Hallock 和 Pomar,2008)。

12.6.4 引起始新世珊瑚成为中光度型的"地域性/区域性"因素与一般因素

一个关键的问题是,珊瑚在中光度区明显富集是否与"地域性/区域性'因素相关或者这种环境倾向是否与更普遍的约束性相关。珊瑚—虫黄藻共生现象提供了石珊瑚形成礁的潜能。基于骨骼稳定同位素和有机基质分析,Stanley 和 Swart(1995),Muscatine 等(2005)和 Stanley(2006)提出珊瑚—虫黄藻共生体在晚三叠世开始形成。Palliani 和 Riding(2000)也指出晚三叠世珊瑚和虫黄藻协同进化,基于沟鞭藻包囊族的 *Suessiacea* 的多样性增加了这种巧合,*Suessiacea* 与虫黄藻和造礁的石珊瑚密切相关。珊瑚虫和沟鞭藻的协同进化,对形成共生组合是必要的,可能出现数次,可能与中生代—新生代虫黄藻珊瑚的盛衰相关(Veron,1995)。Veron(1995)指出,进化灵活性会减少或者抛弃它们对共生体的依赖,这样就使得至少一些珊瑚具有生态生理上的适应性,使它们在多数沉积事件中存活下来。这种共生体组合要求如下的进化:(1)为寄主内的藻类提供无机碳酸盐和虫黄藻光合作用所需其他营养的机制;(2)生物化学防晒物质(一般由共生体生产出来并转移到寄主);(3)抗氧化剂抵御光合作用产生过量的氧气;(4)特殊的机制(在后生动物中不常见)可以吸收和隔绝无机氮气,以便于在贫养水域中存活(Furla 等,2005;Stanley,2006)。

被现代珊瑚所包含的共生藻虫黄藻的祖先可以追溯到始新世早期,随后在始新世以及中中新世开始分化为两支,与全球变冷时期相吻合(Pochon 等,2006)。Pomar 和 Hallock(2007,2008)指出珊瑚建造向上变化到透光的、浅水环境,这种变化可能与现存的共生藻虫黄藻在后—中中新世多样化作用相关。直到晚中新世,在地中海省的浅水区珊瑚并没有形成大型抗浪构造。但是它们同红藻一起,在中光度环境中形成了孤立的建造。这些珊瑚向透光区的扩张可能是温度梯度增强的结果,从而大大地增加了钙化速率。温度梯度增强导致了"珊瑚温度窗"的水深和纬度的减少,同时造成全球大气 CO_2 含量的降低(Pomar 和 Hallock,2007,2008)。近海洋表面高光照强度支撑了较高的光合作用速率,这样不仅促进了过度钙化,而且增加了大面积发生光氧化的风险,导致在安静而炎热的天气发生大量漂白作用(Shick 等,

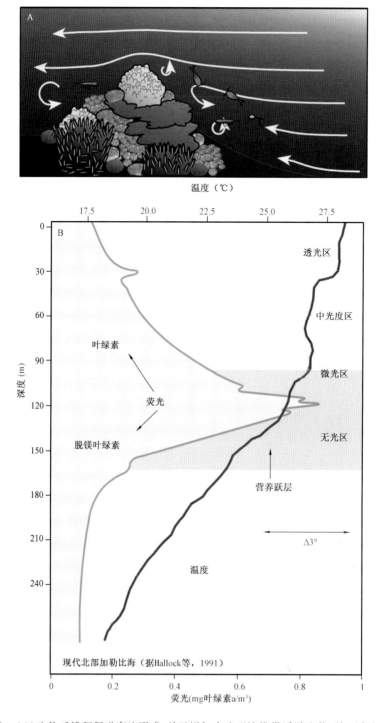

图 12-13 （A）生物丘堆积促进紊流形成，并且增加水流碰撞携带浮游生物，从而有利于食悬浮物质生物的生长。（B）荧光最大值，代表了叶绿素和脱镁叶绿素含量的峰值（叶绿素的降解），典型的是出现在营养跃层，这也与密度跃层相吻合。营养跃层一般出现在透光层的靠下部位，那里有富含营养的（较深）水体，可以通过上升流或者内波向上进行混合，在那里可以获得光能，并不是营养，从而限制了初级生产力

1996；Kleypas 等，1999，2001）。此外，因为较高的 CO_2 浓度提升了光氧化的风险（Wooldridge，2009），所以虫黄藻珊瑚在渐新世—中新世向高光能环境的扩张可能与至新近纪长期以来大气 CO_2 浓度的降低相一致，这种降低有利于文石的过度钙化（见 Pomar 和 Hallock，2008，以及相关文献）。这种向透光区的扩张也可能与获得防晒物质和抗氧化剂机制增强相关。

关键的问题仍然得不到合理的解释，有待于我们进一步开展研究。始新世珊瑚和早期共生藻虫黄藻的分支，都被限制生长在中光度区？出现在这两个陆源物质和碳酸盐主导的沉积环境的珊瑚建造和相关生物之间的相似性，说明这是普遍特征，而不是局部受浅水透光区混浊度或者高硅质碎屑注入的抑制影响。研究结果表明珊瑚被限制生长在中光度区（见图 12-8），这与 Potts 和 Jacobs（2000）的研究结果一致，他们提出石珊瑚—虫黄藻共生体的"祖先"出现在混浊的环境，并且指出珊瑚适应多变环境的生理耐性更大，这种更大的适应性使得珊瑚种群在海洋环境中延续下来。

12.7 结论

结合正色摄影上的相图综合分析、建造相结构（生物相、岩相和界面）反映在倾斜的光镶嵌幕以及剖面记录上详细编图，连同骨骼成分的生态要求以及从岩石结构推断水动力条件等分析，已经成为成功建立理想沉积模式的技术手段。在 Aínsa - Jaca 盆地，在低光能、浪基面以下的以泥质为主的三角洲沉积环境，珊瑚可以形成生物层和生物丘。建造水深位置受生物组合种类的约束，尤其是建造内依赖光能的群落和岩相。对均变论的批判性应用，通过一个过程—结果分析，充分验证了虫黄藻珊瑚在中光度环境的繁盛可以成为常态，而不是仅作为异常情况存在于温室海洋的假设，在向冰室海洋转换阶段亦如此。本次研究结果为现代热带、高光能、浅水珊瑚礁可以适应地球变冷的理论提供了有力的支持。

参 考 文 献

Alamaru, A., Loya, Y., Brokovich, E., Yamd, R. and Shemesh, A. (2009) Carbon and nitrogen utilization in two species of Red Sea corals along a depth gradient: insights from stable isotope analysis of total organic material and lipids. Geochim. Cosmochim. Acta, 73, 5333 – 5342.

Allen, J. R. L. (1964) The Nigerian continental margin: bottom sediments, submarine morphology and geological evolution. Mar. Geol., 1, 284 – 332.

Alvarez, G., Busquets, P., Taberner, C. and Urquiola, M. M. (1994) Facies architecture and coral distribution in a mid Eocene reef tract, South Pyrenean Foreland Basin (NE Spain). Cour. Forsch. - Inst. Senckenberg, 172, 249 – 259.

Alvarez, G., Bosence, D., Busquets, P., Darrell, J. G., Franque`s, J., Gili, E., Pisera, A., Reguant, S., Rosen, B. R., Salas, R., Serra - Kiel, J., Skelton, P. W., Taberner, C., Travé, A. and Valdeperas, F. X. (1995) Bioconstructions of the EoceneSouth Pyrenean Foreland Basin (Vic and Igualada Areas) and of the Upper Cretaceous South Central Pyrenees(Tremp area). VII Int. ymp. on Fossil Cnidaria and Porifera(field trip c), Madrid, 68 pp.

Anderson, G. C. (1969) Subsurface chlorophyll maximumin the Northeast Pacific Ocean. Limnol. Oceanogr., 14, 386 – 391. Anthony, K. R. and Fabricius, K. E. (2000) Shifting roles of heterotrophy and autotrophy in coral energetics under varying turbidity. J. Exp. Mar. Biol. Ecol., 252, 221 – 253.

Baker, A. C., Starger, C. J., McClanahan, T. R. and Glynn, P. W. (2004) Corals' adaptive response to climate change. Nature, 430, 741.

Barbera, C., Simone, L. and Carannante, G. (1978) Depositi circalittorali di piattaforma aperta nel Miocene Campano, Analisi sedimentologica e paleoecologica. Boll. Soc. Geol. It., 97, 821 – 834.

Bassi, D. (1998) Coralline algal facies and their palaeoenvironmentsin the Late Eocene of Northern Italy (Calcare di Nago, Trento). Facies, 39, 179–202.

Bates, R. L. and Jackson, J. A. (1987) Glossary of Geology. American Geological Institute, Alexandria, 788 pp.

Berner, R. A. (2006) GEOCARBSULF: a combined model forPhanerozoic atmospheric O_2 and CO_2. Geochim. Cosmochim. Acta, 70, 5653–5664.

Bilger, R. W. and Atkinson, M. J. (1992) Anomalous masstransfer of phosphate on coral reef flats. Limnol. Oceanogr. ,37, 261–272.

Bosellini, A. (1989) Dynamics of Tethyan carbonate platforms. In: Controls on Carbonate Platform and Basin Development (Eds P. D. Wilson, J. L. Sarg and J. F. Read). SEPM Spec. Publ. ,44, 3–13.

Bosellini, F. R. (1998) Diversity, composition and structure of Late Eocene shelf – edge coral associations (Nago Limestone, orthern Italy). Facies, 39, 203–226.

Braga, J. C., Martin, J. M. and Alcala, B. (1990) Coral reefs in coarse – terrigenous sedimentary environments (Upper Tortonian, Granada Basin, southern Spain). Sed. Geol. , 66, 135–150.

Busquets, P., Ortí, F., Pueyo, J. J., Riba, O., Rosel – Ortiz, L. ,Saez, A., R., S. and Taberner, C. (1985) Evaporite deposition and diagenesis in the Saline (potash) Catalan Basin,Upper Eocene. In: IAS, 6th European Regional Meeting, Excursion Guidebook, Excursion No. 1 (Eds M. D. Mila and J. Rosell), pp. 13–59.

Institut d'Estudis Ilerdencs, Lleida. Callot, P., Odonne, F., Debroas, E. J., Maillard, A., Dhont, D. ,Basile, C. and Hoareau, G. (2009) Three – dimensional architecture of submarine slide surfaces and associated soft – sediment deformation in the Lutetian Sobrarbe deltaic complex (Ainsa, Spanish Pyrenees). Sedimentology, 56, 1226–1249.

Canudo, J. I., Molina, E., Riveline, J., Serra – Kiel, J. and Sucunza, M. (1988) Lesévé nements biostratigraphiques de la zone prépyrénéenne d'Aragon (Espagne), de l'Eocène moyena l'Oligocène infèrieur. Rev. Micropaleontol. , 31, 15–29.

Canudo, J. I., Malagon, J., Melendez, A., Millán, H., Molina, E. and Navarro, J. J. (1991) Las secuencias deposicionales del Eoceno medio y superior de las Sierras externas (Prepirineo meridional aragonés). Geogaceta, 9, 81–84.

Carannante, G., Severi, C. and Simone, L. (1996) Off – shelf transport along foramol (temperate – type) open shelf margins: an example from the Miocene of the Central – southern Apennines (Italy). Bull. Soc. Géol. Fr. , 169, 277–288.

Castelltort, S., Guillocheau, F., Robin, C., Rouby, D., Nalpas,T., Lafont, F. and Eschard, R. (2003) Fold control on the stratigraphic record: a quantified sequence stratigraphic study of the Pico del Aguila anticline in the south – western Pyrenees (Spain). Basin Res. , 15, 527–551.

Cavagnetto, C. and Anadón, P. (1996) Preliminary palynological data on floristic and climatic changes during the Middle Eocene – Early Oligocene of the eastern Ebro Basin, northeast Spain. Rev. Palaeobot. Palynol. , 92, 281–305.

Chan, Y. L., Pochon, X., Fisher, M. A., Wagner, D. T G. ,Concepcion, G. T., Kahng, S. E., Toonen, R. J. and Gates, R. D. (2009) Generalist dinoflagellate endosymbionts and hostgenotype diversity detected from mesophotic (67–100 m depths) coral Leptoseris. BMC Ecology, 9, 21. doi: 10. 1186/1472–6785–9–21.

Chen, C. A., Wang, J. T., Fang, L. S. and Yang, Y. W. (2005)Fluctuating algal symbiont communities in Acropora palifera (Scleractinia: Acroporidae) from Taiwan. Mar. Ecol. Prog. Ser. , 295, 113–121.

Chun – ting, X., Beets, D. J., Guang – xue, L. and Peersman, M. (1995) Sedimentary evolution of modern Huanghe River delta lobe. Chinese J. Oceanol. Limnol. , 13, 325–331.

Correggiari, A., Cattaneo, A. and Trincardi, F. (2005) Themodern Po Delta system: lobe switching and asymmetric prodelta growth. Mar. Geol. , 222–223, 49–74.

Darga, R. (1990) The Eisenrichterstein near Hallthurm,Bavaria: an upper Eocene carbonate ramp (Northern Calcareous Alps). Facies, 23, 17–36.

Darga, R. (1992) Geologie, Paläontologie und Palökologie dersüdostbayerischen unter – priabonien (Ober – Eozän) Riffkalkvorkommendes Eisenrichtersteins bei Hallthurm (Nördliche Kalkalpen) und des Kirchbergs bei Neubeuern (Helvetikum). MünchnerGeowissenschaften. Abh., 23, 1 – 166.

Dreyer, T., Corregidor, J., Arbues, P. and Puigdefàbregas, C. (1999) Architecture of the tectonically influenced Sobrarbe deltaic complex in the Ainsa Basin, northern Spain. Sed. Geol. 127, 127 – 169.

Dunham, R. J. (1962) Classification of carbonate rocks according to their depositional texture. In: Classification of Carbonate Rocks (Ed. W. E. Ham), pp. 108 – 121. American Association of Petroleum Geologists Memoir 1, Tulsa, OK. Eichenseer, H. and Luterbacher, H. (1992) The marine Paleogene of the Tremp Region (NE Spain. Depositional sequences, facies history, biostratigraphy and controlling factors. Facies, 27, 119 – 152.

Embry, A. F. and Klovan, J. E. (1971) A late Devonian reef tract of northeastern Banks Islands. N. W. T. Bull. Can. Petrol. Geol., 19, 730 – 781.

Fautin, D. G. and Buddemeier, R. W. (2004) Adaptive bleaching: a general phenomenon. Hydrobiologia, 530/531, 459 – 467. Brahmaputra: cyclone – dominated sedimentation patterns. Mar. Geol., 149, 133 – 154.

Millán, H., Aurell, M. and Melendez, A. (1994) Synchronous detachment folds and coeval sedimentation in the Prepyrenean external Sierras (Spain): a case study for a tectonic origin of sequences and systems tracts. Sedimentology, 41, 1001 – 1024.

Millán, H., Pueyo, E. L., Aurell, M., Luzón, A., Oliva, B., Martínez, M. B. and Pocoví, A. (2000) Actividad tectónicaregistrada en los depósitos terciarios del frente meridional del Pirineo central. Rev. Sociedad Geoló gica de Espam, 13, 279 – 300.

Morse, J. W., Wang, Q. and Tsio, M. Y. (1997) Influences of temperature and Mg:Ca ratio on $CaCO_3$ precipitates from seawater. Geology, 25, 85 – 87.

Mount, J. F. (1984) Mixing of siliciclastic and carbonate sediments in shallow shelf environments. Geology, 12, 432 – 435.

Muñoz, J. A., McClay, K. and Poblet, J. (1994) Synchronous extension and contraction in frontal thrust sheets of the Spanish Pyrenees. Geology, 22, 921 – 924.

Murillo, V. C., Silva, C. G. and Fernandez, G. B. (2009) Nearshore Sediments and Coastal Evolution of Paraíba do Sul River Delta, Rio de Janeiro, Brazil. J. Coast. Res., 56, 650 – 654.

Muscatine, L., Porter, J. W. and Kaplan, I. R. (1989) Resource partitioning by reef corals as determined from stable isotope composition. Mar. Biology, 100, 185 – 193.

Muscatine, L., Goiran, C., Land, L., Jaubert, J., Cuif, J. P. and Allemand, D. (2005) Stable isotopes ($\delta^{13}C$ and $\delta^{15}N$) of organic matrix from coral skeleton. Proc. Natl. Acad. Sci. USA, 102, 1525 – 1530.

Mutti, E., Remacha, E., Sgavetti, M., Rosell, J., Valloni, R. and Zamorano, M. (1985) Stratigraphy and facies characteristics of the Eocene Hecho Group turbidite systems, south – central Pyrenees. In: IAS, 6th European Regional Meeting, Excursion

Guidebook (Eds M. D. Mila and J. Rosell), pp. 521 – 576, Institut d'Estudis Ilerdencs, Lleida. Mutti, E., Tinterri, R., Benevelli, G., Di Biase, D. and Cavanna, G. (2003) Deltaic, mixed and turbidite sedimentation of ancient foreland basins. Mar. Petrol. Geol., 20, 733 – 755.

Nebelsick, J. H., Rasser, M. W. and Bassi, D. (2005) Facies dynamics in Eocene to Oligocene circumalpine carbonates. Facies, 51, 197 – 216.

Palliani, R. B. and Riding, J. B. (2000) Subdivision of the dinoflagellate cyst Family Suessiaceae and discussion of its evolution. J. Micropaleontol., 19, 133 – 137.

Papazzoni, C. A. and Sirotti, A. (1995) Nummulite biostratigraphy at the Middle/Upper Eocene boundary in the Northern Mediterranean area. Riv. It. Paleont. Stratigr., 101, 63 – 80.

Perrin, C. (2002) Tertiary: the emergence of modern reef ecosystems. In: Phanerozoic Reef Patterns (Eds W. Kiessling and E. Flügel), SEPM Spec. Publ., 72, 587 – 621.

Perry, C. T. (2005) Structure and development of detrital reef deposits in turbid nearshore environments, Inhaca Island, Mozambique. Mar. Geol. , 214, 143 – 161.

Perry, C. T. and Smithers, S. G. (2010) Evidence for the episodic "turn on" and "turn off" of turbid-zone coral reefs during the late Holocene sea-level highstand. Geology, 38,119 – 122.

Poblet, J. and Hardy, S. (1995) Reverse modelling of detachment folds; application to the Pico del Aguila anticline in the south central Pyrenees (Spain). J. Struct. Geol. , 17,1707 – 1724.

Poblet, J. ,Muñoz, J. A. , Travé, A. and Serra-Kiel, J. (1998) Quantifying the kinematics of detachment folds using three dimensionalgeometry: application to the Mediano anticline(Pyrenees, Spain). Geol. Soc. Am. Bull. , 110, 111 – 125.

Pochon, X. , Montoya-Burgos, J. I. , Stadelmann, B. and Pawlowski, J. (2006) Molecular phylogeny, evolutionary rates, and divergence timing of the symbiotic dinoflagellate genus Symbiodinium. Molecul. Phylogen. Evolut. , 38, 20 – 30.

Pomar, L. (2001) Types of carbonate platforms, a genetic approach. Basin Res. , 13, 313 – 334.

Pomar, L. and Hallock, P. (2007) Changes in coral-reef structures through the Miocene in the Mediterranean province: adaptive vs. environmental influence. Geology, 35, 899 – 902.

Pomar, L. and Hallock, P. (2008) Carbonate factories: a conundrum in sedimentary geology. Earth Sci. Rev. , 87,134 – 169.

Postigo Mijarra, J. M. , Barro, E. , Manzaneque, F. G. and Morla, C. (2009) Floristic changes in the Iberian Peninsula and Balearic Islands (south-west Europe) during the Cenozoic. J. Biogeogr. , 36, 2025 – 2043.

Potts, D. C. and Jacobs, I. R. (2000) Evolution of reef-building scleractinian corals' in turbid environments: a aleo-ecological hypothesis. Proceedings 9th International Coral Reef Symposium, 23 – 27 October 2000. Vol. 1, 249 – 254, Bali, Indonesia.

Prothero, D. R. , Ivany, L. C. and Nesbitt, E. A. (2003) From Greenhouse to Icehouse, the Marine Eocene – Oligocene Transition. Columbia University Press, New York, 541 pp.

Reitner, J. and Engeser, T. S. (1987) Skeletal structures and habitats of recent and fossil Acanthochaetes (subclass Tetractinomorpha, Demospongiae, Porifera). Coral Reefs, 6,13 – 18.

Remacha, E. , Oms, O. , Gual, G. , Bolano, F. , Climent, F. ,Fernandez, L. P. , Crumeyrollle, P. , Pettingill, H. , Vicente, J. C. and Suarez, J. (2003) Sand-rich turbidite systems of the Hecho Group from slope to basin plain; facies, stacking patterns, controlling factors and diagnostic features. AAPG

International Conference and Exhibition, Barcelona, Spain, September 21 – 24, Geological Field Trip 12. Remacha, E. , Fernandez, L. P. and Maestro, E. (2005)

The transition between sheet-like lobe and basin-plain turbidites in the Hecho basin (south-central Pyrenees, Spain). J. Sed. Res. , 75, 798 – 819.

Ribes, M. , Coma, R. , Atkinson, M. J. and Kinzie, R. A. (2003)

Particle removal by coral reef communities: picoplankton is a major source of nitrogen. Mar. Ecol. Progr. Series, 257, 13 – 23.

Riding, R. (2002) Structure and composition of organic reef and carbonate mud mounds: concepts and categories. Earth Sci. Rev. , 58, 163 – 231.

Romero, J. , Caus, E. and Rosell, J. (2002) A model for the palaeoenvironmental distribution of larger foraminifera based on late Middle Eocene deposits on the margin of the South Pyrenean basin (NE Spain). Palaeogeogr. Palaeoclimatol. Palaeoecol. , 179, 43 – 56.

Rosen, B. R. , Aillud, G. S. , Bosellini, F. R. , Clack, N. J. , Insalaco, E. , Valldeperas, F. X. and Wilson, M. E. J. (2002) Platy coral assemblages: 200 million years of functional stability in response to the limiting effects of light and turbidity. Proc. 9th Intern. Coral Reef Symp, Bali, vol. 1, 255 – 294.

Rowan, R. (2004) Thermal adaptation in reef coral symbiont. Nature, 430, 742.

Sabatier, F. , Samat, O. , Ullmann, A. and Suanez, S. (2009) Connecting large – scale coastal behaviour with coastal management of the Rhne delta. Geomorphology, 107, 79 – 89.

Sandberg, P. A. (1983) An oscillating trend in Phanerozoic nonskeletal carbonate mineralogy. Nature, 305, 19 – 22.

Sanders, D. and Baron-Szabo, R. C. (2005) Scleractinian assemblages under sediment input: their characteristics and relation to nutrient input concept. Palaeogeogr. Palaeoclimatol. Palaeoecol. , 216, 139 – 181.

Santisteban, C. and Taberner, C. (1988) Sedimentary models of siliciclastic deposits and coral reef interrelations. In: Carbonate-Clastic Transitions (Eds L. J. Doyle and H. H. Roberts). Dev. Sedimentol. , 42, 35 – 76.

Elsevier, Amsterdam. Schlager, W. (1992) Sedimentology and Sequence Stratigraphy of Reefs and Carbonate Platforms. American Association of Petroleum Geologists Continuing Education Course Note Series No. 34, Tulsa, OK, 71 pp.

Schlager, W. (2000) Sedimentation rates and growth potential of tropical, cool water and mud mound carbonate factories. In: Carbonate Platform Systems: Components and Interactions (Eds E. Insalaco, P. W. Skelton and T. J. Palmer). Geol. Soc. London, Spec. Publ. , 178, 217 – 227.

Schlager, W. (2003) Benthic carbonate factories of the Phanerozoic. Int. J. Earth Sci. , 92, 445 – 464.

Schlager, W. (2005) Secular oscillations in the stratigraphic record—an acute debate. Facies, 51, 12 – 16.

Schuster, F. (1996) Paleoecology of Paleocene and Eocene corals from the Kharga and Farafra Oases (Western Desert, Egypt) and the depositional history of the Paleocene Abu Tartur carbonate platform, Kharga Oasis. Tübinger eowissenschaftliche Arbeiten, 31, 96 pp. Tübingen.

Serra-Kiel, J. , Hottinger, L. , Caus, E. , Drobne, K. , Ferarandez, C. , Jauhri, A. K. , Less, G. , Pavlovec, R. , Pignatti, J. , Samso, J. M. , Schaub, H. , Sirel, E. , Strugo, A. , Tambareau, Y. , Tosquella, J. and Zakrevskaya, E. (1998) Larger foraminiferal biostratigraphy of the Tethyan Paleocene and Eocene. Bull. Geol. Soc. France, 169, 281 – 299.

Shick, J. M. , Lesser, M. P. and Jokiel, P. L. (1996) Effects of ultraviolet radiation on corals and other coral reef organisms. Glob. Chang. Biol. , 2, 527 – 545.

Simone, L. and Carannante, G. (1985) Evolution of a carbonate open shelf up to its drowning. Rendiconti Accademia Scienze Fisiche e Matematiche, 53, 1 – 43.

Somoza, L. , Barnolas, A. , Arasa, A. , Maestro, A. , Rees, J. G. and Hernadez – Molina, F. J. (1998) Architectural stacking patterns of the Ebro delta controlled by Holocene highfrequency eustatic fluctuations, delta-lobe switching and subsidence processes. Sed. Geol. , 117, 11 – 32.

Stanley, G. D. , Jr (2006) Photosymbiosis and the evolution of modern coral reefs. Science, 312, 857 – 858.

Stanley, S. M. and Hardie, L. A. (1998) Secular oscillations in the carbonate mineralogy of reef – building and sedimentproducing organisms driven by tectonically forced shifts in seawater chemistry. Palaeogeog. , Palaeoclimat. , Palaeoecol. , 144, 3 – 19.

Stanley, G. D. , Jr and Swart, P. K. (1995) Evolution of the coralzooxanthellae symbiosis during the Triassic: a geochemical approach. Paleobiology, 21, 179 – 199.

Steele, J. H. and Yentsch, C. S. (1960) The vertical distribution of chlorophyll. J. Mar. Biol. Assoc. UK, 39, 217 – 226.

Sternberg, R. W. , Cacchione, D. A. , Paulson, B. , Kineke, G. C. and Drake, D. E. (1996) Observations of sediment transport on the Amazon subaqueous delta. Cont. Shelf Res. , 16, 697 – 715.

Storms, J. E. A. , Hoogendoorn, R. M. , Dam, R. A. C. , Hoitink, A. J. F. and Kroonenberg, S. B. (2005) Late – Holocene evolution of the Mahakam delta, East Kalimantan, Indonesia. Sed. Geol. , 180, 149 – 166.

Sztrákos, K. and Castelltort, S. (2001) Bartonian and Priabonian foraminifera from Arguis sections (Pre – Pyrenean external sierras, Spain). Impact on the correlation of biozones at the Bartonian/Priabonian boundary. Revue Micropaléontol. , 44, 233 – 247.

Ta, O. T. K. , Nguyen, V. L. , Tateishi, M. , Kobayashi, I. , Saito, Y. and Nakamura, T. (2002) Sediment facies and Late Holocene progradation of the Mekong River Delta in Bentre Province, southern Vietnam: an example of evolution from a tide-dominated to a tide-and wave-dominated delta. Sed. Geol. , 152, 313 – 325.

Taberner, C. and Bosence, D. W. J. (1995) An Eocene biodetrital mud-mound from the southern Pyrenean foreland basin, Spain: an ancient analogue for Florida Bay mounds? in: Carbonate Mud-Mounds-Their Origin and Evolution (Eds C. L. V. Monty, D. W. J. Bosence, P. H. Bridges and B. R. Pratt). IAS Spec. Publ., 23, 423–437.

Tanabe, S., Saito, Y., Lan Vu, Q., Hanebuth, T. J. J., Lan Ngo, Q. and Kitamura, Q. A. (2006) Holocene evolution of the Song Hong (Red River) delta system, northern Vietnam. Sed. Geol., 187, 29–61.

Titlyanov, E. A. and Latypov, Y. Y. (1991) Light dependence in Scleractinian distribution in the sublittoral zone of South China Sea Islands. Coral Reefs, 10, 133–138.

Tortora, P. (1999) Sediment distribution on the Ombrone River delta seafloor and related dispersal processes. Geol. Romana, 35, 211–218.

Van Andel, T. J. H. (1967) The Orinoco Delta. J. Sed. Res., 37, 297–310.

Veron, J. E. N. (1995) Corals in Space and Time; the Biogeography and Evolution of the Scleractinians. Cornell University Pres, Sidney, 321 pp.

Wilson, M. E. J. (2005) Equatorial delta-front patch reef development during the Neogene, Borneo. J. Sedim. Res., 75, 116–134.

Wilson, M. E. J. (2008) Global and regional influences on equatorial shallow-marine carbonates during the Cenozoic. Palaeogeogr. Palaeoclimatol. Palaeoecol., 265, 262–274.

Wilson, M. E. J. and Lokier, S. W. (2002) Siliciclastic and volcaniclastic influences on equatorial carbonates: insights from the Neogene of Indonesia. Sedimentology, 49, 583–601.

Wooldridge, S. A. (2009) A new conceptual model for the warm–water breakdown of the coral-algae endosymbiosis. Mar. Freshw. Res., 60, 483–496.

Woolfe, K. J. and Larcombe, P. (1999) Terrigenous sedimentation and coral reef growth: a conceptual framework. Mar. Geol., 155, 331–345.

Wright, V. P. and Burgess, P. M. (2005) The carbonate factory continuum, facies mosaics and microfacies: an appraisal of some of the key concepts underpinning carbonate sedimentology. Facies, 51, 17–23.

Xue, Z., Liu, J. P., DeMaster, D., Van Nguyen, L. and Ta, O. T. K. (2010) Late Holocene Evolution of the Mekong Subaqueous Delta, Southern Vietnam. Mar. Geol., 269, 46–60.

Yentsch, C. S., Yentsch, C. M., Cullen, J. J., Lapointe, B., Phinney, D. A. and Yentsch, S. W. (2002) Sunlight and water transparency: cornerstones in coral research. J. Experim. Mar. Biol. Ecol., 268, 171–183.

Zachos, J., Pagani, M., Sloan, L., Thomas, E. and Billups, K. (2001) Trends, rhythms and aberrations in global climate 65 Ma to present. Science, 292, 686–693.

第13章 非海相盆地的大型沉积结构:可容空间与沉积物供给之间相互作用的响应

PEDRO HUERTA，ILDEFONSO ARMENTEROS，PABLO G. SILVA 著

吴因业 译,赵 霞 校

摘 要 非海相盆地的沉积序列可以认为是可容空间与沉积物供给变化的产物。控制非海相盆地大型沉积结构的可容空间变化在高沉积物供给和低沉积物供给地区有差异。盆地内部单斜和背斜的隆升减少了可容空间,导致沉积层序几何形态、河流、泥坪和浅水湖相碳酸盐沉积大型结构的变化。主要的河流水系记录了高沉积物供给地区,而泥坪和浅湖振荡代表了接收沉积物较少的地区。西班牙Almazán盆地主要河流水系的大型结构的两部分是:(1)带状河道充填,具有低连通性,侧向上过渡到泥岩和蒸发岩为主的泥坪,然后过渡到沼泽与浅水碳酸盐湖泊沉积(主要是A2沉积层序);(2)席状河道充填,具有高连通性,侧向上与边缘泥坪的叠置钙结层相对应(在A3上部)。带状河道充填形成于高可容空间地区,而席状河道充填形成于可容空间减少区。

关键词 可容空间 Almazán盆地 钙结层 河道 河流 古近纪 沉积物供给 席状

13.1 概况

在层序地层学发展早期,以成因标志开展海相盆地沉积序列的划分、制图和对比被认为是很大的进步。这些进步允许进行油藏的表征和识别,发展出沉积盆地纪录的地质时期全球海平面变化曲线(Mitchum 等,1977a;Haq 等,1988)。海平面的变化(包括全球海平面变化和构造引起的局部变化)是控制地层结构的主要因素。在非海相盆地,层序地层学没有这样好的发展起来。了解可容空间与沉积物供给之间怎样变化导致沉积序列的差异,了解怎样识别他们成为层序地层学方法的重要问题。研究河流体系大型沉积结构已经是此类盆地层序地层学模型建立的基础。河流中河道的分布、密度、宽度以及河道带沉积(例如从低密度带状河道到高密度并列的板状河道),已经基于数值的、实验的和野外的资料应用于这些模型(Leeder,1978;Bridge 和 Leeder,1979;Shanley 和 McCabe,1991,1994;Mackey 和 Bridge,1995;Wright 和 Marriott,1996;Kim 和 Paola,2007)。基于基准面变化、可容空间(AS)、海平面和可容空间/沉积供给比率(AS/S),可以解释这种变化。大多数研究(例如 Leeder – Allen – Bridge 或 LAB 模型)解释高密度河道体系,是可容空间减少的结果。可是,某些实验模型显示,高密度河道带的发育和盆地沉降增加有关(Sheets 等,2002;Hickson 等,2005;Strong 等,2005;Kim 和 Paola,2007)。

不同模型之间的差异反映出的主要问题在于:(1)可容空间变化怎样控制地层结构;(2)当沉积物供给或高或低时可容空间不变,怎样记录这种变化?

本文通过研究河流体系的大型沉积结构,提出侧向进入洪泛平原边缘和泥坪环境时,可以纪录沉积物供给的侧向减少。详细的露头研究结合了实验的和数值的盆地模拟。西班牙Almazán盆地巨厚古近系著名的露头(图13-1),提供了对可容空间在发育或局部消失区域的河流体系的结构、侧向位置的泥坪和活动褶皱区的湖泊详细开展沉积学和地层学分析的好机会。预测河流盆地结构的能力对于石油工业和CO_2地下存储是重要的,这基于对盆地沉积主控因素间的相互作用的理解:构造和沉积会产生沉积物供给和可容空间的变化。

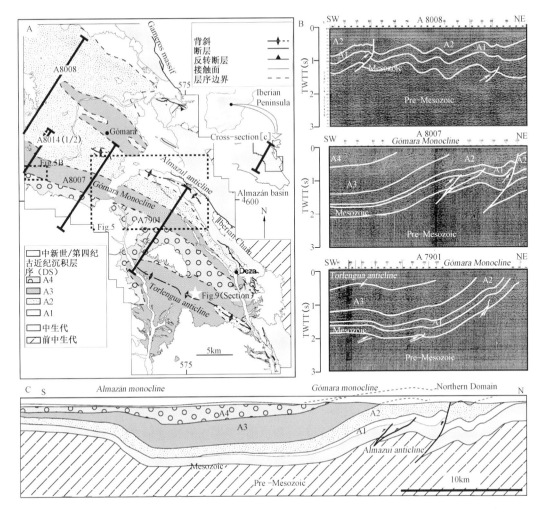

图 13-1 （A）Almazán 盆地古近纪露头剖面图，显示沉积层序、主要构造和垂直于构造的地震测线位置。图 13-5 的位置显示在图中。（B）地震测线（A8008；A8007 和 A7901）。沉积层序边界和断层有图示。TWTT 指双程旅行时间。（C）盆地横剖面，正交于主要构造并平行于测线 A8007

13.2 地质背景

新生代 Almazán 盆地是非海相的背驮式盆地（内部水系，与海洋不连通），充填冲积的、河流的、浅湖的及泥坪沉积物。盆地发育在主要 Cameros 冲断带的上盘（hanging wall），向北移动越过 Ebro 盆地的新生代沉积物，隆升到 Cameros 地块和 Iberian 山脉（Casas-Sainz 等，2000）。在本文古地理一节，下部地貌区位于所研究盆地的南部，局部基准面控制了盆地充填。

除了 Iberian 山脉，还存在显示同沉积活动的盆内构造，例如北部的 Gómara 单斜和 Torlengua 背斜，南部的 Gómara 单斜和 Torlengua 单斜。盆地南部被中新世埋藏，但可以用地震测线和"El Gredal"测井开展研究（图 13-1），而始新世—渐新世剖面在北部有良好的暴露。南倾30°的 Gómara 单斜把上隆的北部区域与平底的盆地主要沉积中心分开。Torlengua 背斜，位于 Gómara 单斜的东南，向西北没于盆地沉积中心之下（图 13-1B，图 13-1C）。所有构造都与东北向的冲断有关（Casas-Sainz 等，2000）。

13.3 地层

Almazán 盆地的古近纪沉积可以划分出四套沉积层序(DS),命名为 A1、A2、A3 和 A4(图 13-1,图 13-2)。建立沉积层序采用了 Mitchum 等(1977b)的标准。层序 A1 和 A2 属于中—上始新统,沉积物厚度大于1300m,分布于盆地北部,向南楔状尖灭(图 13-1C)。上述两套层序形成于主体盆地内部构造活动之前。层序 A3(渐新世?)堆积于盆内构造上隆开始时期(Gómara、Almazán 和 Arcos 单斜及 Torlengua 背斜),显示单斜上部翼部厚度减少,覆盖在背斜中枢之上。单斜上部翼部的厚度减少,在北东到南西过盆地的地震剖面上可以看到(图 13-2,图 13-3)。在 Gómara 单斜,A3 有1100m 厚度,但在上隆的翼部(盆地北部)无法全部测定,由于部分在 A4 时期遭受了剥蚀(图 13-1)。A3 厚度的减少沿盆地南缘也可以在 Almazán 单斜良好标定(Casas-Sainz 等,2002;图 13-3)。在渐新世到早期的中新世,盆地边缘上隆,Gómara 单斜(北部)上翼部遭受剥蚀。这时 A4 堆积在南部,显示同构造的不整合(图 13-1)。

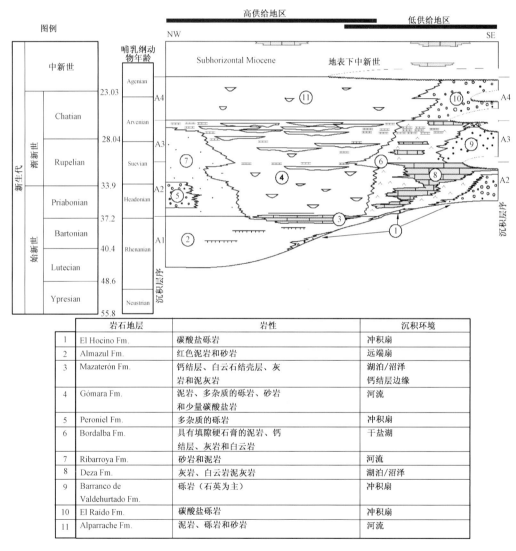

图 13-2 新生代 Almazán 盆地的等时地层剖面和岩石地层描述
高和低沉积供给区显示在图的顶部

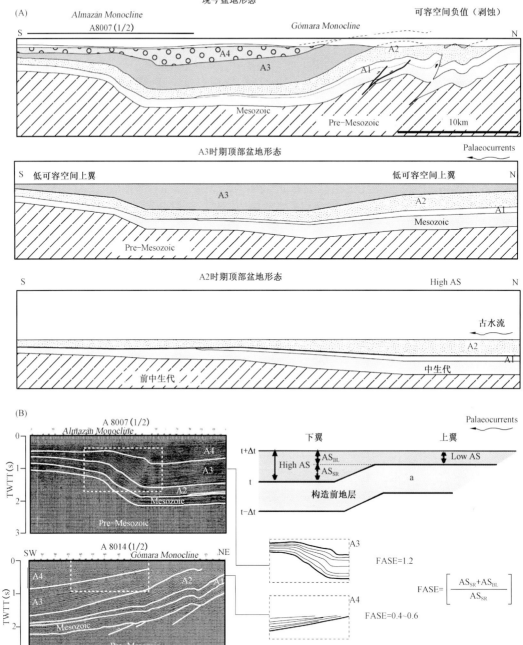

图 13-3 (A)盆地横剖面(位置见图 13-1),图示 A3 和 A2 时期顶部盆地形态的恢复。注意 A1 和 A2 时期高可容空间地区位于盆地北部,等时 A3 时期可容空间减少到负值,这时由于 Gómara 单斜的上隆。
(B)FASE 概念(Patton,2004)的解释,图示 Almazán 单斜(A3)和 Gómara 单斜(A4)的两个实例。
TWTT = 双程旅行时间

古近系 11 套地层记录了河流、湖泊/沼泽、干盐湖和冲积扇环境(图 13-2)。盆地西北部以广泛的河流体系为主,指向西南,具有很大的集水面积,位于 Iberian Ranges 山脉和 Cameros Massif 地块。而且,东南部记录了含石膏的泥坪/干盐湖和碳酸盐湖泊/沼泽体系,侧翼有小型冲积扇,来自 Iberian Chain 山脉的西北集水区。

13.4 可容空间的估算

在非海相盆地,可容空间是陆上的,在一定位置与两套不同时期沉积界面之间的沉积厚度有关(Muto 和 Steel,2000)。在 Almazán 盆地,可容空间的变化从相对意义上是根据上面提到的沉积层序厚度侧向变化来评估的(A1 至 A4;图 13 – 1)。这些变化受同沉积构造活动和相应的构造幅度发育控制。A1 和 A2 的沉积中心位于盆地北部(图 13 – 3A),但后期褶皱,目前定义为 Gómara 单斜的上翼。在 A1 和 A2 的沉积时期,被认为是构造前(单斜前)地层,分别是远端扇和河流沉积,指向西南(图 13 – 4,图 13 – 5)。生长中的单斜上翼 A3 地层减少,被解释为该区可容空间减少的纪录。A3 的地层增长模式,即 Gómara 和 Almazán 单斜上翼薄层和下翼厚层,是由于盆内构造幅度发育(ASSR)在上翼和下翼之间导致新可容空间产生的结果(图 13 – 3B)。而且,两翼之间由于相应基准面上升(ASBL)也造成可容空间的发育(图 13 – 3B)。Patton(2004)的实验揭示了当发育有褶皱可容空间效应("Fold Accommodation Space Efficiency")(FASE) >1 情况下,这种地层生长模式发育,就像 Almazán 单斜的两翼 A3 反映的情况(图 13 – 3B)。根据 Patton(2004),FASE 是产生的总可容空间($AS_{SR} + AS_{BL}$)与构造幅度产生的可容空间(AS_{SR})的比值:

$$\text{FASE} = \left[\frac{AS_{SR} + AS_{BL}}{AS_{SR}} \right]$$

A3 沉积时期,Gómara 河流体系流向西南,几乎垂直于 Gómara 单斜(图 13 – 4,图 13 – 5),上翼相互高度连通的席状河道沉积纪录了下来。相反,Almazán 单斜的上翼,A3 沉积时期,出现泥岩和硬石膏,以干盐湖沉积环境为主,这从测井(例如"El Gredal")和地震相分析可以推断(Huerta,2007)。

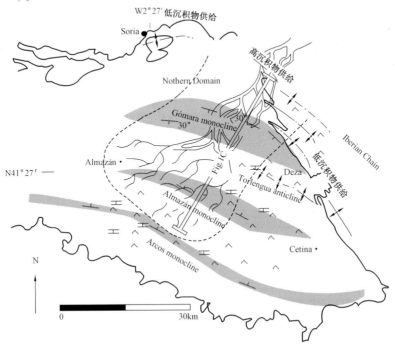

图 13 – 4 Almazán 盆地 A3 沉积期古地理图

盆地内部的单斜是图的灰色区域。河流的分支体系被解释为高沉积物供给区的纪录,泥坪区记录了低沉积物供给;这同时得到了盆地边缘背斜上隆的证据支持,向盆地东部和东南部是覆盖区

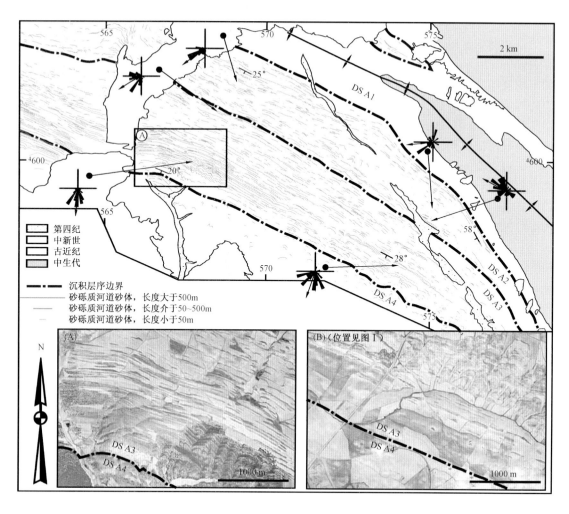

图 13-5 Gómara 单斜古近纪河道和河道带及其沉积层序边界的平面图

河道体按任意长度划分三段。河道体宽度小于 50m,介于 50~500m 和大于 500m。玫瑰图显示古水流垂直于 Gómara 单斜。(A)席状河道体,宽度大于 500m,出现于 A3 上部。(B)叠加钙结层的泥坪(白色层)出现于 A3 顶部。

在 A4 时期,Gómara 单斜的连接北翼隆升,被来自北部的盆地主要河流体系切割。结果,事件性的 A4 层序仅仅限于下翼(南部)有沉积,层序显示远离构造的楔状体。这一更新的地层生长模型在 ASBL 趋于 0 或成为负值而 ASSR 产生($0 < FASE < 1$)(图 13-3)的情况下出现。在最后时期(A4),Gómara 河流体系以带状(ribbon-shaped)河道充填为主,具有低连接度和垂向叠加,显示比 A3 更分散的古水流,但保留了整体南—西方向。

沉积在连接构造(Emerging Structure)的沉积单元经历了可容空间的减少,不同的地层生长模型有纪录(Suppe 等,1992;Casas-Sainz 等,2002;Patton,2004),可以代表不同的沉积相和沉积环境(碎屑岩或碳酸盐岩)(Burbank 和 Vergés,1994;Ford 等,1997;Jordan 等,2002;Castelltort 等,2003)。更多的研究集中在过连接构造的沉积学变化方面。

13.5 沉积物供给的评估

沉积物供给被认为是一定时间段提供的沉积物总量。这一概念在古代沉积体系的野外资料中难以评估。本项研究目标就是作相对定量的,以高和低沉积物供给来划分地区。为了完成这一目标,沉积物供给会从古近纪沉积序列的沉积学特征方面来评估。在 Almazán 盆地,考虑到主要水系(Gómara 河流体系),提供沉积物到盆地是从北东方向,具有一个大的集水区,纪录了高沉积物供给(图 13-4)。大量砾岩河道充填的出现也证明了这一点,砾石尺寸较大(平均 $\phi = 15cm$)。相反,盆地东南地区是低沉积物供给,沉积在泥坪和湖泊环境,因为 Iberian Chain 山脉的 Aragonian 分支的背斜和冲断形成了障碍,具有小型集水区(图 13-1,图 13-4)。后者的环境以分散的小型河道沉积为特征,细粒沉积物为主,存在沼泽/湖泊灰岩。这些沉积体系形成需要低的陆源注入。

13.6 沉积学

在盆地大部分沉积史中,Gómara 河流体系主要从东北到西南(盆地沉积中心)分布水系(图 13-2,图 13-4)。就像野外和航空照片看到的,河流侧向和向远端流到 Bordalba 组,该组表现为在干旱/含盐的泥坪中广泛分布的填隙硬石膏沉淀,以及更远端形成的干盐湖洪泛(Huerta 等,2010)。依据 Hardie 等(1978)和 Warren(1989)的定义,术语"泥坪"指位于短暂盐湖边缘的陆上暴露的细粒沉积物平原。河流体系含有一系列河道,正交穿越活动的 Gómara 单斜。向盆地方向,河道变化为远端分支体系,以朵叶体为终端(参考 Nichols 和 Fisher,2007;Hartley 等,2010)。这些变化表现为:(1)几何形态和河道体的宽度;(2)河道间的连通性;(3)河道沉积的颗粒大小,反映大型结构变化和体系的叠加样式。河道充填可以从正射投影纠正的航空照片进行数字化成图,并用 GIS 软件划分三类宽度(图 13-5)。选择了两块 $1km^2$ 的面积,用于测量河道/洪泛平原的面积比(河道密度),一块在 A2,另一块在 A3(图 13-5)。

13.6.1 河流体系的河道带(Gómara 组)

Gómara 组良好的露头把河道带充填与越岸沉积区分开。河道带充填有一套 Leeder(1993)、Jo 和 Chough(2001)所定义河道模型的结构单元。Gómara 组河道的结构单元在表 13-1 中有总结。根据几何形态、规模、颗粒大小和内部形态,Gómara 组河道带充填划分为两种结构终端成员和一种过渡成员:(1)带状河道和河道带充填;(2)席状河道和河道带充填;(3)席状过渡型(图 13-6)。Gómara 组河道体的平均厚度 A2 是 2~3m,A3 是 3~4m,但是,最大厚度可以达到 20~30m(图 13-7)。带状河道和河道带充填宽度小于 50m,具低的连通性(河道密度 9.3%),充填有粗粒砂岩和砾岩(图 13-6,图 13-8A,和图 13-8B);形成的单一河道具有宽度/厚度比值(w/t)从 3:1 至 15:1。某些河道充填,在 A4 更常见,其垂向叠加的砂体比值为 1:3(w/t)。河道体底部向上凹,顶部平,有时上凸,显示宽翼;内部显示向上变细,单一或多层的充填。这些河道和河道带充填的常见沉积相是 Gp、Sp、Gt、St 和 Sh,沉积相编码依据 Miall(1996)的编码。这些沉积相排成下列结构单元:侧向槽状充填,杂乱地层的蚀穴充填和槽状交错层理(表 13-1)。带状河道和河道带充填在 A2 和 A4 层序常见,并以洪泛平原泥岩为界。

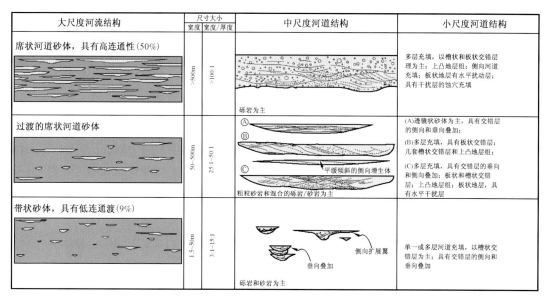

图 13-6　从小尺度到大尺度的河流结构地层综合图

Almazán 盆地充填的三个主要模式代表了大尺度和中尺度：(1)席状河道砂体，具有高连通性；
(2)过渡的席状河道砂体；(3)带状砂体，具有低连通性。结构单元描述见表 13-1

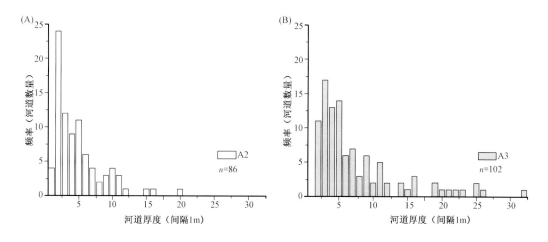

图 13-7　A2(A)和 A3(B)的河道体厚度直方图

厚度最大值在 A3 上部。河道体厚度测量沿剖面 A2 和 A3 进行，厚度间隔 1m(X 轴)。
落在每个数值段的河道数量以 Y 轴为代表。A2 测量了 86 个河道体，A3 测量了 102 个

解释：几何形态、宽度/厚度比和河道的内部结构显示，它们河道化的体系属于低弯度和较少的侧向运动(Friend 等，1979)。低弯度的证据还有古水流的均一和低发散。向上变细是由于河道充填时能量减少引起的(Williams 和 Rust，1969)。侧翼解释为堤岸或决口扇沉积(Marzo 等，1988)。相似的河道充填传统上解释为较少的分流河道或决口河道(Marzo 等，1988；Clemente 和 Pérez Arlucea，1993；Farrell，2001)，也解释为低弯度的网状河，为洪泛平原分隔(Behrensmeyer 和 Tauxe，1982；Nadon，1994；Gibling 等，1998)。网状河与冲积的脊部超高有关，而不是由于坡度(Mohrig 等，2000)。一些砾质河道充填的垂向叠加可以解释为过去废弃的河道再占领，产生加积模型(Gibling，2006)。

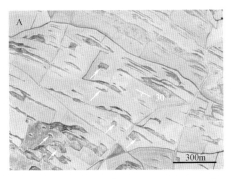

图 13 – 8　Gómara 河流体系的河道体

(A)A2 沉积层序的航空照片,显示带状河道体(箭头),具有低连通性。(B)孤立的带状河道体,具有垂向叠加。(C)厚的分布广泛的河道体,主要由砾岩组成。人比例尺高约 1.8m。(D)砾岩席状河道体,高连通性,Gómara 和 Buberos(A3)位置之间 CL – 101 公路的沟里可以看见。砾岩体底部以点划线标出,顶部有箭头显示。(E)分布广泛的席状河道体,接近于 Gómara 村子,地层南倾 30°

表 13 – 1　出现在 Gómara 组河道充填中简单的结构单元描述和解释

单元	描述	解释
具有侧向和垂向生长的交错层和水平层	从中粒砂到砾石变化。单一层从几厘米厚度到垂向叠加 0.5~1.5m 厚度,侧向延伸几十米。交错层侧向会减少角度,变为水平。纹层上凸和水平	垂向和侧向加积,向下游迁移的丘

— 385 —

续表

单元	描述	解释
板状交错层 30cm~2m	板状砂岩和砾岩，具有板状交错层。层厚度从50cm~1m。砂岩中—粗粒。砾岩层有粒序	向下游直线脊状底形加积，横向坝，流槽沙坝，2D沙丘
透镜状低幅地形层 15~50cm	透镜状砂岩和砾岩体。块状或具有低角度交错层。厚度20~50cm，长度可达10m	低幅地形丘的垂向、侧向和向下游迁移
槽状交错层 10cm~3m / 10cm~4m	砂岩和砾岩体，具有板状、透镜状和河道几何形态。颗粒大小从中砂到砾。槽状交错层位于侵蚀的底面之上。层的规模从几厘米到几米	3D巨型波痕、蚀穴充填和冲刷痕迁移
上凸层 50cm~2m / 50cm~1m	上凸层由几厘米的纹层组成，厚度从50cm~2m，长度大于厚度。中粒—粗粒砂岩组成。纹层倾向砂体边缘	河道中心坝，流槽沙坝
侧向槽状充填 1.5~3m / 75cm~2m	砂质砾岩体，具有河道形态。内部显示侧向倾斜地层，充填垂直于河流方向	解释为侧向河道充填，伴有侧向坝
大型斜线地层组 1~4m	3或6个砂质砾岩组成，具有透镜状形态，倾角5°~20°。侧向叠加相同。内部显示板状和槽状交错层及波痕。单元长度通常几十米	点坝和侧向坝增生体
大型板状体，具有分散水平层 >1m	板状砂砾岩体。可以是块状，或显示水平层和板状交错层。地层的底和顶不清楚，内部具有上凸面和侵蚀的底。厚度和宽度是变化的	低地形的坝和砾岩席的叠加
蚀穴充填，具有分散地层 50cm~2m / 1~2m	河道形态或勺子几何形态，由砾岩组成。内部显示槽状杂乱的交错层。砾岩体宽度2~4m	3D巨波痕的迁移，蚀穴充填

13.6.1.1 席状河道体

席状河道体宽度大于 500m,具有高连通性(河道密度为 50.2%),由中砾和巨砾组成(图 13-6)。一些河道体可以达到 3000m 宽度,宽度/厚度比大于 100,厚度在 5～15m 之间(图 13-6,图 13-8C 至图 13-8E)。这些砾岩体由水平的多层河道充填形成。底部构造不规则,有侵蚀,几十米到几米的冲沟,顶部是平的,具有突变的或渐变的上覆泥岩。砾岩多杂质,砂岩比例较少,碎屑尺寸垂向上几乎不变。砾石的最大直径 10～30cm,最大可达 70cm。沉积相是 Gcm、Gp、Gt、Gch、Sm、Sp、St 和 Sh,可以排列成四种结构单元:(1)大尺度的板状体,具有干扰的水平层;(2)蚀穴充填,具有干扰的地层;(3)槽状交错层组(通常米级);(4)板状交错层组(表 13-1)。古水流测量从河道体底面的交错层、冲刷面和冲沟上完成(图 13-5)。席状体高度连通,以 Gómara 单斜的 A3 上部为主。

解释:席状砾岩体解释为侧向叠加的河道带充填。古水流垂直于露头走向,露头揭示的砾岩倾斜地层,测量的宽度与河道带充填宽度一致。高的宽度/厚度比(>100)显示,河道带活动性强,穿越洪泛平原,类似于 Brahmaputra 河(Coleman,1969),Kosi 河和黄河(Yellow River)(Mackey 和 Bridge,1995)。席状河道带充填的形成原因可能是:(1)连续的沿迁移一侧侵蚀,对岸的沉积作用,尽管没有识别出点坝结构;(2)紧密空间的撕裂和河道带充填的侧向并联。可是,撕裂或不撕裂的证据是含糊其辞的。大的碎屑尺寸可以揭示沉积的高能量和底形建造。综合多层河道充填的结构单元记录了被河道切割的纵向坝向下游迁移,三维底形和蚀穴充填的迁移。这些过程反映了辫状河道体系的沉积特征。相似的结构单元在高能的辫状河已经有识别(Williams 和 Rust,1969;Hjellbakk,1997;Bridge 等,1998;Jo,2003b;Leckie,2003)。除此之外的席状洪泛沉积显示了同样的几何形态和席状河道充填的内部特征(Bridge,2003),如果大尺度的结构没有分析的话,很容易被错误解释。

13.6.1.2 过渡的河道体

过渡河道体充填具有席状几何形态,宽度 50～500m,伴随两个终端成员(图 13-6)。这些河道体由砾岩和砂岩沉积相组成,宽度/厚度比在 25～50。河道体厚度 1～6m,尽管平均厚度在 2～3m。河道体侧向上楔状,周边为泥岩。内部结构单元上可以划分为三类(图 13-6):(1)透镜状为主;(2)侧向增生体为主;(3)多层充填为主。

解释:过渡的河道体以内部形态进行解释(表 13-1)。透镜状低幅地形层、交错层和水平层的透镜体,具有侧向和垂向增生体,解释为低幅坝的河道。常见侧向增生体的河道解释为高弯度河道的点坝沉积(Bluck,1971;Ikeda,1989;Bridge 等,1995)。多层充填的河道体,具有槽状和板状地层组和蚀穴充填,被解释为侧向和向下游的坝体迁移以及辫状模式的河道叠加(Bristow,1987;Bristow 等,1993;Jo,2003a)。

13.6.2 远端洪泛平原和泥坪(Gómara,Bordalba 和 Deza 组)

Gómara 组河流体系的主要河道体侧向上进入洪泛平原,沉积速率随河道带距离的增加而减少。这些低沉积物供给区与 Gómara 组西北部远端洪泛平原一致,侧向上同时又进入 Ribarroya 组的河流体系。向东南方向,Gómara 组渐变为泥坪和干泥坪环境,属于 Bordalba 组纪录的干盐湖体系(Huerta 等,2010)。

远端洪泛平原和泥坪相组合与 Gómara 组河道体的大型结构变化和叠加样式一致:(1)泥岩和沼泽/湖泊灰岩,对应低连通性的带状河道体;(2)叠加的钙结层,具有高连通性的席状河道体。

13.6.2.1 泥岩和沼泽/湖泊灰岩

泥岩和砂岩层,代表 A2 层序 Gómara 组的越岸沉积。Gómara 组的泥岩可以显示填隙硬石膏,与 A2 顶部 Bordalba 组指状相连(图 13-2)。

泥岩形成了板状地层,与板状砂岩交替出现。砂岩底部突变,向上渐变为棕红色上覆泥岩。Bordalba 组的板状砂岩层厚度在 20cm~4m,钙质或硬石膏胶结。内部层理常常被生物扰动中断。Bordalba 组以泥岩和具有填隙硬石膏的泥岩为特征,常见生物扰动、介形类和树叶残余化石(Huerta 等,2010)。Bordalba 组侧向上进入东南部 Deza 组的湖泊/沼泽灰岩,厚度大于 200m,Deza 地区最厚达 450m。灰岩形成块状和板状层,厚度 40cm~2m,灰色或黑灰色。灰岩层与灰色泥灰岩层互层,有时与煤层互层。灰岩主要是粒泥灰岩(Wackestones),含有腹足类、介形类和轮藻类化石,不常见碎屑石英颗粒(<1%)。填隙硬石膏后的透镜状方解石假晶在浅水的、米级规模的、沼泽/湖泊的灰岩层序顶部很常见。在一些情况下,会与本组地层湖泊体系中心区一致,出现丰富的燧石结核。沼泽灰岩在 Deza 组顶部更加常见,显示凝块结构和球粒结构,这与暴露有关。泥岩具有晶洞和孔隙,以及具有平面和曲线的裂缝。从无角砾岩化到球粒—内碎屑结构的几个角砾岩化作用时期(Armenteros 和 Daley,1998),可以被识别出来。存在不规则和透镜状的燧石结核(Armenteros 等,2006)。

解释:具有填隙硬石膏的泥岩沉积在 Gómara 组的洪泛平原,以及 Bordalba 组泥坪和干泥坪。从北西到南东侧向穿过,是 Gómara 组至 Bordalba 组,都以河道充填的减少和泥岩中填隙硬石膏的增加为特征。Bordalba 组板状砂岩层比 Gómara 组颗粒更细。侧向变化被解释为高沉积物供给区向低沉积物供给区的转变。这在区域平面图纪录的构造—地层关系中得到了证实,揭示 Iberian 山脉的 Aragonian 分支在 A2 时期开始上隆,形成向盆地东南的遮蔽区。泥岩记录了洪泛阶段悬浮载荷的沉积作用,广泛分布的填隙硬石膏揭示接近地表的地下水具有高的硫浓度和盆地蒸发条件的存在(Huerta 等,2010)。同样 Deza 组湖泊和沼泽灰岩伴随低的碎屑注入,揭示在自由水位或池塘里有高的碳酸盐生产。黑灰色和灰色沉积相表明,湖泊具有高的有机质生产,有时伴随快速埋藏的煤层沉积,抑制了有机质的氧化作用。在沼泽灰岩中,具有暴露特征和潮湿/干燥结构时,碳酸盐生产的生物化学和化学沉积速率与沉降速率密切相关(Huerta 和 Armenteros,2005;Armenteros 和 Huerta,2006)。古近纪盆地的东南,Deza 组湖泊/沼泽沉积区,位于隆起的碳酸盐中生代岩石附近,发育了一个小的汇水区,具有低的陆源沉积,乃薄层指状冲积,位于源区与湖泊环境之间(Huerta,2007)。这种环境和雨季的出现会有利于湖泊沉积的永久发育,有些情况下会有利于富有机质的堆积和保存,可能显示沉降速率大大超过沉积相的堆积速率。可是,Deza 组的含煤沉积归因于气候(Armenteros 等,2006)。

13.6.2.2 叠加的钙结层

多种多样的叠加米级钙结层局部可以厚达 15m;从结核或底部的棱柱到剖面顶部的块状(阶段Ⅱ至Ⅳ,Machette,1985)(图 13-9A 至 13-9E)。这些钙结层常见于层序 A3 上部,Gómara 组(Gómara 单斜)和 Bordalba 组(Torlengua 背斜),两者与粗粒席状河道体为主的地区一致(图 13-2)。

钙结层由微亮晶方解石镶嵌体组成,常见氧化铁和氧化锰斑块,具有分散的粉砂级石英颗粒,颗粒边缘常见压溶沟,充填有微亮晶方解石。钙结层,少见白云石结层,会形成板状层,具有渐变的底和突变的顶(图 13-9A,13-9B 和 13-9D),尽管有时钙结层剖面的叠加会模糊单一钙结层的识别(图 13-9C);层序上显示碳酸盐含量从下伏棕红色泥岩向上增加。钙结层的顶部被泥岩覆盖,或另一层钙结层覆盖(叠加)。块状的顶部是均一的、连续的碳酸盐块体,胶结程度比结核高,结核被包含在粉状钙结层块体内部。结核钙结层显示原生泥岩的遗迹,剖面上向上变小,直到顶部块状层消失(图13-9E)。一些结核垂向上伸长,根部尖锐。钙结层层序可以是单一层,与泥岩互层或垂向叠加,有时被几十米泥岩层分隔,或在部分情况下,由一个钙结层层序叠加在以前的层序之上。

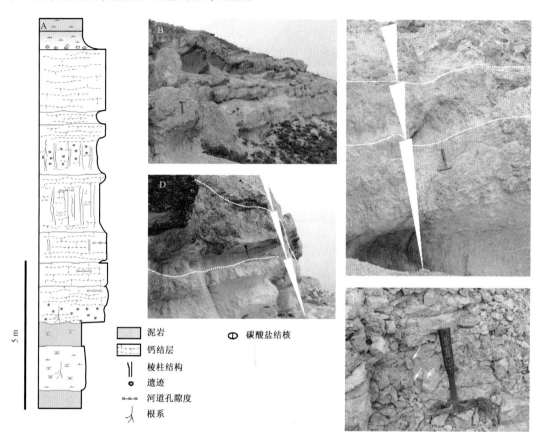

图 13-9 A3 沉积期泥坪地区叠加的钙结层

(A)地层剖面(La Rey,位于图 13-1)显示叠加的钙结层序列。(B)叠加的钙结层露头,锤子作比例尺,长 28cm。
(C)叠加的钙结层序列,从低结核/棱柱钙结层到顶部块状钙结层。钙结层边界是混杂的,以点划线标出。
(D)块状钙结层序列,更加坚硬,层的边界良好界定。(E)块状钙结层,具有氧化铁色斑(红色斑块,箭头)

解释:沉积盆地中的钙结层与地下水位之上的蒸发蒸腾作用有关(A 钙结层),或盆底的成土作用有关(B 钙结层)(Wright 和 Tucker,1991),或两者组合、交替或叠加,区分起来较难(Semeniuk 和 Searle,1985;Zaleha,1997;Williams 和 Krause,1998;Alonso-Zarza,2003;Huerta 和 Armenteros,2005)。垂向上伸长的结核(棱柱结构)和河道孔隙在成土钙结层中常见,铁和锰的氢氧化合物斑块与地下水动力学剖面有关(La Force 等,2002;Alonso-Zarza,2003)。非海相沉积盆地的洪泛平原、泥坪和浅湖常常以低地形幅度和低梯度为特征,有利于地下水接近地表,要加以关注。进一步讲,如果沉积作用与成土作用同步,水成的和成土的过程会叠加出现,就像许多大陆盆地钙结层和边缘的湖泊/沼泽环境的常见证据一样(Armenteros 等,1995;Sanz 等,1995;Wright 和 Platt,1995;Cojan,1999;Freytet 和 Verrecchia,2002;Alonso-Zarza,2003)。

古土壤的水成特征通常发育在基准面上升时期(Kraus,1999)。碳酸盐含量向上增加被解释为沉积速率的减少,可以反映侧向上从活动的沉积区变为较少沉积活动区和低沉积量地区(Bown 和 Kraus,1987;Alonso – Zarza 等,1992;Marriott 和 Wright,1993;Wright 和 Marriott,1996;Kraus,1999;Huerta 和 Armenteros,2005;Alonso – Zarza 和 Wright,2010),处于相对水系较好的环境下水注入盆地较少的阶段。Gómara 和 Bordalba 组纪录的钙结层和叠加钙结层地区被解释为远端洪泛平原缓慢升高的环境,钙结层的发育可能由于相对基准面减少,同时因为剖面上碳酸盐填隙沉淀而地形增加,就像出现在现代 Okawango 河流分支体系的岛一样(Gumbricht 等,2004,2005)。这些地区可容空间减少,沉积物越过而向远端发育。

13.7　讨论

河流体系沉积结构的变化可以解释为可容空间(AS)与沉积物供给的变化。泥坪结构的对比变化反映 AS 在低沉积物供给区的变化。河流体系和泥坪在这里被认为分别是高沉积供给和低沉积物供给区。Gómara 河流体系是盆地的主要水系,具有高沉积物供给。另一方面,Bordalba 组泥坪受到较少沉积物供给的源区供源,具有分散河道沉积和古地理方面的证据(Huerta 等,2010)。

具有低连通性和低宽度/厚度(w/t)比的带状河道沉积被解释为洪泛平原上的网状河道,高加积、高可容空间和高沉积物供给(Gómara 单斜的层序 A2 和 A4,图 13 – 5,图 13 – 8)。这些观察与实验结果一致,即河流体系的高沉积速率出现在大多数沉积物被存储而河道带作为沉积物供给通道时的洪泛平原(Sheets 等,2002)。文献中也有理论模型和野外实例,说明低连通性的带状河道与高沉降、高 AS 或基准面上升有关(Marzo 等,1988;Shanley 和 McCabe,1991;Clemente 和 PérezArlucea,1993;Wright 和 Marriott,1993;Legarreta 和 Uliana,1998;Martinsen 等,1999),以及数值模型(Allen,1978;Leeder,1978;Bridge 和 Leeder,1979;Bridge 和 Mackey,1993)。如果河道深度或单一河道充填厚度可以被解释为评估决口频率的常数,那么 A2 河道充填对 A3 的低厚度(图 13 – 7),就显示 A2 时期较高的决口频率。这种解释与 Bryant 等(1995)的实验是一致的,显示决口频率随沉积速率增加而增加。相反,Heller 和 Paola(1996)预测的高连通性对应高决口频率,具有高沉降速率,没有在 Almazán 盆地 A2 层序观察到(图 13 – 5)。注意,决口频率很难在岩石纪录中进行评估,可能这个参数不是控制河道厚度的唯一参数。A4 河道充填的垂向叠加,具有 1∶3 的 w/t 比值(图 13 – 8B),形成于下翼高 AS 值阶段,在 Gómara 单斜上翼遭受剥蚀。

在临近主要河流体系的周边地区,尤其是南东方向,沉积物供给较低,导致河道带—洪泛平原体系向边缘洪泛带和泥坪过渡。那里的河道沉积不丰富,填隙硬石膏常见,主要是由于这种特殊的盆地背景下减少了水和陆源沉积的供给。厚层浅水湖泊/沼泽碳酸盐发育在这些地区,沉积物供给不足以充填 AS(De Wet 等,1998;Carroll 和 Bohacs,1999;Alonso – Zarza 和 Calvo,2000;Bohacs 等,2000;Huerta 和 Armenteros,2005)。在泥坪地区,水位切断了古地面(图 13 – 10),化学的和生物化学的沉淀出现在池塘(Huerta 和 Armenteros,2005)。A2 层序从带状河道带侧向过渡为泥坪沉积,最后变成沼泽/湖泊碳酸盐,显示沉积物供给的侧向减少(图 13 – 10),可能伴随边缘地区沉降的缩小。这种情况允许主要河流体系在同一位置继续存在,即盆地西北部,这里沉降较高(cf. Mackey 和 Bridge,1995)。如上提及,沉积层序的厚度已经用于评估盆地中央和边缘产生的可容空间。河流体系的位置和 Iberian Chain 山脉的 Aragonian 分支产生了低的沉积物输入,位于盆地的东南部。

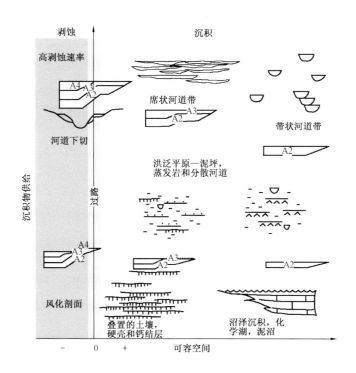

图 13-10 图示可容空间与沉积物供给的关系,以及非海相盆地观察到的不同地层结构
过 Gómara 单斜的沉积层序(A2 至 A4)的几何形态显示该区具有高沉
积物供给(在图的上部单斜),和具有低沉积物供给(在图的下部单斜)

相互连通性好的席状河道和河道带具有高的 w/t 比值,发育在 AS 减少和沉积物供给丰富的阶段(主要在 A3 上部,图 13-5,图 13-10)。可容空间减少出现在 Gómara 单斜初始隆升阶段(A3),在随后发生的切断和剥蚀(AS 负值)之前,对应于 A4 层序的沉积(图 13-5,图 13-10)。在洪泛平原的 AS 减少使沉积物储备减少,增加了系统的潜力。沉积物在该地区通过河道形成过路,向盆地西南远端而去,这时 AS 保持高值,位于单斜的下翼部(图 13-10)。有限的加积条件有利于河道侧向迁移,发育席状河道和河道带,具有高 w/t 比值,位于 Gómara 单斜的上翼部,正如其他盆地和模型解释的一样(Friend,1983;Wright 和 Marriott,1993;Shanley 和 McCabe,1994;Legarreta 和 Uliana,1998;图 13-5,图 13-10)。河道带沉积物丰度、侧向连续性和相互连通性的增加是高沉积物供给和洪泛平原 AS 减少的沉积响应,类似于 Leeder-Allen-Bridge 数值模拟的结果(Allen,1978;Leeder,1978;Bridge 和 Leeder,1979;Bridge 和 Mackey,1993)。这一假说被 Heller 和 Paola 质疑(1996),认为 Leeder-Allen-Bridge 数值模拟的结果来自于河道决口频率是个常数的假定条件,而决口速率的增加会产生高的河道相互连通性,即使在高的沉降情况下。Heller 和 Paola(1996)的数值模拟结果很有意思,可以在解释河流序列时予以考虑。问题是,如上所述,决口频率难以在岩石纪录中予以评估,因此不可能查明不同的决口速率所起的作用。可是,决口速率是内部和外部控制因素的结果,直接与沉积速率成正比(Bryant 等,1995)。因此,在低可容空间时(低沉积速率和高过路),如 A3,决口速率可以是低的,以 A3 观察到的较大河道沉积厚度一致(图 13-7)(如果假定较厚的河道充填形成于低决口速率,如上描述)。Heller 和 Paola(1996)观察的这一结果没有出现在 Almazán 盆地。Gómara 组 A3 高连通性的席状河道体发育在低 AS 条件下,推测与 Gómara 单斜上翼 A3

时期构造背景和厚度减少有关。这一观察支持了 Leeder – Allen – Bridge 模型的结果，但不同于 Sheets 等（2002）和 Hickson 等（2005）报告的实验结果，他们发现在其他条件下低沉降期河道连通性会减少。在 Gómara 组大尺度河流结构观察的相似例子在犹他州的 Mesaverde 组有所描述（Shanley 和 McCabe，1993；Olsen 等，1995），以及西班牙 Pyrenees 的 Castissent 组（Marzo 等，1988；Nijman，1998），或者在 Neuquén 盆地（Legarreta 和 Uliana，1998），都发现高连通性的河道充填形成于低 AS，或沉降条件下。

在相应的泥坪体系，低 AS 时期主要存在叠加的钙结层（图 13 – 9，图 13 – 10）；他们发育在低沉积区（Tandon 等，1998），变得更加成熟，垂向上叠覆遮蔽区（Armenteros 和 Huerta，2006）和/或隆起区（Alonso-Zarza 等，1999），沉积物减少。因此，侧向上延伸和叠加的钙结层出现在低沉积物供给和低 AS 地区。Almazán 盆地内部 A3 沉积期构造的上隆不产生地形显示，类似于 Andean 盆地（Jordan 等，2002），但是这可能形成于海拔较高地区，没有接受重要的碎屑注入。这种海拔背景下地下碳酸盐和硅质沉淀形成的岛的成因相似，会形成钙结层和硅结层，如 Okawango 河流分支体系（McCarthy，1993；McCarthy 和 Ellery，1995）。虽然人们知道气候在非海相盆地是重要的控制因素（Nichols 和 Fisher，2007），就像在 Almazán 盆地（Huerta 和 Armenteros，2004；Armenteros 等，2006；Huerta，2007；Huerta 等，2010），显示岩相有干旱气候期的印记，但 A3 构造隆升对远端洪泛平原和泥坪中叠置硅结层的形成起更加重要的作用。

13.8　结论

Almazán 盆地古近系沉积物可以划分四套层序，A1 到 A4 范围从始新统中部到可能的渐新统中—上部。盆地北部主要是延伸的河流体系（Gómara 和 Alparrache 组），侧向伸展，远端进入干盐湖体系（Bordalba 组；图 13 – 2）。Iberian Chain 山脉的初始隆升和 Almazán 盆地的相应构造（Gómara 单斜，Almazán 单斜，Torlengua 背斜等）在 A3 时期发生了可容空间（AS）和沉积物供给的显著变化。这些变化控制了河流和泥坪/湖泊序列的大尺度沉积结构的改变。

高 AS 值和高沉积物供给被低宽度、带状的河道充填所记录，具有低的相互连通性（沉积界面 A2 和 A4，下翼，图 13 – 10）。当 AS 低值而沉积物供给高值时，发育低的席状河道，具有高的相互连通性（沉积界面 A3；图 13 – 10）。AS 负值有利于河道下切和剥蚀，出现在 Gómara 单斜上翼 A4 时期（图 13 – 10）。

高 AS 值和低沉积物供给会产生泥坪沉积，具有填隙硬石膏的生长，主要在分散的充填砂岩的河道（沉积层序 A2；图 13 – 10）。在极低沉积物供给的地带，仅仅堆积化学的和附属的有机物沉积，此外推测有短期气候变化影响，主要指低幅度物源区为界的盆地东南部。在低沉积物供给和低 AS 条件下，叠置的土壤和钙结层会发育（沉积界面 A3；图 13 – 9，图 13 – 10）。实际上，AS 负值的出现使盆地内部早期沉积的沉积层序遭受剥蚀和风化。可容空间/沉积物供给比值分析可以成功应用于 Almazán 盆地沉积序列，特别是在内部水系为完全大陆环境的沉积盆地。

参 考 文 献

Allen, J. R. L. (1963) The classification of cross – stratified units, with notes on their origin. Sedimentology, 2, 93 – 114.

Allen, J. R. L. (1978) Studies in fluviatile sedimentation: an exploratory quantitative model for architecture of avulsion-controlled alluvial units. Sed. Geol. , 21, 129 – 147.

Allen, J. R. L. (1983) Studies in fluviatile sedimentation: bars, bar complexes and sandstone sheets (low-sinuosity braided streams) in the brownstones (L. Devonian), Welsh Borders. Sed. Geol. , 33, 237 – 293.

Alonso-Zarza, A. M. (2003) Palaeoenvironmental significance of palustrine carbonates and calcretes in the geological record. Earth-Sci. Rev. , 60, 261 – 298.

Alonso-Zarza, A. M. and Calvo, J. P. (2000) Palustrine sedimentation in an episodically subsiding basin: the Miocene of the northern Teruel Graben (Spain). Palaeogeogr. Palaeoclimatol. Palaeoecol. , 160, 1 – 21.

Alonso-Zarza, A. M. and Wright, V. P. (2010) Chapter 5. Calcretes. In: Carbonates in Continental Settings: Facies, Environments, and Processes (Eds A. M. Alonso-Zarza and L. H. Tanner), Dev. Sedimentol. , 226 – 267. Elsevier, Amsterdam. Alonso-Zarza, A. M. , Wright, V. P. , Calvo, J. P. and García del Cura, M. A. (1992) Soil-landscape relationship in the Middle Miocene of the Madrid Basin. Sedimentology, 39, 17 – 35.

Alonso-Zarza, A. M. , Sopeña, A. and Sanchez-Moya, Y. (1999) Contrasting palaeosol development in two different tectonic settings: the Upper Buntsandstein of the Western Iberian Ranges, Central Spain. Terra Nova, 11, 23 – 39.

Armenteros, I. and Daley, B. (1998) Pedogenic modification and structure evolution in palustrine facies as exemplified by the Bembridge Limestone (Late Eocene) of the Isle of Wight, southern England. Sed. Geol. , 119, 275 – 295.

Armenteros, I. and Huerta, P. (2006) The role of clastic sediment influx in the formation of calcrete and palustrine facies: a response to paleographic and climatic conditions in the Southeastern Tertiary Duero basin (northern Spain). In: Paleoenvironmental Record and Applications of Calcretes and Palustrine Carbonates (Eds A. M. Alonso-Zarza and L. H. Tanner), Geol. Soc. Am. Spec. Pap. , 416, 119 – 132.

Armenteros, I. , Bustillo, M. A. and Blanco, J. A. (1995) Pedogenic and groundwater processes in a closed Miocene Basin (Northern Spain). Sed. Geol. , 99, 17 – 36.

Armenteros, I. , Bustillo, M. A. and Huerta, P. (2006) Ciclos climáticos en un sistema lacustre marginal perenne carbonatado-evaporítico. Formación Deza, Eoceno superior, cuenca de Almazán. Geo – temas, 9, 25 – 30.

Behrensmeyer, A. K. and Tauxe, L. (1982) Isochronous fluvial systems in Miocene deposits of Northern Pakistan. Sedimentology, 29, 331 – 352.

Bluck, B. J. (1971) Sedimentation in the meandering River Endrick. Scott. J. Geol. , 7, 93 – 138.

Bohacs, K. M. , Carroll, A. R. , Nede, J. E. and Mankirowicz, P. J. (2000) Lake-basin type, source potential and hydrocarbon character: an integrated sequence-stratigraphic-geochemical framework. In: Lake Basins Through Space and Time (Eds E. H. Gierlowski-Kordesch and K. R. Kelts), AAPG Stud. Geol. , 46, 3 – 33. American Association of Petroleum Geologists, Tulsa, OK.

Bown, T. M. and Kraus, M. J. (1987) Integration of channel and floodplain suites: I, Developmental sequence and lateral relations of alluvial Paleosols. J. Sed. Petrol. , 57, 587 – 601.

Bridge, J. S. (1993) The interaction between channel geometry, water flows, sediment transport and deposition in braided rivers. In: Braided Rivers (Eds J. L. Best and C. S. Bristow), Geol. Soc. Spec. Publ. , 75, 13 – 71.

Bridge, J. S. (2003) Rivers and Floodplains: Forms, Processes and Sedimentary Record. Blackwell Science, Oxford, 491 pp.

Bridge, J. S. and Leeder, M. R. (1979) A simulation model of alluvial stratigraphy. Sedimentology, 26, 617 – 644.

Bridge, J. S. and Mackey, S. D. (1993) A revised alluvial stratigraphy model. In: Alluvial Sedimentation (Eds M. Marzo and C. Puigdefábregas), Int. Assoc. Sedimentol. Spec. Publ. , 17, 319 – 336. Blackwell, Oxford, International.

Bridge, J. S. , Alexander, J. , Collier, R. E. L. , Gawthorpe, R. L. and Jarvis, J. (1995) Ground-penetrating radar and coring used to study the large-scale structure of point-bar deposits in three dimensions. Sedimentology, 42, 839 – 852.

Bridge, J. S. , Collier, R. and Alexander, J. (1998) Large-scale structure of Calamus River deposits (Nebraska,

USA) revealed using ground-penetrating radar. Sedimentology,45, 977–986.

Bristow, C. S. (1987) Brahmaputra River: channel migration and deposition. In: Recent Developments in Fluvial Sedimentology (Eds F. G. Ethridge, R. M. Flores and M. D. Harvey), SEPM Spec. Publ., 39, 63–74. Society of Economic Paleontologists and Mineralogists, Tulsa, OK.

Bristow, C. S., Best, J. L. and Roy, A. G. (1993) Morphology and facies models of channel confluences. In: Alluvial Sedimentation (Eds M. Marzo and C. Puigdefábregas), Int. Assoc. Sedimentol. Spec. Publ., 17, 91–100. Blackwell, Oxford, International.

Bryant, M., Falk, P. and Paola, C. (1995) Experimental study of avulsion frequency and rate of deposition. Geology, 23, 365–368.

Burbank, D. W. and Vergés, J. (1994) Reconstruction of topography and related depositional systems during active thrusting. J. Geophys. Res., 99, 20281–20297.

Cant, D. J. and Walker, R. G. (1978) Fluvial processes and facies sequences in the sandy braided South Saskatchewan River, Canada. Sedimentology, 25, 625–646.

Carroll, A. R. and Bohacs, K. M. (1999) Stratigraphic classification of ancient lakes: balancing tectonic and climatic controls. Geology, 27, 99–102.

Casas-Sainz, A. M., Cortés–García, A. L. and Maestro-González, A. (2000) Intraplate deformation and basin formation during the Tertiary within the northern Iberian plate: origin and evolution of the Almazán basin. Tectonics, 19, 258–289.

Casas-Sainz, A. M., Cortés, A. L. and Maestro, A. (2002) Sequential limb rotation and kink-band migration recorded by growth strata, Almazán Basin, North Spain. Sed. Geol.,146, 25–45.

Castelltort, S., Guillocheau, F., Robin, C., Rouby, D., Nalpas, T., Lafont, F. and Eschard, R. (2003) Fold control on the stratigraphic record: a quantified sequence stratigraphic study of the Pico del Aguila anticline in the south-western Pyrenees (Spain). Basin Res., 15, 527–551.

Clemente, P. and Pérez Arlucea, M. (1993) Depositional architecture of the Cuerda del Pozo Formation, Lower Cretaceous of the extensional Cameros Basin, north-central Spain. J. Sed. Petrol., 63, 437–452.

Cojan, I. (1999) Carbonate rich palaeosols in the Late Cretaceous-Early Palaeogene Series of the Provence basin (France). In: Palaeoweathering, Palaeosurfaces and Continental Deposits (Eds M. Thiry and R. Simon-Coicon), Int. Assoc. Sedimentol. Spec. Publ., 27, 323–335. Blackwell Science.

Coleman, J. M. (1969) Brahmaputra River: channel processes and sedimentation. Sed. Geol., 3, 129–239.

Cowan, E. J. (1991) The large–scale architecture of the fluvial Westwater Canyon Member, Morrison Formation (Upper Jurassic), San Juan Basin, New Mexico. In: The Three-Dimensional Facies Architecture of Terrigenous Clastic Sediments and Its Implications for Hydrocarbon Discovery and Recovery (Eds D. Miall Andrew and N. Tyler), SEPM Concepts Sedimentol. Paleontol., 3, 80–93. SEPM (Society for Sedimentary Geology), Tulsa, OK.

De Wet, C. B., Yocum, D. A. and Mora, C. I. (1998) Carbonate lakes in closed basins; sensitive indicators of climate and tectonics; an example from the Gettysburg Basin (Triassic), Pennsylvania, USA. In: Relative Role of Eustasy, Climate, and Tectonism in Continental Rocks (Eds K. W. Shanley and P. J. McCabe), SEPM Spec. Publ., 59, 191–209.

Díaz-Molina, M. (1993) Geometry and lateral accretion patterns in meander loops; examples from the upper Oligocene-lower Miocene, Loranca Basin, Spain. In: Alluvial Sedimentation (Eds M. Marzo and C. Puigdefábregas), Int. Assoc. Sedimentol. Spec. Publ., 17, 115–131. Blackwell,Oxford, International.

Farrell, K. M. (2001) Geomorphology, facies architecture, and high-resolution, non-marine sequence stratigraphy in avulsion deposits, Cumberland Marshes, Saskatchewan. Sed. Geol., 139, 93–150.

Ford, M., Williams, E. A., Artoni, A., Vergés, J. and Hardy, S. (1997) Progressive evolution of a fault-related fold pair from growth strata geometries, Sant Llorenc de Morunys, SEPyrenees. J. Struct. Geol., 19, 413–441.

Freytet, P. and Verrecchia, E. P. (2002) Lacustrine and palustrine carbonate petrography: an overview. J. Paleolimnol. ,27, 221–237.

Friend, P. F. (1983) Towards the field classification of alluvial architecture or sequence. In: Alluvial Sedimentation (Eds J. D. Collinson and J. Lewin), IAS Spec. Publ., 6, 345–354. Blackwell.

Friend, P. F., Slater, M. J. and Williams, R. C. (1979) Vertical and lateral building of river sandstone bodies, Ebro Basin, Spain. J. Geol. Soc. London, 136(Part 1), 39–46.

Gibling, M. R. (2006) Width and thickness of fluvial channel bodies and valley fills in the geological record: a literature compilation and classification. J. Sed. Res., 76, 731–770.

Gibling, M. R., Nanson, G. C. and Maroulis, J. C. (1998) Anastomosing river sedimentation in the Channel Country of central Australia. Sedimentology, 45, 595–619.

Gumbricht, T., McCarthy, J. and McCarthy, T. S. (2004) Channels, wetlands and islands in the Okavango Delta, Botswana, and their relation to hydrological and sedimentological processes. Earth Surf. Proc. Land., 29, 15–29.

Gumbricht, T., McCarthy, T. S. and Bauer, P. (2005) The micro-topography of the wetlands of the Okavango Delta, Botswana. Earth Surf. Proc. Land., 30, 27–39.

Haq, B. U., Hardenbol, J. and Vail, P. R. (1988) Mesozoic and Cenozoic chronostratigraphy and eustatic cycles. In: Sea Level Changes – An Integrated Approach (Eds C. K. Wilgus, B. S. Hastings, C. G. S. C. Kendall, H. W. Posamentier, C. A. Ross and J. C. Van Wagoner), SEPM Spec. Publ., 42, 71–108.

Hardie, L. A., Smoot, J. P. and Eugster, H. P. (1978) Saline lakes and their deposits: a sedimentological approach. In: Modern and Ancient Lake Sediments (Eds A. Matter and M. E. Tucker), Int. Assoc. Sedimentol. Spec. Publ., 2, 7–41.

Harms, J. C., Southard, J. B. and Walker, R. G. (1982) Structures and Sequences in Clastic Rocks. Society of Economic Paleontologists and Mineralogists, Tulsa, OK, variouslypaginated, 249 pp.

Hartley, A. J., Weissmann, G. S., Nichols, G. J. and Warwick, G. L. (2010) Large distributive fluvial systems: characteristics, distribution, and controls on development. J. Sed. Res.,80, 167–183.

Heller, P. L. and Paola, C. (1996) Downstream changes in alluvial architecture: an exploration of controls on channelstacking patterns. J. Sed. Res., 66, 297–306.

Hickson, T. A., Sheets, B. A., Paola, C. and Kelberer, M. (2005) Experimental test of tectonic controls on three-dimensional alluvial facies architecture. J. Sed. Res., 75, 710–722.

Hjellbakk, A. (1997) Facies and fluvial architecture of a highenergy braided river; the upper Proterozoic Seglodden Member, Varanger Peninsula, Northern Norway. Sed. Geol.,114, 131–161.

Huerta, P. (2007) El Paleógeno de la cuenca de Almazán. Relleno de una cuenca piggyback. PhD Thesis, Universidad de Salamanca, Salamanca, 340 pp.

Huerta, P. and Armenteros, I. (2004) Asociaciones de carbonatos continentales en el Eoceno de la Cuenca de Almazán. Geo–temas, 6, 75–78.

Huerta, P. and Armenteros, I. (2005) Calcrete and palustrine assemblages on a distal alluvial–floodplain: a response to local subsidence (Miocene of the Duero basin, Spain). Sed. Geol., 177, 253–270.

Huerta, P., Armenteros, I., Recio, C. and Blanco, J. A. (2010) Palaeogroundwater evolution in playa–lake environments. Sedimentary facies and stable isotope record (Palaeogene, Almazán basin, Spain). Palaeogeogr. Palaeoclimatol. Palaeoecol.,286, 135–148.

Ikeda, H. (1989) Sedimentary controls on channel migration and origin of point bars in sand–bedded meandering rivers. In: River Meandering (Eds S. Ikeda and G. Parker), WaterResour. Monogr., 12, 51–68. American Geophysical Union, Washington, DC.

Jo, H. R. (2003a) Depositional environments, architecture, and controls of early cretaceous non-marine successions in the northwestern part of Kyongsang Basin, Korea. Sed. Geol.,161, 269–294.

Jo, H. R. (2003b) Non-marine successions in the northwestern part of Kyongsang Basin (Early Cretaceous): fluvial styles and stratigraphic architecture. Geosci. J., 7, 89–106.

Jo, H. R. and Chough, S. K. (2001) Architectural analysis of fluvial sequences in the northwestern part of Kyongsang Basin (Early Cretaceous), SE Korea. Sed. Geol., 144, 307-334.

Jordan, T. E., Muñoz, N., Hein, M., Lowenstein, T., Godfrey, L. and Yu, J. (2002) Active faulting and folding without topographic expression in an evaporite basin, Chile. Geol. Soc. Am. Bull., 114, 1406-1421.

Karpeta, W. P. (1993) Sedimentology and gravel bar morphology in an Archaean braided river sequence; the Witpan Conglomerate Member (Witwatersrand Supergroup) in the Welkom Goldfield, South Africa. In: Braided Rivers (Eds J. L. Best and C. S. Bristow), Geol. Soc. Spec. Publ., 75, 369-388. Geological Society of London, London.

Kim, W. and Paola, C. (2007) Long-period cyclic sedimentation with constant tectonic forcing in an experimental relay ramp. Geology, 35, 331-334.

Kraus, M. J. (1999) Paleosols in clastic sedimentary rocks: their geologic applications. Earth-Sci. Rev., 47, 41-70.

La Force, M., Hansel, C. and Fendorf, S. (2002) Seasonal transformations of manganese in a palustrine emergent wetland. Soil Sci. Soc. Am. J., 66, 1377-1389.

Leckie, D. A. (2003) Modern environments of the Canterbury Plains and adjacent offshore areas, New Zealand – an analog for ancient conglomeratic depositional systems in nonmarine and coastal zone settings. Bull. Can. Petrol Geol., 51, 389-425.

Leeder, M. R. (1978) A quantitative stratigraphic model for alluvium, with special reference to channel deposit density and interconnectedness. In: Fluvial Sedimentology (Ed. A. D. Miall), Can. Soc. Petrol. Geol. Mem., 5, 587-596. Canadian Society of Petroleum Geologists, Calgary, AB.

Leeder, M. R. (1993) Tectonic controls upon drainage basin development, river channel migration and alluvial architecture; implications for hydrocarbon reservoir development and characterization. In: Characterization of Fluvial and Aeolian Reservoirs (Eds P. North Colin and D. J. Prosser), Geol. Soc. Spec. Publ., 73, 7-22. Geological Society of London, London.

Legarreta, L. and Uliana, M. A. (1998) Anatomy of hinterland depositional sequences: Upper Cretaceous fluvial strata, Neuquen basin, west-central Argentina. In: Relative Role of Eustasy, Climate, Tectonism in Continental Rocks (Eds K. W. Shanley and P. J. McCabe), SEPM Spec. Publ., 59, 83-92. Society of Economic Paleontologists and Mineralogists, Tulsa, OK.

Machette, M. N. (1985) Calcic soils of the southwestern United States. In: Soils and Quaternary Geology of the Southwestern United States (Eds D. L. Weide and M. L. Faber), Geol. Soc. Am. Spec. Pap., 203, 1-21. Geological Society of America, Boulder, CO.

Mackey, S. D. and Bridge, J. S. (1995) Three-dimensional model of alluvial stratigraphy: theory and application. J. Sed. Res., B65, 7-31.

Marriott, S. B. and Wright, V. P. (1993) Palaeosols as indicators of geomorphic stability in two Old Red Sandstone alluvial suites, South Wales. J. Geol. Soc. London, 150, 1109-1120.

Martinsen, O. J., Ryseth, A., Helland-Hansen, W., Flesche, H., Torkildsen, G. and Idil, S. (1999) Stratigraphic base level and fluvial architecture: Ericson Sandstone (Campanian), Rock Springs Uplift, SW Wyoming, USA. Sedimentology, 46, 235-263.

Marzo, M., Nijman, W. and Puigdefábregas, C. (1988) Architecture of the Castissent fluvial sheet sandstones, Eocene, South Pyrenees, Spain. Sedimentology, 35, 719-738.

McCarthy, T. S. (1993) The great inland deltas of Africa. J. Afr. Earth Sci., 17, 275-291.

McCarthy, T. S. and Ellery, W. N. (1995) Sedimentation on the distal reaches of the Okavango Fan, Botswana, and its bearing on calcrete and silcrete (ganister) formation. J. Sed. Res. Sect. J. Sediment. Res., A, 65, 77-90.

Miall, A. D. (1996) The Geology of Fluvial Deposits: Sedimentary Facies, Basin Analysis, and Petroleum Geology. Springer-Verlag, Berlin, 582 pp.

Mitchum, R. M. Jr, Vail, P. R. and Sangree, J. B. (1977a) Seismic stratigraphy and global changes of sea level; Part 6, Stratigraphic interpretation of seismic reflection patterns in depositional sequences. In: Seismic Stratigraphy; Applications to Hydrocarbon Exploration (Ed. C. E. Payton), AAPG Mem., 26, 117-133. American Association of Petroleum Geologists, Tulsa, OK.

Mitchum, R. M. Jr, Vail, P. R. and Thompson, S. III (1977b) Seismic stratigraphy and global changes of sea level; Part 2,The depositional sequence as a basic unit for stratigraphic analysis. In: Seismic Stratigraphy; Applications to Hydrocarbon Exploration (Ed. C. E. Payton),AAPGMem., 26, 53 – 62. American Association of Petroleum Geologists, Tulsa, OK.

Mohrig, D., Heller, P. L., Paola, C. and Lyons, W. J. (2000) Interpreting avulsion process from ancient alluvial se-quences: Guadalope-Matarranya system (northern Spain) and Wasatch Formation (western Colorado). Geol. Soc. Am. Bull., 112, 1787 – 1803.

Muto, T. and Steel, R. J. (2000) The accommodation concept in sequence stratigraphy: some dimensional problems and possible redefinition. Sed. Geol., 130, 1 – 10.

Nadon, G. C. (1994) The genesis and recognition of anastomosed fluvial deposits; data from the St. Mary River Formation, southwestern Alberta, Canada. J. Sed. Res., 64, 451 – 463.

Nemec, W. and Postma, G. (1993) Quaternary alluvial fans in southwestern Crete: sedimentation processes and geomorphic evolution. In: Alluvial Sedimentation (Eds M. Marzo and C. Puigdefàbregas), Int. Assoc. Sedimentol. Spec. Publ., 17, 235 – 276.

Nichols, G. and Fisher, J. A. (2007) Processes, facies and architecture of fluvial distributary system deposits. Sed. Geol., 195, 75 – 90.

Nijman, W. (1998) Cyclicity and basin axis shift in a piggyback basin: towards modelling of the Eocene Tremp – Ager Basin, South Pyrenees, Spain. In: Cenozoic Foreland Basins of Western Europe (Eds A. Mascle, C. Puigdefàbregas, H. P. Luterbacher and M. Fernández), Geol. Soc. Spec. Publ., 134,135 – 162.

Olsen, T., Steel, R., Høgseth, K., Skar, T. and Røe, S. -L. (1995) Sequentianl architecture in a fluvial succession: sequence stratigraphy in the Upper Cretaceous Mesaverde Group, Price Canyon, Utah. J. Sed. Res., B65, 265 – 280.

Patton, T. L. (2004) Numerical models of growth-sediment development above an active monocline. Basin Res., 16, 25 – 39.

Sanz, M. E., Alonso-Zarza, A. M. and Calvo, J. P. (1995) Carbonate pond deposits related to semiarid alluvial systems examples from the Tertiary Madrid Basin, Spain. Sedimentology,42, 437 – 452.

Semeniuk, V. and Searle, D. J. (1985) Distribution of calcrete in Holocene coastal sands in relationship to climate, southwestern Australia. J. Sed. Petrol., 55, 86 – 95.

Shanley, K. W. and McCabe, P. J. (1991) Predicting facies architecture through sequence stratigraphy – an example from the Kaiparowits Plateau, Utah. Geology, 19, 742 – 745.

Shanley, K. W. and McCabe, P. J. (1993) Alluvial architecture in a sequence stratigraphic framework: a case history from the Upper Cretaceous of southern Utah, USA. In: Quantitative Description and Modelling of Clastic Hydrocarbon Reservoirs and Outcrop Analogues (Eds S. Flint and I. D. Bryant), Int. Assoc. Sedimentol. Spec. Publ., 15, 21 – 56.

Shanley, K. W. and McCabe, P. J. (1994) Perspectives on the sequence statigraphy of continental Strata. AAPG Bull., 78, 544 – 568.

Sheets, B. A., Hickson, T. A. and Paola, C. (2002) Assembling the stratigraphic record: depositional paterns and timescales in an experimental alluvial basin. Basin Res., 14, 287 – 301.

Strong, N., Sheets, B. A., Hickson, T. A. and Paola, C. (2005) A mass-balance framework for quantifying downstream changes in fluvial architecture. In: Fluvial Sedimentology VII (Eds M. D. Blum, S. B. Marriott and S. Leclair), IAS Spec. Publ. 35, 243 – 253. Blackwell.

Suppe, J., Chou, G. T. and Hook, S. C. (1992) Rates of folding and faulting determined from growth strata. In: Thrust Tectonics (Ed. K. R. McClay), pp. 105 – 121. Chapman and Hall, London.

Tandon, S. K., Andrews, J. E., Sood, A. and Mittal, S. (1998) Shrinkage and sediment supply control on multiple calcrete profile development: a case study from the Maastrichtian of Central India. Sed. Geol., 119, 25 – 45.

Warren, J. K. (1989) Evaporite Sedimentology. Prentice Hall Inc., Old Tappan, NJ, 285 pp.

Williams, G. E. (1971) Flood deposits of the sand-bed ephemeral streams of central Australia. Sedimentology, 17, 1–40.

Williams, C. A. and Krause, F. F. (1998) Pedogenic-phreatic carbonates on a Middle Devonian (Givetian) terrigenous alluvial-deltaic plain, Gilwood Member (Watt Mountain Formation), northcentral Alberta, Canada. Sedimentology, 45, 1105–1124.

Williams, P. F. and Rust, B. R. (1969) Sedimentology of a braided river. J. Sed. Petrol., 39, 649–679.

Wright, V. P. and Marriott, S. B. (1993) The sequence stratigraphy of fluvial depositional systems: the role of floodplain sediment storage. Sed. Geol., 86, 203–210.

Wright, V. P. and Marriott, S. B. (1996) A quantitative approach to soil occurrence in alluvial deposits and its application to the Old Red Sandstone of Britain. J. Geol. Soc., 153, 907–913.

Wright, V. P. and Platt, N. H. (1995) Seasonal wetland carbonate sequences and dynamic catenas – a reappraisal of palustrine limestones. Sed. Geol., 99, 65–71.

Wright, V. P. and Tucker, M. E. (1991) Calcretes: an introduction. In: Calcretes (Eds V. P. Wright and M. E. Tucker), Int. Ass. Sediment., Reprint Series, 2, 1–22. Blackwell ScientificPublications.

Zaleha, M. J. (1997) Siwalik Paleosols (Miocene, northern Pakistan): genesis and controls on their formation. J. Sed. Res., 67, 821–839.

附录1 中国海相碳酸盐岩的图版：北京及周边长城系蓟县系和寒武系露头

(A) 寒武系链条状灰岩；(B) 竹叶状灰岩夹层，注意层面特征；(C) 含叠层石灰岩，注意层间缝和溶孔；(D) 碳酸盐生物丘，注意界面特征。北京西山寒武系露头，2011

(A) 中元古界蓟县系白云岩及其准层序；(B) 中元古界蓟县系白云岩及其藻团块；(C) 中元古界蓟县系白云岩，注意准层序界面和波痕特征；(D) 中元古界蓟县系白云岩，注意波痕特征。北京延庆蓟县系露头，2010

(A) 寒武系鲕粒灰岩；(B) 寒武系竹叶状灰岩与薄板状灰岩；(C) 寒武系灰岩夹泥质条带；(D) 寒武系鲕粒灰岩。河北涞水寒武系露头，2009

(A) 从滨海相石英砂岩"A"向混积岩的变化，越向右碳酸盐岩组分越多；(B) 碳酸盐岩-碎屑岩的混积岩"B"；(C) "C"是"B"的放大，注意混积岩的排列方式。北京延庆长城系露头，2012 (D) 碳酸盐岩沉积"D"，碳酸盐岩-碎屑岩的混积岩"E"，黑色部分是石英砂岩；(E) "F"是"E"的放大，注意混积岩中碳酸盐岩部分的淋滤孔隙；(F) "G"是采集的混积岩样品。北京延庆长城系，2012

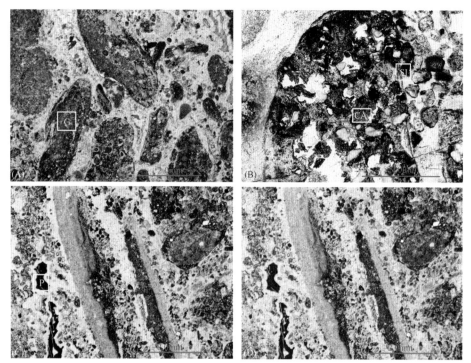

(A) 混积岩薄片观察：滨海沉积环境的混积岩以碳酸盐颗粒"G"为主，部分石英颗粒，粒间充填亮晶白云石，正交光；(B) 颗粒内部主要成分为方解石（"CA"红色部分），右侧看到的主要是陆源碎屑石英颗粒"SI"，单偏光；(C) 滨海沉积环境的混积岩可以在粒间形成孔隙"P"，注意孔隙周边的亮晶白云石，单偏光；(D) 作为对比的正交光照片。北京延庆长城系露头之样品，2012

附录2 中国陆相碳酸盐岩的图版：四川盆地侏罗系露头

(A) 湖泊沉积环境的侏罗系灰岩；(B) 侏罗系灰岩含泥质夹层，注意层面特征；(C) 姜结石灰岩，反映滨浅湖岸线附近的沉积特征；(D) 富介壳发育层的湖相碳酸盐岩，注意介壳直径大小约1cm。四川川东北侏罗系露头，2012

(A) 湖泊沉积环境的侏罗系灰岩，注意单层厚度约20~30cm；(B) 侏罗系灰岩，含丰富的介壳；(C) 介壳灰岩，注意介壳直径大小约1cm；(D) 质较纯的湖相碳酸盐岩，单层厚度约20~30cm，注意人为比例尺。四川川东北侏罗系露头，2012

(A) 侏罗系灰岩，夹有富有机质的泥岩；(B) 侏罗系灰岩，局部呈球状或结核状，风化较严重；(C) 湖相灰岩，注意局部重结晶严重，见有沥青；(D) 团块状石灰岩，注意界面特征。四川川东北侏罗系露头，2012